Recovery and Treatment of Solid Waste

Recovery and Treatment of Solid Waste

Editors

Rossana Bellopede
Lorena Zichella

MDPI • Basel • Beijing • Wuhan • Barcelona • Belgrade • Manchester • Tokyo • Cluj • Tianjin

Editors
Rossana Bellopede
Politecnico di Torino
Italy

Lorena Zichella
Politecnico di Torino
Italy

Editorial Office
MDPI
St. Alban-Anlage 66
4052 Basel, Switzerland

This is a reprint of articles from the Special Issue published online in the open access journal *Materials* (ISSN 1996-1944) (available at: https://www.mdpi.com/journal/materials/special_issues/recovery_waste).

For citation purposes, cite each article independently as indicated on the article page online and as indicated below:

LastName, A.A.; LastName, B.B.; LastName, C.C. Article Title. *Journal Name* **Year**, *Volume Number*, Page Range.

ISBN 978-3-0365-7710-4 (Hbk)
ISBN 978-3-0365-7711-1 (PDF)

Contents

materials

MDPI

Article

Chemical–Electrochemical Process Concept for Lead Recovery from Waste Cathode Ray Tube Glass

Árpád Imre-Lucaci [1], Melinda Fogarasi [2,*], Florica Imre-Lucaci [3,*] and Szabolcs Fogarasi [1,3,*]

[1] Department of Chemical Engineering, Faculty of Chemistry and Chemical Engineering, Babeş-Bolyai University, 11 Arany Janos Street, 400028 Cluj-Napoca, Romania; arpad.imre@ubbcluj.ro
[2] Department of Food Engineering, University of Agricultural Sciences and Veterinary Medicine of Cluj-Napoca, Calea Mănăstur 3-5, 400372 Cluj-Napoca, Romania
[3] Interdisciplinary Research Institute on Bio-Nano-Sciences, Babeş-Bolyai University, 42 Treboniu Laurian Street, 400271 Cluj-Napoca, Romania
* Correspondence: melinda.nagy@usamvcluj.ro (M.F.); florica.imre@ubbcluj.ro (F.I.-L.); szabolcs.fogarasi@ubbcluj.ro (S.F.); Tel.: +40-264-593-833 (S.F.)

Abstract: This paper presents a novel approach for the recovery of lead from waste cathode-ray tube (CRT) glass by applying a combined chemical-electrochemical process which allows the simultaneous recovery of Pb from waste CRT glass and electrochemical regeneration of the leaching agent. The optimal operating conditions were identified based on the influence of leaching agent concentration, recirculation flow rate and current density on the main technical performance indicators. The experimental results demonstrate that the process is the most efficient at 0.6 M acetic acid concentration, flow rate of 45 mL/min and current density of 4 mA/cm^2. The mass balance data corresponding to the recycling of 10 kg/h waste CRT glass in the identified optimal operating conditions was used for the environmental assessment of the process. The General Effect Indices (GEIs), obtained through the Biwer Heinzle method for the input and output streams of the process, indicate that the developed recovery process not only achieve a complete recovery of lead but it is eco-friendly as well.

Keywords: cathode ray tube; lead recovery; lead leaching and electrodeposition; environmental assessment

Citation: Imre-Lucaci, Á.; Fogarasi, M.; Imre-Lucaci, F.; Fogarasi, S. Chemical–Electrochemical Process Concept for Lead Recovery from Waste Cathode Ray Tube Glass. *Materials* **2021**, *14*, 1546. https://doi.org/10.3390/ma14061546

Academic Editors: Rossana Bellopede and Lorena Zichella

Received: 17 February 2021
Accepted: 18 March 2021
Published: 22 March 2021

1. Introduction

In the recent years there has been a major concern to limit the risks associated with the manufacture of electrical and electronic equipment, the management of waste electrical and electronic equipment (WEEE) in order to minimize the negative impact on the environment [1–3]. In this regard, the European Union Regulations on Electronic Waste, WEEE Directive 2002/96/EC, EU WEEE DIRECTIVE 2012/19/EU and RoHs 2002/95/EC, regulate the responsibility of the member countries for the collection, use, recycling and recovery of electronic waste [4]. Rapid growth in production and technical development in electronics involves the accelerated replacement of outdated electronics equipment and accumulation of large amounts of harmful WEEE [5,6], including waste cathode-ray tube (CRT) glass from old televisions and computer monitors [7].

A CRT is composed of two different types of glass, from which one is used for the funnel and neck sections, characterized by high levels of lead oxide and another used for the screen which is typically a non-leaded glass that contains high levels of barium oxide [8]. Recycling of lead from waste CRT glass is an important issue because lead is classified as a neurotoxin that can accumulate in the soft tissues and bones, causing serious health issues [9–11]. The high content of lead oxide (23%) in CRT funnel glasses is an important factor that limits its landfill storage and the recycling process, as result many states have passed bans on putting CRTs in landfills or incinerators [12,13]. Additionally, CRT funnel glass, due to its composition, is unsuitable for applications where metal oxides could leach into food products or ground water [14].

Conventional CRT glass recycling is carried out in a closed-loop, where waste glass, after an appropriate removal of metal and luminophore contaminants, is utilized during manufacturing of new CRTs [15,16]. However, the above-mentioned recycling method is insufficient, as technology develops and modern liquid crystal display (LCD), plasma or light-emitting diode (LED) screens are introduced, the demand for CRT glass decreases [17]. As a result, it is necessary to provide the industry with new technical solutions for the processing of waste glass which led to other products than CRTs [18]. The most common applications of the waste glass are related to the manufacturing of different products like conventional ceramics, aggregates and cements [19,20]. Some of the technologies involve high temperature treatment of the CRT waste glass leading to ceramic or glass composites used in the construction industry, mainly to manufacture bricks and roof tiles [21,22]. The CRT waste glass cullet can be also used in the metallurgical industry to produce ferro-silicates in the form of slag [23]. Other methods have also been developed for the use of waste CRT glass in the production of floor coverings and chemical resistant compounds [24].

Unfortunately, the amount of waste material that can be recycled in the above mentioned technologies is limited due to the fact that waste CRT glass is used without preliminary separation of harmful components, involving higher environmental risks, operation and maintenance costs [25,26]. In order to overcome this drawback, associated with waste CRT glass recycling, there have been attempts to remove the hazardous components like lead employing different hydrometallurgical and pyrometallurgical processes [27–29]. These processes present some major disadvantages like insufficient lead removal efficiency and polluting byproducts, which can be more harmful than the treated waste material [30]. Moreover, many of the studies presented in the literature lack a comprehensive overview, by not assessing both the technical performance and environmental impact of the processes, which is necessary to draw global conclusions [29,31].

In view of the above discussion, the recovery of lead from waste CRT glasses was achieved by acetic acid leaching of Pb coupled with the simultaneous electrowinning of a high purity Pb deposit and regeneration of the leaching agent. The novel process concept defined and assessed in the current paper ensures high technical performance with a low environmental impact, based on the influence of acetic acid concentration, recirculation flow rate and current density on different key performance indicators and GEIs values.

2. Materials and Methods

2.1. Thermal Treatment of CRT Glass Samples

In a preliminary step the CRT glass samples with the composition presented in Table 1 were ground to a fine powder (3.5–5.5 μm) in order to promote the reaction between the CRT glasses samples and Na_2CO_3. All tests were carried out in an electrical furnace at 1000 °C with duration of 30 min, combining 10 g of CRT glass with 24 g of Na_2CO_3. After the thermal treatment the cooled samples were washed with 100 mL of distilled water in order to selectively remove the soluble silicates and hydroxides.

Table 1. Cathode-ray tube (CRT) glass composition.

Component	SiO_2	TiO_2	Fe_2O_3	Al_2O_3	CaO	MgO	Na_2O	K_2O	PbO	Other Trace Elements
Concentration, wt.%	55.6	0.2	0.2	2.96	3.85	1.65	6.25	6.75	22.08	0.46

2.2. Lead Dissolution Process

The solid material obtained from the washing step was dissolved, over a period of one hour, in 150 mL of acetic acid (CH_3COOH) solution using a chemical reactor equipped with a stirrer operated isothermally at 80 °C. In order to determine the optimal concentration of acetic acid for the dissolution process, the experiments were performed at different concentrations of CH_3COOH in the range of 0.2–1 M. Glacial acetic acid of analytical purity and bidistilled water were used to obtain the leaching solutions. Finally, the solution was filtered in order to separate it from the precipitated solid residues which contained mainly SiO_2. The lead concentration in the samples taken during the dissolution was determined using an atomic absorption spectrophotometer.

2.3. Electrochemical Process Description

The installation used for lead electroextraction consisted of a storage tank of the processed solution, connected in series with a divided electrochemical reactor (ER), operated in galvanostatic mode. Recirculation of the solution between the storage tank and ER was performed with a peristaltic pump. The cathode was made of stainless-steel plates and the anode was made of graphite. Two $Ag/AgCl/KCl_{sat}$ reference electrodes were used to measure the cathodic and anodic potentials. All tests were performed at 22 °C for two hours using a 150 mL of electrolyte containing 9.25 g/L Pb^{2+}, 3.91 g/L $Mg(CH_3COO)_2$ and 7.24 g/L $Ca(CH_3COO)_2$ which correspond to the final composition of the leaching solution. The experiments were performed at different flow rates (15, 30 and 45 mL/min) and current densities (4, 8 and 12 mA/cm^2). Experiments also involved the use of a computer-controlled DC power supply, and LabVIEW software for process control and data acquisition. The obtained Pb deposit was dissolved in concentrated HNO_3 to determine the amount and purity of lead deposited. The concentration of lead in the solutions at the end of the experiment was determined using an atomic absorption spectrophotometer.

2.4. Performance Indicators of the Lead Recovery Process

The performance of the dissolution and electro-extraction processes was evaluated on the basis of technical performance indicators:

- **Dissolution degree (%)** was defined as the ratio between the amount of dissolved lead and the initial amount of lead in the processed samples.
- **Efficiency of CH_3COOH utilization (%)** is the ratio of the amount of CH_3COOH consumed in the dissolution process and the initial amount of CH_3COOH in the solution.
- **Specific acetic acid consumption (kg CH_3COOH/kg Pb)** indicates the amount of CH_3COOH consumed to dissolve one kilogram of Pb from the processed waste.
- **Extraction degree (%)** was calculated as the ratio of the quantity of electrodeposited lead and the initial amount of lead in the electrolyte.
- **Current efficiency (%)** was defined as the ratio of the amount of electricity used to form the cathode deposit and the total amount of electricity consumed in the process.
- **Specific energy consumption for the cathodic process (kWh/kg Pb)** indicates the amount of energy used to form one kilogram of Pb deposit.
- **Specific energy consumption for the anode process (kWh/kg CH_3COOH)** indicates the amount of energy required to produce one kilogram of CH_3COOH.

3. Results and Discussions

3.1. Dissolution of Lead from Pretreated CRT Glass

In order to determine the optimal operating conditions for the dissolution process, the evolution of the dissolution degree at different acetic acid concentrations was quantified. It can be seen from Figure 1 that the dissolution degree increases over time at all CH_3COOH concentrations, the final value being almost three times higher than the initial one. The results also show that the concentration of the leaching agent has a decisive influence on the dissolution rate, since at the concentration of 1 M CH_3COOH the dissolution degree is three times higher than at 0.2 M CH_3COOH. However, the dissolution degree values increase only with 30% between 0.6 and 1 M CH_3COOH while between 0.2 and 0.6 M CH_3COOH they increase 110%, which means that above 0.6 M CH_3COOH there is no significant gain in efficiency.

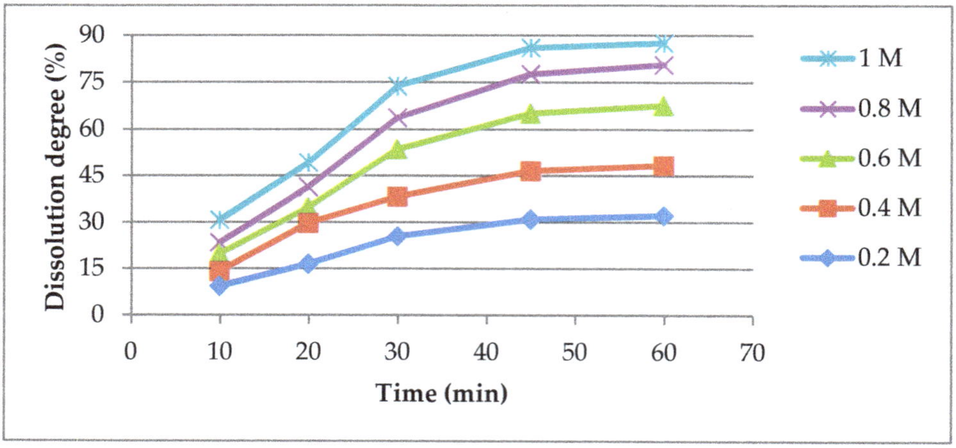

Figure 1. Dissolution degree vs. time at different CH_3COOH concentrations.

In addition to increasing the dissolution degree with increasing CH_3COOH concentration, it is important to increase the conversion of the leaching agent in order to exploit the full potential of the leaching solution. To highlight this aspect, the efficiency of CH_3COOH utilization was determined, which indicates how much of the leaching agent was converted under different experimental conditions compared to what could theoretically be used for lead dissolution, taking into account the initial amount of CH_3COOH in the solution. The results from Figure 2 show that the efficiency of CH_3COOH utilization is diminished by increasing the concentration of CH_3COOH, the maximum value being reached at a concentration of 0.2 M acetic acid.

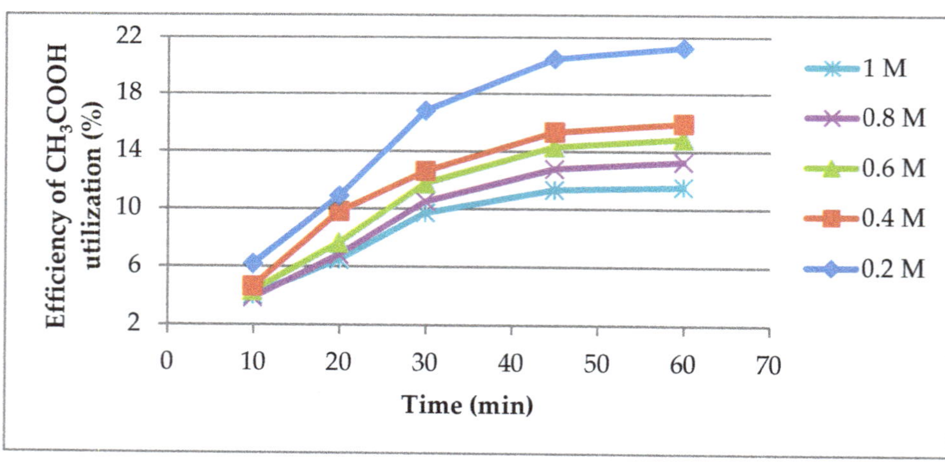

Figure 2. Efficiency of CH_3COOH utilization vs. time at different CH_3COOH concentrations.

This tendency can be attributed to a partial order of reaction regarding CH_3COOH concentration which means that the amount of acetic acid transformed in the leaching reaction does not increases proportional with the increase in initial CH_3COOH concentration. The concentration profiles of Pb^{2+} shown in Figure 3 sustain the above explanation, considering that the total amount of lead dissolved increases only 171% by increasing the initial concentration of CH_3COOH by 400%.

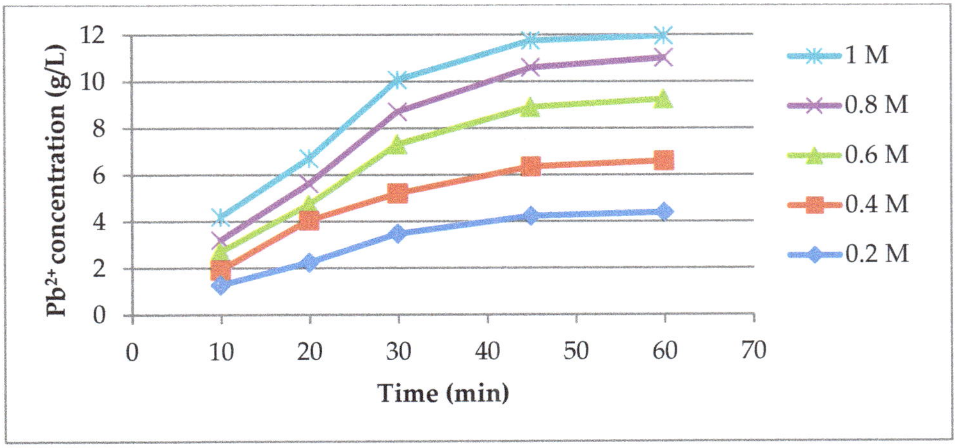

Figure 3. Pb^{2+} concentration profile at different CH_3COOH concentrations.

Considering that the above evaluated performance indicators give contradictory conclusions regarding the optimal CH_3COOH concentrations, the specific acetic acid consumption for the leaching process was determined which gives a comprehensive overview on the efficiency of the process. The specific acetic acid consumption values, Figure 4, show that the lowest concentration (0.2 M) of CH_3COOH allows the most efficient use of the amount of leaching agent present in the system. In contrast, according to Figure 3, the final Pb^{2+} concentration is the lowest at 0.2 M of CH_3COOH which would not ensure the most favorable conditions for the electrodeposition process of lead. Therefore, the intermediate concentration of 0.6 M CH_3COOH would be a better option in comparison to 0.2 M CH_3COOH, considering that the obtained final Pb^{2+} concentration (9.25 g/L) represents 80% of the maximum achievable concentration under these dissolution conditions.

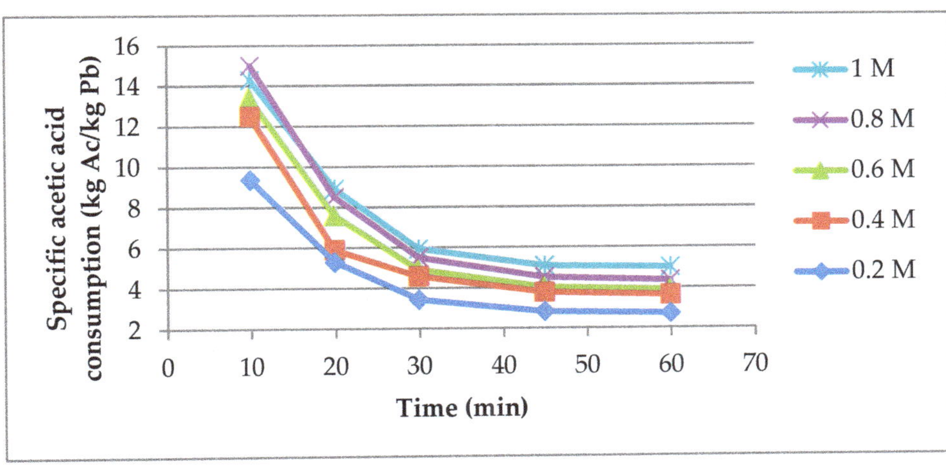

Figure 4. Specific acetic acid consumption vs. time at different CH_3COOH concentrations.

Additionally, taking into account the ecological aspects, to ensure a relatively advanced removal of lead from CRT waste, but with a reasonable yield, the 0.6 M CH_3COOH value can be considered as the optimal concentration for the dissolution process.

3.2. Lead Electrodeposition and CH_3COOH Regeneration

The electrochemical recovery of lead from the leach solutions involved as main reactions the simultaneous deposition of Pb at the cathode and CH_3COOH regeneration at the anode. Considering the position of Pb in the electrochemical series of metals, its formation at the cathode is accompanied by the hydrogen evolution reaction.

Cathode:

$$Pb^{2+} + 2e^- \rightarrow Pb \quad (1)$$

$$2H^+ + 2e^- \rightarrow H_2 \quad (2)$$

Anode:

$$2H_2O \rightarrow 4H^+ + O_2 + 4e^- \quad (3)$$

Chemical reaction:

$$H^+ + CH_3COO^- \rightleftarrows CH_3COOH \quad (4)$$

According to the results, Figure 5, the extraction degree values increase with the increase in electrolyte flow rate and current density reaching the maximum value at 45 mL/min and 12 mA/cm^2. Additionally, the results reveal the fact that the extraction degree is more strongly dependent on current density than electrolyte flow rate. It can be observed, Figure 5, that the extraction degree values almost double with the increase in current density by three times while for the same increase in flow rate at constant current density increases the extraction degree by only 32%.

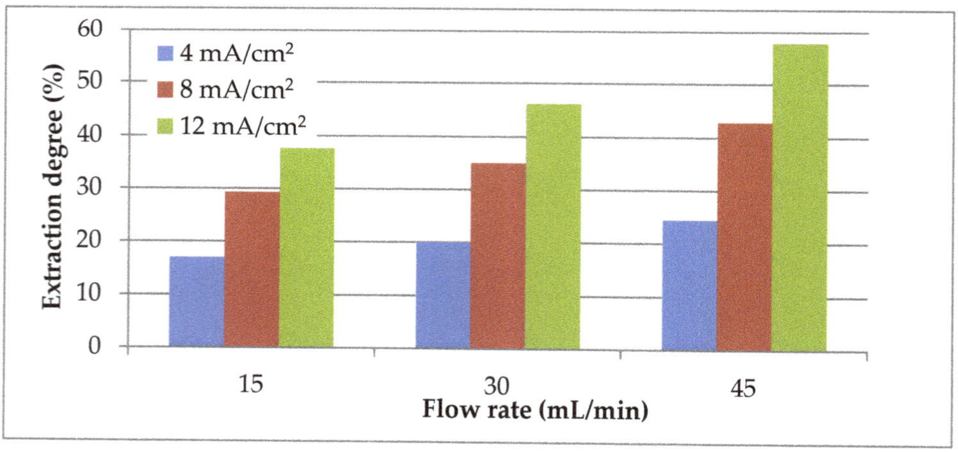

Figure 5. Influence of electrolyte flow rate and current density on lead extraction degree.

As can be seen in Figure 6, in contrast to the evolution of extraction degree, the cathodic current efficiency decreases as the current density increases, while the increase in electrolyte flow rate has a positive impact. This can be explained by the fact that high current densities favor the hydrogen discharge reaction, which leads to a decrease in the cathodic current efficiency by 20–23% between the current densities of 4 and 12 mA/cm^2. However, the experimental data show that the impact of the secondary cathodic reaction is even lower as the electrolyte flow rate is higher. In view of this tendency the maximum cathodic current efficiency (38.08%) was obtained at the highest flow rate (45 mL/min) and the lowest current density.

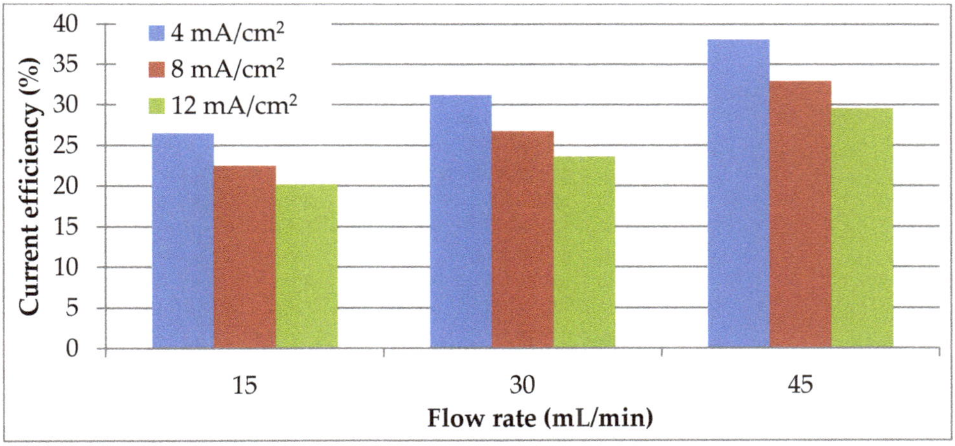

Figure 6. Evolution of current efficiency with electrolyte flow rate at different current densities.

Since the most important performance criterion in the performance of an electrochemical process is the specific energy consumption, this parameter was evaluated for both lead deposition and regeneration of the leaching agent. From Figure 7 it can be seen that the specific energy consumption for the cathodic process depends more strongly on the current density than on the electrolyte flow rate and varies in the opposite direction with the two operating parameters. Increasing the current density by three times increases the specific energy consumption of the electrodeposition process by 197–231%, while the same variation of the electrolyte flow reduces it by about 100%. The beneficial impact of flow rate increase can be attributed to the more intensive transportation of Pb^{2+} ions to the cathode surface which reduces the corresponding mass transport potential.

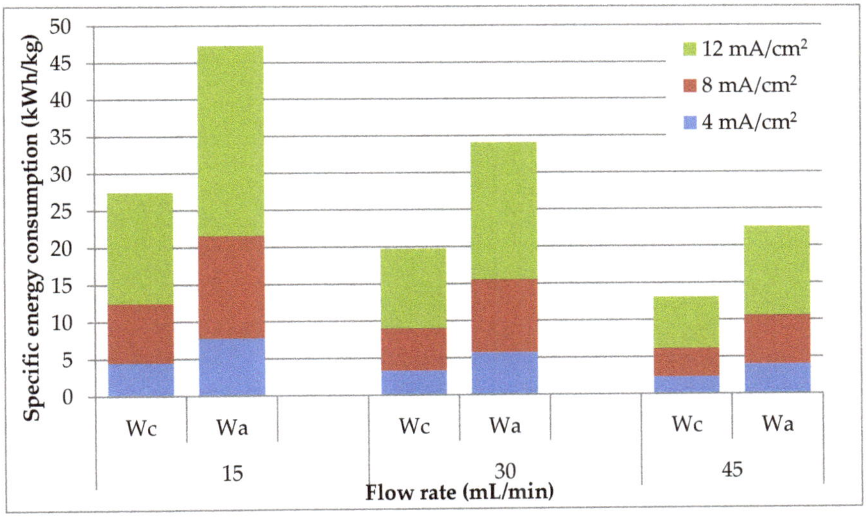

Figure 7. Specific energy consumption values for lead electrodeposition (Wc) and CH_3COOH regeneration (Wa) at different electrolyte flow rates and current densities.

Similar conclusions can be reached related to the influence of the operating parameters on the evolution of the specific energy consumption for acetic acid regeneration. In contrast,

Figure 7 reveals that the specific energy consumption of the anodic process is almost twice as high as for the cathodic one. This is due to the lower molar mass of acetic acid than lead, which leads to the generation of a lower amount of CH_3COOH by consuming the same amount of electricity.

The influence of the operating conditions on the performance of the lead electrodeposition process at the cathode and the CH_3COOH regeneration at the anode is also quantified by the thermodynamic parameters, Table 2, of the electrochemical process. Electrode potentials confirm that lead deposition and oxygen discharge are the main electrochemical reactions, and their increase with increasing current density indicates the negative impact of the current density on the electrode over potentials. As can be seen from Table 2, increasing the current density also leads to higher ohmic drops, which leads to the increase in the cell voltage. In contrast, increasing flow rates reduce both electrode potentials and cell voltage, indifferent of the current density value, due to increased transport of electrochemically active species to the reaction surface.

Table 2. Thermodynamic parameters of the electrochemical process.

Flow Rate, mL/min	E_b, V			ε_c, V			ε_a, V		
	i, mA/cm^2			i, mA/cm^2			i, mA/cm^2		
	4	8	12	4	8	12	4	8	12
15	2.30	3.46	5.81	−0.2	−0.34	−0.74	1.62	1.96	1.98
30	2.01	2.94	4.87	−0.15	−0.21	−0.73	1.50	1.75	2.32
45	1.71	2.42	3.94	−0.07	−0.36	−0.66	1.43	1.67	2.47

E_b—cell voltage; ε_c—cathode potential; ε_a—anode potential; i—current density.

Based on the above discussions, the optimal operating conditions were obtained at a flow rate of 45 mL/min and a current density of 4 mA/cm^2, due to the fact that the specific energy consumption for both main electrochemical processes, Figure 7, attain the lowest values.

3.3. Environmental Assessment of the Lead Recovery Process

The environmental assessment was performed using the Biwer–Heinzle method [32,33] which is easily applicable in the early phases of process development and reveals the contribution of each input and output substance to the overall environmental impact of the lead recovery process.

In accordance with the Biwer–Heinzle method (Figure 8) the environmental factors were obtained from 6 impact groups which contained 14 impact categories. All of the components involved in the lead recovery process were allocated to a class *A*, *B* or *C* in each impact category (*A* = 1—highly toxic substances, *B* = 0.3—less toxic substances, *C* = 0—non-toxic substances) [32]. Next, the input and output environmental indices were determined by combining the obtained environmental factors with the mass indices resulting from the mass balance data corresponding with the processing of 10 kg/h CRT glass in the identified optimal conditions. Finally, the overall environmental impact of the lead recovery process was evaluated based on the General Effect Indices (GEIs) calculated by dividing the sum of environmental indices by the total mass indices [33].

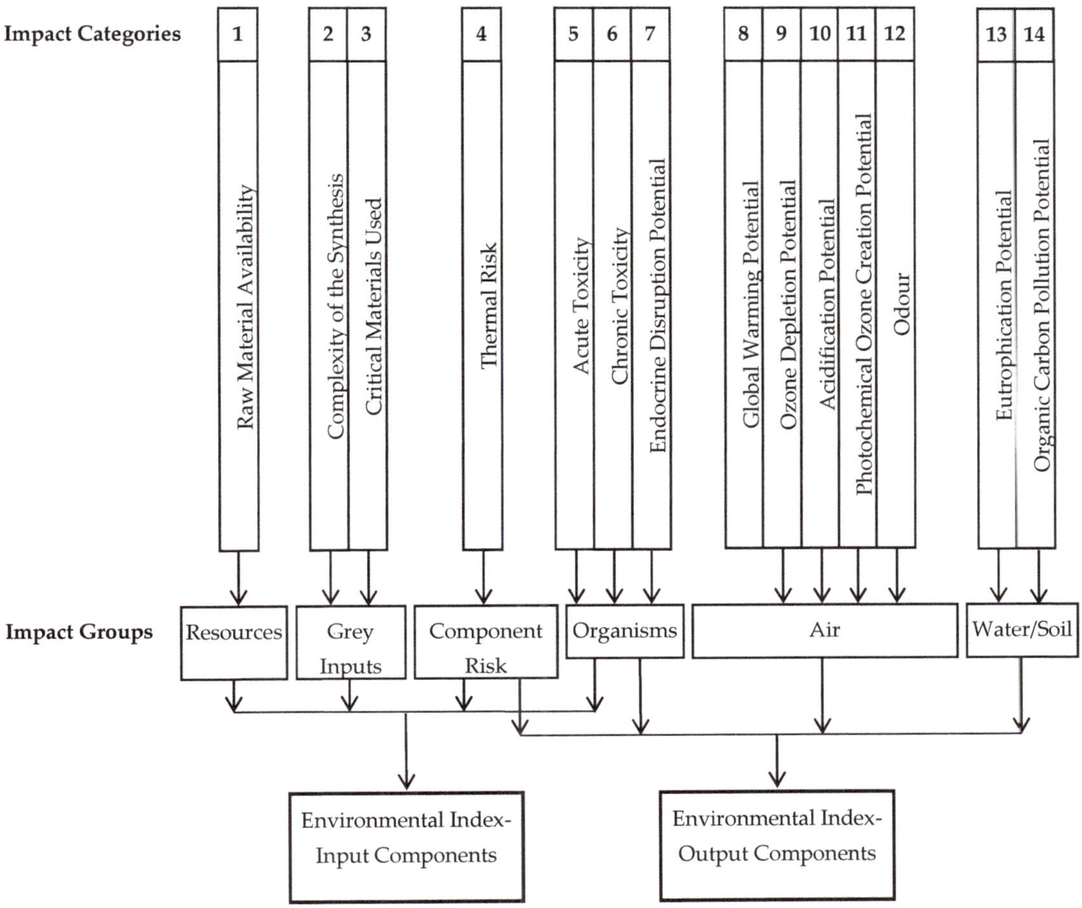

Figure 8. Schematic representation of the Biwer–Heinzle method.

Among the input materials (Table 3), water has the lowest environmental impact considering that it was allocated to class *C* in all seven impact categories. In contrast, CH_3COOH was assigned to class *B* for its acute toxicity, thermal risk and raw material availability associated with their production. The processed waste CRT glass was assigned to class *B* in impact category 5 and 6 because Pb can cause serious health issues. For the same reason, in the case of the output streams, Pb was assigned to class *B* in impact category 5 and 6 together. An output stream with similar environmental impact is the waste acetic acid solution which was also assigned to class *B* in impact categories 4, 5, 11 and 14 regarding its impact on air, soil and water pollution. Considering the global warming potential of CO_2, it was assigned to class *B* in impact category 8. The other output streams, H_2O and SiO_2 have the lowest environmental impact, considering that they were allocated to class *C* in all 11 impact categories, being valuable secondary products of the developed process together with the resulting lead acetate, calcium acetate and magnesium acetate.

The *GEIs* values from Table 4 indicate that the environmental impact of the output streams is lower than for the input streams, which means that the developed process lowers the environmental impact of waste CRT glass through the recovery of lead.

Table 3. Input impact assessment.

Input			Impact Categories							Environmental Factors	Environmental Index
Streams	**Quantity (kg/h)**	**Mass Index**	**1**	**2**	**3**	**4**	**5**	**6**	**7**	***EF***	***EI***
Waste CRT	10	4.96	C	C	C	C	B	B	C	0.075	0.372
Sodium carbonate	7.8	3.87	C	C	C	B	C	C	C	0.075	0.290
Acetic acid	1.55	0.77	B	C	C	B	B	C	C	0.225	0.173
Water	50	24.81	B	C	C	B	B	B	C	0	0.000
Total:	**69.35**	**34.41**								Environmental Index, EI_{inputs}:	**0.835**
										General Effect Index, GEI_{inputs}:	**0.024**

Table 4. Output impact assessment.

Output			Impact Categories											Environmental Factors	Environmental Index
Streams	**Quantity (kg/h)**	**Mass Index**	**4**	**5**	**6**	**7**	**8**	**9**	**10**	**11**	**12**	**13**	**14**	***EF***	***EI***
Lead	2.02	**1.00**	C	B	B	C	C	C	C	C	C	C	C	0.075	0.075
Lead acetate	0.09	0.05	C	B	B	C	C	C	C	C	C	C	C	0.075	0.004
Calcium acetate	1.21	0.60	C	B	C	C	C	C	C	C	C	C	C	0.075	0.045
Magnesium acetate	0.62	0.31	C	B	C	C	C	C	C	C	C	C	C	0.075	0.023
Slicon dioxide	1.27	0.63	C	C	C	C	C	C	C	C	C	C	C	0	0.000
Waste acetic acid	1.16	0.57	B	B	C	C	C	C	C	B	C	C	B	0.3	0.172
CO_2	3.24	1.61	C	C	C	C	B	C	C	C	C	C	C	0.075	0.121
Gases (O_2, H_2,...)	3.19	1.58	C	C	C	C	B	C	C	C	C	C	C	0.075	0.119
Wastewater	56.55	28.06	C	C	C	C	C	C	C	C	C	C	C	0	0.000
Total:	**69.35**	**34.41**												Environmental Index, $EI_{outputs}$:	**0.558**
														General Effect Index, GEI_{ouputs}:	**0.016**

Since the input and output streams have *GEI*s values close to the minimum possible (0), according to the Biwer–Heinzle method, it means that globally the process has low environmental impact. Nevertheless, caution and special protective measures must be applied when handling concentrated CH_3COOH solutions.

4. Conclusions

The results demonstrate that the developed combined chemical–electrochemical process can be efficiently applied for the recovery of Pb from waste CRT glass in the form of metallic Pb. It was found that the Pb dissolution process from the pre-treated waste CRT glass samples is most effective at a concentration of 0.6 M CH_3COOH, a consideration which ensures equilibrium between the yield of dissolution process and leaching agent consumption. Based on the specific energy consumption, it can be concluded that the electrochemical process is carried out with the highest performance at a flow rate of 45 mL/min and a current density of 4 mA/cm^2, leading to the formation of a high purity Pb deposit (99.98 wt.%) and CH_3COOH regeneration. In the identified operating conditions, the amount of metallic lead recovered in one hour of processing represents ~10% of the lead present in the treated waste material the rest is in form of dissolved lead acetate.

The environmental impact assessment of the Pb recovery process was performed successfully in the early phases of process development by applying the Biwer–Heinzle method and the corresponding mass balance data for the treatment of 10 kg/h waste CRT glass. Based on the GEIs values obtained for the input and output streams, it can be stated as an overall conclusion that the novel approach for the recovery of lead from waste CRT glass proved to be a promising alternative with low environmental impact. Still, further studies are recommended in order to model, simulate and scale up the process for higher production and assess its economic performance.

Author Contributions: Conceptualization S.F., M.F. and Á.I.-L.; methodology S.F. and F.I.-L.; formal analysis M.F. and Á.I.-L.; investigation, F.I.-L., S.F., M.F. and Á.I.-L.; writing—original draft preparation, F.I.-L., S.F.; writing—review and editing, S.F., M.F. and Á.I.-L.; supervision, M.F., S.F. All authors have read and agreed to the published version of the manuscript.

Funding: This research received no external funding.

Institutional Review Board Statement: Not applicable.

Informed Consent Statement: Not applicable.

Data Availability Statement: Data sharing is not applicable to this article.

Acknowledgments: We are grateful for the administrative and financial support offered by the Babes-Bolyai University and University of Agricultural Sciences and Veterinary Medicine Cluj-Napoca.

Conflicts of Interest: The authors declare no conflict of interest.

References

1. Wang, J.; Xu, Z. Disposing and recycling waste printed circuit boards: Disconnecting, resource recovery, and pollution control. *Environ. Sci. Technol.* **2015**, *49*, 721–733. [CrossRef]
2. Siddiqi, M.M.; Naseer, M.N.; Abdul Wahab, Y.; Hamizi, N.A.; Badruddin, I.A.; Hasan, M.A.; Zaman Chowdhury, Z.; Akbarzadeh, O.; Johan, M.R.; Kamangar, S. Exploring E-Waste Resources Recovery in Household Solid Waste Recycling. *Processes* **2020**, *8*, 1047. [CrossRef]
3. Bizzo, W.A.; Figueiredo, R.A.; de Andrade, V.F. Characterization of Printed Circuit Boards for Metal and Energy Recovery after Milling and Mechanical Separation. *Materials* **2014**, *7*, 4555–4566. [CrossRef] [PubMed]
4. Ismail, H.; Hanafiah, M.M. An overview of LCA application in WEEE management: Current practices, progress and challenges. *J. Clean. Prod.* **2019**, *232*, 79–93. [CrossRef]
5. Suresh, S.S.; Bonda, S.; Mohanty, S.; Nayak, S.K. A review on computer waste with its special insight to toxic elements, segregation and recycling techniques. *Process Saf. Environ. Prot.* **2018**, *116*, 477–493. [CrossRef]
6. Fogarasi, S.; Imre-Lucaci, F.; Ilea, P.; Agachi, P.S.; Imre-Lucaci, A. Dissolution of base metals from WPCBs using $Na_2S_2O_8$ solution. *Studia Univ. Babes-Bolyai Chem.* **2015**, *60*, 205–214.
7. Liu, T.; Song, W.; Zou, D.; Li, L. Dynamic mechanical analysis of cement mortar prepared with recycled cathode ray tube (CRT) glass as fine aggregate. *J. Clean. Prod.* **2018**, *174*, 1436–1443. [CrossRef]
8. Kumar, A.; Holuszko, M.; Espinosa, D.C.R. E-waste: An overview on generation, collection, legislation and recycling practices. *Resour. Conserv. Recycl.* **2017**, *122*, 32–42. [CrossRef]
9. Wang, G.; Sang, L.; Tariq, M.; Lu, C.; Zhang, W.; Lin, K.; Huang, B. Systematic facile study of singleton e-waste recycling site to unveil the potential bio-indicator for atmospheric heavy metals by using tree leaves. *Process Saf. Environ. Prot.* **2020**, *143*, 304–312. [CrossRef]
10. Liu, M.; Jia, L.; Zhao, Z.; Han, Y.; Li, Y.; Peng, Q.; Zhang, Q. Fast and robust lead (II) removal from water by bioinspired amyloid lysozyme fibrils conjugated with polyethyleneimine (PEI). *Chem. Eng. J.* **2020**, *390*, 124667. [CrossRef]
11. Petrella, A.; Spasiano, D.; Race, M.; Rizzi, V.; Cosma, P.; Liuzzi, S.; De Vietro, N. Porous Waste Glass for Lead Removal in Packed Bed Columns and Reuse in Cement Conglomerates. *Materials* **2018**, *12*, 94. [CrossRef] [PubMed]
12. Liu, T.; Wei, H.; Zou, D.; Zhou, A.; Jian, H. Utilization of waste cathode ray tube funnel glass for ultra-high performance concrete. *J. Clean. Prod.* **2020**, *249*, 119333. [CrossRef]
13. Cabrera, M.; Perez, P.; Rosales, J.; Agrela, F. Feasible Use of Cathode Ray Tube Glass (CRT) and Recycled Aggregates as Unbound and Cement-Treated Granular Materials for Road Sub-Bases. *Materials* **2020**, *13*, 748. [CrossRef] [PubMed]
14. Yuan, W.; Li, J.; Zhang, Q.; Saito, F. Innovated application of mechanical activation to separate lead from scrap cathode ray tube funnel glass. *Environ. Sci. Technol.* **2012**, *46*, 4109–4114. [CrossRef]
15. Qi, Y.; Xiao, X.; Lu, Y.; Shu, J.; Wang, J.; Chen, M. Cathode ray tubes glass recycling: A review. *Sci. Total Environ.* **2019**, *650*, 2842–2849. [CrossRef] [PubMed]

16. Karuppannan Gopalraj, S.; Kärki, T. A Study to Investigate the Mechanical Properties of Recycled Carbon Fibre/Glass Fibre-Reinforced Epoxy Composites Using a Novel Thermal Recycling Process. *Processes* **2020**, *8*, 954. [CrossRef]
17. Akcil, A.; Agcasulu, I.; Swain, B. Valorization of waste LCD and recovery of critical raw material for circular economy: A review. *Resour. Conserv. Recycl.* **2019**, *149*, 622–637. [CrossRef]
18. Yu, M.; Liu, L.; Li, J. An overall Solution to Cathode-Ray Tube (CRT) Glass Recycling. *Procedia Environ. Sci.* **2016**, *31*, 887–896. [CrossRef]
19. You, L.; Jin, D.; Guo, S.; Wang, J.; Dai, Q.; You, Z. Leaching evaluation and performance assessments of asphalt mixtures with recycled cathode ray tube glass: A preliminary study. *J. Clean. Prod.* **2021**, *279*, 123716. [CrossRef]
20. Jurczak, R.; Szmatula, F.; Rudnicki, T.; Korentz, J. Effect of Ground Waste Glass Addition on the Strength and Durability of Low Strength Concrete Mixes. *Materials* **2021**, *14*, 190. [CrossRef]
21. Hu, B.; Hui, W. Lead recovery from waste CRT funnel glass by high-temperature melting process. *J. Hazard. Mater.* **2018**, *343*, 220–226. [CrossRef] [PubMed]
22. Long, W.-J.; Li, H.-D.; Ma, H.; Fang, Y.; Xing, F. Green alkali-activated mortar: Sustainable use of discarded cathode-ray tube glass powder as precursor. *J. Clean. Prod.* **2019**, *229*, 1082–1092. [CrossRef]
23. Lv, J.; Yang, H.; Jin, Z.; Ma, Z.; Song, Y. Feasibility of lead extraction from waste Cathode-Ray-Tubes (CRT) funnel glass through a lead smelting process. *Waste Manag.* **2016**, *57*, 198–206. [CrossRef]
24. Mrowiec, K.; Kubica, S.; Kuczyńska, H. Cathode ray tube (CRT) waste glass recycling. *CHEMIK* **2011**, *65*, 1212–1217.
25. Lu, X.; Ning, X.A.; Chen, D.; Chuang, K.H.; Shih, K.; Wang, F. Lead extraction from Cathode Ray Tube (CRT) funnel glass: Reaction mechanisms in thermal reduction with addition of carbon (C). *Waste Manag.* **2018**, *76*, 671–678. [CrossRef]
26. Xing, M.; Wang, J.; Fu, Z.; Zhang, D.; Wang, Y.; Zhang, Z. Extraction of heavy metal (Ba, Sr) and high silica glass powder synthesis from waste CRT panel glasses by phase separation. *J. Hazard. Mater.* **2018**, *347*, 8–14. [CrossRef]
27. Bursi, E.; Ponzoni, C.; Lancellotti, I.; Vassura, I.; Ferroni, L.; Barbieri, L. Cathode ray tube (CRT) lead glass: Lead leaching study after a chelating agent treatment. *Environ. Eng. Manag. J.* **2015**, *14*, 1503–1509. [CrossRef]
28. Veit, H.M.; Oliveira, E.d.; Richter, G. Thermal processes for lead removal from the funnel glass of CRT monitors. *Rem Rev. Esc. Minas* **2015**, *68*, 287–294. [CrossRef]
29. Xing, M.; Fu, Z.; Wang, Y.; Wang, J.; Zhang, Z. Lead recovery and high silica glass powder synthesis from waste CRT funnel glasses through carbon thermal reduction enhanced glass phase separation process. *J. Hazard. Mater.* **2017**, *322*, 479–487. [CrossRef]
30. Yao, Z.; Ling, T.-C.; Sarker, P.K.; Su, W.; Liu, J.; Wu, W.; Tang, J. Recycling difficult-to-treat e-waste cathode-ray-tube glass as construction and building materials: A critical review. *Renew. Sustain. Energy Rev.* **2018**, *81*, 595–604. [CrossRef]
31. Mingfei, X.; Yaping, W.; Jun, L.; Hua, X. Lead recovery and glass microspheres synthesis from waste CRT funnel glasses through carbon thermal reduction enhanced acid leaching process. *J. Hazard. Mater.* **2016**, *305*, 51–58. [CrossRef] [PubMed]
32. Heinzle, E.; Biwer, A.; Eissen, M.; Kholiq, M.A. Evaluation of biotechnological processes in early steps of their development in terms of ecology, safety and health. *Chem.-Ing.-Tech.* **2006**, *78*, 301–305. [CrossRef]
33. Heinzle, E.; Biwer, A.P.; Cooney, C.L. *Development of Sustainable Bioprocesses: Modeling and Assessment*; John Wiley & Sons, Ltd.: Hoboken, NJ, USA, 2007; pp. 1–294. [CrossRef]

materials

MDPI

Article

Phosphate Cements Based on Calcined Dolomite: Influence of Calcination Temperature and Silica Addition

Cristina Andreea Vijan, Alina Badanoiu *, Georgeta Voicu and Adrian Ionut Nicoara

Department of Science and Engineering of Oxide Materials and Nanomaterials, Faculty of Applied Chemistry and Materials Science, University Politehnica of Bucharest, 1-7 Gheorghe Polizu Street, 011061 Bucharest, Romania; cris_deea_vajan@yahoo.com (C.A.V.); georgeta.voicu@upb.ro (G.V.); adrian.nicoara@upb.ro (A.I.N.)
* Correspondence: alina.badanoiu@upb.ro

Abstract: The aim of this study is to assess the possibility of obtaining phosphate cements based on dolomite calcined at various temperatures with/without quartz sand addition. A lower calcination temperature of dolomite (1200 °C) determines a high increase in the system temperature when calcined dolomite is mixed with KH_2PO_4 (MKP) solution and also a rapid expansion of the paste. The increase in calcination temperature up to 1400 °C reduces the oxides reactivity; however, for lower dosages of MKP, the expansion phenomenon is still recorded. The increase in MKP dosage increases the compressive strength due to the formation of K-struvite. The mixing of dolomite with sand, followed by thermal treatment at 1200 °C, modifies its composition and reactivity; the compressive strength of phosphate cements obtained by mixing this solid precursor with MKP increases up to 28 days of curing. We assessed the nature of hydrates formed in the phosphate systems studied by X-ray diffraction in order to explain the hardening processes and the mechanical properties of these systems. The microstructure and elemental composition of hardened cement pastes were assessed by scanning electronic microscopy with energy-dispersive spectroscopy. The phosphate cements based on calcined magnesite or dolomite were used to immobilize an industrial hazardous waste with high chromium content. The partial substitution of calcined magnesite/dolomite with this waste determines an important decrease in compressive strengths. Nevertheless, the leaching tests confirm an adequate immobilization of chromium in some of the matrices studied (for a waste dosage corresponding to 0.5 wt % Cr).

Keywords: phosphate cements; calcined dolomite; quartz sand; calcined magnesite; temperature; K-struvite; industrial waste; chromium

Citation: Vijan, C.A.; Badanoiu, A.; Voicu, G.; Nicoara, A.I. Phosphate Cements Based on Calcined Dolomite: Influence of Calcination Temperature and Silica Addition. *Materials* **2021**, *14*, 3838. https://doi.org/10.3390/ma14143838

Academic Editor: Rossana Bellopede

Received: 7 June 2021
Accepted: 6 July 2021
Published: 9 July 2021

1. Introduction

Magnesium phosphate cement (MPC) hardens due to an acid–base reaction between magnesia (MgO) and phosphate acid or a phosphate salt solution [1–4]. The usual source of magnesium oxide is magnesite (magnesium carbonate), which is thermally treated at increasing temperatures to obtain caustic calcined magnesite/magnesia (CCM), dead burned magnesite/magnesia (DBM), and fused magnesia (FM). CCM has numerous applications such as hydrometallurgy, steel industry, ceramic and cement manufacture, fertilizers, water treatment, etc. [5,6]. DBM and FM are mainly used in the manufacturing process of refractory materials [5], but DBM is also a key ingredient in MPC manufacture.

In the 2014 "Report on critical raw materials for EU", magnesite was identified as a critical raw material [5]. Taking into account this aspect, it is important to find alternative sources of raw materials for the manufacture of MPC.

Dolomite is a sedimentary rock containing calcium and magnesium carbonates [7,8]. Dolomite is an important material in various industries such as the pharmaceutical industry, metallurgy, the production of paper, inorganic binders, concrete, fertilizer, refractory bricks, water treatment, absorption of heavy metals, etc. [8–10].

The calcination of dolomite is used to transform this mineral into magnesium and calcium oxides. Depending on the experimental conditions, i.e., chemical composition of dolomite, presence and amount of impurities, grain size distribution, decomposition temperature and atmosphere (air, carbon dioxide, nitrogen etc.), the thermal decomposition of dolomite in MgO and CaO can proceed in one or several steps (endothermic processes) [8,11–13].

Yu et al. [14] studied the possibility of using dolomite as raw material to produce magnesium phosphate cement. According to these authors, mixing fine dolomite with coarse quartz sand and thermal treatment at relatively low temperatures (1100–1250 °C) substantially reduces the amount of free lime, and the MgO obtained has an adequate reactivity vs. phosphate salt ($NH_4H_2PO_4$). The compressive strengths of phosphate cements pastes prepared with this type of calcined dolomite can reach 22 MPa after 3 h of hardening and 63 MPa after 7 days, with sufficient soundness [14].

MPCs can be used for the immobilization of various types of wastes with heavy metals content such as Ni, Pb, Cr, Cd, etc. [15–19]. Heavy metals are toxic and can cause serious health problems. Chromium (especially Cr (VI)) has an important toxic effect; it can produce skin irritation and ulcerations as well as liver and kidney deficiency and, if inhaled, it increases lung cancer risk [20].

Deng et al. [16] studied the influence of Cr^{3+} (brought in the system by $Cr(NO_3)_3{\cdot}9H_2O$) on the compressive strength, microstructure, as well as leaching toxicity of solidified forms into MPCs based on calcined magnesite and KH_2PO_4. According to these authors, the presence of Cr^{3+} changed the system's pH and affected the morphology of hydration products; however, the MPCs leaching toxicity was less than the one assessed for other matrices i.e., geopolymer, calcium aluminum cement, and alkali-activated slag binders.

Therefore, we assessed in this paper the possibility of producing phosphate cements by replacing calcined magnesite with dolomite thermally treated at various temperatures, with/without quartz sand addition. We also evaluated the efficiency of magnesium phosphate cements (MPC) based on calcined magnesite and magnesium and calcium phosphate cements (MCPC) based on calcined dolomite in order to immobilize an industrial waste with high chromium content. To the best of our knowledge, the immobilization of chromium in phosphate cements based on calcined dolomite has been first reported in this paper.

2. Materials and Methods

The precursors used in this study were as follows:

- Calcined magnesite (M), industrial product (Tremag, Tulcea, Romania), obtained by the calcination of magnesite at 1500 °C; the residue on sieve 90 microns mesh was 7.93%.
- Calcined dolomites, obtained by thermal treatment of natural dolomite (Rodbungrup, Bucharest, Romania) at 1200 °C (D_{12}) and 1400 °C (D_{14}) for 3 h. The natural dolomite had a content of 47% $CaCO_3$ and 37.5% $MgCO_3$ and a residue on 90 microns mesh of 24.83%. After the thermal treatment, the calcined dolomites were ground up to a fineness corresponding to total passing through a 90 microns sieve.
- Calcined mixture of dolomite and quartz sand (D_{12S}); the sand (Societe Nouvelle du Litoral, Leucate, France) had a fineness corresponding to total passing through a 200 microns sieve. The dolomite to quartz sand ratio was 1.5, and the thermal treatment was performed at 1200 °C for 1 h, based on the results reported by Yu et.al. [14]. The rate of heating was 10 °C/minute, and the cooling was performed in the oven.
- Potassium dihydrogen phosphate (KH_2PO_4—MKP), chemical reagent Sigma-Aldrich (Darmstadt, Germany).
- Setting retarder—borax (B)—chemical reagent Sigma-Aldrich (Darmstadt, Germany).
- Industrial waste with high chromium content (2.71%) in the form of $CaCrO_4.2H_2O$ [21]; $CaCO_3$, $Mg(OH)_2$, and $Ca(OH)_2$ were also detected by X-ray diffraction in this waste.

The waste also contains Si, Al, Fe, and S as well as very small amounts of As, Ba, Cu, Hg, Mn, V, W, Zn and Zr [21].

The compositions of the magnesium (and calcium) phosphate cements are presented in Table 1.

Table 1. Compositions of phosphate cements based on calcined magnesite (M) and calcined dolomite (D) with/without chromium waste.

Sample	Calcined Magnesite (M) wt %	Calcined Dolomite (D) wt %	KH_2PO_4 (MKP) wt %	Borax * (B) wt %	Sand wt %	Cr Waste ** % wt %	M or D to MKPRatio (wt)	Water to Solid Ratio (wt)	Calcination Temperature (°C)
M_MKP_4_B3.3	80	-	20	3.3	-	-	4	0.2	1500
M_MKP_4_B3.3_Cr 1	50.4	-	12.6	3.3	-	37	4	0.35	1500
M_MKP_4_B3.3_Cr_0.5	64.8	-	16.2	3.3	-	19	4	0.35	1500
D_{12}_W	-	100	-	-	-	-	-	0.8	1200
D_{12}_MKP_4	-	80	20	-	-	-	4	0.67	1200
D_{12}_MKP_B3.3_4	-	80	20	3.3	-	-	4	0.67	1200
D_{12}_MKP_2.5	-	71.43	28.57	-	-	-	2.5	0.55	1200
D_{12S}_MKP_2	-	40.2	33	-	26.8	-	2	0.2	1200
D_{14}_MKP_4	-	80	20	-	-	-	4	0.3	1400
D_{14}_MKP_2	-	67	33		-	-	2	0.2	1400
D_{14}_MKP_2_Cr1	-	42	21	-	-	37	2	0.35	1400
D_{14}_MKP_2_Cr0.5	-	54	27	-	-	19	2	0.35	1400
D_{12S}_MKP_2_Cr0.5	-	32.4	27	-	21.6	19	2	0.2	1200

* Borax dosage was calculated with reference to calcined magnesite or calcined dolomite. ** Cr waste was dosed to bring in the system 0.5 wt % Cr and 1 wt % Cr; Cr waste substitutes the oxide + phosphate salt mixture.

The specimens were obtained by the mixing of solid component (calcined magnesite or calcined dolomite) with potassium dihydrogen phosphate, water, and in some cases borax; the resulting paste was poured in rectangular molds (15 mm × 15 mm × 60 mm- width × heigh × length). The curing of specimens was performed in the mold the first 24 h and then, after demolding, in air at 20 ± 2 °C.

The reactive CaO and MgO content (available for the reaction with water) of calcined dolomite was determined according to the method presented in the standard SR EN 459-2 [22]. The available (unbound) CaO and MgO and corresponding hydroxides are dissolved in a sucrose solution and titrated with hydrochloric acid.

A Shimadzu XRD 6000 (Shimadzu, Kyoto, Japan), CuKα (λ = 1.5406 Å), 2θ ranging between 10 and 60, a 0.02 step size, and a 2 deg./min scan speed was used for X ray diffraction analyses.

The microstructure of pastes was assessed by Scanning Electron Microscopy (SEM) using an FEI Inspect F50 (Thermo Fisher—former FEI, Eindhoven, Nederland) electronic microscope equipped with a Schottky emission electron beam with a resolution of 1.2 nm at 30 kV and 3 nm at 1 kV (BSE). In this analysis, the freshly fractured samples were visualized in a vacuum mode using a 30 kV acceleration voltage and spot 3.5.

A differential thermal analyzer Shimadzu DTG-TA 51H (Shimadzu, Kyoto, Japan) was used for complex thermal analysis (DTA-TG); the analyses were performed in air, with a heating rate of 10 °C/minute, in the temperature range 20–1000 °C.

Prismatic specimens (15 mm× 15 mm × 60 mm- width × heigh × length), cured for 1 up to 28 days in air at 20 ± 2 °C, were employed for the assessment of compressive strength using a Matest testing machine (Matest, Treviolo, Italy). For the calculation of average compressive strength, a minimum of 6 compressive strength values were considered. The outliers (±10%) were not considered in calculation.

The chromium leaching test was performed according to the method presented in standard SR EN 12457-4: 2003 [23]. The MPC and MCPC specimens, hardened in air at 20 ± 2 °C for 28 days, were triturated and sieved; the particles smaller than 10 mm were mixed with water (water to solid ratio was 10). The resulting suspension was stirred for 24 h at a rate of 10 rpm by means of an orbital shaker (Heidolph Instrument Gmbh&Co.KG, Schwabach, Germany); next, the suspension was filtered, and the leachate was mixed with nitric acid to achieve a pH lower than 2. An atomic absorption spectrometer (Analytik, Jena, Germany) was used to assess the heavy metals concentration in the leachate.

3. Results

The X-ray diffraction patterns of natural dolomite presented in Figure 1 confirm the presence of dolomite ($CaMg(CO_3)_2$) along with a small amount of calcite ($CaCO_3$).

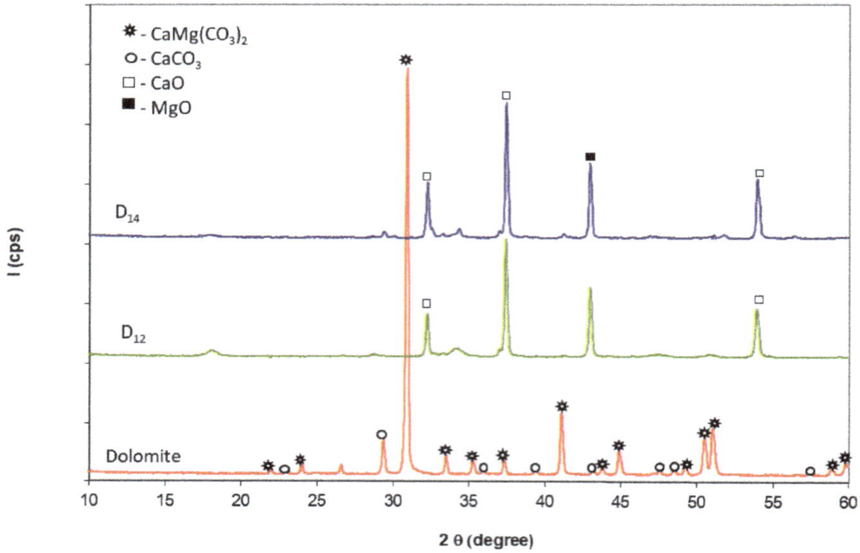

Figure 1. XRD patterns of dolomite and dolomite calcined at 1200 °C (D_{12}) and 1400 °C (D_{14}) for 3 h.

The complex thermal analysis of dolomite (Figure 2) shows on the DTA curve an endothermic process with a shoulder at approximately 650 °C and maximum at 800 °C with a corresponding weight loss of 46.35% (assessed on TG curve). This process, which ends at 850 °C, can be attributed to the decarbonation of the magnesium carbonate and the calcium carbonate with CaO and MgO formation [8,11–13].

In correlation with previous results, the thermal treatment of dolomite at 1200 °C and 1400 °C leads to the transformation of calcium magnesium carbonate (083-1530) and calcium carbonate (072-1652) into magnesium oxide (004-0829) and calcium oxide (082-1690) (Figure 1). There are no significant differences between the XRD patterns of dolomite thermally treated at these two temperatures.

The mixing of dolomite with quartz sand and thermal treatment at 1200 °C for 1 h determines, as expected, the formation of calcium and/or magnesium silicates (see Figure 3). The XRD patterns presented in Figure 3 also show the presence of SiO_2 along with MgO. The intensities of XRD peaks of CaO are much smaller (as compared with those assessed on D_{12} and D_{14} XRD patterns—Figure 1) due to its partial consumption in the reaction with SiO_2.

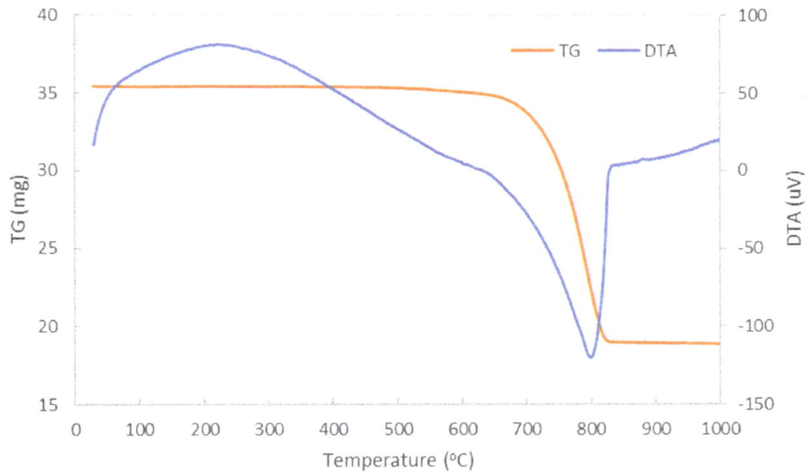

Figure 2. TG-DTA curves of dolomite.

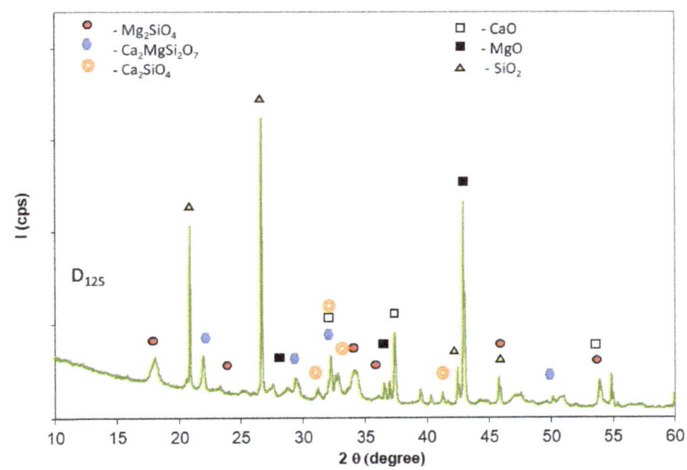

Figure 3. XRD patterns of dolomite with sand calcined at 1200 °C (D_{12S}) for 1 h.

The amount of reactive calcium and magnesium oxides assessed by the method presented in SR EN 459-2 [22] in the dolomite thermally treated for 3 h at 1200 °C was 54.13%, and for the dolomite calcined at 1400 °C, it was 22%. The decrease of oxides' reactivity when the thermal temperature increases is due, as in the case of thermal treatment of limestone, to the increase of oxides' crystals sizes correlated with the decrease of porosity, when the material is thermally treated at a higher temperature [24–27].

The mixing with water (W) or KH_2PO_4 solution (MKP) determines, for the specimens based on the dolomite calcined at 1200 °C, an intense and rapid heat release (Table 2), due to the hydration of MgO and CaO (in the case of D_{12}_W) and to the reaction with the MKP (in the case of D_{12}_MKP_4 and D_{12}_MKP_2.5). It has been noticed that a decrease of MKP content determines a slower heat release corresponding to the exothermic processes specific for the setting and hardening of these phosphate systems i.e., the maximum temperature assessed on pastes is reached after a longer time (see D_{12}_MKP_4 as compared to D_{12}_MKP_2.5—Table 2).

Table 2. Maximum temperature (T_{max}) and corresponding time (t_{max}) for the studied binders.

Sample	T_{max} * (°C)	t_{max} ** (min)	Obs.
D_{12}_W	98	4	Expansion
D_{12}_MKP_4	90	26	Expansion
D_{12}_MKP_2.5	90	10	Expansion

* T_{max}—maximum temperature (°C) of paste assessed after the mixing of precursors; ** t_{max}—time (minutes) corresponding to Tmax.

Figure 4 shows the XRD patterns of sample D_{12}_W after 3 days of hardening. It can be observed the presence of $Ca(OH)_2$ (084-1276) resulting from the hydration of CaO and the presence of $Mg(OH)_2$ (084-2163) resulting from the hydration of MgO—both exothermic processes that explain the significant temperature increase; the presence of MgO peaks on the XRD patterns confirms the smaller reactivity vs. water of this oxide as compared to CaO [24,27,28], which correlates with the thermal treatment temperature and plateau.

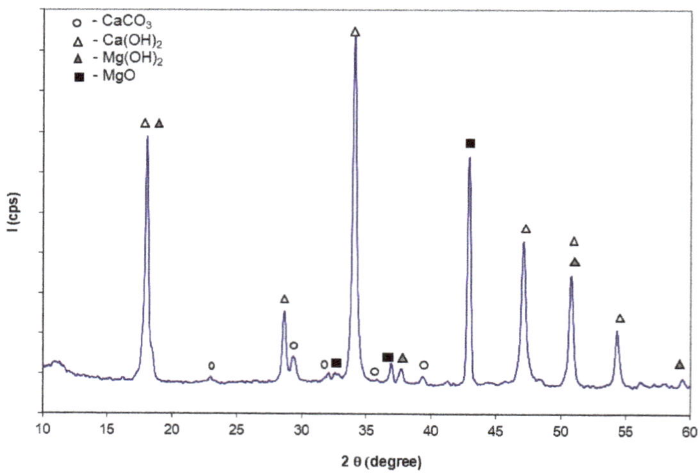

Figure 4. XRD patterns of paste obtained by mixing of water with dolomite calcined at 1200 °C/3 h (D_{12}_W).

On the XRD patterns of cements based on dolomite with/without borax and MKP (Figure 5a) can be noticed the presence of calcium and magnesium hydroxides as well as the presence of hydroxyapatite (HAp)—which resulted in the reaction of calcium with phosphate, which was brought into the system by the potassium dihydrogen phosphate (MKP). The formation of HAp is also facilitated by the basicity of this system (the addition of MKP to the calcined dolomite + water mixture shifts the pH value at 8–9).

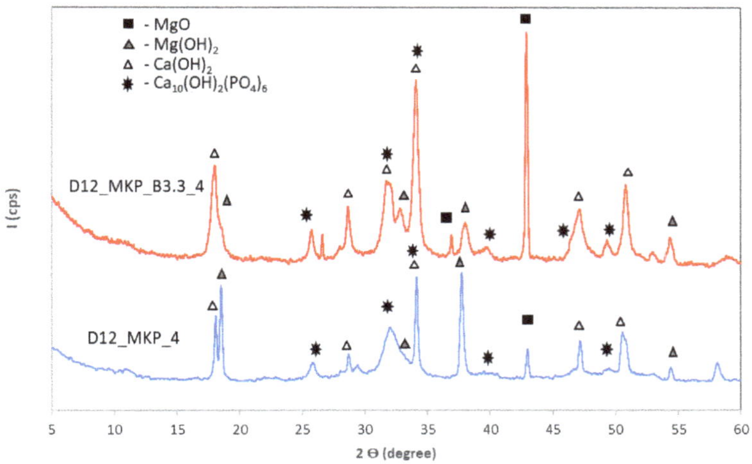

(a)

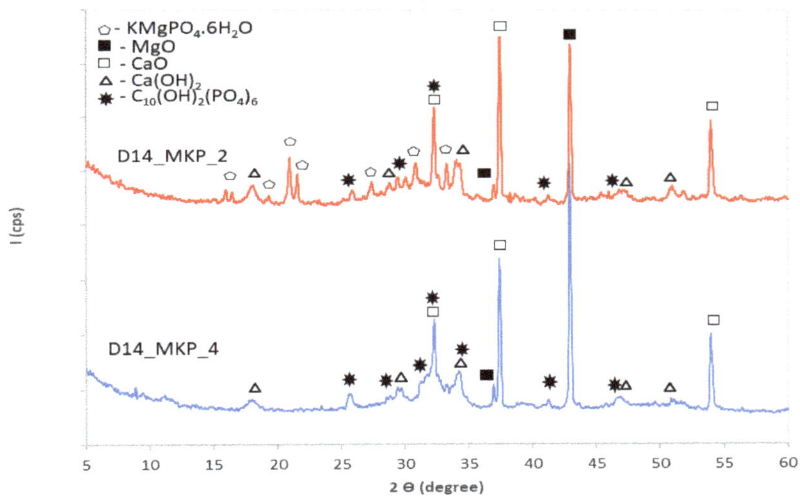

(b)

Figure 5. XRD patterns of the paste obtained by mixing water with potassium dihydrogen phosphate (with/without borax) and dolomite calcined for 3 h at (**a**) 1200 °C; (**b**) 1400 °C.

Due to the high reactivity vs. water or MKP solution of dolomite calcinated at 1200 °C, phosphate cements based on dolomite calcined at 1400 °C were also obtained.

Figure 5b shows the XRD patterns of phosphate cements based on dolomite calcined at 1400 °C prepared with different dosages of potassium dihydrogen phosphate (MKP). In the case of the specimens with a higher dosage of KH_2PO_4 (D_{14}_MKP_2) along with XRD peaks specific for magnesium oxide (004-0829), calcium oxide (082-1690), calcium hydroxide (084-1276) and HAp (084-1998) (which are also present on XRD patterns of D_{14}_MKP_4) appear also XRD peaks specific for K-struvite ($KMgPO_4.6H_2O$)(020-0685). Due to the higher reactivity of CaO as compared to MgO [24,27,28], HAp is the first reaction product formed in this system; if there are still available phosphate groups in the solution, K-struvite is formed by their reaction with magnesium. The presence of K-struvite contributes to the increase of mechanical strength [29,30].

These results are in correlation with the values of compressive strengths, which are presented in Table 3. The higher compressive strengths were assessed for the MPC based on calcined magnesite (M_MKP_4_B3.3) in which the main reaction product is K-struvite [30]. The specimens based on dolomite calcined at 1400 °C have recordable strengths only when the MgO/KH_2PO_4 ratio is 2, i.e., when K-struvite is detected in the hardened paste.

Table 3. Compressive strengths versus time. Influence of Cr waste presence.

Specimens	Compressive Strength (MPa)			
	1 Day	3 Days	7 Days	28 Days
M_MKP_4_B3.3	16.87	21.62	26.5	27.1
M_MKP_4_B3.3_Cr0.5	0.6	1	1.15	1.5
M_MKP_4_B3.3_Cr1	0	0	0	0
D_{14}_MKP_4	0	0	0	0
D_{14}_MKP_2	7.1	10.8	12.2	0
D_{14}_MKP_2_Cr0.5	1.5	1.7	2	0
D_{14}_MKP_2_Cr1	0	0	0	0
D_{12S}_MKP_2	2.8	6.3	7.2	9.2
D_{12S}_MKP_2_Cr0.5	0	3.5	5.4	5.4

In order to assess the influence of chromium waste on the composition of hardened phosphate cements, pastes with various amounts of waste were prepared (Table 1). The XRD patterns of the pastes based on calcined magnesite (M) and dolomite calcined at 1400 °C (D_{14}) with a dosage of chromium waste corresponding to 0.5 wt % Cr are presented in Figure 6.

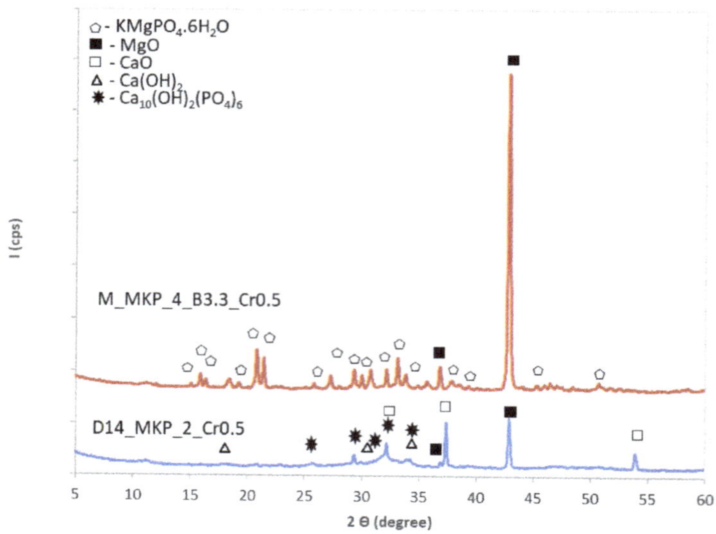

Figure 6. XRD patterns of the specimens based on calcined magnesite or dolomite calcined at 1400 °C with chromium waste corresponding to 0.5 wt % Cr.

For the paste based on magnesite (M_MKP_4_B3.3_Cr0.5), one can assess through this method the presence of MgO and K-struvite ($KMgPO_4.6H_2O$). The substitution of calcined magnesite with chromium waste determines an important decrease of the compressive strengths (Table 3), which could be due both to the smaller amount of K-struvite formed in the system (chromium waste substitute calcined magnesite and MKP) as well as the increase of water dosage (from 0.2 to 0.35—see Table 1) necessary to improve the workability of fresh paste.

For the specimen based on calcined dolomite (D_{14}), the presence of chromium waste seems to inhibit the K-struvite formation (see also Figure 5b). This explains the decrease of compressive strengths values as compared with those recorded for specimen D_{14}_MKP_2, with the increase of Cr content (see Table 3). However, after 7 days of hardening, the compressive strength of specimens based on D_{14} (D_{14}_MKP_2 and D_{14}_MKP_2_Cr0.5) dramatically decrease, which is most probably due to a delayed hydration of free CaO and MgO.

Therefore, in order to reduce the free lime content and to obtain magnesium oxide with an adequate reactivity, while keeping the same thermal treatment temperature of 1200 °C, a mixture of dolomite and quartz sand was thermally treated at this temperature for 1 h, based on the method proposed by Yu et al. [14].

The XRD pattens of the phosphate cement based on dolomite + sand calcined at 1200 °C—D_{12S} (Figure 7) show the presence of hydrates i.e., K-struvite and $Ca(OH)_2$ along with MgO, SiO_2, and Mg_2SiO_4 assessed in D_{12S} (see Figure 3). The presence of chromium waste does not change the nature of the reaction products (hydrates) assessed by this method (Figure 7).

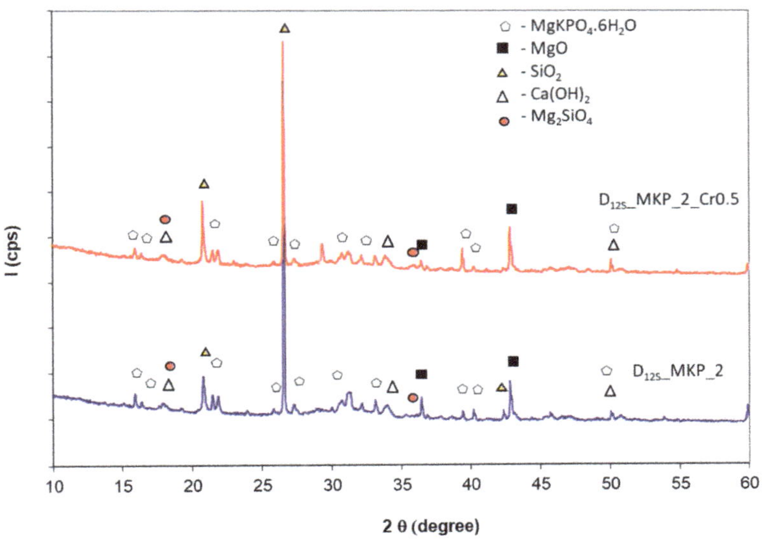

Figure 7. XRD patterns of the pastes obtained by mixing potassium dihydrogen phosphate and dolomite + sand calcined for 1 h at 1200 °C, with/without chromium waste.

In correlation with the above presented data, the compressive strengths of phosphate cements based on dolomite + sand calcined at 1200 °C (D_{12S}_MKP_2) are lower in comparison to the ones assessed for the phosphate cement based on magnesite (M_MKP_B3.3_4); however, these values steadily increase up to 28 days (Table 3). This compressive strength evolution can be related to the formation of K-struvite (assessed by XRD) and to the presence of sand grains, which act as aggregates (Figure 8).

Figure 9 presents the SEM images and elemental compositions assessed by EDX on various areas of D_{12S}_MKP_2 cement paste. As can be seen from Figure 9a, in area 1, the atomic ratio of K:Mg:P is 13.25:13.14:14.14 confirming the presence of K-struvite in these specimens; the elemental compositions in area 3 (Figure 9a) and area 1 (Figure 9b) show the presence of Ca together with Mg, K, and P, which suggest a complex composition of these hydrates [14].

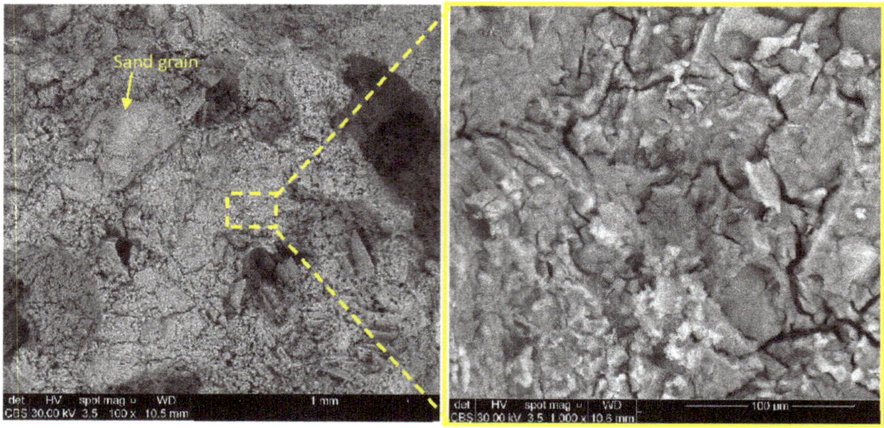

Figure 8. BSE images of D_{12S}_MKP_2.

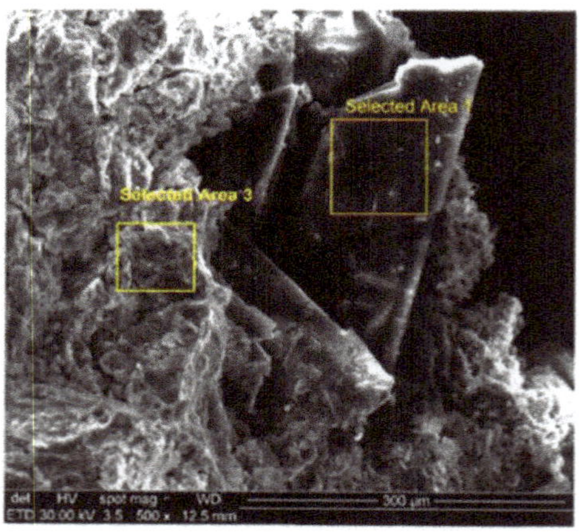

Element	Area 1 (atomic %)	Area 3 (atomic %)
Mg	13.14	4.29
Si	0.45	0.18
P	14.14	9.18
K	13.25	8.78
Ca	0.43	5.3
C	3.76	4.71
O	54.83	67.56

a

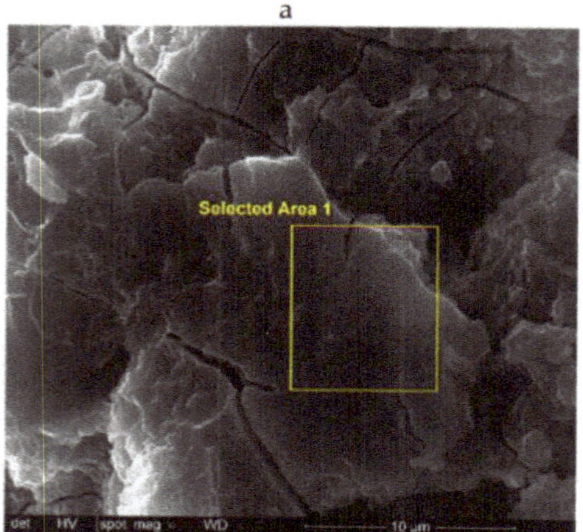

Element	Area 1 (atomic %)
Mg	7.49
Si	1.05
P	13.10
K	12.83
Ca	5.34
O	60.19

b

Figure 9. SEM images and EDX analyses of D_{12S}_MKP_2 cement paste at different magnifications: (**a**) ×500; (**b**) ×10,000.

For the cement paste based on D_{12S}, the presence of chromium waste (in a dosage corresponding to 0.5 wt % Cr) determines a reduction of compressive strengths in comparison to the cement without Cr (D_{12S}_MKP_2); however, these values increase up to 28 days (Table 3). The decrease of compressive strengths can be explained by the lower amount of MgO (sand partially substitutes the dolomite) available in this system for the formation of K-struvite.

For a better understanding of the correlation between the morphology/composition and properties of D_{12S}_MKP_2_Cr0.5, SEM and EDX analyses were performed on this specimen (Figure 10). As it can be noticed, the sand grains are embedded in a continuous matrix (Figure 10a) in which are present plate-like and prismatic crystals intermixed with agglomerates of small grains (Figure 10b). The coherence of this matrix seems to be much lower in these specimens in comparison to the one without Cr (Figure 9), which can explain the lower mechanical strength values.

a b

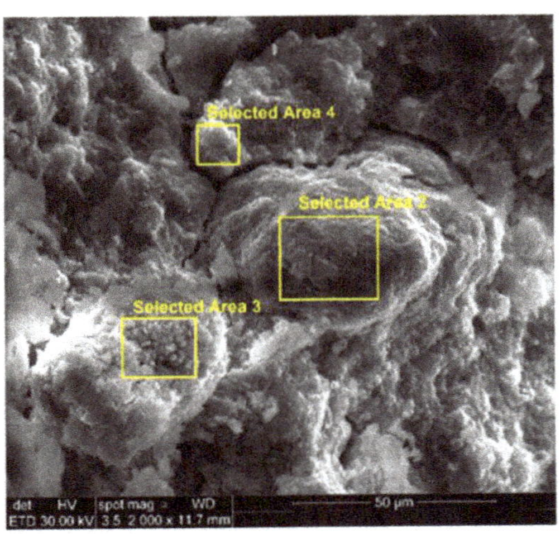

Element	Area 2 atomic %	Area 3 atomic %	Area 4 atomic %
Mg	10.25	4.32	11.25
Al	2.26	-	-
Si	3.47	0.78	0.84
P	2.51	3.07	7.65
K	4.04	2.63	5.88
Ca	5.68	11.49	8.92
Cr	1.04	-	0.1
Fe	2.48	0.15	0.22
C	0.02	3.85	1.54
O	68.25	73.71	63.6

c

Figure 10. SEM images and EDX analyses of D_{12S}_MKP_2_Cr0.5 paste, at different magnifications: (**a**) ×500; (**b**) ×2000; (**c**) ×2000.

The EDX analysis performed in three areas on SEM image presented in Figure 10c shows the presence of Cr, Al, and Fe (from waste [21]) mainly in area 2, which suggests the presence of a waste grain embedded in a layer (matrix) with Ca, K, and P content.

To assess the efficiency of the studied MPC and CMPC to immobilize Cr, a leaching test (described in SR EN 12457-4: 2003 [23]) was performed for phosphate cement pastes hardened for 28 days. The results are presented in Figure 11.

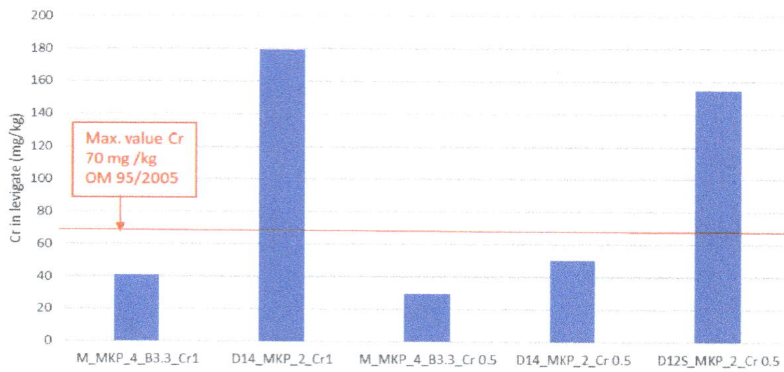

Figure 11. Cr content in levigates.

In Figure 11, one can observe that the MPCs based on calcined magnesite are effective for the immobilization of chromium even for a high waste content (M_MKP_4_B3.3_Cr1); the Cr content determined in leachate is below the limit stipulated by the Romanian Ministerial Order OM 95/2009 [31] for both phosphate cement pastes based on calcined magnesite.

The good immobilization of Cr in the MPCs based on calcined magnesite (M) can be explained by the presence of K-struvite, which could play an important role [16,32]. Rouff [32] reported Cr adsorption or/and substitution in the struvite ($NH_4.H_2PO_4.6H_2O$), which precipitates from concentrate solutions of $MgCl_2.6H_2O$ and $(NH_4)_2HPO_4$, with $Cr(NO_3)_3.9H_2O$ or Na_2CrO_4 additions.

For the phosphate cements based on calcined dolomite, only the one based on dolomite calcined at 1400 °C with a waste content corresponding to 0.5 wt % Cr (D_{14}_MKP_2_Cr0.5) fulfills the legal requirement. The reduced efficiency in immobilization of Cr in this type of cement can be due to the inhibition of K-struvite formation (suggested by XRD data—Figure 6). Nevertheless, the presence of hydroxyapatite (HAp) in this composition could contribute to the immobilization of Cr [33,34], explaining the low amount of chromium levigated in the CMPC with a lower chromium waste content (0.5 wt. %).

The high Cr content assessed in the levigate of the CMPC based on D_{12S} can be due to a low amount of K-struvite formed in this cement, due to partial substitution of dolomite with quartz sand.

4. Conclusions

In this study, we obtained and studied the properties of magnesium phosphate and calcium magnesium phosphate cements based on calcined magnesite and calcined dolomite (at different temperatures) with/without sand addition and potassium dihydrogen phosphate.

The following conclusions can be drawn:

- The thermal treatment of dolomite at 1200 °C and 1400 °C leads to the decomposition of calcium magnesium carbonate into CaO and MgO. The increase in calcination temperature, from 1200 to 1400 °C reduces the reactivity of calcium and magnesium oxides vs. water or phosphate (MKP) solution; for the phosphate cements based on dolomite calcined at 1200 °C, an important increase in paste temperature during the setting and paste's expansion was noticed due to the high reactivity of oxides (CaO and MgO); the increase of thermal treatment temperature at 1400 °C determines a decrease of the oxides' reactivity, and for a higher KH_2PO_4 dosage (corresponding to $D_{14}/KH_2PO_4 = 2$ weight ratio), the pastes have measurable compressive strength at early ages. Nevertheless, for all specimens based on dolomite calcined at 1400 °C, the compressive strengths dramatically decrease after 7 days of hardening, which is most probably due to a delayed hydration of CaO and MgO.

- The main compounds observed in hardened phosphate binders based on calcined dolomite were calcium and magnesium hydroxides; in the case of specimens with a lower dosage of MKP (corresponding to $D_{14}/MKP = 4$ weight ratio)—along with $Ca(OH)_2$ and $Mg(OH)_2$, a new compound was detected by XRD—hydroxyapatite (HAp); HAp results from the reaction of calcium with phosphate, which is brought into the system by MKP. For a higher dosage of MKP (corresponding to $D_{14}/MKP = 2$ weight ratio) on the XRD patterns, peaks specific for the K-struvite ($KMgPO_4.6H_2O$) compound are also present.
- In order to obtain a solid precursor for CMPC synthesis by the calcination of dolomite at relatively low temperature, a mixture of dolomite and quartz sand was thermally treated at 1200 °C for 1 h. The compressive strengths of resulting CMPC are lower in comparison with those assessed on phosphate cement based on calcined magnesite; however, they steadily increase up to 28 days. The lower values of compressive strengths assessed on these compositions are mainly due to the lower content of MgO available in this type of calcined dolomite (D_{12S}) for the formation of K-struvite.
- The partial substitution of calcined magnesite and calcined dolomite with an industrial waste product with chromium content (corresponding to a Cr dosage of 0.5 wt % and 1 wt %) led to a significant decrease of compressive strength. In the case of MPC based on calcined magnesite, this decrease can be explained by the reduction of K-struvite amount due to the replacement of calcined magnesite and KH_2PO_4 with chromium waste, as well as by the increase of the water-to-solid ratio (necessary to obtain an adequate workability). For the CMPC based on dolomite calcined at 1400 °C (D_{14}), the replacement of active components (D_{14} and KH_2PO_4) with chromium waste further inhibits the formation of K-struvite, and the compressive strengths decrease with the increase of the waste content. In the case of CMPC based on D_{12S}, the lower compressive strengths assessed for specimen with chromium waste are also explained by the lower amount of K-struvite formed in this system (due to the decrease of MgO available for the reaction with KH_2PO_4).
- Phosphate cements based on calcined magnesite with a waste content corresponding to 1 and 0.5 wt % Cr can effectively reduce the Cr leaching for pastes cured for 28 days; in the case of phosphate cements based on calcined dolomite (D_{14}), it was found that for a waste dosage corresponding to 0.5 wt % Cr, after 28 days of curing, the concentration of leached Cr is below the limit value imposed by the legislation currently in force. Nevertheless, considering the evolution of mechanical strength vs. time for this phosphate cement, it seems necessary to extend the evaluation of leached chromium for longer curing times (over 28 days). Although the CMPCs based on D_{12S} developed adequate compressive strengths even at longer hardening times (28 days), the amount of chromium leached exceeded the limit imposed by the current legislation most probably due to the decrease of K-struvite content. The chromium immobilization in this type of CMPC can be improved if the appropriate amount of D_{12S} (or chromium waste) is selected.

Author Contributions: Conceptualization, C.A.V. and A.B.; Investigation, C.A.V., A.I.N.; G.V.; Methodology: C.A.V. and A.B.; Supervision, A.B. and G.V.; Writing—original draft, C.A.V. and A.B.; Writing—review and editing, A.B. and G.V. All authors have read and agreed to the published version of the manuscript.

Funding: Cristina Andreea Vijan acknowledges the financial support of the Operational Programme Human Capital of the Ministry of European Funds through the Financial Agreement 51668/09.07.2019, SMIS code 124705.

Institutional Review Board Statement: Not applicable.

Informed Consent Statement: Not applicable.

Data Availability Statement: Not applicable.

Conflicts of Interest: The authors declare no conflict of interest.

References

1. Xu, B.; Ma, H.; Li, Z. Influence of magnesia-to-phosphate molar ratio on microstructures, mechanical properties and thermal conductivity of magnesium potassium phosphate cement paste with large water-to-solid ratio. *Cem. Concr. Res.* **2015**, *68*, 1–9. [CrossRef]
2. Liu, R.; Yang, Y.; Sun, S. Effect of M/P and Borax on the Hydration Properties of Magnesium Potassium Phosphate Cement Blended with Large Volume of Fly Ash. *J. Wuhan Univ. Technol. Mater. Sci. Ed.* **2018**, *33*, 1159–1167. [CrossRef]
3. Yue, L.; Bing, C. Factors that affect the properties of magnesium phosphate cement. *Constr. Build. Mater.* **2013**, *47*, 977–983.
4. Jiang, Z.; Zhang, L.; Geng, T.; Lai, Y.; Zheng, W.; Huang, M. Study on the Compressive Properties of Magnesium Phosphate Cement Mixing with Eco-Friendly Coir Fiber Considering Fiber Length. *Material* **2020**, *13*, 3194. [CrossRef]
5. European Commission. Report on Critical Raw Materials for EU. Available online: http://ec.europa.eu/DocsRoom/documents/11911/attachments/1/translations/en/renditions/native (accessed on 23 May 2021).
6. Alvarado, E.; Torres-Martinez, L.M.; Fuentes, A.F.; Quintana, P. Preparation and characterization of MgO powders obtained from different magnesium salts and the mineral dolomite. *Polyhedron* **2000**, *19*, 2345–2351. [CrossRef]
7. Subagjo; Wulandari, W.; Adinata, P.M.; Fajrin, A. Thermal decomposition of dolomite under CO_2-air atmosphere. In *AIP Conference Proceedings, Proceedings of the 1st International Process Metallurgy Conference (IPMC 2016), Bandung, Indonesia, 10–11 November 2016*; AIP publishing: Dresden, Germany, 2017; Volume 1805. [CrossRef]
8. Olszak-Humienik, M.; Jablonski, M. Thermal behavior of natural dolomite. *J. Therm. Anal. Calorim.* **2015**, *119*, 2239–2248. [CrossRef]
9. Wulandari, W.; Subagjo; Mursito, A.T.; Juanjaya, F.J.; Alwi, M.F. Performance of Dolomite Calcination in a Bench-Scale Rotary Kiln. In *MATEC Web of Conferences, Proceedings of the The 24th Regional Symposium on Chemical Engineering (RSCE 2017), Semarang, Indonesia, 15–16 November 2017*; EDP Science: Les Ulis, France, 2018; Volume 156, pp. 704–708.
10. Roques, H.; Nugroho-Jeudy, L.; Lebugle, A. Phosphorus removal from wastewater by half-burned dolomite. *Wat. Res.* **1991**, *25*, 959–965. [CrossRef]
11. Engler, P.; Santana, M.W.; Mittleman, M.L.; Balazs, D. Non isothermal, In situ XRD analysis of dolomite decomposition. *Thermochim. Acta* **1989**, *140*, 67–76. [CrossRef]
12. Mako, E. The effect of quartz content on the mechanical activation of dolomite. *J. Eur. Ceram. Soc.* **2007**, *27*, 535–540. [CrossRef]
13. Gunasekaran, S.; Anbalagan, G. Thermal decomposition of natural dolomite. *Bull. Mater. Sci.* **2007**, *30*, 339–344. [CrossRef]
14. Yu, J.; Qian, J.; Wang, F.; Qin, J.; Dai, X.; You, C.; Jia, X. Study of using dolomite ores as raw materials to produce magnesium phosphate cement. *Constr. Build. Mater.* **2020**, *253*, 119147. [CrossRef]
15. Singh, D.; Wagh, A.S.; Cunnane, J.C.; Mayberry, J.L. Chemically bonded phosphate ceramics for low-level mixed-waste stabilization. *J. Environ. Sci. Health* **1997**, *32*, 527–541. [CrossRef]
16. Deng, Q.; Lai, Z.; Yan, T.; Wu, J.; Liu, M.; Lu, Z.; Lv, S. Effect of Cr (III) on hydration, microstructure of magnesium phosphate cement, and leaching toxicity evaluation. *Environ. Sci. Pollut. Res.* **2021**, *28*, 15290–15304. [CrossRef]
17. Wang, Y.S.; Dai, J.G.; Wang, L.; Tsang, D.C.W.; Poon, C.S. Influence of lead on stabilization/solidification by ordinary Portland cement and magnesium phosphate cement. *Chemosphere* **2018**, *190*, 90–96. [CrossRef] [PubMed]
18. Sua, Y.; Yanga, J.; Liuc, D.; Zhena, S.; Lind, N.; Zhoue, Y. Effects of municipal solid waste incineration fly ash on solidification/stabilization of Cd and Pb by magnesium potassium phosphate cement. *J. Environ. Chem. Eng.* **2016**, *4*, 259–265. [CrossRef]
19. Vijan, C.A.; Badanoiu, A.; Nicoara, A.I.; Barcan, I. Effect of lead and nickel on the hardening processes and properties of phosphate cements. *Rev. Rom. Mater.* **2020**, *50*, 510–520.
20. ATSDR, Agency for Toxic Substances and Disease Registry. Available online: https://www.atsdr.cdc.gov/csem/chromium/signs_and_symptoms.html (accessed on 29 June 2021).
21. Oproiu, C.; Parvan, M.; Voicu, G.; Badanoiu, A.I. Inertization of an industrial waste rich in chromium in composite portland cements. *Rev. Rom. Mater.* **2018**, *48*, 458–466.
22. SR EN 459-2. *2001-Building Lime. Part 2. Test Methods*; ASRO: Bucharest, Romania, 2001.
23. SR EN 12457-4:2003. *Characterization of Waste—Leaching—Compliance Test for Leaching of Granular Waste Materials and Sludges—Part 4: One Stage Batch Test at a Liquid to Solid Ratio of 10 l/kg for Materials with Particle Size Below 10 mm (without or with Size Reduction)*; ASRO: Bucharest, Romania, 2003.
24. Teoreanu, I. *The Basics of Inorganic Binder Technology (in Romanian)*; Editura Didactică și Pedagogică: Bucharest, Romania, 1993.
25. Schorcht, F.; Kourti, I.; Scalet, B.M.; Roudier, S.; Sancho, L.D. Best Available Techniques (BAT) Reference Document for the Production of Cement, Lime and Magnesium Oxide (Integrated Pollution Prevention and Control). Available online: https://ec.europa.eu/jrc/en/publication/reference-reports/best-available-techniques-bat-reference-document-production-cement-lime-and-magnesium-oxide (accessed on 23 May 2021).

26. Stork, M.; Meindetsma, W.; Overgaag, M.; Neelis, M. *A Competitive and Efficient Lime Industry. Cornerstone for a Sustainable Europe*; Technical Report; Ecofys by order of EuLA: Brussels, Belgium, 2014; pp. 1–60.
27. Hashimoto, H.; Komaki, E.; Hayashi, F.; Uematsu, T. Partial Decomposition of Dolomite in CO_2. *J. Solid State Chem.* **1980**, *33*, 181–188. [CrossRef]
28. Wells, L.S.; Taylor, K. Hydration of magnesia in dolomitic hydrated limes and putties. *J. Res. Natl. Bur. Stand.* **1937**, *19*, 215–236. [CrossRef]
29. Wagn, A.S. *Chemically Bonded Phosphate Ceramics*; Elsevier: Amsterdam, The Netherlands, 2016.
30. Vijan, C.A.; Badanoiu, A. The influence of potassium phosphate and fly ash addition on the setting time and mechanical strengths of magnesium phosphate cement. *Univ. Politeh. Buchar. Sci. Bull. Ser. B* **2020**, *82*, 21–32.
31. Romanian Ministerial Order no. 95/2005 on the Establishment of Acceptance Criteria and Preliminary Acceptance Procedures for Landfill Waste and the National List of Waste Accepted in Each Landfill Class (According to 2003/33/EC). Available online: https://www.ecotic.ro/wp-content/uploads/2015/07/1e62400fcaa487767564187df6343e669b6f36ad.pdf (accessed on 23 May 2021).
32. Rouff, A.A. The use of TG/DSC–FT-IR to assess the effect of Cr sorption on struvite stability and composition. *J. Therm. Anal. Calorim.* **2012**, *110*, 1217–1223. [CrossRef]
33. Nayak, A.; Bhushan, B. Hydroxyapatite as an advanced adsorbent for removal of heavy metal ions from water: Focus on its applications and limitations. *Mater. Today Proc.* **2021**. [CrossRef]
34. Campisi, S.; Evangelisti, C.; Postole, G.; Gervasini, A. Combination of Interfacial Reduction of Hexavalent Chromium and Trivalent Chromium Immobilization on Tin-Functionalized Hydroxyapatite Material. *Appl. Surf. Sci.* **2021**, *539*, 148227. [CrossRef]

materials

MDPI

Review

Materials for Production of High and Ultra-High Performance Concrete: Review and Perspective of Possible Novel Materials

Markssuel Teixeira Marvila [1], Afonso Rangel Garcez de Azevedo [2,*], Paulo R. de Matos [3,4], Sergio Neves Monteiro [5] and Carlos Maurício Fontes Vieira [1]

1 LAMAV—Advanced Materials Laboratory, UENF—State University of the Northern Rio de Janeiro, Av. Alberto Lamego, 2000, Campos dos Goytacazes 28013-602, Brazil; markssuel@hotmail.com (M.T.M.); veira@uenf.br (C.M.F.V.)

2 LECIV—Civil Engineering Laboratory, UENF—State University of the Northern Rio de Janeiro, Av. Alberto Lamego, 2000, Campos dos Goytacazes 28013-602, Brazil

3 Department of Civil Engineering, UFSC—Federal University of Santa Catarina, Rua João Pio Duarte Silva, 205, Florianópolis 88040-900, Brazil; paulorm.matos@gmail.com

4 Coordenadoria Acadêmica, UFSM—Federal University of Santa Maria, Rodovia Taufik Germano, 3013, Cachoeira do Sul 96503-205, Brazil

5 Department of Materials Science, IME—Military Institute of Engineering, Square General Tibúrcio, 80, Rio de Janeiro 22290-270, Brazil; snevesmonteiro@gmail.com

* Correspondence: afonso.garcez91@gmail.com

Citation: Marvila, M.T.; de Azevedo, A.R.G.; de Matos, P.R.; Monteiro, S.N.; Vieira, C.M.F. Materials for Production of High and Ultra-High Performance Concrete: Review and Perspective of Possible Novel Materials. *Materials* **2021**, *14*, 4304. https://doi.org/10.3390/ma14154304

Academic Editors: Rossana Bellopede and Lorena Zichella

Received: 6 July 2021
Accepted: 29 July 2021
Published: 31 July 2021

Abstract: This review article proposes the identification and basic concepts of materials that might be used for the production of high-performance concrete (HPC) and ultra-high-performance concrete (UHPC). Although other reviews have addressed this topic, the present work differs by presenting relevant aspects on possible materials applied in the production of HPC and UHPC. The main innovation of this review article is to identify the perspectives for new materials that can be considered in the production of novel special concretes. After consulting different bibliographic databases, some information related to ordinary Portland cement (OPC), mineral additions, aggregates, and chemical additives used for the production of HPC and UHPC were highlighted. Relevant information on the application of synthetic and natural fibers is also highlighted in association with a cement matrix of HPC and UHPC, forming composites with properties superior to conventional concrete used in civil construction. The article also presents some relevant characteristics for the application of HPC and UHPC produced with alkali-activated cement, an alternative binder to OPC produced through the reaction between two essential components: precursors and activators. Some information about the main types of precursors, subdivided into materials rich in aluminosilicates and rich in calcium, were also highlighted. Finally, suggestions for future work related to the application of HPC and UHPC are highlighted, guiding future research on this topic.

Keywords: high-performance concrete; ultra-high-performance concrete; fibers; mechanical strength; alkali-activated cement; geopolymers

1. Introduction

With the advancement of cement science and technology and due to the necessity for slender and bolder structures, conventional concrete (CC), also named as normal strength concrete [1], no longer meets the requirements for the execution of these works. In this context, other concretes and other cementitious mixtures, with properties superior to CC, emerged to meet this need. This is the case for high-strength concrete (HSC) [2], high-performance concrete (HPC) [1,3] and, more recently, ultra-high performance concrete (UHPC) [3,4].

Another important aspect of using HPC and UHPC is the environmental issue, as these concretes provide lower values of binder intensity (*bi*). This index measures the total amount of binder needed to provide one unit of a given performance indicator, for example, compressive strength in MPa, as per Equation (1). In this sense, the application of these types of concrete provides an ecological gain, due to the reduction in cement consumption to reach the same level of compressive strength [5,6].

$$bi = \frac{b}{\sigma} \quad (1)$$

where b is the total consumption of binder materials ($kg \cdot m^{-3}$) and σ is the compressive strength (MPa) at 28 days.

It is very common to confuse these terms, due to the absence of technical standards that provide satisfactory definitions for these materials. Some standards for concrete structures, such as NBR 8953 [7], NBR 6118 [8], ACI 363 [9], ACI 318 [10], and BS EN 1992 [11], differentiate two classes of concrete, as a function of characteristic compressive strength (f_{ck}). For instance, class I concretes have a compressive strength between 20 and 50 MPa, while class II concretes have a strength between 55 and 100 MPa. It can be considered that class I concretes be associated with the CC while class II concretes are related to HSC.

While the definition of HSC is based only on compressive strength, the definition of HPC and UHPC is related to performance, which includes not only mechanical strength but also workability, aesthetics, finish, integrity, and durability [12,13]. Despite not being unanimous, a satisfactory definition for the HPC is that this concrete presents a compressive strength equivalent to the HSC, that is, f_{ck} above 50 or 55 MPa [2,14], but presents a workability equivalent to self-compacting concrete (SCC), that is, a spread between 455 to 810 mm by the slump-flow test [1,15,16]. Regarding the water/binder factor (w/b), some authors suggest that this factor should be less than 0.40 [17–19], contrary to what is observed in conventional concretes, where the w/b factor ranges from 0.45 to 0.65 [20–22]. Regarding cement consumption, it is usual for it to be between 400 to 700 kg/m^3 [23–26], while for CC, the cement consumption is usually between 260 to 380 kg/m^3 [25,27,28].

The UHPC, on the other hand, presents even greater requirements. Some authors suggest a minimum strength of 120 MPa [13,29], while others stipulate a minimum of 150 MPa [30,31], with a fluidity equivalent or greater than HPC, in addition to low porosity. In theory, a concrete that presents a strength above class II could be considered a concrete with a mechanical performance superior to HSC or HPC. In other words, these materials would compose a kind of strength class III, where the UHPC fits. These properties are achieved using a w/c ratio between 0.2 to 0.3 [32,33] associated with a very high cement consumption, around 800 to 1000 kg/m^3 [34,35]. In addition, UHPC is usually produced without the use of coarse aggregates, which theoretically turns it into a mortar, allowing the evaluation of workability to be carried out by the flow table test. Some authors recommend that flow table measurements be greater than 260 mm, without the application of blows to the material [36]. All this information is summarized in Table 1, which presents the main characteristics of the concretes discussed in this literature review.

Table 1. Properties of different types of concrete.

Concrete	Abbreviation	Compressive Strength (MPa)	w/b Ratio	Workability (mm)	Cement Consumption (kg/m^3)
Conventional	CC	20–50	0.45–0.65	NA	260–380
High Strength	HSC	55–100	NA	NA	400–700
High Performance	HPC	55–100 50–100	<0.4	455–810 (slump flow)	400–700
Ultra-High Performance	UHPC	>100 >120 >150	0.2–0.3	>260 (flow table without drops)	800–1000

NA—Not applicable.

Another way to understand the difference between the concretes presented in Table 1, from the point of view of compressive strength. This is one of the reasons for the need to apply fibers in UHPC, which will be further discussed in this paper. In addition to conventional (i.e., steel and polymeric) fibers, the search for alternative natural fibers, such as basalt, sisal, and banana fibers has grown in recent years (detailed in Section 3.3) mainly due to their low cost and good fiber-matrix bond. Finally, despite the great interest in the application of fibers in HPC and UHPC, few standards are available to specify it [37].

The main applications of HPC and UHPC initially were in the construction of high-rise residential and/or commercial buildings, mainly between the 60 and 80 s. Examples are the Lake Point Tower and Water Tower Place buildings, in Chicago—the USA, built, respectively, in 1965 and 1970 [23,38]. Later, these classes of concrete started to be applied more frequently in infrastructure works, through the construction of bridges, viaducts, and other special buildings of art [39–41]. Since then, HPC and UHPC have been also applied to road pavements, industrial floors, and underground buildings [42,43]. More broadly, these types of concretes can be applied in any works requiring high compression loads as well as structures subjected to aggressive environments that need high durability and in cases of emergency or recovery works [12,13,44]. However, the cost must be taken into account when analyzing the feasibility of the building construction, since in general high-performance concrete is more expensive than conventional concrete [45,46].

In this context, the objective of this literature review is to present some concepts related to the main materials used for the production of HPC and UHPC. The materials named in this research as "classics" will be highlighted, as they are the same ones traditionally used to elaborate the CC. In addition, it is also an objective of this review to identify new materials used to obtain (ultra) high-performance concretes; in particular through the inclusion of fibers and the use of alkali-activated cement.

2. Classic Components Used for HPC and UHPC Production

In this section, the main materials that make up the structure of HPC and UHPC will be described. These materials are named classics as they are also used for CC production. These are ordinary Portland cement (OPC), mineral additions, aggregates, and chemical additives.

2.1. Ordinary Portland Cement (OPC)

The OPC used for the production of HPC and UHPC can be the same for the production of CC. However, several authors recommend the use of OPC with higher clinker contents and lower amounts of mineral additions. This is necessary because the production of HPC/UHPC uses mineral additions with greater reactivity than those usually considered in the production of commercial cement, such as blast furnace slag and fly ash [13]. Different denominations are used for these special OPCs. According to the main international standards, these cement are known as CP-I. CP-V-ARI cement according to the Brazilian standards NBR 16697 [47]. CEM I cement according to the European standard EN 197-1 [48]. Type I cement according to the American ASTM C150 [49]. All these cement have in common their high clinker content, usually above 90% or 95%, and also high fineness. However, it should be noted that CP-I cement is not a commercial product, which is why research-based on NBR 16697 [47] focuses on the use of CP-V-ARI.

In particular, the use of one of these types of OPCs, the Brazilian CP-V-ARI, is cited as an example. Silva et al. (2020) [50] studied the effect of high temperatures on the mechanical performance of HPC/UHPC containing recycled aggregates and CP-V-ARI. This cement had a compressive strength at 7 days of 37.4 MPa and at 28 days of 43.7 MPa. The reference concrete compositions presented a compressive strength of 65.2 MPa, proving that it is an HPC. Roberti et al. (2021) [13] studied the autogenous shrinkage effects and the fresh and hardened state properties of an HPC. They used cement CP-V-ARI, with a strength of 38 MPa at 7 days and 45.8 MPa at 28 days. The evaluated compositions showed compressive strength from 68.3 to 84.3 MPa. Viana et al. (2020) [51] used CP-V-ARI to evaluate the influence of the incorporation of carbon nanotubes in HPC, obtaining compressive strength for concrete around 80 MPa at 28 days of cure. De Matos et al. (2020) [52] used CP-V-ARI to produce UHP cement pastes with 28-day strength of around 130 MPa. Pilar et al. (2021) [53] studied the rheological behavior of HPC/UHP using cement type CP-V-ARI, however, the authors did not report the results of compressive strength since the focus of the work was to evaluate the properties of the fresh state.

It is also important to mention other investigations on special processing and aggregates. Sohail et al. (2021) [1] studied the durability characteristics of HPC and obtained concrete with compressive strength of a little less than 100 MPa at 28 days. Storm et al. (2021) [54] evaluated the ways in which different fibers were pulled out in high-performance concrete. Li and Zhang (2021) [55] studied the thermal stresses in UHPC containing polypropylene and steel fibers, obtaining a compressive strength of 141.5 MPa. Liu et al. (2021) [56] evaluated the application of steel slag as a complementary material in UHPC, obtaining strength between 120 to 150 MPa. Suescum-Morales et al. (2021) [57] evaluated the effect of temperatures on high-performance concrete performing a microstructural analysis. Rashid et al. (2020) [58] evaluated the effects of using magnetite sand in UHPC, obtaining compressive strength of 134 MPa at 28 days. Olawuui et al. (2021) [59] evaluated the development of initial and long-term strength of HPC containing polymer incorporation, obtaining a strength of approximately 80 MPa at 28 days. Zhang et al. (2021) [60] valuated the fragmentation strength and mechanical properties of UHPC at high temperatures, obtaining strength around 130 MPa.

On the use of type I OPC, it is worth mentioning the following works. Choi et al. (2021) [61] evaluated the effect of TiO_2 as a filler in concrete, obtaining compressive strength above 150 MPa at 28 days, typical of UHPC. Khan et al. (2020) [62] developed an ultra-high performance concrete for shielding from nuclear radiation, with a strength greater than 160 MPa. Kim et al. (2021) [63] used an OPC with a strength of 42.5 MPa to evaluate the benefits of curved steel fibers in the pullout strength of HPC. Yoo et al. (2021) [64] evaluated the effect of glass powder on the mechanical properties of UHPC, obtaining compressive strength greater than 200 MPa. Manigandan et al. (2021) [26] evaluated the use of treated banana fibers in the compressive strength of HPC, obtaining a value of approximately 52 MPa. Other researchers such as Yoo et al. (2021) [65], Kareem et al. (2021) [66], and Bae and Pyo (2020) [67] also used type I cement in their research with HPC and UHPC.

Regarding the chemical composition of the OPCs considered for the production of HPC and UHPC, it is observed that there are no major differences between the OPCs used for the production of CC, as shown in Table 2. The typical OPC base is essentially the same: between 60 to 72% of CaO, between 14 and 22% of SiO_2, resulting in approximately 80% of the cement composed of CaO + SiO_2. One can note that virtually all the cement presented in Table 2 had appreciable values (up to ~4%) of loss on ignition (associated with the CO_2 thermal decomposition) and/or MgO. This can be explained by the presence of carbonaceous fillers, e.g., calcite ($CaCO_3$) and/or dolomite ($CaMg(CO_3)_2$), or periclase (MgO). However, from a mineralogical point of view, it is relevant that the cement used for the production of HPC must present higher amounts of alite (C_3S) and belite (C_2S), which are the constituents of the OPC responsible for the formation of C-S-H, the main strength product of concrete [30]. Therefore, the CaO content must be analyzed coupled with the loss of ignition value, since this element can be present either as calcium silicate (i.e., C_3S and C_2S) or calcium carbonate filler. The content of C_3A must be reduced because this mineral is incompatible with the conventional water-reducing chemical additives used since it mainly forms ettringite which adsorbs the additive molecules and tends to increase the additive content required to reach proper flowability [68,69]. This is the reason why cement must have the lowest possible amounts of Fe_2O_3 and Al_2O_3.

Table 2. Chemical composition of Ordinary Portland Cement (OPC) used for HPC and UHPC.

CaO (%)	SiO_2 (%)	Fe_2O_3 (%)	Al_2O_3 (%)	SO_3 (%)	MgO (%)	Loss on Ignition (%)	Reference
62.91	20.34	4.58	4.47	2.58	1.24	3.27	[70]
61.33	21.01	3.12	6.40	2.30	3.02	-	[61]
63.62	19.70	2.93	-	-	1.28	-	[71]
66.45	17.84	3.58	4.26	4.10	2.14	-	[72]
63.07	19.38	3.28	4.58	3.50	2.79	1.54	[3]
64.62	20.18	3.24	4.98	3.15	1.98	2.59	[12]
62.60	20.60	3.20	5.10	3.60	3.00	-	[73]
67.97	16.19	3.79	3.59	4.05	1.71	0.51	[56]
71.22	14.80	3.48	4.54	4.11	-	4.02	[50]
68.91	15.74	4.80	3.18	3.80	2.00	1.00	[1]
62.90	18.90	2.80	3.70	3.10	4.20	3.20	[53]
62.15	20.95	3.80	4.85	2.00	3.10	-	[74]

Regarding other important properties of OPCs, it is noteworthy that the density of OPC for HPC and UHPC and for conventional applications is the same, around 3.10 to 3.15 g/cm^3. This is due to the low content of additions, since silica fume and fly ash, for example, have a specific mass around 2.2 to 2.4 g/cm^3. The fineness, in general, is greater, due to the use of high initial strength OPCs, which are generally thinner, and due to the low amount of water used, generally less than necessary for the hydration reaction. So, greater fineness helps in a considerable degree of hydration, which is why OPCs need to be thinner. The same happens with the percentage of material retained in the #200 sieve, considered an important parameter for post-production OPC grinding control. This information is summarized in Table 3, where it is observed that the fineness of the OPC varies in values above 3500 cm^2/g by the Blaine method. In general, the OPC used for conventional concrete presents a fineness close to 3000 cm^2/g. Regarding the percentage retained in the #200 mesh sieve, it is observed that the maximum value observed was 2%, much lower than other types of OPC used for CC production, such as CP-II, CP-III, CP-cement V, cement type II and III and CEM II. This is necessary because the larger specific area allows these cement to react faster than those used for conventional applications.

Table 3. Physical properties of OPC used in HPC and UHPC.

Density (g/cm^3)	Blaine Fineness (cm^2/g)	Retained in Sieve #200 (%)	Reference
-	3600	-	[71]
3.10	3600	-	[72]
3.12	4430	0.20	[13]
3.15	3500	-	[73]
-	-	2.00	[56]
3.06	-	1.38	[51]
3.09	4070	-	[53]
3.15	-	1.80	[12]

Based on this, it is possible to establish that the OPC used in HPC and UHPC must be chemically rich in calcium and silica, mineralogically rich in C_3S and C_2S, with the least number of mineral additions possible, aiming at the incorporation of more reactive pozzolans than those used in OPC commercial.

2.2. Mineral Additions

Mineral additions are supplementary cementitious materials used together with OPC, with the aim of providing differentiated technological performance to cementitious products. Initially emerged as a need to improve some properties of concrete, it now contributes considerably to environmental aspects [75,76]. The need to replace Portland clinker with low-impact binders is today a common desire, but without losing the properties provided by these binders [77]. This is because ordinary cement is the world's second most used material, second only to water. In addition, OPC production releases great amounts of CO_2 mainly due to limestone decarbonation, necessary for clinker production [78]. Another important factor is the high energy demand required to reach a burning temperature of around 1500 °C. As a consequence, some authors point out that for every 1 ton of clinker produced, about 0.8–1.0 tons of CO_2 are emitted into the atmosphere [76,79,80]. A portion of this emission is mitigated through CO_2 uptake by mortar and concrete carbonation [78,81,82]; however, this topic will not be further addressed in this paper.

As a result, some researchers have proposed the replacement of clinker by agro-industrial residues or by-products. This is the case of blast furnace slag, fly ash, silica fume, and other pozzolans such as agricultural ashes. It was observed a reduction in production costs, due to the reduction in the consumption of clinker, which generally has a higher cost than aggregates. For example, Li and Jiang (2020) [83] observed a 21% cost reduction when used 60% slag and 10% limestone in OPC replacement for the same concrete strength class. Zhang et al. (2021) [84] demonstrated that concrete mix design can be optimized simultaneously for environmental, economic, and mechanical objectives with silica fume incorporation. In addition, there is a contribution to sustainable development, and the achievement of concrete with greater mechanical strength, especially with the use of pozzolanic materials [85–87].

Mineral additions can be classified according to their reaction with the clinker as inert. This, in general, only contributes to a physical filling effect improving the packaging. For example, this happens with the application of limestone filler as well as in cement, such as blast furnace slag, and in pozzolanic products, such as metakaolin, fly ash, agro-industrial ash, and silica fume [85,88,89]. Cementing additions are materials, usually based on calcium, that present an accelerated reaction capacity in the presence of alkaline media, as is the case of $Ca(OH)_2$ of portlandite present in OPC [90,91]. They can also be activated with the use of other alkaline hydroxides, such as NaOH and KOH, through alkaline activation, dispensing in this case, the presence of OPC [92,93]. This type of reaction will be described in Section 4.

However, the additions with the greatest potential for the production of HPC/UHPC are the pozzolanic ones that in addition to the physical packing effect, described above, have a chemical effect. Pozzolans are materials based on silica or silica and alumina which, isolated, do not present any binding power, but when finely ground, and in the presence of clinker and water, develop binding powers [94,95]. This happens through the so-called pozzolanic reaction. It is observed that portlandite, formed through the hydration of the silicates present in the OPC, reacts with the amorphous silica and alumina present in the pozzolan and forms C-S-H and C-A-H. As aforementioned, the compound responsible for the strength of hardened OPC is C-S-H. Thus, the conversion of CH to C-S-H provided by pozzolans contributes significantly to the strength of the formed product [94,96].

It is observed that mineral additions have certain requirements for their efficiency in clinker replacement. From a physical point of view, they need to present a high specific surface, measured by the Blaine fineness, for example, to increase the contact area of mineral additions with OPC [96]. Mineralogically, these materials need to be predominantly amorphous, which means that they present structural disorders. Otherwise, the additions are not reactive, as crystalline materials have an organized structure and hardly change their structure under normal conditions of temperature and pressure due to hydration reactions.

From a chemical point of view, there is a wide range of mineral additions used. According to the bibliography [97–99], there are reports of the use of materials rich in calcium, such as furnace slag, limestone filer, and class F ash as well as materials rich in silica or aluminosilicates, such as fly ash, silica fume, and metakaolin, which make up the class of pozzolans.

One of the most used additions is fly ash, a by-product of thermoelectric plants that burn coal to produce energy. Ash is trapped in the combustion gas exhaust system and, when pulverized, it acquires pozzolanic characteristics [95,100]. Another great advantage of fly ash is the spherical shape of the material, which can promote a rolling effect between grains and improve workability, providing a reduction in the amount of water and consequently contributing to mechanical strength. This material is mineralogically amorphous and has a chemical composition of approximately 45–60% SiO_2, 30–32% Al_2O_3, in addition to Fe_2O_3 and CaO in variable but detectable contents [101–103]. As thermoelectric plants are the main energy sources in the world, with the exception of some countries such as Brazil, the availability of fly ash is high, which justifies the wide application of the material as pozzolans [103].

In the research by Mohan et al. (2021) [101], in which the authors used two different types of fly ash as additions for HPC/UHPC, the proportion of additions used was approximately 50% of the OPC mass. The authors obtained a concrete with 69 MPa compressive strength at 28 days. The two fly ashes used, named 1 (59.32% SiO_2, 29.95% Al_2O_3, 4.32% Fe_2O_3, 1.28% CaO) and 2 (60.56% SiO_2, 32.67% Al_2O_3, 4.44% Fe_2O_3, 1.41% CaO), presented a fineness by the Blaine method of 5636 and 6210 cm^2/g, respectively. Comparing with the OPC used by the authors, with a fineness of 3300 cm^2/g, it is observed that their additions were finer than this binder.

Other studies with the application of HPC/UHPC that used fly ash are highlighted below, to justify that this material is one of the most used additions in high-performance concrete: Sujay et al. (2020) [104] studied the effect of the application of steel fibers in high-performance concrete containing mineral addition of fly ash replacing 15% of the OPC mass, obtaining compressive strength at 28 days of approximately 55 MPa. Bahedh and Jaafar (2018) [105] studied the application of fly ash (69.41% SiO_2, 28.20% Al_2O_3, 5.30% Fe_2O_3, 6.47% CaO) to replace OPC in percentages of 0–40% for the production of UHPC by molding in an autoclave, with the application of pressure in the production stage of the specimens. The authors observed that the use of 40% ash allowed to obtain a compressive strength at 28 days of 120 MPa, while the reference composition presented a strength of 80 MPa at that same aging period. This information is illustrated in Figure 1.

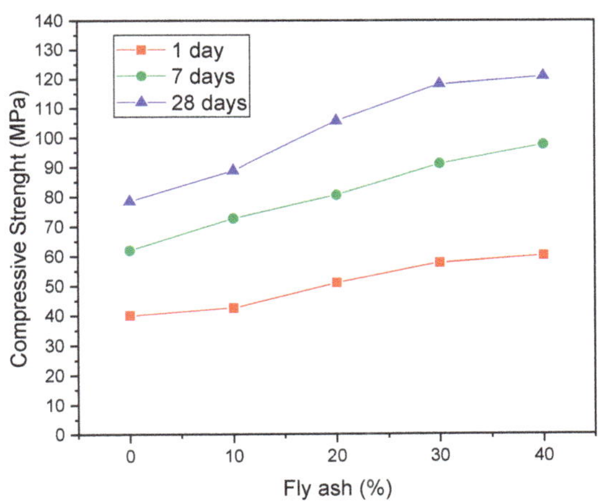

Figure 1. Influence of fly ash content on the compressive strength of ultrahigh performance concrete [105].

Zhang et al. (2020) [106] studied the effects of replacing 25% OPC by fly ash (50.35% SiO_2, 29.65% Al_2O_3, 6.61% Fe_2O_3, 5.85% CaO) in HPC subjected to situations of high temperatures. The authors obtained a compressive strength at 28 days of approximately 55 MPa for the composition containing ash, while the reference composition showed a strength of less than 50 MPa, not being characterized as HPC. At all ages evaluated, the compositions containing fly ash showed a better mechanical performance, which is directly related to the pozzolanic effect.

Choudhary et al. (2021) [107] evaluated the effect of various mineral additions on the abrasion and mechanical strength properties of high-performance concrete. Although the best results were obtained with the use of silica fume (95.58% SiO_2, 0.71% Al_2O_3, 0.81% Fe_2O_3, 0.90% CaO), which will be discussed below, the results obtained with fly ash (58.19% SiO_2, 26.93% Al_2O_3, 4.27% Fe_2O_3, 0.90% CaO) were also positive. For example, obtaining compressive strength at 28 days of 55 MPa. One of the reasons that explain the better performance of silica fume is the fineness of the material, 9550 cm^2/g by the Blaine method, while fly ash showed 3530 and cement 2860 cm^2/g. Another explanation is the chemical composition since silica fume has a higher SiO_2 content than fly ash.

As already emphasized, silica fume is another pozzolanic mineral additive used in HPC/UHPC, with mechanical properties superior to those obtained by using fly ash. In the research by Choudhary et al. (2021) [107], cited in the previous paragraph, the compositions with silica fume obtained compressive strength at 28 days of 75 MPa, while the same compositions containing fly ash obtained 55 MPa. Some relevant information helps to explain this difference in behavior.

Silica fume is a by-product obtained in the production process of plain silicon or in silicon iron alloys. The process is carried out in large metal furnaces, by reducing the quartz in the presence of coal, or iron in the case of alloys, at very high temperatures, around 2000 °C. During the heating step, silicon monoxide (SiO) is eliminated as a gas, oxidizing and condensing into extremely small spherical particles of amorphous silica (SiO_2) [108,109]. In terms of chemical composition, silicas fume have a SiO_2 content above 95%, another characteristic that contributes to their pozzolanic effect [102,107].

This produced SiO_2 material is trapped in the furnace exhaust gas filtration systems, being removed and used as a pozzolan. The average diameter of the material is 0.1 mm, about 100 times smaller than the average diameter of the OPC particles, presenting a specific surface by Blaine fineness in the order of 9000 to 10,000 cm^2/g [109,110]. This contributes to the greater reactivity of this addition, in addition to contributing to granular packing and helping in pozzolanic reactions in regions that other conventional pozzolans cannot, such as at the paste-aggregate interface. However, it should be noted that silica fume, unlike fly ash, has a higher commercial value, as the production numbers of the silicon industry are much lower than the production of fly ash in thermoelectric plants [110,111].

Some researches that studied the application of silica fume are highlighted below. Wu et al. (2019) [112] studied the changes in rheological and mechanical properties with the use of silica fume (95.2% SiO_2) in ultra-high performance concrete. The authors obtained strength of 120 MPa at 28 days of cure, with the use of 20% silica fume (94.8% SiO_2) to replace OPC. Smarzewski et al. (2019) [113] used silica fume to evaluate the mechanical properties of UHPC, obtaining a strength of 95 MPa for the reference composition and 110 MPa for the composition containing 20% of the additive. Pedro et al. (2017) [114] and Pedro et al. (2018) [115] evaluated the mechanical properties and durability of HPC produced with recycled aggregate and silica fume (94.2% SiO_2). The authors obtained compressive strength of 76.70 MPa after 28 days with the compositions containing silica fume in the first study and observed that the performance in the durability tests was superior for the compositions containing 20% silica fume, in all evaluated conditions (strength to carbonation, strength to chloride attack, water absorption by immersion, and by capillarity) in the second study.

Chen et al. (2018) [116] studied the application of fly ash to replace OPC in percentages of 0–30% for the production of UHPC by molding in an autoclave, with an application of pressure in the production stage of the specimens. The authors observed that the use of 20% ash allowed to obtain a compressive strength at 28 days of 125 MPa, while the reference composition presented a strength of 105 MPa at the same age, as shown in Figure 2. This change in strength was attributed by the authors to the greater formation of C-S-H, which is directly attributed to the pozzolanic reaction promoted by the material, as well as to the likely improvement in granular packing, related to the physical effect. In addition to the aforementioned research, several other studies used silica fume as a mineral addition [68,117–121].

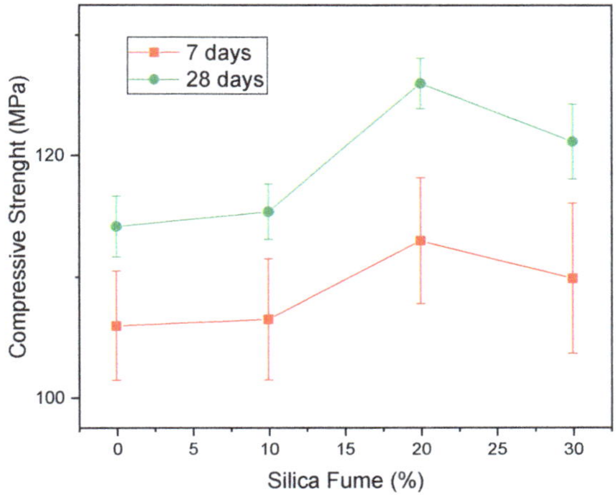

Figure 2. Compressive strength of Ultra-High Performance (UHPC) containing silica fume [116].

Other incorporations are also used in the production of HPC and UHPC, although in much smaller proportions than fly ash and silica fume. Several countries, such as China, Japan, Brazil, the USA, India, and Germany, have a great potential for using blast furnace slag, obtained through the steel industries, a strong industrial sector in these countries. Blast furnace slag is generated as a by-product of the production of pig iron with a high calcium content [122,123]. The same can be highlighted in the ceramic industry, a strong sector in countries such as China, Brazil, Italy, and Spain, responsible for the production of ceramic waste and metakaolin, two pozzolans with a high content of silica and alumina [124–126]. Several countries, such as China, the USA, Argentina, and Brazil, also present the possibility of using agro-industrial ash, such as rice husk ash and sugarcane bagasse ash, creating alternatives for the application of renewable forms of pozzolans [127–129].

On the use of slag, the following works stand out. Shen et al. (2020) [130] and Shen et al. (2020) [131] studied the use of blast furnace slag in HPC obtaining a strength of 52 and 66.9 MPa, respectively. Cheah et al. (2019) [132] and Ma et al. (2018) [133] studied the use of blast furnace slag together with fly ash for the production of high-performance concrete, obtaining strength at 28 days of approximately 50 MPa. On the application of ceramic waste, it is worth mentioning the following research works. Kannan et al. (2017) [134] obtained a concrete of 51.5 MPa at 28 days with the replacement of 10% OPC by ceramic waste. Xu et al. (2021) [135] obtained a UHPC of 120 MPa at 28 days using 15% ceramic waste as pozzolan replacing OPC. Salami et al. (2020) [136] studied the application of metakaolin as a mineral additive in HPC obtaining a compressive strength of 60 MPa. Song et al. (2019) [137], Shehab et al. (2017) [138], and Tafraoui et al. (2016) [139] studied the use of metakaolin as an additive in UHPC containing different types of fibers. All authors obtained compressive strength at 28 days above 100 MPa.

Regarding the use of agro-industrial ash, the research by Le and Ludwing (2016) [140] stands out, which evaluated the use of fly ash, silica fume, and rice husk ash together for the production of UHPC, obtaining a strength of 110 MPa at 28 days. Le et al. (2015) [141] evaluated the durability of HPC containing rice husk ash as a pozzolan. Shaaban et al. (2021) [142] evaluated the rheological and hardened state properties of HPC containing this same additive, obtaining a strength of 60 MPa at 28 days. Finally, the research by Gar et al. (2017) [143] studied the use of sugarcane bagasse ash as pozzolan in HPC, obtaining strength at 28 days of 52 MPa.

2.3. Aggregates

Aggregates are defined as inert, granular materials without defined shape and volume, used in concrete for economic and technological reasons [144]. They are classified according to their origin as artificial (or industrialized), natural, or recycled. Crushed stones are cited as an example of artificial aggregates. Other examples are natural sand washed from the river and construction and demolition waste used after recycled [145,146].

As for their own weight, aggregates are classified as light, such as expanded clay, conventional, such as crushed stone, or heavy, such as hematite aggregates [144,147]. From a granulometric point of view, they are classified into coarse, those whose grains pass through the 152 mm opening sieve and are retained in the 4.75 mm sieve, and in giblets, those whose grains pass through the 4.75 mm opening sieve and remain retained in the 0.075 mm aperture sieve [145,148]. The particle size distribution of the materials plays a major role in the fresh and hardened performance of HPC and UHPC. While the aggregates correspond to the macroscale components, the binder materials (i.e., OPC and mineral additions) correspond to the microscale fraction of concrete. In addition, the very small silica fume particles can improve the particle packing of the binder fraction, leading to higher compactness due to physical effects besides the pozzolanic contribution [149]. By optimizing the particle size distribution and mix proportions, one can achieve maximum particle packing, therefore improving the fresh and hardened properties of concrete. This strategy has been widely used for HPC and UHPC design over the last years [150–152].

Dealing specifically with the aggregates used in HPC, it is observed that there is no specificity regarding the use of these materials when compared to conventional concrete. However, care must be taken with regard to the particle size of the materials, in order to find the packing of all aggregates, following a continuous distribution, which presents the smallest possible void volume. This characteristic is even more important in the production of UHPC, which usually do not have coarse aggregates in their composition, due to the possible presence of micro-cracks caused by crushing or because the strength of the coarse aggregate is usually inferior to that of the cement matrix, making the aggregate the fragile point of the material. Owing to the above-described reasons, it is common for authors to use more than one type of fine aggregate or different combinations of fine and large aggregate to obtain the best packing. Arunothayan et al. (2021) [70], for example, uses three different types of sand as fine aggregate for HPC production used in 3D printing applications. With the different combinations proposed, the authors obtained compressive strength ranging from 110.1 to 152.5 MPa.

In addition to aspects related to packaging, some physical parameters need to be analyzed. Regarding the coarse aggregate used for HPC, the following points are worth mentioning. The content of fines, passing through the 75 mm sieve, should be limited to 1%, as this material is generally attributed to silt and clay particles, which can increase the aggregate water absorption. The D/d shape ratio, which relates to the largest size and smallest aggregate size, should be limited to 3 to avoid anisotropy in concrete. Water absorption should also be limited to 7%, as should abrasion wear strength measured by the Los Angeles method, which should be restricted to 50%.

On the D/d ratio, Zhao et al. (2021) [153] evaluated the influence of three coarse aggregate geometry on the mechanical properties of HPC. The authors used lamellar, irregular, and rounded aggregates. The strength results obtained by the authors are highlighted in Figure 3, where it is possible to observe that the best results are obtained with irregular aggregates. This is attributed by the authors to greater adherence between the cementitious paste phases and the aggregates, which improves the behavior of HPC.

Regarding the physical characteristics of small aggregates, some information is pertinent. For example, the number of fines is also limited, but to a total of 3%. The limit water absorption is 7%, while the swelling coefficient must be as low as possible, to avoid excessive volume increase. There are still some recommendations related to concrete durability problems, which apply to both coarse and fine aggregates. The content of chlorides and sulfates, for example, should be limited to 0.1% of the chemical composition of the aggregates.

This is necessary to avoid the occurrence of oxidation in the concrete reinforcement and to avoid the formation of late ettringite in the concrete, causing an expansive reaction that destabilizes the material volume and generates internal stresses [154]. Problems related to alkali-aggregate reaction (AAR) must also be verified through the mortar bar test. The maximum expansion allowed in this test is 0.05% after 3 months and 0.10% after 6 months. In AAR, a gel is formed that absorbs water and tends to increase the volume of the concrete, which can generate cracking and disaggregation of the aggregate paste [148,155].

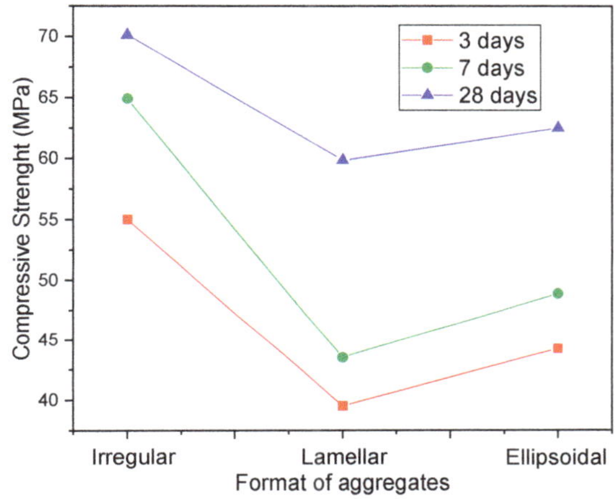

Figure 3. Compressive strength results as a function of aggregate shape [153].

Analyzing the researches published with HPC and UHPC, it is observed that the authors traditionally use quartz sand washed from the river [1,3,13,70] or quartz sand from dunes [62,156] as the main fine aggregate. As a coarse aggregate, it is common to use crushed stones [25] and gravel [157]. The stones used are limestone [114,115,158] and diabase [159], mainly.

There are also researches using heavy aggregates such as barite, magnetite, and hematite for HPC in nuclear protection applications [147,160,161]. Figure 4, for example, shows the compressive strength results obtained using 3 types of fine aggregate: silica sand (97.32% SiO_2), barite (58.69% BaO), and hematite (71.71% Fe_2O_3). The silica sand used has a specific mass of 2.7 g/cm^3, while the barite has 3.0 to 4.4 g/cm^3. Although the compressive strength results are better for the composition with silica sand, the radiation absorption values of cobalt and cesium were much higher for heavy concrete manufactured with barite, justifying its use in this type of application.

There are also researches that use light aggregates, mainly expanded clay, for the production of HPC. Angelin et al. (2020) [162] evaluated the packing of lightweight concrete containing expanded clay and rubber as aggregates, obtaining strength at 28 days of 58.5 MPa. Lu et al. (2021) [163] also obtained compressive strength results compatible with HPC using expanded clay aggregates. However, Garcia et al. (2021) [164] reported that the use of expanded clay in high-performance concretes is problematic, due to defects arising from the calcination of these aggregates. As a result, it is usual to increase OPC consumption and use a high number of additives, in addition to reducing the w/c factor, which is only possible using considerable amounts of plasticizer additive. This makes concrete too expensive and is therefore not recommended.

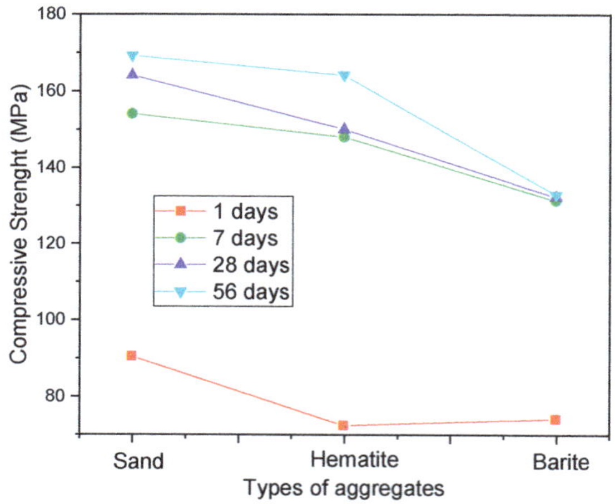

Figure 4. Compressive strength due to different types of aggregate [147].

Research that use recycled aggregates for the production of HPC and UHPC are also mentioned. These researches use aggregate from concrete waste [114,115] and ceramic waste [157,165]. Using these residues, which present the particle size curve within the normative limits of the ASTM C33 standard, the authors obtained a compression strength after 28 days of curing of 77.3 MPa, proving the feasibility of using recycled aggregates, as long as they meet the stipulated granulometry parameters.

Based on this, it is observed that the choice of aggregates for application in HPC and UHPC should be more carefully selected for application in CC, due to the need to obtain greater packaging. It is possible to use not only coarser and finer but also conventional and heavy aggregates, as well as lighter and recycled aggregates for the production of HPC and UHPC.

2.4. Chemical Additives

Chemical additives are materials used in the production of concrete aiming to improve properties of interest. They are generally used in quantities of up to 5% of the OPC mass [166]. The main types of chemical additives are air-incorporated additives, tack modifying additives, water-reducing additives, and shrinkage-mitigating additives, especially used in UHPC. As the name suggests, air-incorporated additives add air bubbles to concrete voids, improving the plasticity and workability of the material, but potentially compromising compressive strength. They are used in situations where concrete is subjected to ice and thaw, acting as a kind of reservoir where water can migrate, since when freezing the water expands, which could cause the concrete to crack [166–168].

Most used shrinkage mitigating additives are amphyphylic molecules, which have a hydrophilic end and a hydrophobic end. When interacting with the water present in the hydration of OPC, or any other polar solvent, the molecules of these additives are mainly absorbed in the liquid-vapor interface by electrostatic repulsions due to the interaction of hydrogen bonds [169,170]. In this way, there is a reduction in the surface tension of the water in the pores of the concrete, allowing a reduction of up to 50% in concrete shrinkage. The reduction is possible because shrinkage-mitigating additives continue to act in the pore system even with hardened concrete, reducing the effects of water surface tension that contribute to drying shrinkage [170,171]. The main examples of such additives are propylene glycol, general glycol ethers, and polyethylene glycol [169].

Set-modifying additives serve to delay or accelerate the setting time and hardening time of concrete. They do not significantly change the final strength of concrete, however, they do change the strength at early ages [172]. The water-reducing additives, on the other hand, serve to reduce the w/c ratio without losing the workability of the concrete. They are the main additives used in the production of HPC and UHPC [173,174]. In addition to these additives, there are polyfunctionals that have two or functions simultaneously, generally modifying not only the set but reducing the amount of water.

As mentioned, water-reducing additives are the most used for the production of HPC and UHPC. These additives are subdivided into three generations. Depending on the amount of water they reduce in the concrete, these are: (i) 1st generation of superplasticizers that reduce from 6 to 12% of water; (ii) 2nd generation of superplasticizers that reduce 12 to 20% of water; and (iii) 3rd generation of superplasticizers that can reduce water above 20%, reaching a reduction of up to 45% of the mixing water [174,175]. In HPC and UHPC, 3rd generation superplasticizers are preferably used. The importance of water reduction is related to Abrams' Law, which indicates that the lower the w/c ratio, the greater the mechanical strength, using the same material parameters [176]. Logically, the reduction of the w/c factor impairs the workability of the concrete, which is why chemical additives are used.

The water reduction promoted by 1st and 2nd generations of superplasticizer reducers can be explained through the electrostatic dispersion effect. This effect occurs because the additive involves a system of OPC particle charges of the same sign [177]. Due to the effect of electrostatic repulsion, the superplasticizer will disperse the cement particles, making less water necessary to reach a given workability [178,179].

The 3rd generation of superplasticizers works due to the steric effect or due to the combination of the electrostatic repulsion effect with the steric effect [180]. This effect, which occurs mainly in additives based on polycarboxylate (PCE), the main additive used in HPC and UHPC. PCE features a long main chair, with shorter branches and side chains, increasing floor space in an OPC particulate system, resulting in much greater water reduction than 1st and 2nd generation plasticizers [181–183].

As examples of some additives used in HPC, the following works stand out. Ibragimov and Fediuk (2019) [184] evaluated the influence of different types of superplasticizers on the mechanical properties of concrete. The authors used five different superplasticizer additives: the first is a copolymer based on polyoxyethylene derived from unsaturated carboxylic acids (1st generation); the second is based on sodium salts of polymethylene naphthalenesulfonic acids (2nd generation); the third is a polyfunctional consisting of naphthalenesulfonate and an organic accelerator; the fourth additive used is a superplasticizer based on polyoxyethylene derivatives of polymethacrylic acid (PAA); finally a copolymer based on polyether carboxylates (PCE). The strength results obtained by the authors are shown in Table 4. It is observed that the additives that contributed the most to the compressive strength were the PAA and PCE, which is why they are the most used in the literature.

Table 4. Compressive strength results obtained with the use of various water reducing additives. Source: [184].

Composition	Compressive Strength after 1 Day (MPa)	Compressive Strength after 3 Days (MPa)	Compressive Strength after 28 Days (MPa)
Reference	7.1 (100%)	19.57 (100%)	40.53 (100%)
1st generation	12.32 (174%)	31.89 (163%)	57.55 (142%)
1st generation (naphthalene)	12.81 (180%)	32.42 (166%)	51.70 (128%)
Polyfunctional	13.76 (194%)	29.01 (148%)	46.15 (114%)
Polyoxyethylene derivatives of polymethacrylic acid (PAA)	22.53 (331%)	46.38 (237%)	62.81 (155%)
Copolymer based on polyether carboxylates (PCE)	19.62 (276%)	45.01 (230%)	65.72 (162%)

Benaicha et al. (2019) [185] analyzed the effects of superplasticizer additives on the rheological and strength properties of HPC. They used PCE-type superplasticizers in different percentages, obtaining a compression strength of 73.49 MPa at 28 days with the use of 0.3% of the superplasticizers. Cheah et al. (2020) [186] evaluated the changes in the mechanical and microstructural properties of HPC produced with PCE-type superplasticizers containing a ternary mixture of OPC, blast furnace slag, and silica fume. The authors obtained results compatible with the behavior of HPC, obtaining compressive strength at 28 days of around 80 MPa. Cheah et al. (2019) [132] evaluated the performance of HPC containing fly ash, blast furnace slag, and PCE-type superplasticizers. The compression results obtained were consistent with HPC applications. Finally, the work by Guan et al. (2021) [179] reported the durability effects of HPC sulfates produced with PAA-type superplasticizers. The authors did not perform mechanical tests but emphasize that the applied concrete presents behavior for high-performance applications.

Thus, based on what has been presented, it is observed that the use of superplasticizers is essential to obtain an HPC and UHPC with an adequate behavior. The most used additives are PCE, 3rd generation of superplasticizers that work by the combined principle of electrostatic repulsion with steric effect.

3. HPC and UHPC Containing Fibers: Composite Materials

High-performance fiber reinforced concrete (HPRFC) or ultra-high-performance fiber reinforced concrete (UHPFRC), emerged as a need to improve ductility and tensile strength properties in HPC and UHPC [25,70]. Research with this type of material is relatively recent, starting from the 1990s, about 30 years ago [187]. The researchers observed that, although the mechanical behavior related to compression was increasing, the same did not happen with tension, which compromised the concept of high performance [188].

In this context, a new perspective of HPC and UHPC emerged: using the concepts of composites applied to concrete materials. Composites are materials that use two different components to obtain a material with superior properties, these two components are the matrix phase, the majority phase, responsible for involving and protecting the second phase, which is the reinforcement, dispersed throughout the matrix. The matrix of a composite can be metallic, polymeric, or ceramic. In the case of high-performance concrete, the matrix used is cementitious, essentially composed of HPC and UHPC [13,188,189]. Despite the great potential of using fibers to reinforce HPC and UHPC, few standards were developed to date, such as the French standard [37].

This concrete, used as a matrix, has the same components highlighted in Section 2, that is, OPC, mineral additions, preferably pozzolanic of high fineness, such as silica fume and fly ash, low w/c, requiring the use of superplasticizer chemical additives, and fine and coarse aggregates following the same specification detailed in Section 2.2.

The reinforcement phase of a composite can be in the form of filler particles or in the form of fibers. In the case of HPC and UHPC discussed in this section, the reinforcement phase is composed of synthetic fibers that can be metallic, polymeric, and ceramic, or natural fibers, of mineral, animal, and vegetable origin [12,13,190]. Figure 5 shows the behavior of tenacity and elongation at the break of the main fibers used for application in HPC and UHPC, while Table 5 shows the main properties of these fibers. The main advantages of applying fibers in HPC and UHPC are the high gain in mechanical strength, both in compression and in traction, and the increase in ductility. Disadvantages include the reduction in the workability of concrete, which can be solved using 3rd generation of superplasticizer reducers, and the higher cost of the material, especially with the use of synthetic fibers. These issues will be detailed in this section.

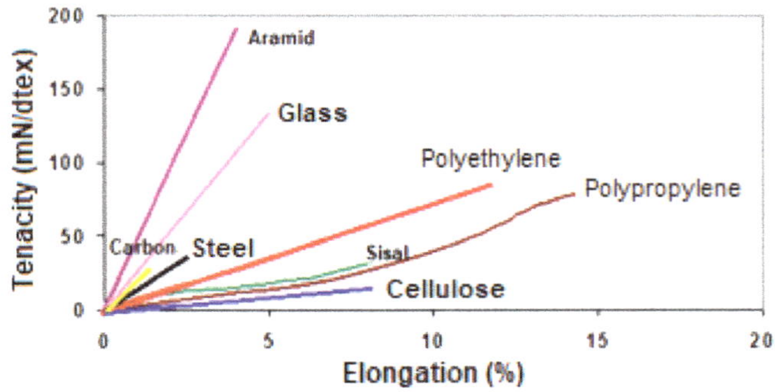

Figure 5. Stress x strain diagram of fibers used in HPC and UHPC [191,192].

Table 5. Properties of fibers used in HPC and UHPC. Source: [193–199].

Fibers	Diameter (mm)	Density (g/cm^3)	Modulus of Elasticity (GPa)	Tensile Strength (MPa)	Elongation (%)
Matrix (comparison)	-	2.7	10–45	3.5–8	0.02
Steel	5–500	7.84	200	500–2580	0.5–3.5
Carbon	5	1.9	65–135	2600	1
Glass	9–15	2.6	70–80	2000–4000	2–4.5
Polypropylene	20–200	0.9	164	500–750	9
Polyethylene	20–200	0.95	14–19.5	200–300	7.5
Asbesto	0.02–0.4	2.6–3.4	196	33000	2–3
Kevlar	10	1.45	5–17	3600	2.1–4
Cellulose	-	1.2	4	300–500	-
Sisal	10–50	1.5	15–20	800	7

3.1. Steel Fibers

Through the consulted database, it is observed that the majority of research carried out with HPC and UHPC, mainly use steel fibers, produced through essentially ferrous metallic alloys, containing between 0.008 and 2.11% of carbon. These fibers present positive properties, such as high ductility and tensile strength, in addition to compatibility with concrete, which is typically observed in the use of steel bars in reinforced concrete [73,200–202]. However, they are prone to corrosion, which is why several authors have studied the effects of this pathology on the behavior of HPC. According to Shin and Yoo (2020) [203], Yoo et al. (2020) [204], Lv et al. (2021) [190], and Ngo et al. (2021) [205], corrosion reduces the strength of the cementitious composite and decreases its ductility to levels even worse than its behavior without reinforcement. Therefore, corrosion must be severely avoided.

On the mechanical properties obtained by the application of steel fibers, Gou et al. (2021) [206], studied the mechanical properties of HPC containing different fractions of steel fibers in an orderly and disorderly manner in the cementitious matrix. The authors observed, as can be seen in Figure 6a, that the use of 1.0, 1.2, and 1.5% of disordered fibers presents compressive strength after 28 days of cure, superior to equivalent compositions with oriented fibers. However, this pattern of behavior is changed with the use of 1.8% of fibers because, as the fiber content is high, the workability of the cement mortar reduces, resulting in an agglomerate of steel fibers, which causes the formation of inter-defects in the concrete and, consequently, the loss of strength.

Regarding the flexural tensile strength at 28 days, Figure 6b, it was observed that the oriented fibers had better results, regardless of the amount of fiber evaluated. This is attributed by the authors to the formation of stress transfer bridges and crack propagation. This effect is obtained in greater intensity when the fibers are properly oriented. That is also why the results improve considerably with the use of higher amounts of fiber, where the composition with 1.8% oriented showed a result of 40 MPa, for example.

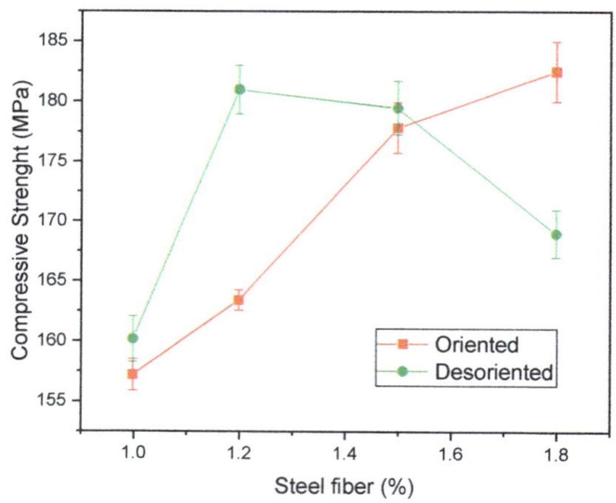

(**a**)

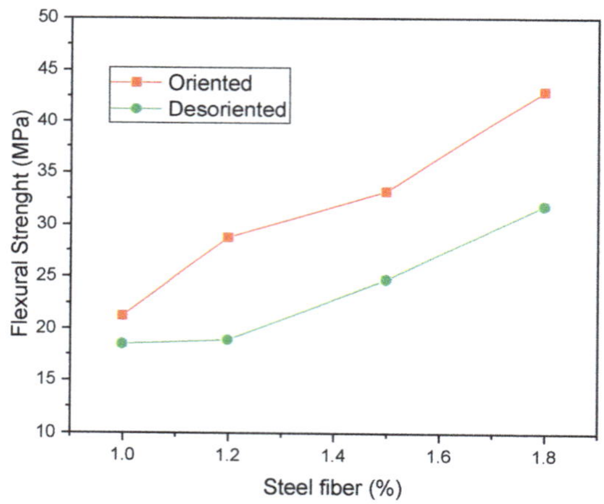

(**b**)

Figure 6. (**a**) Compressive strength, (**b**) flexural tensile strength fibers [206].

Other important works using steel fibers in HPC are now cited. Ashkezari et al. (2020) [201] studied the experimental relationships between the volume fraction of steel fiber and the mechanical properties of HPC. Bao and Pyo (2020) [67] evaluated the mechanical behavior and electrical properties of high-performance concrete containing fiber incorporation for application in railway sleepers. Park et al. (2021) [207] evaluated and verified the orientation and distribution of steel fibers in high-performance concrete pillars using computerized microtomography. Kim et al. (2021) [63] and Kim et al. (2020) [208] studied the benefits of applying curved steel fibers with a radius of 10–50 mm on the properties of high-performance concretes. Dingqiang et al. (2021) [209] evaluated the influence of the use of straight and short (6 mm), straight and long (13 mm) and hook-end fibers on the mechanical properties of UHPC, performing experimental tests and computational analyses. The best results were obtained using 2% steel fibers with a hook, obtaining approximately 150 MPa compressive strength and 45 MPa flexural tensile strength at 28 days.

In this way, it was concluded that the steel fiber is compatible with the cementitious matrix for UHPC applications.

3.2. Other Synthetic Fibers

Other synthetic fibers used for HPC and UHPC applications are carbon, glass, and polymeric materials such as polypropylene and polyethylene. Carbon fibers, composed of thousands of unified filaments, have the advantage of greater adhesion to the cement matrix due to their high specific area.

Relevant researches with the application of carbon fibers in HPC and UHPC may cite the following: Afzal and Khushnood (2021) [210] evaluated the influence of carbon fibers on the performance of UHPC exposed to high temperatures; Jung et al. (2020) [211] reported on the changes in the structural behavior of HPC containing carbon fibers; Liu et al. (2020) [197] measured the strength gain of HPC containing carbon fibers at early and intermediate ages; Zhou et al. (2020) [212] evaluated the behavior of HPC e UHPC containing several types of fiber, including carbon, exposed to high temperatures.

Glass fibers are produced using borosilicate glass (E-glass) or soda-lime-silica (A-glass), with the advantages of being able to apply thin panels and elements without the occurrence of corrosion. However, they can present durability problems, since the cementitious matrix is highly alkaline, which can degrade the fiber and promote the embrittlement of HPC and UHPC [12,64]. Some important works with this fiber are worth mentioning. Bilisk and Ozdemir (2021) [213] studied the three-dimensional configuration of glass fibers in cementitious composites composed of HPC and UHPC. Al-Khafaji et al. (2021) [214] evaluated the behavior of sustainable HPC with ecological materials and fiberglass application. Kumar et al. (2021) [215] studied the application of UHPC containing glass fiber, and other fibers, for structural reinforcement of concrete and reported corrosion problems. Ali et al. (2020) [216] investigated the behavior of beams subjected to bending of UHPC containing recycled aggregates and fiberglass.

Regarding the application of glass fibers, the work by Mohamed et al. (2021) [217] in which the authors applied glass fibers at various levels of UHPC with a w/c ratio of 0.12 and 0.14, varying the curing age in 7, 14, and 28 days, as shown in Figure 7, is worth mentioning. The authors observed that after using 1.5% fibers, the compressive strength results did not change. This happens because the composite has reached the fiber saturation point, which is the point at which the matrix is not wettable to absorb the fibers used. In concrete, this can cause a drop in strength or a loss of workability.

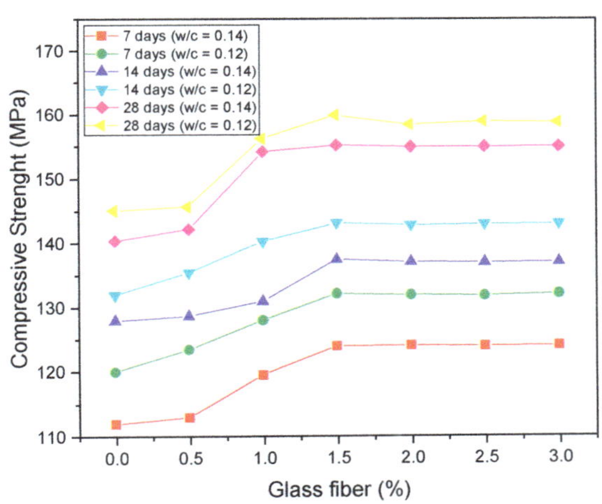

Figure 7. Compressive strength of HPC containing glass fibers [217].

The most used polymer fibers are polypropylene (PP) and polyethylene (PE). The main advantages are the low density of fibers when compared to others already mentioned in this review. As for disadvantages, the mechanical behavior is mentioned, which is inferior to steel, glass, and carbon fibers. Hussain et al. (2020) [218], for example, compared the mechanical behavior of CC and HPC containing 1% steel, glass, and PP fibers, as shown in Figure 8; as for HPC, steel fibers are the best performers, followed by glass. PP fibers have the worst results, even though they are superior to the reference concrete. However, as these fibers are lighter, their application should not be discarded, mainly because they help to reduce HPC and UHPC retraction.

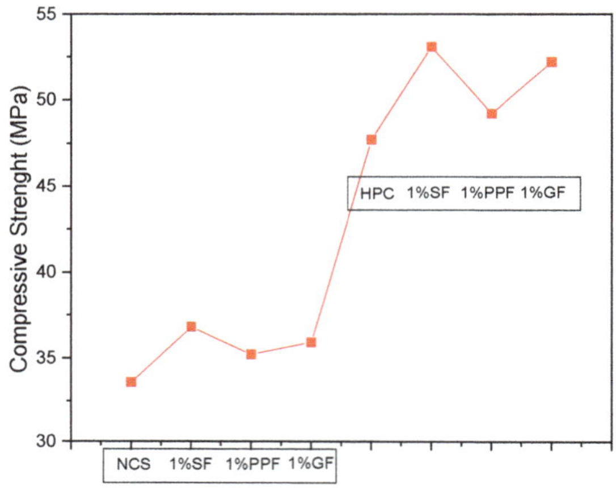

Figure 8. Compressive strength of HPC containing fibers: S (steel), G (glass), P (po polypropylene) and R (reference) [218].

Other important research that used polymeric fibers are now cited. Behafarnia and Behravan (2014) [219] evaluated the application of PP fibers in HPC to be used in water tunnels; Zhu et al. (2017) [220] evaluated the effect of the degree of aggregate saturation on the freeze-thaw strength of HPC e UHPC with light PP fibers; Li et al. (2018) [221] and Li and Zhang (2021) evaluated the influence of aggregate size and inclusion of PP and steel fibers on the hot permeability of UHPC at high temperature; Yan et al. (2021) [12] carried out experimental research on the increase of HPC ductility using basalt fiber, PP fiber, and glass fiber; Shen at al. (2020) [222] analyzed the effect of the length of PP fibers on the crack strength of high-performance concrete at an early age; Yoo and Kim (2019) [223] evaluated the influence of PE fibers on the mechanical and impact strength properties of HPC; and finally, He et al. (2017) [224] evaluated PE fiber coating mechanisms to improve the adhesion and mechanical properties of UHPC.

3.3. Natural Fibers

Natural fibers, such as fibers of mineral origin like basalt [12,212,225,226], and fibers of plant origin, like sisal [72,227], as well as banana [228,229], can also be used as reinforcement of HPC and UHPC. Their main advantages are the fact that they are renewable and eco-friendly resources. The main disadvantages are the great variability of properties and the possibility of degradation in an alkaline environment, especially in the case of fibers of vegetable origin that need previous treatments to be applied in cementitious matrices [72,227,230]. Another important advantage of natural fibers, especially vegetable fibers, is related to the fiber's high adherence to the OPC matrix. The adhesion mechanisms of this type of fiber are related to the fibrous structure of the cellulose [72]. Some authors have carried out studies on the improvement of the adhesion properties of these fibers through treatments with alkaline materials, which attack the fiber surface, increasing the roughness and adhesion with the matrix [227,230].

4. HPC and UHPC Produced with Alkali-Activated Cement

Another type of HPC and UHPC that has emerged as a great potential for application in civil construction is based on alkali-activated cement [231]. This type of cement, unlike OPC, is composed of two materials: a powdery and amorphous component named as a precursor, and another liquid composed of an aqueous solution of a dilute alkali metal, named as an activator [232,233]. The types of precursors and activators will be presented in detail in the following sections; however, in summary, they must result in composites with similar characteristics to OPC-based HPC and UHPC: adequate flowability, high mechanical strength, and good durability.

From the point of view of aggregates, the same as conventional HPC and UHPC are used, namely: washed sand from rivers or crushed sand, as fine, as well as gravel or rolled pebbles, as coarse. Thus, the materials used as precursors and activators will be discussed in this topic, since these are the components that differentiate concretes produced from alkali-activated cement.

4.1. Precursors: Overview

Precursors, as highlighted, are very fine materials and are predominantly amorphous, which can have two basic chemical compositions: (i) rich in calcium or (ii) rich in aluminosilicates. To classify the material, the $CaO/(SiO_2 + Al_2O_3)$ mass ratio is calculated. If this ratio is greater than 1, the material is considered rich in calcium, as is the case with blast furnace slag and some types of fly ash. If the ratio is less than 1, the material is considered rich in aluminosilicates, giving rise to a subclass of alkali-activated materials known as geopolymers [234,235]. This is the case of metakaolin, ceramic waste, glass waste, and most types of fly ash.

The use of precursors, which totally replaces the application of OPC, presents as main benefits the possibility of applying residues and agro-industrial by-products from different sectors, producing environmental and economic advantages. This is also related to the fact that the OPC industry is highly polluting. Alkali-activated precursors and cement are considered alternative and eco-friendly binders [232,236,237], usually resulting in concretes with lower CO_2 emissions than OPC-based concrete (around 10%, according to [238]). The accurate determination of the environmental impact associated with these binders—and consequently compare it with those associated with OPC—would require a detailed analysis which will not be addressed in this paper; nonetheless, further details can be found in [239–241].

4.2. Precursors: Rich in Aluminosilicates

Metakaolin is one of the main precursors rich in aluminosilicates according to some authors [233,234,242]. This is due to its high reactivity and the way in which the material is obtained, which can be originated from ceramic wastes [243], or more commonly, due to calcination of clays rich in kaolinite mineral, known as kaolin [244,245]. The commercial production of metakaolin occurs with the calcination of kaolin at temperatures ranging from 500 to 800 °C, depending on the degree of crystallization and purity of the material [246]. The kaolinite present in the material undergoes a dehydroxylation reaction at around 550 °C, becoming metakaolinite. Figure 9 presents a scheme for producing metakaolin from traditional kaolin [233,247]. Burning at temperatures below 400 °C is not suitable for producing the precursor. Likewise, burning at temperatures above 950 °C is not suitable either, due to the formation of mullite that does not have the ability to be alkali activated owing to its high crystallinity and because it is not soluble in an alkaline medium [248,249].

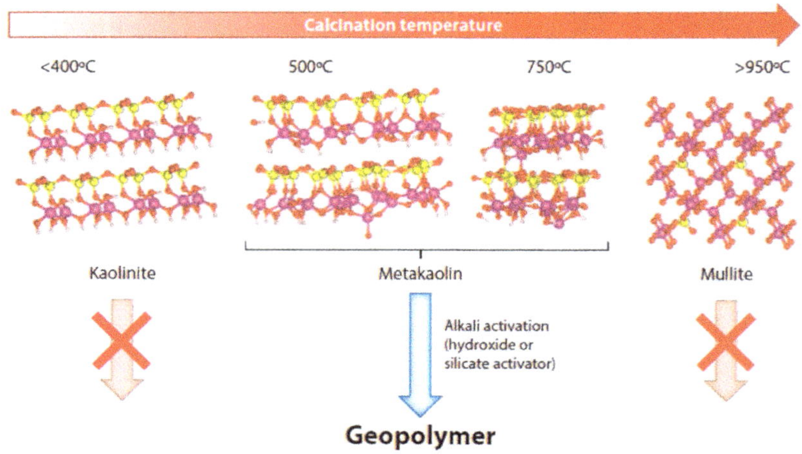

Figure 9. Metakaolin manufacturing process through kaolin calcination [247].

It is interesting to note that the use of metakaolin is environmentally advantageous because metakaolin synthesis emits about 5 to 6 times less CO_2 than the OPC production process [250]. Furthermore, kaolin can be extracted not only from mineral sources but depending on the composition it can be obtained from industrial paper waste [233]. The different sources will influence the reactivity of the metakaolin obtained, but they are ecologically more viable solutions. Another advantage is that, due to the high reactivity of metakaolin, the structure formed by alkaline activation is extremely resistant, forming a better-defined gel microstructure [233].

Regarding the disadvantages, the price of metakaolin is emphasized. Although lower than the price of OPC, it is superior to other options of precursors rich in aluminosilicates, such as fly ash [249]. Another disadvantage is related to the tendency of efflorescence of the compounds obtained by the alkaline activation of metakaolin, in general, related to the inefficiency of the chemical reaction, which can generate a whitish appearance in the concrete obtained [251,252]. The high retraction tendency of metakaolin is also mentioned, due to the chemical composition of the material, with higher levels of aluminum oxide [253].

Some examples of metakaolin-based alkali-activated cement are now presented. Hasnaoui et al. (2021) [254] evaluated the behavior of geopolymeric concretes produced with metakaolin activated by hydroxide and sodium silicate with recycled fine and coarse aggregates. The authors obtained compressive strength above 50 MPa at 28 days. Gomes et al. (2020) [255] carried out the evaluation of the mechanical properties of geopolymeric concretes produced with metakaolin activated by silicate and sodium hydroxide, containing conventional fine and coarse aggregate and steel fibers. The authors analyzed the results obtained through concepts of fracture mechanics, obtaining values compatible with application in HPC. Dias and Silva (2019) [256] evaluated the effects of the mass ratio of Na_2O/SiO_2 and K_2O/SiO_2 on the compressive strength of geopolymers produced with metakaolin without the use of silicates as activators and without the use of coarse aggregates. Albidah et al. (2020) [257] reported on the application of geopolymer concrete produced from metakaolin activated by sodium hydroxide. The authors used two types of coarse aggregates of different particle sizes, in addition to sand as a fine aggregate. The authors also used steel fiber to reinforce the geopolymer matrix, obtaining compressive strength results of approximately 58 MPa at 28 days of cure. Finally, Rocha et al. (2018) [258] evaluated the mechanical properties of geopolymers produced with metakaolin activated by hydroxide and silicate, both sodium and potassium. The authors used river washed sand as fine aggregate. The compositions evaluated by the authors reached approximately

40 MPa with 1 day of cure, showing compressive strength higher than 80 MPa after 28 days of evaluation. These results are compatible with the stipulated values for HPC.

After metakaolin, the second most used precursor rich in aluminosilicates is fly ash. As highlighted in Section 2.2, this material is obtained through the burning of mineral coal for energy production in thermoelectric plants, being one of the most produced alternative binders around the world, with an annual production of more than 900 million tons in 2019 [249]. Its characteristic is fine spherical particles, with a chemical composition based on aluminum, silicon, calcium, iron, magnesium, and carbon residues [259]. Another relevant feature is the particle size in the fly ash, ranging from <1 mm to more than 100 mm, which indicates a high specific surface area and void filling capacity. The main disadvantage of using fly ash in alkali-activated cement is the difficulty in synthesizing hardened products, due to the material curing, which must be done in thermal curing at a temperature of 65 to 90 °C to increase the material's reactivity, which it is very low at ambient temperatures [246,260–262].

In the works by Luo et al. (2021) [263], the authors compared the interfacial transition zone (ITZ) behavior of two different types of paste, one based on OPC, and another based on alkali-activated cement produced with fly ash. The results obtained showed that the ITZ of the alkali-activated paste is more strongly adhered to the aggregates than the cementitious paste, contributing to the high mechanical strength of the material. Moghaddam et al. (2021) [264] evaluated the mechanical behavior of geopolymeric concrete produced with fly ash, using rubber as aggregates and incorporating steel fibers. The strength values obtained were equivalent with HPC applications. Finally, the work by Pasupathy et al. (2021) [265], investigated the durability performance of geopolymeric concretes produced with fly ash in saline environments. The authors used two particle sizes of coarse aggregate and sand as fine aggregate. As an activator solution, a combination of sodium hydroxide and silicate was used. In addition to the greater durability of geopolymer concrete, when compared to OPC-based concrete, the authors observed that the compressive strength values were compatible with HPC applications.

Other precursors rich in aluminosilicates with high application potential are calcined illite-smectite clays [266] or calcined feldspars [267,268]. In addition to these, there are volcanic ash, natural pozzolans, and metallurgical slag with low amounts of calcium [269–272]. Recently, some authors have proposed the application of industrial waste, such as chamotte, waste from the ceramic industry [243], and magnesium phosphate from the chemical industry for the production of ammonium [273]. In the studies mentioned above, the alternative precursors shown compatible performance when compared with conventional precursors (e.g., metakaolin), indicating its potential applications in HPC and UHPC. However, further investigations are required to confirm their use for these applications.

4.3. Precursors: Rich in Calcium

Regarding calcium-rich precursors, the main materials used are steel residues or by-products, mainly blast furnace slag. The alkaline activation of these materials results in products similar to those obtained during the hydration of OPC such as C-S-H, but with the potential to be even stronger [236,274]. Regarding calcium-rich precursors, the main materials used are steel residues or by-products, mainly blast furnace slag. The alkaline activation of these materials results in products similar to those obtained during the hydration of OPC such as C-S-H, but with the potential to be even more resistant [275]. In fact, blast furnace slag is already reactive in water, but with very low kinetics. Thus, the presence of alkaline compounds only accelerates the material's hardening reaction [276]. This, in fact, is what motivates the use of slag as a substitute for clinker. Indeed, in the presence of this binder, which during the hydration step forms a solution rich in calcium, the slag can be activated.

The products obtained by the alkaline activation of blast furnace slag in the presence of sodium hydroxide or sodium silicate are composed of hydrated calcium silicates, however, showing substitution of chemical species Si by Al. Figure 10 allows us to understand the alkaline activation process of the slag, comparing it to the hydration process of OPC. It is observed that the OPC hydration reaction produces large calcium chains linked to Si tetrahedrons and with the presence of chemically linked interstitial water. This structure is called the Dreiketten structure [232,233,275]. In the case of alkaline activation of blast furnace slag, very similar calcium chains are formed. However, some Si tetrahedra, attached to the calcium structure, are replaced by Al tetrahedrons, which allows the formation of cross-links between different Dreiketten chains, giving greater rigidity and strength to the formed compounds. The occurrence of cross-links chemically unbalances the compounds formed by the alkaline activation of the slag, which is why the presence of alkaline ions, preferable metals such as Na^{+1} and K^{+1}, is necessary to promote the balance of charges. Other metallic ions can also occupy the interstice of the chains, as is the case of Al^{+3} e Ca^{+2}. In addition to these ions, there is the presence of chemically bound interstitial water in the structure of this material, which is called C-A-S-H or tobermorite [277,278], to differentiate it from hydrated OPC products, usually nomenclated as C-S-H. The microscopic appearance of tobermorite is illustrated in Figure 11.

The main disadvantages of the application of blast furnace slag as a precursor in alkali-activated cement are related to the loss of workability in the material in the fresh state. Marvila et al. (2021) [236] evaluated the rheological aspects of HPC produced with alkali cement activated on the basis of blast furnace slag and sodium hydroxide solution. The authors observed that in smaller amounts of sodium, up to 7.5% Na_2O, the obtained material behaved rheologically similar to materials based on OPC. However, at levels above 10% of Na_2O, the materials behaved rheologically as fluids with high initial yield stress and dynamic viscosity, impairing the applicability of the material. However, this characteristic is not so relevant, because, as aforementioned, blast furnace slag is reactive to lower levels of alkalinity, enabling the application of the material with lower amounts of sodium.

It is worth highlighting some recent studies that proved the viability of using blast furnace slag as a precursor of alkali-activated cement. For example, Chen et al. (2021) [279] verified the effects of alkaline solution dosage on the properties of materials produced with alkali-activated blast furnace slag, obtaining compressive strength values above 60 MPa after 1 day of cure and above 100 MPa at 28 days, with results equivalent to the production of UHPC. He et al. (2021) [280] also produced materials based on alkali-activated cement from blast furnace slag, evaluating the influence of hydrated lime as an activated agent for the material. The results obtained were above 40 MPa after three days of curing and above 70 MPa after 28 days, with respect to compressive strength. Therefore, the feasibility of applying blast furnace slag as a precursor for alkali-activated materials is proven. Other examples of calcium-rich precursors with potential for application in alkali-activated cement are gypsum desulfurization waste (FGD), cellulosic paper sludge residues and marble residues, which are also amorphous and fine.

Another possibility of alkali-activated cement widely used by some authors is the combined application of calcium-rich and aluminosilicate-rich precursors, as is the case reported by Neupane and Hadigheh (2021) [281]. These authors produced an HPC with alkali-activated cement using silica fume and blast furnace slag, also considering two types of fine aggregate (medium and fine sand) and two types of coarse aggregate, with different particle sizes. The compressive strength results obtained were higher than 30 MPa after 7 days of curing and higher than 50 MPa after 28 days, proving the application of the material as an HPC. Kotop et al. (2021) [282] evaluated the engineering properties of concrete obtained from alkali-activated cement of fly ash and calcium-rich slag, using the application of nanoclays and carbon nanotubes. The compressive strength results obtained reached 60 MPa at 28 days, proving the viability of this type of material for HPC application. The work by Mahmood et al. (2020) [283] also evaluated the mechanical properties of concrete produced with alkali-activated cement-based on fly ash and blast furnace slag, using coarse and fine aggregates and alkaline solution based on hydroxide and sodium silicate. The results obtained for compressive strength at 28 days are above 50 MPa.

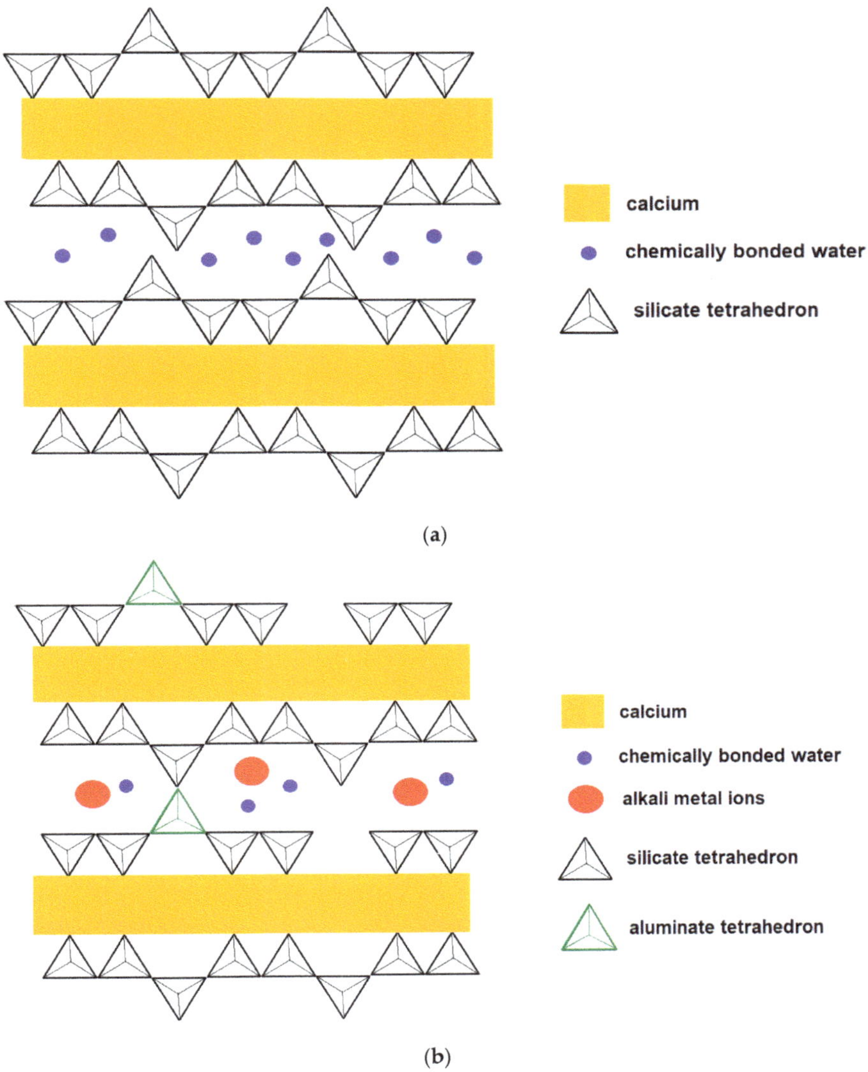

Figure 10. (**a**) Structure of the C-S-H; (**b**) Structure of C-A-S-H.

4.4. Activator Solution

The activating solution is produced by dissolving hydroxide and/or silicate of some alkaline metal in water as well as varying parameters of molarity and mass concentration. In general, the most used activators are sodium and potassium hydroxide, in addition to sodium silicate [258]. They play the same role as water in the hydration reaction of OPC, that is, starting the chemical reaction process of hardening the binder.

Some authors highlight the importance of using activated agents obtained by dissolving silicates since, in the alkaline activation of precursors rich in aluminosilicates, alumina dissolves before silica [232]. Thus, the use of silicates increases the reaction kinetics because it already provides, in a faster way, the necessary silica species for the activation reaction [233,275]. However, it favors the occurrence of efflorescence. In the case of alkaline activation of calcium-rich precursors, the need for the formation of tobermorite (C-A-S-H).

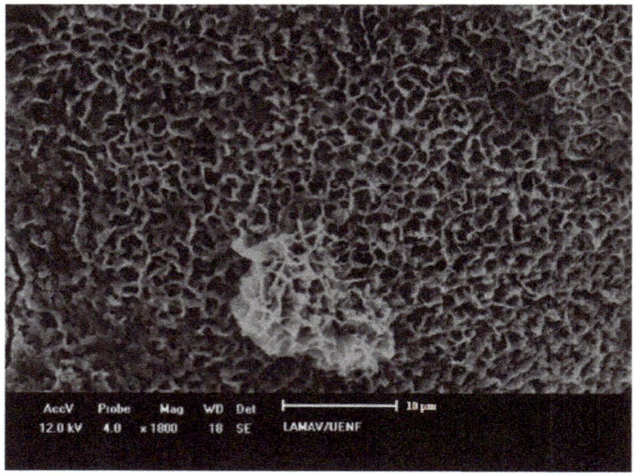

Figure 11. Scanning electron microscopy illustrating tobermorite obtained by activating alkali cement activated on a blast furnace slag base [236].

One of the disadvantages of using silicates is the high cost associated with this material. This highlights the need for alternative activated agents, which, in addition to reducing the cost of alkali-activated cement, contribute to the sustainable development of this type of HPC and UHPC, as indicated by Mendes et al. (2021) [284]. It is suggested, for example, the use of activated rice husk ash, glass waste, and silica fume.

5. Conclusions and Suggestion for Future Work

The main objective of this review article is to evaluate relevant concepts related to precursor materials used for the production of HPC and UHPC. Although there is no normative definition, these concretes can be understood as materials with high mechanical strength, good workability parameters, and high durability. After consulting the bibliography, a limit of 50 MPa for HPC and 100 MPa for UHPC was established, analyzing the compressive strength at 28 days.

Initially dealing with classic HPC and UHPC, which generally uses the same construction materials as conventional concrete, it was observed that the main type of OPC used is one that is richer in clinker and with few mineral additions. This is necessary so that pozzolans with superior quality and reactivity than those used in OPC production are used in the production of HPC and UHPC. From a chemical and mineralogical point of view, it is preferable to use OPCs rich in C_3S and C_2S, that is, with low levels of oxides of Al_2O_3 and Fe_2O_3.

Regarding mineral additives, the most used are pozzolanic ones, such as fly ash and silica fume, although there are HPC and UHPC applications using blast furnace slag. Silica fume is the ideal pozzolan due to its high specific surface, high amorphism content, and high presence of SiO_2 (>95%). This promotes the occurrence of pozzolanic reactions in a more intense way. Regarding aggregates, greater care must be taken when applying them in HPC and UHPC than with aggregates used for CC. These precautions are related to the degree of packaging, chemical composition, and shape of the grains.

For the production of HPC and UHPC, it is essential to use chemical additives, especially shrinkage mitigators and superplasticizers that allow the reduction of the w/c factor without loss of workability. The 3rd generation of superplasticizers, which work due to the electrical repulsion effect and the steric effect, are the most used for these applications.

Due to the low tensile strength and lack of ductility of HPC and UHPC, some authors proposed the inclusion of fibers for the formation of composites. The main type used is steel, followed by carbon, glass, polymeric and natural, such as sisal. The fiber improves mechanical properties but reduces workability properties due to the densification of the cement matrix. As a result, fiber contents must be studied to avoid harming the behavior of HPC and UHPC.

Another new perspective on the application of HPC and UHPC emerged due to the environmental problems generated by the production of OPC clinker. Thus, the use of alkali-activated cement, with the possibility of using waste and by-products as a binder, became a reality. The alkali-activated cement is produced through a precursor, rich in calcium (usually with >20% CaO) such as blast furnace slag, or rich in aluminosilicates, giving rise to geopolymers. The main precursors of this last group are metakaolin and low-calcium fly ash. The main difference between the high-calcium and low-calcium precursors is that the former presents a higher reaction rate due to the higher solubility of calcium in comparison with silica and alumina, generally leading to higher mechanical strength at early ages compared with the latter. In addition, high-calcium precursors form calcium aluminosilicate hydrate (C-A-S-H) as the binding phase, while low-calcium precursors form 3D sodium/potassium aluminosilicate hydrate (N/K-A-S-H) frameworks. The last component used is the alkaline solution, generally based on sodium and/or potassium hydroxides or silicates. The results obtained, both in the fresh and hardened state, are compatible with HPC and UHPC applications. Besides, alkali-activated HPC and UHPC composites tend to present higher durability when compared with OPC-based composites.

In general, HPC and UHPC produced with OPC (and mineral admixtures) are easier to produce in practice because they do not require the manipulation of highly alkaline materials. In turn, those produced with alkali-activated materials generally have a lower environmental impact.

Finally, some perspectives for future work are highlighted:

- Further standardization of fiber application methodologies in HPC and UHPC;
- Development of HPC and UHPC with other natural, renewable, and more economical fibers, such as piassava, açaí, guaruman, and pineapple fibers;
- Development of alkali-activated cement dosage methodologies for application in HPC and UHPC;
- Application of other agro-industrial residues and by-products as precursors of alkali-activated cement, such as sugarcane bagasse ash and rice husk ash;
- Research on mechanisms to improve the workability and aspects of alkali-activated cement without loss of mechanical strength;
- Development of activated agents that are more ecological than sodium and/or potassium hydroxides and silicates used for the application of alkali-activated cement, such as those based on glass residue and rice husk ash.

Author Contributions: Conceptualization, A.R.G.d.A. and M.T.M.; methodology, A.R.G.d.A., C.M.F.V., M.T.M. and P.R.d.M.; validation, S.N.M. and P.R.d.M.; formal analysis, M.T.M.; investigation, A.R.G.d.A. and M.T.M.; resources, M.T.M.; data curation, S.N.M. and C.M.F.V.; writing—original draft preparation, M.T.M., A.R.G.d.A. and P.R.d.M.; writing—review and editing, M.T.M., A.R.G.d.A. and P.R.d.M.; supervision, S.N.M. and C.M.F.V.; funding acquisition, S.N.M. and C.M.F.V. All authors have read and agreed to the published version of the manuscript.

Funding: CNPq: an FAPERJ, Proc. No. E-26/010.001953/2019 and E-26/210.150/2019.

Institutional Review Board Statement: Not applicable.

Informed Consent Statement: Not applicable.

Data Availability Statement: All the data is available within the manuscript.

Acknowledgments: The authors thank CAPES, CNPq and FAPER for the financial support.

Conflicts of Interest: The authors declare no conflict of interest.

References

1. Sohail, M.G.; Kahraman, R.; Al Nuaimi, N.; Gencturk, B.; Alnahhal, W. Durability Characteristics of High and Ultra-High Performance Concretes. *J. Build. Eng.* **2021**, *33*, 101669. [CrossRef]
2. Phan, D.H.H.; Patel, V.I.; Liang, Q.Q.; Al Abadi, H.; Thai, H.-T. Simulation of Uniaxially Compressed Square Ultra-High-Strength Concrete-Filled Steel Tubular Slender Beam-Columns. *Eng. Struct.* **2021**, *232*, 111795. [CrossRef]
3. Ding, M.; Yu, R.; Feng, Y.; Wang, S.; Zhou, F.; Shui, Z.; Gao, X.; He, Y.; Chen, L. Possibility and Advantages of Producing an Ultra-High Performance Concrete (UHPC) with Ultra-Low Cement Content. *Constr. Build. Mater.* **2021**, *273*, 122023. [CrossRef]
4. Zhang, T.; Zhang, Y.; Xiao, Z.; Yang, Z.; Zhu, H.; Ju, J.W.; Yan, Z. Development of a Novel Bio-Inspired Cement-Based Composite Material to Improve the Fire Resistance of Engineering Structures. *Constr. Build. Mater.* **2019**, *225*, 99–111. [CrossRef]
5. Azmee, N.M.; Shafiq, N. Ultra-High Performance Concrete: From Fundamental to Applications. *Case Stud. Constr. Mater.* **2018**, *9*, e00197. [CrossRef]
6. Damineli, B.L.; Kemeid, F.M.; Aguiar, P.S.; John, V.M. Measuring the Eco-Efficiency of Cement Use. *Cem. Concr. Compos.* **2010**, *32*, 555–562. [CrossRef]
7. Brazilian Association of Technical Standards (ABNT). *ABNT NBR 8953: Concrete for Structural Use—Density, Strength and Consistence Classification*; Brazilian Association of Technical Standards (ABNT): Rio de Janeiro, Brazil, 2015. (In Portuguese)
8. Brazilian Association of Technical Standards (ABNT). *ABNT NBR 6118: Design of Concrete Structures—Procedure*; Brazilian Association of Technical Standards (ABNT): Rio de Janeiro, Brazil, 2014. (In Portuguese)
9. American Concrete Institute. *High-Strength Concrete (ACI 363R)*; American Concrete Institute: New York, NY, USA, 2005.
10. ACI Committee 318. *Building Code Requirements for Structural Concrete*; American Concrete Institute: New York, NY, USA, 2014.
11. EUROCODES British Standards. *Eurocode BS EN 1992-2:2005—Eurocode 2: Design of Concrete Structures—Part 2: Concrete Bridges—Design and Detailing Rules*; EUROCODES British Standards: London, UK, 2005.
12. Yan, P.; Chen, B.; Afgan, S.; Aminul Haque, M.; Wu, M.; Han, J. Experimental Research on Ductility Enhancement of Ultra-High Performance Concrete Incorporation with Basalt Fibre, Polypropylene Fibre and Glass Fibre. *Constr. Build. Mater.* **2021**, *279*, 122489. [CrossRef]
13. Roberti, F.; Cesari, V.F.; de Matos, P.R.; Pelisser, F.; Pilar, R. High- and Ultra-High-Performance Concrete Produced with Sulfate-Resisting Cement and Steel Microfiber: Autogenous Shrinkage, Fresh-State, Mechanical Properties and Microstructure Characterization. *Constr. Build. Mater.* **2021**, *268*, 121092. [CrossRef]
14. Sohail, M.G.; Wang, B.; Jain, A.; Kahraman, R.; Ozerkan, N.G.; Gencturk, B.; Dawood, M.; Belarbi, A. Advancements in Concrete Mix Designs: High-Performance and Ultrahigh-Performance Concretes from 1970 to 2016. *J. Mater. Civ. Eng.* **2018**, *30*, 04017310. [CrossRef]
15. Cu, Y.T.H.; Tran, M.V.; Ho, C.H.; Nguyen, P.H. Relationship between Workability and Rheological Parameters of Self-Compacting Concrete Used for Vertical Pump up to Supertall Buildings. *J. Build. Eng.* **2020**, *32*, 101786. [CrossRef]
16. Buttignol, T.E.T.; Sousa, J.L.A.O.; Bittencourt, T.N. Ultra High-Performance Fiber-Reinforced Concrete (UHPFRC): A Review of Material Properties and Design Procedures. *Rev. IBRACON Estrut. Mater.* **2017**, *10*, 957–971. [CrossRef]
17. Bangi, M.R.; Horiguchi, T. Effect of Fibre Type and Geometry on Maximum Pore Pressures in Fibre-Reinforced High Strength Concrete at Elevated Temperatures. *Cem. Concr. Res.* **2012**, *42*, 459–466. [CrossRef]
18. Ding, Y.; Zhang, C.; Cao, M.; Zhang, Y.; Azevedo, C. Influence of Different Fibers on the Change of Pore Pressure of Self-Consolidating Concrete Exposed to Fire. *Constr. Build. Mater.* **2016**, *113*, 456–469. [CrossRef]
19. Zeng, J.-J.; Ye, Y.-Y.; Gao, W.-Y.; Smith, S.T.; Guo, Y.-C. Stress-Strain Behavior of Polyethylene Terephthalate Fiber-Reinforced Polymer-Confined Normal-, High- and Ultra High-Strength Concrete. *J. Build. Eng.* **2020**, *30*, 101243. [CrossRef]
20. Meddah, M.S.; Zitouni, S.; Belâabes, S. Effect of Content and Particle Size Distribution of Coarse Aggregate on the Compressive Strength of Concrete. *Constr. Build. Mater.* **2010**, *24*, 505–512. [CrossRef]
21. Al-Yousuf, A.; Pokharel, T.; Lee, J.; Gad, E.; Abdouka, K.; Sanjayan, J. Effect of Fly Ash and Slag on Properties of Normal and High Strength Concrete Including Fracture Energy by Wedge Splitting Test: Experimental and Numerical Investigations. *Constr. Build. Mater.* **2021**, *271*, 121553. [CrossRef]
22. Yaphary, Y.L.; Lam, R.H.W.; Lau, D. Reduction in Cement Content of Normal Strength Concrete with Used Engine Oil (UEO) as Chemical Admixture. *Constr. Build. Mater.* **2020**, *261*, 119967. [CrossRef]
23. Peng, G.-F.; Yang, W.-W.; Zhao, J.; Liu, Y.-F.; Bian, S.-H.; Zhao, L.-H. Explosive Spalling and Residual Mechanical Properties of Fiber-Toughened High-Performance Concrete Subjected to High Temperatures. *Cem. Concr. Res.* **2006**, *36*, 723–727. [CrossRef]
24. Wu, F.; Xu, L.; Chi, Y.; Zeng, Y.; Deng, F.; Chen, Q. Compressive and Flexural Properties of Ultra-High Performance Fiber-Reinforced Cementitious Composite: The Effect of Coarse Aggregate. *Compos. Struct.* **2020**, *236*, 111810. [CrossRef]
25. Paschalis, S.A.; Lampropoulos, A.P. Developments in the Use of Ultra High Performance Fiber Reinforced Concrete as Strengthening Material. *Eng. Struct.* **2021**, *233*, 111914. [CrossRef]
26. Manigandan, S.; Praveenkumar, T.R.; Al-Mohaimeed, A.M.; Brindhadevi, K.; Pugazhendhi, A. Characterization of Polyurethane Coating on High Performance Concrete Reinforced with Chemically Treated Ananas Erectifolius Fiber. *Prog. Org. Coat.* **2021**, *150*, 105977. [CrossRef]
27. Karimipour, A.; Edalati, M. Shear and Flexural Performance of Low, Normal and High-Strength Concrete Beams Reinforced with Longitudinal SMA, GFRP and Steel Rebars. *Eng. Struct.* **2020**, *221*, 111086. [CrossRef]

28. Zhang, Y.; Zhu, P.; Wang, X.; Wu, J. Shear Properties of the Interface between Ultra-High Performance Concrete and Normal Strength Concrete. *Constr. Build. Mater.* **2020**, *248*, 118455. [CrossRef]
29. Xiong, M.-X.; Xiong, D.-X.; Liew, J.Y.R. Behaviour of Steel Tubular Members Infilled with Ultra High Strength Concrete. *J. Constr. Steel Res.* **2017**, *138*, 168–183. [CrossRef]
30. Zhang, T.; Zhu, H.; Zhou, L.; Yan, Z. Multi-Level Micromechanical Analysis of Elastic Properties of Ultra-High Performance Concrete at High Temperatures: Effects of Imperfect Interface and Inclusion Size. *Compos. Struct.* **2021**, *262*, 113548. [CrossRef]
31. Shi, C.; Wu, Z.; Xiao, J.; Wang, D.; Huang, Z.; Fang, Z. A Review on Ultra High Performance Concrete: Part I. Raw Materials and Mixture Design. *Constr. Build. Mater.* **2015**, *101*, 741–751. [CrossRef]
32. Yoo, D.-Y.; Banthia, N. Mechanical Properties of Ultra-High-Performance Fiber-Reinforced Concrete: A Review. *Cem. Concr. Compos.* **2016**, *73*, 267–280. [CrossRef]
33. Zhang, D.; Tan, K.H. Effect of Various Polymer Fibers on Spalling Mitigation of Ultra-High Performance Concrete at High Temperature. *Cem. Concr. Compos.* **2020**, *114*, 103815. [CrossRef]
34. Li, Y.; Tan, K.H.; Yang, E.-H. Synergistic Effects of Hybrid Polypropylene and Steel Fibers on Explosive Spalling Prevention of Ultra-High Performance Concrete at Elevated Temperature. *Cem. Concr. Compos.* **2019**, *96*, 174–181. [CrossRef]
35. Wang, D.; Shi, C.; Wu, Z.; Xiao, J.; Huang, Z.; Fang, Z. A Review on Ultra High Performance Concrete: Part II. Hydration, Microstructure and Properties. *Constr. Build. Mater.* **2015**, *96*, 368–377. [CrossRef]
36. Wu, Z.; Shi, C.; Khayat, K.H. Investigation of Mechanical Properties and Shrinkage of Ultra-High Performance Concrete: Influence of Steel Fiber Content and Shape. *Compos. Part B Eng.* **2019**, *174*, 107021. [CrossRef]
37. NF P18-470: Bétons Fibrés à Ultra-Hautes Performances–Spécification, Performance, Production et Conformité. Available online: https://www.boutique.afnor.org/norme/nf-p18-470/betons-betons-fibres-a-ultra-hautes-performances-specification-performance-production-et-conformite-/article/866839/fa0633532016 (accessed on 6 July 2021).
38. Schipporeit, G.C.; Heinrich, J.C. Lake Point Tower Chicago. *Lux. City Apartm.* **2018**, *2*, 265–275.
39. Kathirvel, P.; Sreekumaran, S. Sustainable Development of Ultra High Performance Concrete Using Geopolymer Technology. *J. Build. Eng.* **2021**, *39*, 102267. [CrossRef]
40. Li, S.; Zhao, T.; Alam, M.S.; Cheng, Z.; Wang, J. Probabilistic Seismic Vulnerability and Loss Assessment of a Seismic Resistance Bridge System with Post-Tensioning Precast Segmental Ultra-High Performance Concrete Bridge Columns. *Eng. Struct.* **2020**, *225*, 111321. [CrossRef]
41. Xue, J.; Briseghella, B.; Huang, F.; Nuti, C.; Tabatabai, H.; Chen, B. Review of Ultra-High Performance Concrete and Its Application in Bridge Engineering. *Constr. Build. Mater.* **2020**, *260*, 119844. [CrossRef]
42. Tsai, C.-T.; Kung, G.T.-C.; Hwang, C.-L. Use of High Performance Concrete on Rigid Pavement Construction for Exclusive Bus Lanes. *Constr. Build. Mater.* **2010**, *24*, 732–740. [CrossRef]
43. Akhnoukh, A.K.; Elia, H. Developing High Performance Concrete for Precast/Prestressed Concrete Industry. *Case Stud. Constr. Mater.* **2019**, *11*, e00290. [CrossRef]
44. Zhang, Y.; Zhu, P.; Liao, Z.; Wang, L. Interfacial Bond Properties between Normal Strength Concrete Substrate and Ultra-High Performance Concrete as a Repair Material. *Constr. Build. Mater.* **2020**, *235*, 117431. [CrossRef]
45. Makul, N. Cost-Benefit Analysis of the Production of Ready-Mixed High-Performance Concrete Made with Recycled Concrete Aggregate: A Case Study in Thailand. *Heliyon* **2020**, *6*, e04135. [CrossRef]
46. Dong, Y. Performance Assessment and Design of Ultra-High Performance Concrete (UHPC) Structures Incorporating Life-Cycle Cost and Environmental Impacts. *Constr. Build. Mater.* **2018**, *167*, 414–425. [CrossRef]
47. ABNT ABNT NBR. *16697: Portland Cement—Requirements*; Brazilian Association of Technical Standards (ABNT): Rio de Janeiro, Brazil, 2018. (In Portuguese)
48. BSI Standards Publication. *EN 197-1 Cement: Composition, Specifications and Conformity Criteria for Common Cements*; BSI Standards Publication: London, UK, 2011.
49. ASTM International. *ASTM_C150/C150-07: Standard Specification for Portland Cement*; ASTM International: West Conshohocken, PA, USA, 2007.
50. Beatriz da Silva, J.; Pepe, M.; Toledo Filho, R.D. High Temperatures Effect on Mechanical and Physical Performance of Normal and High Strength Recycled Aggregate Concrete. *Fire Saf. J.* **2020**, *117*, 103222. [CrossRef]
51. Viana, T.M.; Bacelar, B.A.; Coelho, I.D.; Ludvig, P.; Santos, W.J. Behaviour of Ultra-High Performance Concretes Incorporating Carbon Nanotubes under Thermal Load. *Constr. Build. Mater.* **2020**, *263*, 120556. [CrossRef]
52. de Matos, P.R.; Sakata, R.D.; Gleize, P.J.P.; de Brito, J.; Repette, W.L. Eco-Friendly Ultra-High Performance Cement Pastes Produced with Quarry Wastes as Alternative Fillers. *J. Clean. Prod.* **2020**, *269*, 122308. [CrossRef]
53. Pilar, R.; Schankoski, R.A.; Ferron, R.D.; Repette, W.L. Rheological Behavior of Low Shrinkage Very High Strength Self-Compacting Concrete. *Constr. Build. Mater.* **2021**, *286*, 122838. [CrossRef]
54. Storm, J.; Pise, M.; Brands, D.; Schröder, J.; Kaliske, M. A Comparative Study of Micro-Mechanical Models for Fiber Pullout Behavior of Reinforced High Performance Concrete. *Eng. Fract. Mech.* **2021**, *243*, 107506. [CrossRef]
55. Li, Y.; Zhang, D. Effect of Lateral Restraint and Inclusion of Polypropylene and Steel Fibers on Spalling Behavior, Pore Pressure, and Thermal Stress in Ultra-High-Performance Concrete (UHPC) at Elevated Temperature. *Constr. Build. Mater.* **2021**, *271*, 121879. [CrossRef]

56. Liu, G.; Schollbach, K.; Li, P.; Brouwers, H.J.H. Valorization of Converter Steel Slag into Eco-Friendly Ultra-High Performance Concrete by Ambient CO_2 Pre-Treatment. *Constr. Build. Mater.* **2021**, *280*, 122580. [CrossRef]
57. Suescum-Morales, D.; Ríos, J.D.; De La Concha, A.M.; Cifuentes, H.; Jiménez, J.R.; Fernández, J.M. Effect of Moderate Temperatures on Compressive Strength of Ultra-High-Performance Concrete: A Microstructural Analysis. *Cem. Concr. Res.* **2021**, *140*, 106303. [CrossRef]
58. Rashid, R.S.M.; Salem, S.M.; Azreen, N.M.; Voo, Y.L.; Haniza, M.; Shukri, A.A.; Yahya, M.-S. Effect of Elevated Temperature to Radiation Shielding of Ultra-High Performance Concrete with Silica Sand or Magnetite. *Constr. Build. Mater.* **2020**, *262*, 120567. [CrossRef]
59. Olawuyi, B.J.; Babafemi, A.J.; Boshoff, W.P. Early-Age and Long-Term Strength Development of High-Performance Concrete with SAP. *Constr. Build. Mater.* **2021**, *267*, 121798. [CrossRef]
60. Zhang, D.; Liu, Y.; Tan, K.H. Spalling Resistance and Mechanical Properties of Strain-Hardening Ultra-High Performance Concrete at Elevated Temperature. *Constr. Build. Mater.* **2021**, *266*, 120961. [CrossRef]
61. Choi, H.-J.; Park, J.-J.; Yoo, D.-Y. Benefits of TiO_2 Photocatalyst on Mechanical Properties and Nitrogen Oxide Removal of Ultra-High-Performance Concrete. *Constr. Build. Mater.* **2021**, *285*, 122921. [CrossRef]
62. Khan, M.U.; Ahmad, S.; Naqvi, A.A.; Al-Gahtani, H.J. Shielding Performance of Heavy-Weight Ultra-High-Performance Concrete against Nuclear Radiation. *Prog. Nucl. Energy* **2020**, *130*, 103550. [CrossRef]
63. Kim, J.-J.; Yoo, D.-Y.; Banthia, N. Benefits of Curvilinear Straight Steel Fibers on the Rate-Dependent Pullout Resistance of Ultra-High-Performance Concrete. *Cem. Concr. Compos.* **2021**, *118*, 103965. [CrossRef]
64. Yoo, D.-Y.; You, I.; Zi, G. Effects of Waste Liquid–Crystal Display Glass Powder and Fiber Geometry on the Mechanical Properties of Ultra-High-Performance Concrete. *Constr. Build. Mater.* **2021**, *266*, 120938. [CrossRef]
65. Yoo, D.-Y.; Jang, Y.S.; Chun, B.; Kim, S. Chelate Effect on Fiber Surface Morphology and Its Benefits on Pullout and Tensile Behaviors of Ultra-High-Performance Concrete. *Cem. Concr. Compos.* **2021**, *115*, 103864. [CrossRef]
66. Kareem, R.S.; Dang, C.N.; Hale, W.M. Flexural Behavior of Concrete Beams Cast with High-Performance Materials. *J. Build. Eng.* **2021**, *34*, 101912. [CrossRef]
67. Bae, Y.; Pyo, S. Effect of Steel Fiber Content on Structural and Electrical Properties of Ultra High Performance Concrete (UHPC) Sleepers. *Eng. Struct.* **2020**, *222*, 111131. [CrossRef]
68. Kadri, E.-H.; Aggoun, S.; De Schutter, G. Interaction between C3A, Silica Fume and Naphthalene Sulphonate Superplasticiser in High Performance Concrete. *Constr. Build. Mater.* **2009**, *23*, 3124–3128. [CrossRef]
69. Prince, W.; Edwards-Lajnef, M.; Aïtcin, P.C. Interaction between Ettringite and a Polynaphthalene Sulfonate Superplasticizer in a Cementitious Paste. *Cem. Concr. Res.* **2002**, *32*, 79–85. [CrossRef]
70. Arunothayan, A.R.; Nematollahi, B.; Ranade, R.; Bong, S.H.; Sanjayan, J.G.; Khayat, K.H. Fiber Orientation Effects on Ultra-High Performance Concrete Formed by 3D Printing. *Cem. Concr. Res.* **2021**, *143*, 106384. [CrossRef]
71. Zhou, Y.; Huang, J.; Yang, X.; Dong, Y.; Feng, T.; Liu, J. Enhancing the PVA Fiber-Matrix Interface Properties in Ultra High Performance Concrete: An Experimental and Molecular Dynamics Study. *Constr. Build. Mater.* **2021**, *285*, 122862. [CrossRef]
72. Ren, G.; Yao, B.; Huang, H.; Gao, X. Influence of Sisal Fibers on the Mechanical Performance of Ultra-High Performance Concretes. *Constr. Build. Mater.* **2021**, *286*, 122958. [CrossRef]
73. Nguyen, D.-L.; Thai, D.-K.; Nguyen, H.T.T.; Nguyen, T.-Q.; Le-Trung, K. Responses of Composite Beams with High-Performance Fiber-Reinforced Concrete. *Constr. Build. Mater.* **2021**, *270*, 121814. [CrossRef]
74. Navabi, D.; Javidruzi, M.; Hafezi, M.R.; Mosavi, A. The High-Performance Light Transmitting Concrete and Experimental Analysis of Using Polymethylmethacrylate Optical Fibers in It. *J. Build. Eng.* **2021**, *38*, 102076. [CrossRef]
75. Lothenbach, B.; Scrivener, K.; Hooton, R.D. Supplementary Cementitious Materials. *Cem. Concr. Res.* **2011**, *41*, 1244–1256. [CrossRef]
76. Scrivener, K.L.; John, V.M.; Gartner, E.M. Eco-Efficient Cements: Potential Economically Viable Solutions for a Low-CO_2 Cement-Based Materials Industry. *Cem. Concr. Res.* **2018**, *114*, 2–26. [CrossRef]
77. Celik, F.; Akcuru, O. Rheological and Workability Effects of Bottom Ash Usage as a Mineral Additive on the Cement Based Permeation Grouting Method. *Constr. Build. Mater.* **2020**, *263*, 120186. [CrossRef]
78. Sanjuán, M.Á.; Andrade, C.; Mora, P.; Zaragoza, A. Carbon Dioxide Uptake by Cement-Based Materials: A Spanish Case Study. *Appl. Sci.* **2020**, *10*, 339. [CrossRef]
79. Gartner, E. Industrially Interesting Approaches to "Low-CO_2" Cements. *Cem. Concr. Res.* **2004**, *34*, 1489–1498. [CrossRef]
80. Meyer, C. The Greening of the Concrete Industry. *Cem. Concr. Compos.* **2009**, *31*, 601–605. [CrossRef]
81. Sanjuán, M.Á.; Andrade, C.; Mora, P.; Zaragoza, A. Carbon Dioxide Uptake by Mortars and Concretes Made with Portuguese Cements. *Appl. Sci.* **2020**, *10*, 646. [CrossRef]
82. Pade, C.; Guimaraes, M. The CO_2 Uptake of Concrete in a 100 Year Perspective. *Cem. Concr. Res.* **2007**, *37*, 1348–1356. [CrossRef]
83. Li, C.; Jiang, L. Utilization of Limestone Powder as an Activator for Early-Age Strength Improvement of Slag Concrete. *Constr. Build. Mater.* **2020**, *253*, 119257. [CrossRef]
84. Zhang, J.; Huang, Y.; Ma, G.; Nener, B. Mixture Optimization for Environmental, Economical and Mechanical Objectives in Silica Fume Concrete: A Novel Frame-Work Based on Machine Learning and a New Meta-Heuristic Algorithm. *Resour. Conserv. Recycl.* **2021**, *167*, 105395. [CrossRef]

85. Rodríguez, E.D.; Bernal, S.A.; Provis, J.L.; Payá, J.; Monzó, J.M.; Borrachero, M.V. Structure of Portland Cement Pastes Blended with Sonicated Silica Fume. *J. Mater. Civ. Eng.* **2012**, *24*, 1295–1304. [CrossRef]
86. de Matos, P.R.; Sakata, R.D.; Prudêncio, L.R. Eco-Efficient Low Binder High-Performance Self-Compacting Concretes. *Constr. Build. Mater.* **2019**, *225*, 941–955. [CrossRef]
87. Juenger, M.C.G.; Siddique, R. Recent Advances in Understanding the Role of Supplementary Cementitious Materials in Concrete. *Cem. Concr. Res.* **2015**, *78*, 71–80. [CrossRef]
88. Aydın, S. A Ternary Optimisation of Mineral Additives of Alkali Activated Cement Mortars. *Constr. Build. Mater.* **2013**, *43*, 131–138. [CrossRef]
89. Cao, L.; Liu, C.; Tian, H.; Jia, D.; Wang, D.; Xu, Y.; Guo, J. Adsorption Interaction between Cement Hydrates Minerals with Fluid Loss Additive Investigated by Fluorescence Technique. *Constr. Build. Mater.* **2019**, *223*, 1106–1111. [CrossRef]
90. Deboucha, W.; Leklou, N.; Khelidj, A.; Oudjit, M.N. Hydration Development of Mineral Additives Blended Cement Using Thermogravimetric Analysis (TGA): Methodology of Calculating the Degree of Hydration. *Constr. Build. Mater.* **2017**, *146*, 687–701. [CrossRef]
91. Yoon, H.N.; Seo, J.; Kim, S.; Lee, H.K.; Park, S. Characterization of Blast Furnace Slag-Blended Portland Cement for Immobilization of Co. *Cem. Concr. Res.* **2020**, *134*, 106089. [CrossRef]
92. Ke, X.; Bernal, S.A.; Provis, J.L.; Lothenbach, B. Thermodynamic Modelling of Phase Evolution in Alkali-Activated Slag Cements Exposed to Carbon Dioxide. *Cem. Concr. Res.* **2020**, *136*, 106158. [CrossRef]
93. Park, S.; Park, H.M.; Yoon, H.N.; Seo, J.; Yang, C.-M.; Provis, J.L.; Yang, B. Hydration Kinetics and Products of MgO-Activated Blast Furnace Slag. *Constr. Build. Mater.* **2020**, *249*, 118700. [CrossRef]
94. Kaminskas, R.; Kubiliute, R.; Prialgauskaite, B. Smectite Clay Waste as an Additive for Portland Cement. *Cem. Concr. Compos.* **2020**, *113*, 103710. [CrossRef]
95. Terzić, A.; Pezo, L.; Mijatović, N.; Stojanović, J.; Kragović, M.; Miličić, L.; Andrić, L. The Effect of Alternations in Mineral Additives (Zeolite, Bentonite, Fly Ash) on Physico-Chemical Behavior of Portland Cement Based Binders. *Constr. Build. Mater.* **2018**, *180*, 199–210. [CrossRef]
96. Kubiliute, R.; Kaminskas, R.; Kazlauskaite, A. Mineral Wool Production Waste as an Additive for Portland Cement. *Cem. Concr. Compos.* **2018**, *88*, 130–138. [CrossRef]
97. Marvila, M.T.; Alexandre, J.; Azevedo, A.R.G.; Zanelato, E.B.; Xavier, G.C.; Monteiro, S.N. Study on the Replacement of the Hydrated Lime by Kaolinitic Clay in Mortars. *Adv. Appl. Ceram.* **2019**, *118*, 373–380. [CrossRef]
98. Marvila, M.T.; Azevedo, A.R.G.; Alexandre, J.; Colorado, H.; Pereira Antunes, M.L.; Vieira, C.M.F. Circular Economy in Cementitious Ceramics: Replacement of Hydrated Lime with a Stoichiometric Balanced Combination of Clay and Marble Waste. *Int. J. Appl. Ceram. Technol.* **2020**, *18*, 192–202. [CrossRef]
99. Marvila, M.T.; Alexandre, J.; de Azevedo, A.R.G.; Zanelato, E.B. Evaluation of the Use of Marble Waste in Hydrated Lime Cement Mortar Based. *J. Mater. Cycles Waste Manag.* **2019**, *21*, 1250–1261. [CrossRef]
100. Marshdi, Q.S.R. Benefits of Using Mineral Additives, as Components of the Modern Oil-Well Cement. *Case Stud. Constr. Mater.* **2018**, *8*, 455–458. [CrossRef]
101. Mohan, M.K.; Pillai, R.G.; Santhanam, M.; Gettu, R. High-Performance Cementitious Grout with Fly Ash for Corrosion Protection of Post-Tensioned Concrete Structures. *Constr. Build. Mater.* **2021**, *281*, 122612. [CrossRef]
102. Jing, R.; Liu, Y.; Yan, P. Uncovering the Effect of Fly Ash Cenospheres on the Macroscopic Properties and Microstructure of Ultra High-Performance Concrete (UHPC). *Constr. Build. Mater.* **2021**, *286*, 122977. [CrossRef]
103. Herath, C.; Gunasekara, C.; Law, D.W.; Setunge, S. Performance of High Volume Fly Ash Concrete Incorporating Additives: A Systematic Literature Review. *Constr. Build. Mater.* **2020**, *258*, 120606. [CrossRef]
104. Sujay, H.M.; Nair, N.A.; Sudarsana Rao, H.; Sairam, V. Experimental Study on Durability Characteristics of Composite Fiber Reinforced High-Performance Concrete Incorporating Nanosilica and Ultra Fine Fly Ash. *Constr. Build. Mater.* **2020**, *262*, 120738. [CrossRef]
105. Bahedh, M.A.; Jaafar, M.S. Ultra High-Performance Concrete Utilizing Fly Ash as Cement Replacement under Autoclaving Technique. *Case Stud. Constr. Mater.* **2018**, *9*, e00202. [CrossRef]
106. Zhang, D.; Yang, Q.; Mao, M.; Li, J. Carbonation Performance of Concrete with Fly Ash as Fine Aggregate after Stress Damage and High Temperature Exposure. *Constr. Build. Mater.* **2020**, *242*, 118125. [CrossRef]
107. Choudhary, R.; Gupta, R.; Nagar, R.; Jain, A. Mechanical and Abrasion Resistance Performance of Silica Fume, Marble Slurry Powder, and Fly Ash Amalgamated High Strength Self-Consolidating Concrete. *Constr. Build. Mater.* **2021**, *269*, 121282. [CrossRef]
108. Yu, L.; Li, Y.; Liu, T.; Qin, Z.; Tan, H.; Zhang, H.; Chen, Z.; Ni, H. Mechanical and Microstructural Characterization of Geopolymers Synthesized from FCC Waste Catalyst and Silica Fume. *Ceram. Int.* **2021**, *47*, 15186–15194. [CrossRef]
109. Brescia-Norambuena, L.; González, M.; Avudaiappan, S.; Saavedra Flores, E.I.; Grasley, Z. Improving Concrete Underground Mining Pavements Performance through the Synergic Effect of Silica Fume, Nanosilica, and Polypropylene Fibers. *Constr. Build. Mater.* **2021**, *285*, 122895. [CrossRef]
110. Ren, J.; Hu, L.; Dong, Z.; Tang, L.; Xing, F.; Liu, J. Effect of Silica Fume on the Mechanical Property and Hydration Characteristic of Alkali-Activated Municipal Solid Waste Incinerator (MSWI) Fly Ash. *J. Clean. Prod.* **2021**, *295*, 126317. [CrossRef]

111. Das, S.K.; Mustakim, S.M.; Adesina, A.; Mishra, J.; Alomayri, T.S.; Assaedi, H.S.; Kaze, C.R. Fresh, Strength and Microstructure Properties of Geopolymer Concrete Incorporating Lime and Silica Fume as Replacement of Fly Ash. *J. Build. Eng.* **2020**, *32*, 101780. [CrossRef]
112. Wu, Z.; Khayat, K.H.; Shi, C. Changes in Rheology and Mechanical Properties of Ultra-High Performance Concrete with Silica Fume Content. *Cem. Concr. Res.* **2019**, *123*, 105786. [CrossRef]
113. Smarzewski, P. Influence of Silica Fume on Mechanical and Fracture Properties of High Performance Concrete. *Procedia Struct. Integr.* **2019**, *17*, 5–12. [CrossRef]
114. Pedro, D.; de Brito, J.; Evangelista, L. Evaluation of High-Performance Concrete with Recycled Aggregates: Use of Densified Silica Fume as Cement Replacement. *Constr. Build. Mater.* **2017**, *147*, 803–814. [CrossRef]
115. Pedro, D.; de Brito, J.; Evangelista, L. Durability Performance of High-Performance Concrete Made with Recycled Aggregates, Fly Ash and Densified Silica Fume. *Cem. Concr. Compos.* **2018**, *93*, 63–74. [CrossRef]
116. Chen, T.; Gao, X.; Ren, M. Effects of Autoclave Curing and Fly Ash on Mechanical Properties of Ultra-High Performance Concrete. *Constr. Build. Mater.* **2018**, *158*, 864–872. [CrossRef]
117. Ali, K.; Qureshi, M.I.; Saleem, S.; Khan, S.U. Effect of Waste Electronic Plastic and Silica Fume on Mechanical Properties and Thermal Performance of Concrete. *Constr. Build. Mater.* **2021**, *285*, 122952. [CrossRef]
118. Abo-El-Enein, S.A.; El-Sayed, H.A.; Ali, A.H.; Mohammed, Y.T.; Khater, H.M.; Ouda, A.S. Physico-Mechanical Properties of High Performance Concrete Using Different Aggregates in Presence of Silica Fume. *HBRC J.* **2014**, *10*, 43–48. [CrossRef]
119. Hamid, R.; Yusof, K.M.; Zain, M.F.M. A Combined Ultrasound Method Applied to High Performance Concrete with Silica Fume. *Constr. Build. Mater.* **2010**, *24*, 94–98. [CrossRef]
120. Tanyildizi, H.; Çevik, A. Modeling Mechanical Performance of Lightweight Concrete Containing Silica Fume Exposed to High Temperature Using Genetic Programming. *Constr. Build. Mater.* **2010**, *24*, 2612–2618. [CrossRef]
121. Tanyildizi, H.; Coskun, A. Performance of Lightweight Concrete with Silica Fume after High Temperature. *Constr. Build. Mater.* **2008**, *22*, 2124–2129. [CrossRef]
122. de Matos, P.R.; Oliveira, J.C.P.; Medina, T.M.; Magalhães, D.C.; Gleize, P.J.P.; Schankoski, R.A.; Pilar, R. Use of Air-Cooled Blast Furnace Slag as Supplementary Cementitious Material for Self-Compacting Concrete Production. *Constr. Build. Mater.* **2020**, *262*, 120102. [CrossRef]
123. Salvador, R.P.; Rambo, D.A.S.; Bueno, R.M.; Silva, K.T.; Figueiredo, A.D. de On the Use of Blast-Furnace Slag in Sprayed Concrete Applications. *Constr. Build. Mater.* **2019**, *218*, 543–555. [CrossRef]
124. Andrade Neto, J.d.S.; Santos, T.A.; de Andrade Pinto, S.; Dias, C.M.R.; Ribeiro, D.V. Effect of the Combined Use of Carbon Nanotubes (CNT) and Metakaolin on the Properties of Cementitious Matrices. *Constr. Build. Mater.* **2021**, *271*, 121903. [CrossRef]
125. Sousa, M.I.C.; da Silva Rêgo, J.H. Effect of Nanosilica/Metakaolin Ratio on the Calcium Alumina Silicate Hydrate (C-A-S-H) Formed in Ternary Cement Pastes. *J. Build. Eng.* **2021**, *38*, 102226. [CrossRef]
126. Miranda de Lima, A.J.; Iwakiri, S.; Satyanarayana, K.G.; Lomelí-Ramírez, M.G. Preparation and Characterization of Wood-Cement Particleboards Produced Using Metakaolin, Calcined Ceramics and Residues of *Pinus* spp. *J. Build. Eng.* **2020**, *32*, 101–140. [CrossRef]
127. Mosaberpanah, M.A.; Umar, S.A. Utilizing Rice Husk Ash as Supplement to Cementitious Materials on Performance of Ultra High Performance Concrete: A Review. *Mater. Today Sustain.* **2020**, *7–8*, 100030. [CrossRef]
128. Anjos, M.A.S.; Araújo, T.R.; Ferreira, R.L.S.; Farias, E.C.; Martinelli, A.E. Properties of Self-Leveling Mortars Incorporating a High-Volume of Sugar Cane Bagasse Ash as Partial Portland Cement Replacement. *J. Build. Eng.* **2020**, *32*, 101694. [CrossRef]
129. Jahanzaib Khalil, M.; Aslam, M.; Ahmad, S. Utilization of Sugarcane Bagasse Ash as Cement Replacement for the Production of Sustainable Concrete—A Review. *Constr. Build. Mater.* **2021**, *270*, 121371. [CrossRef]
130. Shen, D.; Jiao, Y.; Gao, Y.; Zhu, S.; Jiang, G. Influence of Ground Granulated Blast Furnace Slag on Cracking Potential of High Performance Concrete at Early Age. *Constr. Build. Mater.* **2020**, *241*, 117839. [CrossRef]
131. Shen, D.; Jiao, Y.; Kang, J.; Feng, Z.; Shen, Y. Influence of Ground Granulated Blast Furnace Slag on Early-Age Cracking Potential of Internally Cured High Performance Concrete. *Constr. Build. Mater.* **2020**, *233*, 117083. [CrossRef]
132. Cheah, C.B.; Tiong, L.L.; Ng, E.P.; Oo, C.W. The Engineering Performance of Concrete Containing High Volume of Ground Granulated Blast Furnace Slag and Pulverized Fly Ash with Polycarboxylate-Based Superplasticizer. *Constr. Build. Mater.* **2019**, *202*, 909–921. [CrossRef]
133. Ma, L.; Zhao, Y.; Gong, J. Restrained Early-Age Shrinkage Cracking Properties of High-Performance Concrete Containing Fly Ash and Ground Granulated Blast-Furnace Slag. *Constr. Build. Mater.* **2018**, *191*, 1–12. [CrossRef]
134. Kannan, D.M.; Aboubakr, S.H.; EL-Dieb, A.S.; Reda Taha, M.M. High Performance Concrete Incorporating Ceramic Waste Powder as Large Partial Replacement of Portland Cement. *Constr. Build. Mater.* **2017**, *144*, 35–41. [CrossRef]
135. Xu, K.; Huang, W.; Zhang, L.; Fu, S.; Chen, M.; Ding, S.; Han, B. Mechanical Properties of Low-Carbon Ultrahigh-Performance Concrete with Ceramic Tile Waste Powder. *Constr. Build. Mater.* **2021**, *287*, 123036. [CrossRef]
136. Salimi, J.; Ramezanianpour, A.M.; Moradi, M.J. Studying the Effect of Low Reactivity Metakaolin on Free and Restrained Shrinkage of High Performance Concrete. *J. Build. Eng.* **2020**, *28*, 101053. [CrossRef]
137. Song, Q.; Yu, R.; Shui, Z.; Wang, Y.; Rao, S.; Wu, S.; He, Y. Physical and Chemical Coupling Effect of Metakaolin Induced Chloride Trapping Capacity Variation for Ultra High Performance Fibre Reinforced Concrete (UHPFRC). *Constr. Build. Mater.* **2019**, *223*, 765–774. [CrossRef]

138. Shehab El-Din, H.K.; Eisa, A.S.; Abdel Aziz, B.H.; Ibrahim, A. Mechanical Performance of High Strength Concrete Made from High Volume of Metakaolin and Hybrid Fibers. *Constr. Build. Mater.* **2017**, *140*, 203–209. [CrossRef]
139. Tafraoui, A.; Escadeillas, G.; Vidal, T. Durability of the Ultra High Performances Concrete Containing Metakaolin. *Constr. Build. Mater.* **2016**, *112*, 980–987. [CrossRef]
140. Le, H.T.; Ludwig, H.-M. Effect of Rice Husk Ash and Other Mineral Admixtures on Properties of Self-Compacting High Performance Concrete. *Mater. Des.* **2016**, *89*, 156–166. [CrossRef]
141. Le, H.T.; Siewert, K.; Ludwig, H.-M. Alkali Silica Reaction in Mortar Formulated from Self-Compacting High Performance Concrete Containing Rice Husk Ash. *Constr. Build. Mater.* **2015**, *88*, 10–19. [CrossRef]
142. Shaaban, M. Properties of Concrete with Binary Binder System of Calcined Dolomite Powder and Rice Husk Ash. *Heliyon* **2021**, *7*, e06311. [CrossRef]
143. Setayesh Gar, P.; Suresh, N.; Bindiganavile, V. Sugar Cane Bagasse Ash as a Pozzolanic Admixture in Concrete for Resistance to Sustained Elevated Temperatures. *Constr. Build. Mater.* **2017**, *153*, 929–936. [CrossRef]
144. Pu, Y.; Li, L.; Wang, Q.; Shi, X.; Luan, C.; Zhang, G.; Fu, L.; El-Fatah Abomohra, A. Accelerated Carbonation Technology for Enhanced Treatment of Recycled Concrete Aggregates: A State-of-the-Art Review. *Constr. Build. Mater.* **2021**, *282*, 122671. [CrossRef]
145. Nedeljković, M.; Visser, J.; Šavija, B.; Valcke, S.; Schlangen, E. Use of Fine Recycled Concrete Aggregates in Concrete: A Critical Review. *J. Build. Eng.* **2021**, *38*, 102196. [CrossRef]
146. Sahoo, S.; Singh, B. Punching Shear Capacity of Recycled-Aggregate Concrete Slab-Column Connections. *J. Build. Eng.* **2021**, 102430. [CrossRef]
147. Azreen, N.M.; Rashid, R.S.M.; Mugahed Amran, Y.H.; Voo, Y.L.; Haniza, M.; Hairie, M.; Alyousef, R.; Alabduljabbar, H. Simulation of Ultra-High-Performance Concrete Mixed with Hematite and Barite Aggregates Using Monte Carlo for Dry Cask Storage. *Constr. Build. Mater.* **2020**, *263*, 120161. [CrossRef]
148. Trottier, C.; Ziapour, R.; Zahedi, A.; Sanchez, L.; Locati, F. Microscopic Characterization of Alkali-Silica Reaction (ASR) Affected Recycled Concrete Mixtures Induced by Reactive Coarse and Fine Aggregates. *Cem. Concr. Res.* **2021**, *144*, 106426. [CrossRef]
149. Sanjuán, M.Á.; Argiz, C.; Gálvez, J.C.; Moragues, A. Effect of Silica Fume Fineness on the Improvement of Portland Cement Strength Performance. *Constr. Build. Mater.* **2015**, *96*, 55–64. [CrossRef]
150. de Larrard, F.; Sedran, T. Optimization of Ultra-High-Performance Concrete by the Use of a Packing Model. *Cem. Concr. Res.* **1994**, *24*, 997–1009. [CrossRef]
151. Liu, S.; Zhu, M.; Ding, X.; Ren, Z.; Zhao, S.; Zhao, M.; Dang, J. High-Durability Concrete with Supplementary Cementitious Admixtures Used in Corrosive Environments. *Crystals* **2021**, *11*, 196. [CrossRef]
152. Chu, S.H.; Kwan, A.K.H. Mixture Design of Self-Levelling Ultra-High Performance FRC. *Constr. Build. Mater.* **2019**, *228*, 116761. [CrossRef]
153. Zhao, Y.; Duan, Y.; Zhu, L.; Wang, Y.; Jin, Z. Characterization of Coarse Aggregate Morphology and Its Effect on Rheological and Mechanical Properties of Fresh Concrete. *Constr. Build. Mater.* **2021**, *286*, 122940. [CrossRef]
154. De Grazia, M.T.; Goshayeshi, N.; Gorga, R.; Sanchez, L.F.M.; Santos, A.C.; Souza, D.J. Comprehensive Semi-Empirical Approach to Describe Alkali Aggregate Reaction (AAR) Induced Expansion in the Laboratory. *J. Build. Eng.* **2021**, *40*, 102298. [CrossRef]
155. Munhoz, G.S.; Dobrovolski, M.E.G.; Pereira, E.; Medeiros-Junior, R.A. Effect of Improved Autogenous Mortar Self-Healing in the Alkali-Aggregate Reaction. *Cem. Concr. Compos.* **2021**, *117*, 103905. [CrossRef]
156. Amin, M.; Tayeh, B.A.; Agwa, I.S. Effect of Using Mineral Admixtures and Ceramic Wastes as Coarse Aggregates on Properties of Ultrahigh-Performance Concrete. *J. Clean. Prod.* **2020**, *273*, 123073. [CrossRef]
157. Zareei, S.A.; Ameri, F.; Bahrami, N.; Shoaei, P.; Musaeei, H.R.; Nurian, F. Green High Strength Concrete Containing Recycled Waste Ceramic Aggregates and Waste Carpet Fibers: Mechanical, Durability, and Microstructural Properties. *J. Build. Eng.* **2019**, *26*, 100914. [CrossRef]
158. Pedro, D.; de Brito, J.; Evangelista, L. Mechanical Characterization of High Performance Concrete Prepared with Recycled Aggregates and Silica Fume from Precast Industry. *J. Clean. Prod.* **2017**, *164*, 939–949. [CrossRef]
159. Suzuki, M.; Seddik Meddah, M.; Sato, R. Use of Porous Ceramic Waste Aggregates for Internal Curing of High-Performance Concrete. *Cem. Concr. Res.* **2009**, *39*, 373–381. [CrossRef]
160. Ouda, A.S. Development of High-Performance Heavy Density Concrete Using Different Aggregates for Gamma-Ray Shielding. *Prog. Nucl. Energy* **2015**, *79*, 48–55. [CrossRef]
161. Gökçe, H.S.; Yalçınkaya, Ç.; Tuyan, M. Optimization of Reactive Powder Concrete by Means of Barite Aggregate for Both Neutrons and Gamma Rays. *Constr. Build. Mater.* **2018**, *189*, 470–477. [CrossRef]
162. Angelin, A.F.; Cecche Lintz, R.C.; Osório, W.R.; Gachet, L.A. Evaluation of Efficiency Factor of a Self-Compacting Lightweight Concrete with Rubber and Expanded Clay Contents. *Constr. Build. Mater.* **2020**, *257*, 119573. [CrossRef]
163. Lu, Y.; Hu, X.; Yang, X.; Xiao, Y. Comprehensive Tests and Quasi-Brittle Fracture Modeling of Light-Weight Foam Concrete with Expanded Clay Aggregates. *Cem. Concr. Compos.* **2021**, *115*, 103822. [CrossRef]
164. Burbano-Garcia, C.; Hurtado, A.; Silva, Y.F.; Delvasto, S.; Araya-Letelier, G. Utilization of Waste Engine Oil for Expanded Clay Aggregate Production and Assessment of Its Influence on Lightweight Concrete Properties. *Constr. Build. Mater.* **2021**, *273*, 121677. [CrossRef]

165. Juan-Valdés, A.; Rodríguez-Robles, D.; García-González, J.; de Rojas Gómez, M.I.S.; Ignacio Guerra-Romero, M.; De Belie, N.; Morán-del Pozo, J.M. Mechanical and Microstructural Properties of Recycled Concretes Mixed with Ceramic Recycled Cement and Secondary Recycled Aggregates. A Viable Option for Future Concrete. *Constr. Build. Mater.* **2021**, *270*, 121455. [CrossRef]
166. Shah, H.A.; Yuan, Q.; Zuo, S. Air Entrainment in Fresh Concrete and Its Effects on Hardened Concrete-a Review. *Constr. Build. Mater.* **2021**, *274*, 121835. [CrossRef]
167. Dąbrowski, M.; Glinicki, M.A.; Dziedzic, K.; Antolik, A. Validation of Sequential Pressure Method for Evaluation of the Content of Microvoids in Air Entrained Concrete. *Constr. Build. Mater.* **2019**, *227*, 116633. [CrossRef]
168. Nowak-Michta, A. Impact Analysis of Air-Entraining and Superplasticizing Admixtures on Concrete Compressive Strength. *Procedia Struct. Integr.* **2019**, *23*, 77–82. [CrossRef]
169. Tran, N.P.; Gunasekara, C.; Law, D.W.; Houshyar, S.; Setunge, S.; Cwirzen, A. A Critical Review on Drying Shrinkage Mitigation Strategies in Cement-Based Materials. *J. Build. Eng.* **2021**, *38*, 102210. [CrossRef]
170. Zhan, P.; He, Z. Application of Shrinkage Reducing Admixture in Concrete: A Review. *Constr. Build. Mater.* **2019**, *201*, 676–690. [CrossRef]
171. Mo, L.; Fang, J.; Huang, B.; Wang, A.; Deng, M. Combined Effects of Biochar and MgO Expansive Additive on the Autogenous Shrinkage, Internal Relative Humidity and Compressive Strength of Cement Pastes. *Constr. Build. Mater.* **2019**, *229*, 116877. [CrossRef]
172. Kubissa, W.; Jaskulski, R.; Grzelak, M. Torrent Air Permeability and Sorptivity of Concrete Made with the Use of Air Entraining Agent and Citric Acid as Setting Retardant. *Constr. Build. Mater.* **2021**, *268*, 121703. [CrossRef]
173. Sathyan, D.; Anand, K.B. Influence of Superplasticizer Family on the Durability Characteristics of Fly Ash Incorporated Cement Concrete. *Constr. Build. Mater.* **2019**, *204*, 864–874. [CrossRef]
174. Bravo, M.; de Brito, J.; Evangelista, L.; Pacheco, J. Superplasticizer's Efficiency on the Mechanical Properties of Recycled Aggregates Concrete: Influence of Recycled Aggregates Composition and Incorporation Ratio. *Constr. Build. Mater.* **2017**, *153*, 129–138. [CrossRef]
175. Zeyad, A.M.; Almalki, A. Influence of Mixing Time and Superplasticizer Dosage on Self-Consolidating Concrete Properties. *J. Mater. Res. Technol.* **2020**, *9*, 6101–6115. [CrossRef]
176. Marvila, M.T.; Azevedo, A.R.G.; Monteiro, S.N. Verification of the Application Potential of the Mathematical Models of Lyse, Abrams and Molinari in Mortars Based on Cement and Lime. *J. Mater. Res. Technol.* **2020**, *9*, 7327–7334. [CrossRef]
177. Colombo, A.; Geiker, M.R.; Justnes, H.; Lauten, R.A.; De Weerdt, K. On the Effect of Calcium Lignosulfonate on the Rheology and Setting Time of Cement Paste. *Cem. Concr. Res.* **2017**, *100*, 435–444. [CrossRef]
178. Ali Khan, R.; Gupta, C.; Alam, S. Strength and Durability of Self-Curing Concrete Developed Using Calcium Lignosulfonate. *J. King Saud Univ. Eng. Sci.* **2021**. [CrossRef]
179. Guan, J.; Liu, X.; Lai, G.; Luo, Q.; Qian, S.; Zhan, J.; Wang, Z.; Cui, S. Effect of Sulfonation Modification of Polycarboxylate Superplasticizer on Tolerance Enhancement in Sulfate. *Constr. Build. Mater.* **2021**, *273*, 122095. [CrossRef]
180. Ren, C.; Hou, L.; Li, J.; Lu, Z.; Niu, Y. Preparation and Properties of Nanosilica-Doped Polycarboxylate Superplasticizer. *Constr. Build. Mater.* **2020**, *252*, 119037. [CrossRef]
181. Tang, X.; Zhao, C.; Yang, Y.; Dong, F.; Lu, X. Amphoteric Polycarboxylate Superplasticizers with Enhanced Clay Tolerance: Preparation, Performance and Mechanism. *Constr. Build. Mater.* **2020**, *252*, 119052. [CrossRef]
182. Ma, Y.; Shi, C.; Lei, L.; Sha, S.; Zhou, B.; Liu, Y.; Xiao, Y. Research Progress on Polycarboxylate Based Superplasticizers with Tolerance to Clays—A Review. *Constr. Build. Mater.* **2020**, *255*, 119386. [CrossRef]
183. Ma, B.; Qi, H.; Tan, H.; Su, Y.; Li, X.; Liu, X.; Li, C.; Zhang, T. Effect of Aliphatic-Based Superplasticizer on Rheological Performance of Cement Paste Plasticized by Polycarboxylate Superplasticizer. *Constr. Build. Mater.* **2020**, *233*, 117181. [CrossRef]
184. Ibragimov, R.; Fediuk, R. Improving the Early Strength of Concrete: Effect of Mechanochemical Activation of the Cementitious Suspension and Using of Various Superplasticizers. *Constr. Build. Mater.* **2019**, *226*, 839–848. [CrossRef]
185. Benaicha, M.; Hafidi Alaoui, A.; Jalbaud, O.; Burtschell, Y. Dosage Effect of Superplasticizer on Self-Compacting Concrete: Correlation between Rheology and Strength. *J. Mater. Res. Technol.* **2019**, *8*, 2063–2069. [CrossRef]
186. Cheah, C.B.; Chow, W.K.; Oo, C.W.; Leow, K.H. The Influence of Type and Combination of Polycarboxylate Ether Superplasticizer on the Mechanical Properties and Microstructure of Slag-Silica Fume Ternary Blended Self-Consolidating Concrete. *J. Build. Eng.* **2020**, *31*, 101412. [CrossRef]
187. Yu, R.; van Beers, L.; Spiesz, P.; Brouwers, H.J.H. Impact Resistance of a Sustainable Ultra-High Performance Fibre Reinforced Concrete (UHPFRC) under Pendulum Impact Loadings. *Constr. Build. Mater.* **2016**, *107*, 203–215. [CrossRef]
188. Mobasher, B.; Bakhshi, M.; Barsby, C. Backcalculation of Residual Tensile Strength of Regular and High Performance Fiber Reinforced Concrete from Flexural Tests. *Constr. Build. Mater.* **2014**, *70*, 243–253. [CrossRef]
189. Gurusideswar, S.; Shukla, A.; Jonnalagadda, K.N.; Nanthagopalan, P. Tensile Strength and Failure of Ultra-High Performance Concrete (UHPC) Composition over a Wide Range of Strain Rates. *Constr. Build. Mater.* **2020**, *258*, 119642. [CrossRef]
190. Lv, L.-S.; Wang, J.-Y.; Xiao, R.-C.; Fang, M.-S.; Tan, Y. Influence of Steel Fiber Corrosion on Tensile Properties and Cracking Mechanism of Ultra-High Performance Concrete in an Electrochemical Corrosion Environment. *Constr. Build. Mater.* **2021**, *278*, 122338. [CrossRef]
191. Pillai, C.K.S.; Paul, W.; Sharma, C.P. Chitin and Chitosan Polymers: Chemistry, Solubility and Fiber Formation. *Prog. Polym. Sci.* **2009**, *34*, 641–678. [CrossRef]

192. Ahmed, D.; Hongpeng, Z.; Haijuan, K.; Jing, L.; Yu, M.; Muhuo, Y. Microstructural Developments of Poly (p-Phenylene Terephthalamide) Fibers during Heat Treatment Process: A Review. *Mater. Res.* **2014**, *17*, 1180–1200. [CrossRef]
193. Mu, R.; Xing, P.; Yu, J.; Wei, L.; Zhao, Q.; Qing, L.; Zhou, J.; Tian, W.; Gao, S.; Zhao, X.; et al. Investigation on Reinforcement of Aligned Steel Fiber on Flexural Behavior of Cement-Based Composites Using Acoustic Emission Signal Analysis. *Constr. Build. Mater.* **2019**, *201*, 42–50. [CrossRef]
194. Zhang, Z.; Shao, X.-D.; Zhu, P. Direct Tensile Behaviors of Steel-Bar Reinforced Ultra-High Performance Fiber Reinforced Concrete: Effects of Steel Fibers and Steel Rebars. *Constr. Build. Mater.* **2020**, *243*, 118054. [CrossRef]
195. Wei, J.; Meyer, C. Degradation Mechanisms of Natural Fiber in the Matrix of Cement Composites. *Cem. Concr. Res.* **2015**. [CrossRef]
196. Luna-Galiano, Y.; Leiva, C.; Villegas, R.; Arroyo, F.; Vilches, L.; Fernández-Pereira, C. Carbon Fiber Waste Incorporation in Blast Furnace Slag Geopolymer-Composites. *Mater. Lett.* **2018**, *233*, 1–3. [CrossRef]
197. Liu, Y.; Tian, W.; Wang, M.; Qi, B.; Wang, W. Rapid Strength Formation of On-Site Carbon Fiber Reinforced High-Performance Concrete Cured by Ohmic Heating. *Constr. Build. Mater.* **2020**, *244*, 118344. [CrossRef]
198. Guo, P.; Bao, Y.; Meng, W. Review of Using Glass in High-Performance Fiber-Reinforced Cementitious Composites. *Cem. Concr. Compos.* **2021**, *120*, 104032. [CrossRef]
199. Rodier, L.; da Costa Correia, V.; Savastano Junior, H. Elaboration of Eco-Efficient Vegetable Fibers Reinforced Cement-Based Composites Using Glass Powder Residue. *Cem. Concr. Compos.* **2020**, *110*, 103599. [CrossRef]
200. Wang, Y.; Shao, X.; Cao, J.; Zhao, X.; Qiu, M. Static and Fatigue Flexural Performance of Ultra-High Performance Fiber Reinforced Concrete Slabs. *Eng. Struct.* **2021**, *231*, 111728. [CrossRef]
201. Ashkezari, G.D.; Fotouhi, F.; Razmara, M. Experimental Relationships between Steel Fiber Volume Fraction and Mechanical Properties of Ultra-High Performance Fiber-Reinforced Concrete. *J. Build. Eng.* **2020**, *32*, 101613. [CrossRef]
202. Ashkezari, G.D.; Razmara, M. Thermal and Mechanical Evaluation of Ultra-High Performance Fiber-Reinforced Concrete and Conventional Concrete Subjected to High Temperatures. *J. Build. Eng.* **2020**, *32*, 101621. [CrossRef]
203. Shin, W.; Yoo, D.-Y. Influence of Steel Fibers Corroded through Multiple Microcracks on the Tensile Behavior of Ultra-High-Performance Concrete. *Constr. Build. Mater.* **2020**, *259*, 120428. [CrossRef]
204. Yoo, D.-Y.; Shin, W.; Chun, B. Corrosion Effect on Tensile Behavior of Ultra-High-Performance Concrete Reinforced with Straight Steel Fibers. *Cem. Concr. Compos.* **2020**, *109*, 103566. [CrossRef]
205. Ngo, T.T.; Tran, N.T.; Kim, D.J.; Pham, T.C. Effects of Corrosion Level and Inhibitor on Pullout Behavior of Deformed Steel Fiber Embedded in High Performance Concrete. *Constr. Build. Mater.* **2021**, *280*, 122449. [CrossRef]
206. Gou, H.; Zhu, H.; Zhou, H.; Yang, Z. Reinforcement Mechanism of Orientally Distributed Steel Fibers on Ultra-High-Performance Concrete. *Constr. Build. Mater.* **2021**, *281*, 122646. [CrossRef]
207. Park, T.; Her, S.; Jee, H.; Yoon, S.; Cho, B.; Hwang, S.-H.; Bae, S. Evaluation of Orientation and Distribution of Steel Fibers in High-Performance Concrete Column Determined via Micro-Computed Tomography. *Constr. Build. Mater.* **2021**, *270*, 121473. [CrossRef]
208. Kim, J.-J.; Jang, Y.S.; Yoo, D.-Y. Enhancing the Tensile Performance of Ultra-High-Performance Concrete through Novel Curvilinear Steel Fibers. *J. Mater. Res. Technol.* **2020**, *9*, 7570–7582. [CrossRef]
209. Dingqiang, F.; Yu, R.; Kangning, L.; Junhui, T.; Zhonghe, S.; Chunfeng, W.; Shuo, W.; Zhenfeng, G.; Zhengdong, H.; Qiqi, S. Optimized Design of Steel Fibres Reinforced Ultra-High Performance Concrete (UHPC) Composites: Towards to Dense Structure and Efficient Fibre Application. *Constr. Build. Mater.* **2021**, *273*, 121698. [CrossRef]
210. Afzal, M.T.; Khushnood, R.A. Influence of Carbon Nano Fibers (CNF) on the Performance of High Strength Concrete Exposed to Elevated Temperatures. *Constr. Build. Mater.* **2021**, *268*, 121108. [CrossRef]
211. Jung, M.; Park, J.; Hong, S.-G.; Moon, J. Micro- and Meso-Structural Changes on Electrically Cured Ultra-High Performance Fiber-Reinforced Concrete with Dispersed Carbon Nanotubes. *Cem. Concr. Res.* **2020**, *137*, 106214. [CrossRef]
212. Zhou, A.; Qiu, Q.; Chow, C.L.; Lau, D. Interfacial Performance of Aramid, Basalt and Carbon Fiber Reinforced Polymer Bonded Concrete Exposed to High Temperature. *Compos. Part A Appl. Sci. Manuf.* **2020**, *131*, 105802. [CrossRef]
213. Bilisik, K.; Ozdemir, H. Multiaxis Three-Dimensional (3D) Glass Fiber Preform/Cementitious Matrix Concrete Composites: Experimental Characterizations by Panel Test. *Cem. Concr. Compos.* **2021**, *119*, 104020. [CrossRef]
214. Al-Khafaji, A.F.; Myers, J.J.; Alghazali, H.H. Evaluation of Bond Performance of Glass Fiber Rebars Embedded in Sustainable Concrete. *J. Clean. Prod.* **2021**, *282*, 124516. [CrossRef]
215. Rajesh Kumar, K.; Shyamala, G.; Adesina, A. Structural Performance of Corroded Reinforced Concrete Beams Made with Fiber-Reinforced Self-Compacting Concrete. *Structures* **2021**, *32*, 1145–1155. [CrossRef]
216. Ali, B.; Qureshi, L.A.; Khan, S.U. Flexural Behavior of Glass Fiber-Reinforced Recycled Aggregate Concrete and Its Impact on the Cost and Carbon Footprint of Concrete Pavement. *Constr. Build. Mater.* **2020**, *262*, 120820. [CrossRef]
217. Mohammed, B.H.; Sherwani, A.F.H.; Faraj, R.H.; Qadir, H.H.; Younis, K.H. Mechanical Properties and Ductility Behavior of Ultra-High Performance Fiber Reinforced Concretes: Effect of Low Water-to-Binder Ratios and Micro Glass Fibers. *Ain Shams Eng. J.* **2021**, *12*, 1557–1567. [CrossRef]
218. Hussain, I.; Ali, B.; Akhtar, T.; Jameel, M.S.; Raza, S.S. Comparison of Mechanical Properties of Concrete and Design Thickness of Pavement with Different Types of Fiber-Reinforcements (Steel, Glass, and Polypropylene). *Case Stud. Constr. Mater.* **2020**, *13*, e00429. [CrossRef]

219. Behfarnia, K.; Behravan, A. Application of High Performance Polypropylene Fibers in Concrete Lining of Water Tunnels. *Mater. Des.* **2014**, *55*, 274–279. [CrossRef]
220. Zhu, C.; Niu, J.; Li, J.; Wan, C.; Peng, J. Effect of Aggregate Saturation Degree on the Freeze–Thaw Resistance of High Performance Polypropylene Fiber Lightweight Aggregate Concrete. *Constr. Build. Mater.* **2017**, *145*, 367–375. [CrossRef]
221. Li, Y.; Tan, K.H.; Yang, E.-H. Influence of Aggregate Size and Inclusion of Polypropylene and Steel Fibers on the Hot Permeability of Ultra-High Performance Concrete (UHPC) at Elevated Temperature. *Constr. Build. Mater.* **2018**, *169*, 629–637. [CrossRef]
222. Shen, D.; Liu, X.; Zeng, X.; Zhao, X.; Jiang, G. Effect of Polypropylene Plastic Fibers Length on Cracking Resistance of High Performance Concrete at Early Age. *Constr. Build. Mater.* **2020**, *244*, 117874. [CrossRef]
223. Yoo, D.-Y.; Kim, M.-J. High Energy Absorbent Ultra-High-Performance Concrete with Hybrid Steel and Polyethylene Fibers. *Constr. Build. Mater.* **2019**, *209*, 354–363. [CrossRef]
224. He, S.; Qiu, J.; Li, J.; Yang, E.-H. Strain Hardening Ultra-High Performance Concrete (SHUHPC) Incorporating CNF-Coated Polyethylene Fibers. *Cem. Concr. Res.* **2017**, *98*, 50–60. [CrossRef]
225. Wang, D.; Ju, Y.; Shen, H.; Xu, L. Mechanical Properties of High Performance Concrete Reinforced with Basalt Fiber and Polypropylene Fiber. *Constr. Build. Mater.* **2019**, *197*, 464–473. [CrossRef]
226. Smarzewski, P. Influence of Basalt-Polypropylene Fibres on Fracture Properties of High Performance Concrete. *Compos. Struct.* **2019**, *209*, 23–33. [CrossRef]
227. Naraganti, S.R.; Pannem, R.M.R.; Putta, J. Impact Resistance of Hybrid Fibre Reinforced Concrete Containing Sisal Fibres. *Ain Shams Eng. J.* **2019**, *10*, 297–305. [CrossRef]
228. Elbehiry, A.; Elnawawy, O.; Kassem, M.; Zaher, A.; Uddin, N.; Mostafa, M. Performance of Concrete Beams Reinforced Using Banana Fiber Bars. *Case Stud. Constr. Mater.* **2020**, *13*, e00361. [CrossRef]
229. Elbehiry, A.; Elnawawy, O.; Kassem, M.; Zaher, A.; Mostafa, M. FEM Evaluation of Reinforced Concrete Beams by Hybrid and Banana Fiber Bars (BFB). *Case Stud. Constr. Mater.* **2021**, *14*, e00479. [CrossRef]
230. da Gloria, M.Y.R.; Toledo Filho, R.D. Innovative Sandwich Panels Made of Wood Bio-Concrete and Sisal Fiber Reinforced Cement Composites. *Constr. Build. Mater.* **2021**, *272*, 121636. [CrossRef]
231. Sabarish, K.V.; Paul, P.; Bhuvaneshwari; Jones, J. An Experimental Investigation on Properties of Sisal Fiber Used in the Concrete. *Mater. Today Proc.* **2020**, *22*, 439–443. [CrossRef]
232. Marvila, M.T.; de Azevedo, A.R.G.; Vieira, C.M.F. Reaction Mechanisms of Alkali-Activated Materials. *Rev. Ibracon Estrut. Mater.* **2021**, 14–28. [CrossRef]
233. Provis, J.L.; Bernal, S.A. Geopolymers and Related Alkali-Activated Materials. *Annu. Rev. Mater. Res.* **2014**, *44*, 299–327. [CrossRef]
234. Marvila, M.T.; Azevedo, A.R.G.; Delaqua, G.C.G.; Mendes, B.C.; Pedroti, L.G.; Vieira, C.M.F. Performance of Geopolymer Tiles in High Temperature and Saturation Conditions. *Constr. Build. Mater.* **2021**, *286*, 122994. [CrossRef]
235. Azevedo, A.R.G.; Marvila, M.T.; Rocha, H.A.; Cruz, L.R.; Vieira, C.M.F. Use of Glass Polishing Waste in the Development of Ecological Ceramic Roof Tiles by the Geopolymerization Process. *Int. J. Appl. Ceram. Technol.* **2020**, *17*, 2649–2658. [CrossRef]
236. Marvila, M.T.; de Azevedo, A.R.G.; de Matos, P.R.; Monteiro, S.N.; Vieira, C.M.F. Rheological and the Fresh State Properties of Alkali-Activated Mortars by Blast Furnace Slag. *Materials* **2021**, *14*, 2069. [CrossRef] [PubMed]
237. Ramos, G.A.; de Matos, P.R.; Pelisser, F.; Gleize, P.J.P. Effect of Porcelain Tile Polishing Residue on Eco-Efficient Geopolymer: Rheological Performance of Pastes and Mortars. *J. Build. Eng.* **2020**, *32*, 101699. [CrossRef]
238. Turner, L.K.; Collins, F.G. Carbon Dioxide Equivalent (CO_2-e) Emissions: A Comparison between Geopolymer and OPC Cement Concrete. *Constr. Build. Mater.* **2013**, *43*, 125–130. [CrossRef]
239. Van Den Heede, P.; De Belie, N. Environmental Impact and Life Cycle Assessment (LCA) of Traditional and "green" Concretes: Literature Review and Theoretical Calculations. *Cem. Concr. Compos.* **2012**, *34*, 431–442. [CrossRef]
240. Chen, C.; Habert, G.; Bouzidi, Y.; Jullien, A.; Ventura, A. LCA Allocation Procedure Used as an Incitative Method for Waste Recycling: An Application to Mineral Additions in Concrete. *Resour. Conserv. Recycl.* **2010**, *54*, 1231–1240. [CrossRef]
241. Habert, G.; d'Espinose de Lacaillerie, J.B.; Roussel, N. An Environmental Evaluation of Geopolymer Based Concrete Production: Reviewing Current Research Trends. *J. Clean. Prod.* **2011**, *19*, 1229–1238. [CrossRef]
242. Longhi, M.A.; Walkley, B.; Rodríguez, E.D.; Kirchheim, A.P.; Zhang, Z.; Wang, H. New Selective Dissolution Process to Quantify Reaction Extent and Product Stability in Metakaolin-Based Geopolymers. *Compos. Part B Eng.* **2019**, *176*, 107172. [CrossRef]
243. Azevedo, A.R.G.; Vieira, C.M.F.; Ferreira, W.M.; Faria, K.C.P.; Pedroti, L.G.; Mendes, B.C. Potential Use of Ceramic Waste as Precursor in the Geopolymerization Reaction for the Production of Ceramic Roof Tiles. *J. Build. Eng.* **2020**, *29*, 101156. [CrossRef]
244. Kovářík, T.; Rieger, D.; Kadlec, J.; Křenek, T.; Kullová, L.; Pola, M.; Bělský, P.; Franče, P.; Říha, J. Thermomechanical Properties of Particle-Reinforced Geopolymer Composite with Various Aggregate Gradation of Fine Ceramic Filler. *Constr. Build. Mater.* **2017**, *143*, 599–606. [CrossRef]
245. Kamseu, E.; Catania, V.; Djangang, C.; Sglavo, V.M.; Leonelli, C. Correlation between Microstructural Evolution and Mechanical Properties of A-Quartz and Alumina Reinforced K-Geopolymers during High Temperature Treatments. *Adv. Appl. Ceram.* **2012**, *111*, 120–128. [CrossRef]
246. Fernández-Jiménez, A.; Monzó, M.; Vicent, M.; Barba, A.; Palomo, A. Alkaline Activation of Metakaolin–Fly Ash Mixtures: Obtain of Zeoceramics and Zeocements. *Microporous Mesoporous Mater.* **2008**, *108*, 41–49. [CrossRef]
247. White, C.E.; Provis, J.L.; Proffen, T.; Riley, D.P.; van Deventer, J.S.J. Density Functional Modeling of the Local Structure of Kaolinite Subjected to Thermal Dehydroxylation. *J. Phys. Chem. A* **2010**, *114*, 4988–4996. [CrossRef] [PubMed]

248. Zhang, Z.; Wang, H.; Provis, J.L.; Bullen, F.; Reid, A.; Zhu, Y. Quantitative Kinetic and Structural Analysis of Geopolymers. Part 1. The Activation of Metakaolin with Sodium Hydroxide. *Thermochim. Acta* **2012**, *539*, 23–33. [CrossRef]
249. Wu, Y.; Lu, B.; Bai, T.; Wang, H.; Du, F.; Zhang, Y.; Cai, L.; Jiang, C.; Wang, W. Geopolymer, Green Alkali Activated Cementitious Material: Synthesis, Applications and Challenges. *Constr. Build. Mater.* **2019**, *224*, 930–949. [CrossRef]
250. Sindhunata; van Deventer, J.S.J.; Lukey, G.C.; Xu, H. Effect of Curing Temperature and Silicate Concentration on Fly-Ash-Based Geopolymerization. *Ind. Eng. Chem. Res.* **2006**, *45*, 3559–3568. [CrossRef]
251. Longhi, M.A.; Zhang, Z.; Rodríguez, E.D.; Kirchheim, A.P.; Wang, H. Efflorescence of Alkali-Activated Cements (Geopolymers) and the Impacts on Material Structures: A Critical Analysis. *Front. Mater.* **2019**, *6*, 1–13. [CrossRef]
252. Longhi, M.A.; Zhang, Z.; Walkley, B.; Rodríguez, E.D.; Kirchheim, A.P. Strategies for Control and Mitigation of Efflorescence in Metakaolin-Based Geopolymers. *Cem. Concr. Res.* **2021**, *144*, 106431. [CrossRef]
253. Kuenzel, C.; Vandeperre, L.J.; Donatello, S.; Boccaccini, A.R.; Cheeseman, C. Ambient Temperature Drying Shrinkage and Cracking in Metakaolin-Based Geopolymers. *J. Am. Ceram. Soc.* **2012**, *95*, 3270–3277. [CrossRef]
254. HASNAOUI, A.; GHORBEL, E.; WARDEH, G. Performance of Metakaolin/Slag-Based Geopolymer Concrete Made with Recycled Fine and Coarse Aggregates. *J. Build. Eng.* **2021**, *42*, 102813. [CrossRef]
255. Gomes, R.F.; Dias, D.P.; Silva, F. de A. Determination of the Fracture Parameters of Steel Fiber-Reinforced Geopolymer Concrete. *Theor. Appl. Fract. Mech.* **2020**, *107*, 102568. [CrossRef]
256. Dias, D.P.; de Andrade Silva, F. Effect of Na_2O/SiO_2 and K_2O/SiO_2 Mass Ratios on the Compressive Strength of Non-Silicate Metakaolin Geopolymeric Mortars. *Mater. Res. Express* **2019**, *6*, 075514. [CrossRef]
257. Albidah, A.; Altheeb, A.; Alrshoudi, F.; Abadel, A.; Abbas, H.; Al-Salloum, Y. Bond Performance of GFRP and Steel Rebars Embedded in Metakaolin Based Geopolymer Concrete. *Structures* **2020**, *27*, 1582–1593. [CrossRef]
258. da Silva Rocha, T.; Dias, D.P.; França, F.C.C.; de Salles Guerra, R.R.; de Oliveira Marques, L.R.d.C. Metakaolin-Based Geopolymer Mortars with Different Alkaline Activators (Na^+ and K^+). *Constr. Build. Mater.* **2018**, *178*, 453–461. [CrossRef]
259. Ahmaruzzaman, M. A Review on the Utilization of Fly Ash. *Prog. Energy Combust. Sci.* **2010**, *36*, 327–363. [CrossRef]
260. Cai, J.; Li, X.; Tan, J.; Vandevyvere, B. Thermal and Compressive Behaviors of Fly Ash and Metakaolin-Based Geopolymer. *J. Build. Eng.* **2020**, *30*, 101307. [CrossRef]
261. Sun, Z.; Vollpracht, A.; van der Sloot, H.A. PH Dependent Leaching Characterization of Major and Trace Elements from Fly Ash and Metakaolin Geopolymers. *Cem. Concr. Res.* **2019**, *125*, 105889. [CrossRef]
262. Cai, J.; Pan, J.; Li, X.; Tan, J.; Li, J. Electrical Resistivity of Fly Ash and Metakaolin Based Geopolymers. *Constr. Build. Mater.* **2020**, *234*, 117868. [CrossRef]
263. Luo, Z.; Li, W.; Wang, K.; Castel, A.; Shah, S.P. Comparison on the Properties of ITZs in Fly Ash-Based Geopolymer and Portland Cement Concretes with Equivalent Flowability. *Cem. Concr. Res.* **2021**, *143*, 106392. [CrossRef]
264. Charkhtab Moghaddam, S.; Madandoust, R.; Jamshidi, M.; Nikbin, I.M. Mechanical Properties of Fly Ash-Based Geopolymer Concrete with Crumb Rubber and Steel Fiber under Ambient and Sulfuric Acid Conditions. *Constr. Build. Mater.* **2021**, *281*, 122571. [CrossRef]
265. Pasupathy, K.; Singh Cheema, D.; Sanjayan, J. Durability Performance of Fly Ash-Based Geopolymer Concrete Buried in Saline Environment for 10 Years. *Constr. Build. Mater.* **2021**, *281*, 122596. [CrossRef]
266. Buchwald, A.; Hohmann, M.; Posern, K.; Brendler, E. The Suitability of Thermally Activated Illite/Smectite Clay as Raw Material for Geopolymer Binders. *Appl. Clay Sci.* **2009**, *46*, 300–304. [CrossRef]
267. González-García, D.M.; Téllez-Jurado, L.; Jiménez-Álvarez, F.J.; Balmori-Ramírez, H. Structural Study of Geopolymers Obtained from Alkali-Activated Natural Pozzolan Feldspars. *Ceram. Int.* **2017**, *43*, 2606–2613. [CrossRef]
268. Tian, L.; Feng, W.; Ma, H.; Zhang, S.; Shi, H. Investigation on the Microstructure and Mechanism of Geopolymer with Different Proportion of Quartz and K-Feldspar. *Constr. Build. Mater.* **2017**, *147*, 543–549. [CrossRef]
269. Yankwa Djobo, J.N.; Elimbi, A.; Tchakouté, H.K.; Kumar, S. Mechanical Activation of Volcanic Ash for Geopolymer Synthesis: Effect on Reaction Kinetics, Gel Characteristics, Physical and Mechanical Properties. *RSC Adv.* **2016**, *6*, 39106–39117. [CrossRef]
270. Xu, H.; van Deventer, J.S.J. Factors Affecting the Geopolymerization of Alkali-Feldspars. *Mining Metall. Explor.* **2002**, *19*, 209–214. [CrossRef]
271. Djobo, Y.J.N.; Elimbi, A.; Dika Manga, J.; Djon Li Ndjock, I.B. Partial Replacement of Volcanic Ash by Bauxite and Calcined Oyster Shell in the Synthesis of Volcanic Ash-Based Geopolymers. *Constr. Build. Mater.* **2016**, *113*, 673–681. [CrossRef]
272. Djobo, J.N.Y.; Elimbi, A.; Tchakouté, H.K.; Kumar, S. Volcanic Ash-Based Geopolymer Cements/Concretes: The Current State of the Art and Perspectives. *Environ. Sci. Pollut. Res.* **2017**, *24*, 4433–4446. [CrossRef] [PubMed]
273. Mo, L.; Lv, L.; Deng, M.; Qian, J. Influence of Fly Ash and Metakaolin on the Microstructure and Compressive Strength of Magnesium Potassium Phosphate Cement Paste. *Cem. Concr. Res.* **2018**, *111*, 116–129. [CrossRef]
274. Marvila, M.T.; Azevedo, A.R.G.; Zanelato, E.B.; Lima, T.E.S.; Delaqua, G.C.G.; Vieira, C.M.F.; Pedroti, L.G.; Monteiro, S.N. *Study of Pathologies in Alkali-Activated Materials Based on Slag*; Springer International Publishing: Cham, Switzerland, 2021; pp. 523–531.
275. Provis, J.L. Geopolymers and Other Alkali Activated Materials: Why, How, and What? *Mater. Struct. Constr.* **2014**, *47*, 11–25. [CrossRef]
276. Duxson, P.; Fernández-Jiménez, A.; Provis, J.L.; Lukey, G.C.; Palomo, A.; van Deventer, J.S.J. Geopolymer Technology: The Current State of the Art. *J. Mater. Sci.* **2007**, *42*, 2917–2933. [CrossRef]

277. Provis, J.L.; van Deventer, J.S.J. Geopolymerisation Kinetics. 2. Reaction Kinetic Modelling. *Chem. Eng. Sci.* **2007**, *62*, 2318–2329. [CrossRef]
278. Provis, J.L.; van Deventer, J.S.J. Geopolymerisation Kinetics. 1. In Situ Energy-Dispersive X-ray Diffractometry. *Chem. Eng. Sci.* **2007**, *62*, 2309–2317. [CrossRef]
279. Chen, W.; Li, B.; Wang, J.; Thom, N. Effects of Alkali Dosage and Silicate Modulus on Autogenous Shrinkage of Alkali-Activated Slag Cement Paste. *Cem. Concr. Res.* **2021**, *141*, 106322. [CrossRef]
280. He, J.; Bai, W.; Zheng, W.; He, J.; Sang, G. Influence of Hydrated Lime on Mechanical and Shrinkage Properties of Alkali-Activated Slag Cement. *Constr. Build. Mater.* **2021**, *289*, 123201. [CrossRef]
281. Neupane, K.; Hadigheh, S.A. Sodium Hydroxide-Free Geopolymer Binder for Prestressed Concrete Applications. *Constr. Build. Mater.* **2021**, *293*, 123397. [CrossRef]
282. Kotop, M.A.; El-Feky, M.S.; Alharbi, Y.R.; Abadel, A.A.; Binyahya, A.S. Engineering Properties of Geopolymer Concrete Incorporating Hybrid Nano-Materials. *Ain Shams Eng. J.* **2021**, 65–73. [CrossRef]
283. Mahmood, A.H.; Foster, S.J.; Castel, A. High-Density Geopolymer Concrete for Port Kembla Breakwater Upgrade. *Constr. Build. Mater.* **2020**, *262*, 120920. [CrossRef]
284. Mendes, B.C.; Pedroti, L.G.; Fontes, M.P.F.; Ribeiro, J.C.L.; Vieira, C.M.F.; Pacheco, A.A.; de Azevedo, A.R.G. Technical and Environmental Assessment of the Incorporation of Iron Ore Tailings in Construction Clay Bricks. *Constr. Build. Mater.* **2019**, *227*, 116669. [CrossRef]

materials

Article

Technological Characterization of PET—Polyethylene Terephthalate—Added Soil-Cement Bricks

Tulane Rodrigues da Silva [1,2], Daiane Cecchin [2], Afonso Rangel Garcez de Azevedo [3,*], Izabella Valadão [4], Jonas Alexandre [3], Flavio Castro da Silva [2], Markssuel Teixeira Marvila [1], Murali Gunasekaran [5], Fabio Garcia Filho [6] and Sergio Neves Monteiro [6]

1 LAMAV—Advanced Materials Laboratory, UENF—State University of the Northern Rio de Janeiro, Av. Alberto Lamego, 2000, Campos dos Goytacazes 28013-602, Brazil; tulanerodrigues@gmail.com (T.R.d.S.); m.marvila@ucam-campos.br (M.T.M.)
2 Department of Agricultural and Environmental Engineering, Federal Fluminense University—UFF, R. Passo da Pátria, 156, Niterói 24210-240, Brazil; daianececchin@id.uff.br (D.C.); flaviocastro@id.uff.br (F.C.d.S.)
3 LECIV—Civil Engineering Laboratory, UENF—State University of the Northern Rio de Janeiro, Av. Alberto Lamego, 2000, Campos dos Goytacazes 28013-602, Brazil; jonasuenf@gmail.com
4 Department of Civil Engineering, Federal Fluminense University—UFF, R. Passo da Pátria, 156, Niterói 24210-240, Brazil; izabellavaladao@id.uff.br
5 School of Civil Engineering, SASTRA Deemed University, Thanjavur 613401, India; murali@civil.sastra.edu
6 Military Engineering Institute, IME—Materials Science Program, Praça Gen. Tibúrcio, 80, Urca, Rio de Janeiro 22290-270, Brazil; fabiogarciafilho@gmail.com (F.G.F.); snevesmonteiro@gmail.com (S.N.M.)
* Correspondence: afonso@uenf.br

Abstract: The ever-growing consumption and improper disposal of non-biodegradable plastic wastes is bringing worrisome perspectives on the lack of suitable environmentally correct solutions. Consequently, an increasing interest in the circular economy and sustainable techniques is being raised regarding the management of these wastes. The present work proposes an eco-friendly solution for the huge amount of discarded polyethylene terephthalate (PET) wastes by addition into soil-cement bricks. Room temperature molded 300 × 150 × 70 mm bricks were fabricated with mixtures of clay soil and ordinary Portland cement added with up to 30 wt.% of PET waste particles. Granulometric analysis of soil indicated it as sandy and adequate for brick fabrication. As for the PET particles, they can be considered non-plastic and sandy. The Atterberg consistency limits indicated that addition of 20 wt.% PET waste gives the highest plasticity limit of 17.3%; moreover, with PET waste addition there was an increase in the optimum moisture content for the compaction and decrease in specific weight. Standard tests showed an increase in compressive strength from 0.83 MPa for the plain soil-cement to 1.80 MPa for the 20 wt.% PET-added bricks. As for water absorption, all bricks displayed values between 15% and 16% that attended the standards and might be considered an alternative for non-structural applications, such as wall closures in building construction.

Keywords: building materials; PET; compressive strength; water absorption

Citation: da Silva, T.R.; Cecchin, D.; de Azevedo, A.R.G.; Valadão, I.; Alexandre, J.; da Silva, F.C.; Marvila, M.T.; Gunasekaran, M.; Garcia Filho, F.; Monteiro, S.N. Technological Characterization of PET—Polyethylene Terephthalate—Added Soil-Cement Bricks. *Materials* **2021**, *14*, 5035. https://doi.org/10.3390/ma14175035

Academic Editor: Rossana Bellopede

Received: 3 August 2021
Accepted: 25 August 2021
Published: 3 September 2021

1. Introduction

The scarcity of natural resources and the generation of solid wastes without adequate disposal is of worldwide concern, which makes their reuse feasible in the civil construction sector, besides encouraging sustainable development and a circular economy [1–4]. The solid waste problem is of concern mainly in urbanized regions and developing countries where collection and disposal services have difficulties dealing with increasing amounts of waste [2]. As a result, waste is either disposed in open, uncontrolled dumps, accounting for 61% of the landfill sector's CO_2 emissions, or burned in the open dumps, accounting for 40% of global waste [5]. Sustainable solid waste management has become a necessity for industries seeking to promote industrialization and sustainable development. Government regulations have become more stringent around the world, representing an accelerating factor in the adoption of reverse logistics initiatives, including in countries that have been facing difficulties in recycling processes, thus giving more space for the use of this waste within its life cycle [4].

Global plastic production is growing rapidly and by 2030 the world may produce about 619 million tons of plastic per year [6]. A study by Spósito et al. [7] emphasizes that post-consumer polyethylene terephthalate (PET) products have generated a growing interest regarding their recycling potential and their negative impacts on the environment, such as pollution and long degradation time. According to WWF [8], phasing out single-use plastics has the potential to reduce plastic demand by up to 40% by the year 2030. Thus, the current scenario requires a more sustainable route for the recycling of this waste, which if not realized will increase the environmental imbalance due to its non-biodegradability in nature [8,9].

The growing demand for sustainable products has encouraged several studies that search for alternative techniques regarding the reuse of waste in construction materials, such as mortars with sugarcane bagasse [10], cement pastes with açai fiber [11,12], blast furnace slag [13,14], construction and demolition waste [15], ceramic materials with rice ash water treatment plant sludge [16], pulp and paper industry sludge [17], construction and demolition waste [18], and agricultural waste [19], as well as concrete with plastic waste [20]. Therefore, research also highlights the reuse of plastic waste in construction materials [19,20] such as paving [21], mortar [7], concrete [19,20,22,23], fired clay blocks, and bricks [24], as well as unfired blocks and bricks [25,26], thus showing PET (Figure 1a) as an addition in the production of these materials.

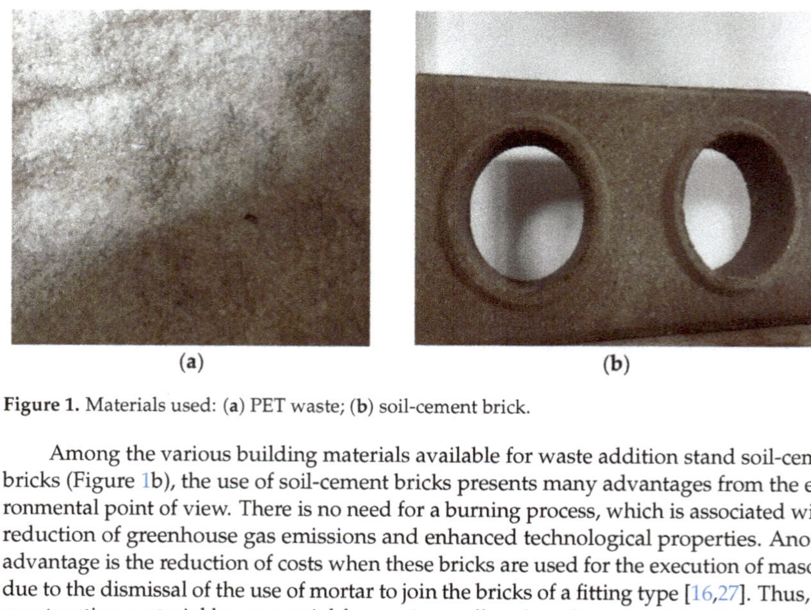

Figure 1. Materials used: (**a**) PET waste; (**b**) soil-cement brick.

Among the various building materials available for waste addition stand soil-cement bricks (Figure 1b), the use of soil-cement bricks presents many advantages from the environmental point of view. There is no need for a burning process, which is associated with a reduction of greenhouse gas emissions and enhanced technological properties. Another advantage is the reduction of costs when these bricks are used for the execution of masonry due to the dismissal of the use of mortar to join the bricks of a fitting type [16,27]. Thus, this construction material has potential for use in small and medium sized buildings without structural function, in addition to having a low financial cost [28,29].

The soil-cement bricks allow the incorporation of waste in their composition and reduce costs up to 40% compared to traditional masonry, especially in low-income housing. In this way, the brick can be considered more sustainable in relation to the traditional brick [30]. In this context, it is also possible to verify the use of waste in soil-cement bricks as shown França et al. [28], who studied the durability of soil-cement bricks with incorporation limestone waste. The authors used 30%, 40% and 50% of waste for the manufactured soil-cement mixtures. The results verified that the incorporation of waste rock was technologically feasible. The parameters studied for compressive strength, water absorption and durability showed superior performance of bricks with waste incorporation. Reis et al. [31] evaluated the incorporation of quartzite mining tailings in soil-cement bricks. The authors tested additions of quartzite tailings at 0%, 15% and 30%. The results showed that the compressive strength of the soil-cement bricks decreased with the addition of quartzite waste. However, the authors observed that the results of compressive strength and water absorption performed at 7 and 28 days demonstrated the possibility of using the waste without compromising the physical and mechanical properties required by the standard, thus verifying that the soil-cement brick is a viable technique for the disposal of this type of waste. Kongkajun et al. [32] evaluated soil-cement bricks with incorporation of construction waste (clay bricks) and fiber-cement industry sludge. The authors used 15 wt.% of Portland cement, 15 wt.% of sand and 70 wt.% of laterite to produce the bricks. The clay brick waste was added from 10% to 50 wt.% of laterite in the control samples. Sludge, on the other hand, was added at 5% and 10 wt.% to replace the total weight of the mixture in the control samples. The maximum obtainable substitution of laterite for clay brick waste was 50 wt.% in the mixture. Maximum compressive strength was achieved for the 10% replacement of laterite with clay bricks. Partial replacement of laterite with clay bricks improved the compressive strength of soil-cement bricks for load-bearing brick application. Although the incorporation of silt caused a reduction in the compressive strength of the brick samples compared to the samples prepared from the control sample, they still exceeded the Thailand community product standard. Increasing the percentage of sludge from 0% to 10 wt.% resulted in a significant decrease in thermal conductivity of 45% compared to the control formula. When using the sludge and clay bricks, the thermal conductivity and density of the bricks were further reduced, while their compressive strength and water absorption values were still satisfactory.

Kouamé et al. [33] verified the influence of shea butter wastes on the physical properties of cement-stabilized soil bricks. The authors used three local clay raw materials consisting mainly of kaolinite, quartz and micaceous phases, as well as shea butter waste and cement. In the mixtures tested, the amount of cement was kept constant (5%), while the amount of shea butter waste varied from 2% to 10%, replacing the soil. The results obtained by the authors showed that the presence of pores due to the shea butter waste influences the reduction of density and thermal conductivity of the bricks. A 25% decrease in thermal conductivity was verified for the samples with clay F, 16% for the samples with clay K, and 22% with clay Y. The authors concluded that the bricks showed good stiffness related to the presence of cementation phases. Therefore, for the samples with clay F and clay Y, the replacement rate of 6% by the shea butter waste was sufficient as compared to 8% for the formulations with K to obtain a physical property. Thus, they found that the shea butter waste offered good thermal insulation and good stiffness properties with a lower amount of cement used.

Vilela et al. [34] evaluated the incorporation of mining waste in soil-cement bricks for soil substitution at 10%, 20%, 30% and 40% of waste. As for the mechanical strength, the authors found that all treatments showed values above the required standards, with the minimum standard value being set at 2.0 MPa. The treatment with 10% mining waste presented the best results. The thermal conductivity showed a direct link with the density of the bricks since the increase in density (bricks with 40% mining waste) led to a material with a lower heat dissipation property. The study showed that the addition of mining waste to soil-cement bricks met all the required standards [34].

In this sense, the use of plastic waste can also be considered in studies such as [7], which showed that hydrated mortars produced with PET bottle waste replaced the fine aggregate in the mixture and suffered changes in properties in their fresh and hardened states. Reference [21] investigated the effects of PET waste on hardened properties in high strength concrete and the investigation showed the interference of high temperatures on concrete properties, in this case, the occurrence of material fragmentation and the release of greenhouse gases. Reference [22] evaluated PET blends for sidewalk sub-bases, highlighting bottles and food packaging taken from collection points and crushed into mixtures with two main constituents of waste materials or construction and demolition by-products: concrete aggregate and crushed brick.

Studies by [24] found that compared to normal concrete, high strength concrete has a failure mode and that the lack of ductility can be solved by using different types of plastic fibers. Akinyele et al. (2020) evaluated the incorporation of PET into fired blocks varied by 0%, 5%, 10%, 15% and 20% and found changes in the samples with respect to high temperature, compressive strength and water absorption. Reference [23] investigated concrete with added waste plastic by varying it at 0%, 5%, 10%, 15% and 20% and found that the concrete showed failure in shear, while in the hardened state it showed gradual reduction in the strength of the material as more granular plastic was added to the concrete mix.

According to [27], the addition of plastic waste in pressed blocks with a variation of 0%, 1%, 3%, and 7% showed that the compressive strength of the stabilized earth block without additives was low and that there was an initial increase in compressive strength with increasing plastic waste. The optimum compressive strength for the study was obtained for blocks containing 1% crushed plastic waste, whose particle sizes were less than 6.3 mm. The increase in compressive strength was 244.4% when compared to the block without the addition of plastic waste.

This paper aimed to evaluate the influence of the incorporation of polyethylene terephthalate (PET) waste in the properties of soil-cement bricks. The study emphasizes mainly the analysis of characterization of the materials used, since this type of brick has particularities for its manufacturing. Therefore, characterization of the soil was performed, as well as the PET waste, to see the relationship of both in the mixtures. The compaction curves were also highlighted, since most studies have difficulties regarding the optimum moisture content used in the production of mixtures for this type of masonry. Moreover, this work also presents the PET waste as an innovation since it is still minimally used and not often discussed in research on the subject of soil-cement.

2. Materials and Methods

2.1. Materials and Mixtures

The sandy-clay soil used in the experiment was collected from a deposit located in Campos dos Goytacazes, Rio de Janeiro, Brazil, at a depth of 1.0 m on average, after removing the top-soil layer with a high percentage of organic matter. The samples were separated in plastic bags and kept hermetically sealed in order to maintain the humidity of the material. After collection, the soil was divided into different portions for homogeneity of the moisture of the samples. Then, the soil was crushed to reduce its volume, thus standardizing its granulometry. After this step, the soil was sieved, using a sieve with an aperture of 4.75 mm according to [35]. The distribution of the grain sizes of the collected soil samples and mixtures was performed by sieving, according to the procedures by [36]. The consistency limits (Atterberg) of the soil and mixtures were tested, determining the plastic properties of the samples according to [36,37]. Normal Proctor compaction tests were performed for soil and mixtures according to [38,39].

The PET waste used in this research presented particles of virgin material supplied by industries, with physical breakdowns and without contaminants as well as recycled PET treated post-consumer waste, i.e., mixed, with a uniform granulated aspect and white color. As a soil stabilizer, ordinary Portland cement type CPV-ARI was used, which has a high initial strength and is widely used in materials that require rapid hardening [16,27]. By considering that the bricks produced are molded, the use of this type of cement facilitates the transportation of the machine to the curing sites, avoiding its disintegration.

To perform the morphological analysis of the PET incorporated into the bricks, scanning electron microscopy (SEM) model FEG Quanta 250, belonging to the Laboratory of Electron Microscopy (LME) of the Military Institute of Engineering, Rio de Janeiro, Brazil (IME) was used. For this, the PET sample was coated with gold film and analyzed at

60 × magnification. To make the bricks, the mixture used was in the proportion of 1 part cement to 6 parts of soil (1:6). The bricks with 10% addition of PET in relation to the soil, 20% addition of PET in relation to the soil and 30% addition of PET in relation to the soil were analyzed in addition to the reference mixture, which was without any addition of waste (Table 1).

Table 1. Compositions (vol.%) of the soil-cement mixtures.

Mixtures	Soil (vol.%)	Cement (vol.%)	PET (vol.%)
0%	90	10	0
10%	80	10	10
20%	70	10	20
30%	60	10	30

Before manufacturing, the materials used in each mixture were weighed with the aid of a digital scale. After weighing, the mixture (soil, waste and cement) was homogenized with the aid of a mechanical mixer. The addition of water was done with a sprayer in order to distribute the water in the mixture in a controlled and uniform way, thus avoiding the formation of lumps of soil concentrated by the excess of water. The amount of water used was calculated according to the optimum moisture found in the Normal Proctor compaction tests, as shown in studies by [27,33].

2.2. Experimental Procedures

Compaction tests were performed for the soil and the mixture with PET according to [37,38]. They related the moisture content to be used with the specific weight of the soil samples and the mixtures with PET. To perform the compaction test, a cylindrical mold (Normal Proctor) was used, with a base attached to a metal socket which has a drop control apparatus [40,41].

The molding process of the bricks was performed according to [35] at the company Arte Cerâmica Sardinha, located in Campos dos Goytacazes. A hydraulic press, Model 7000 Turbo II, of the manufacturer Ecomáquinas, Navegantes, Santa Catarina, Brazil, was used, which has a molding capacity of up to 2 bricks at a time and a compression pressure of approximately 15 tons. The bricks were made in the size of 30 × 15 cm (length × width) and variable height of 7 cm ± 1 cm, with 2 holes and a useful area of 80%.

After making the bricks, the curing process was carried out by water sprinkling for 28 days. The process occurred in a humid chamber, where the specimens were placed on pallet racks covered with plastic sheeting, and during the 28 days constant humidity was added to the bricks through the water pump sprayer to help in the hydration process of the cement. The compressive strength was analyzed after 7 and 28 days of curing. For the statistical analysis, 7 samples of each treatment were used. The dimensions of the samples were measured with a calliper, according to the ABNT standards [40,41]. Then the bricks were cut in half (Figure 2a) in the transverse direction, having the upper part of the socket removed, as recommended by the ABNT standard [41]. The cuts were made with the aid of an electric saw (Figure 2b). Next, the cut halves were overlapped and joined by cement paste and then the capping was performed.

(**a**) (**b**)

Figure 2. Brick capping step: (**a**) Bricks cut in half; (**b**) upper part of socket removed.

The cement used to make the capping paste was the same CPV-ARI used in the manufacture of the bricks. After capping, the bricks were submerged in water for 24 h to ensure their saturation before rupture, which was performed in the press Model 100 T Manual Hydraulic Press DIG.110/220, Solotest, São Paulo, Brazil.

The water absorption test of the bricks was performed according to the ABNT standard [42,43] at 28 days after manufacturing. Three bricks of each mixture were used, which were dried in a SOLAB oven, SL-100 model, with temperatures ranging between 105 °C and 110 °C until reaching constant mass, for 48 h and then stored. After this period, the weight of each of the bricks was measured with the aid of a digital balance brand Marte, Model AD5002, São Paulo, Brazil to obtain the dry mass (g), as recommended by the ABNT standard [43]. Then, the bricks were immersed in water for 24 h, and after this saturation step, they were dried superficially with the help of a clean and dry cloth and then weighed again to obtain the wet mass (g). The water absorption of the bricks was obtained, according to the ABNT standard [43]:

$$A(\%) = \frac{(m1 - m2)}{m2} \times 100 \quad (1)$$

where: A—is the water absorption (%); m1—is the dry mass of the specimen (g); m2—is the mass of the saturated specimen (g).

Analysis of Variance (ANOVA) was used to verify the existence of significant differences between the results obtained. Statistical differences were confirmed by means of the comparison of means test, using Tukey's method ($p < 0.05$). The experimental design used was the Completely Randomized Design for the two variables analyzed: water absorption and compressive strength. For water absorption, three specimens were used for each treatment (28 days of curing): 0%, 10%, 20% and 30%. For compressive strength, seven specimens were used for each of the four treatments (7 days and 28 days of curing).

3. Results and Discussion

3.1. Characterization of Materials and Mixtures

3.1.1. Granulometric Analysis

The soil used for making the bricks, according to its particle distribution, showed 57.1% sand, 24.3% clay and 18.6% silt fraction, which is classified as sandy clay soil according to the ABNT standard [44]. This soil is considered suitable for making bricks as recommended by [45], who state that soils with sandy characteristics are the most suitable for the confection of soil-cement bricks, and also by the ABCP [46]. Indeed, the results of the granulometric analysis of the soil indicated that 100% passed through the ABNT

4.8 mm sieve (n°. 4) and 10% to 50% passed through the ABNT 0.075 mm sieve (n° 200), as observed in the granulometric curve presented in Figure 3.

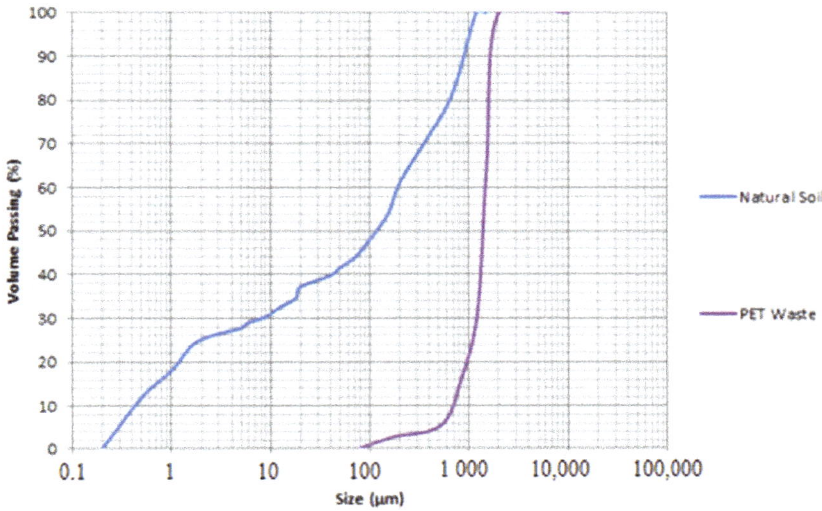

Figure 3. Natural soil and PET waste granulometric analysis.

The particle size distribution of PET particles was 99.9% sandy, 85.5% coarse, 11.5% medium and 2.9% fine particles, as shown in Figure 3. Since PET did not present consistency limits nor hygroscopicity, it can be considered non-plastic and sandy, according to the ABNT standard [44].

3.1.2. Consistency Limits

The consistency limits of the soil used presented the plasticity index (PI) values in Table 2. Thus, it was possible to verify that the values were adequate, since the PI for the production of soil-cement bricks should be up to 18%, as shown in studies [45,46]. In addition, the results reached the reference values for the manufacture of soil-cement bricks, i.e., soils with the maximum limit of 45% for liquidity limit (LL) and 18% for plasticity limit (PL), as well as those recommended by the ABNT standard [35].

Table 2. Sample consistency limits.

PET (%)	LL (%)	PL (%)	PI (%)
0%	40.1	22.4	17.7
10%	34.0	20.0	14.0
20%	36.7	19.4	17.3
30%	33.8	18.4	15.3

* Liquidity Limit (LL), Plasticity Limit (PL) and Plasticity Index (PI).

Differently from the results of the soil (reference sample), the values of the PET-Soil cement mixtures showed decreased LL, PL and PI compared to natural soil. This occurred due to the addition of PET waste (sandy fraction), showing that with an increase in the percentage of addition, the greater the amount of water needed in the mixture to get out of its plastic state. This can be related to the workability of the mixture, i.e., the greater the amount of water added, the lower the workability of the mixture [47,48].

3.1.3. Moisture Content and Compaction Energy

The optimum compaction moisture results with the respective maximum densities of the mixtures are presented in Table 3.

Table 3. Optimum compaction humidity and maximum mixture densities.

PET	Specific Weight (kN/m³)	Optimum Humidity (%)
0%	17.8	14.5
10%	16.6	15.0
20%	15.6	15.3
30%	13.9	16.5

The results in this table showed that with the increase of PET waste, there was an increase in the optimum moisture content of compaction and a decrease in the maximum specific weight. As for the decrease in maximum specific weight, this is due to the density value of the PET waste being lower than that of the natural soil. Moreover, the sample mass decreased as water was added to the compaction process because the waste did not have plastic properties. With this, as the water addition increased, it was possible to verify that the workability was decreasing, therefore this is also related to the compaction strength, as reported in study by Akinyele and Ajede [23]. The compaction curves of the mixtures are presented in Figure 4. Moreover, the curves proved (especially between 0% and 30%) the difference between the mixtures in relation to the addition of waste, i.e., the higher the percentage of addition, the greater the difficulty of reaching the optimum moisture content with increasing percentage of PET.

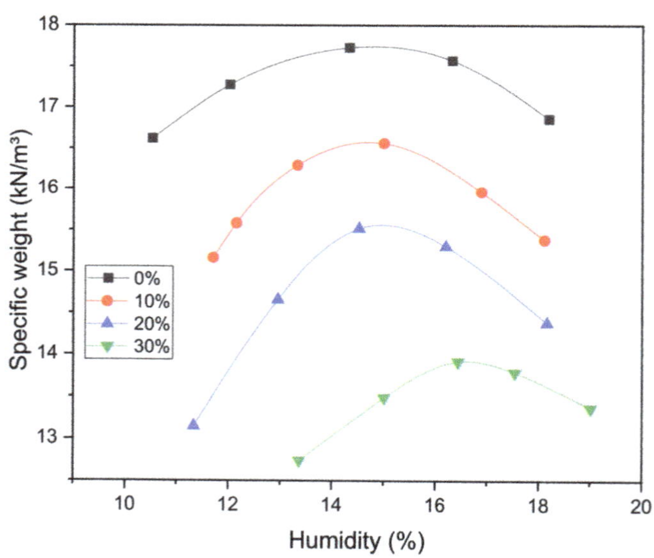

Figure 4. Compaction curves and optimum moisture of mixtures. Percentages (%) indicate the amount of water content in each mixture (0%, 10%, 20% and 30%).

3.1.4. Microstructural Characterization

A morphological analysis in Figure 5 by SEM showed that PET presents high surface area and irregularities in the shape and size of its particles. These are typical of different cutting and crushing processes of the waste, as shown in studies by Siqueira and Holanda [49]. As the particles present an irregular distribution, smaller sized particles can occupy the free spaces left by larger sized particles, forming particle packing [50].

Figure 5. Scanning Electron Microscopy (SEM) of PET 60× magnification.

3.2. Technological Characterization of Bricks

3.2.1. Compressive Strength

The compressive strength analysis of the bricks produced with different levels of in-corporation of PET waste showed a significant difference between the percentages, as well as between the ages. At seven days, the bricks made with 20% and 30% addition of PET waste showed the highest values of compressive strength, values that did not differ statistically between them. At 28 days, the highest average value was observed for the 20% addition (1.80 MPa), followed by 30% (1.45 MPa) and then the lowest values (0% and 10%) with no differences between them, as shown in Table 4.

Table 4. Compressive strength of samples at 7 and 28 days.

PET (%)	Compressive Strength 7 Days	Standard Deviation 7 Days	Compressive Strength 28 Days	Standard Deviation 28 Days
0	0.68 Ba	0.132	0.83 Ca	0.115
10	0.88 Ba	0.053	0.89 Ca	0.043
20	1.56 Ab	0.270	1.80 Aa	0.111
30	1.39 Aa	0.120	1.45 Ba	0.118

* Val. * Values followed by the same letter, uppercase in the column and lowercase in the row, do not differ at the level of 5% by Tukey's test.

According to the results, it was possible to observe that an increase in compressive strength occurred in bricks with different PET mixtures (20% and 30%). The samples with 20% PET reached the average value of compressive strength required by the ABNT standard NBR 15270-1 [51], which recommends 1.5 MPa for the total average of the samples. The mixtures with more PET had more water content. Since the PET particles were larger and the material was not porous, more water was available for cement hydration, which was probably the main reason why higher strengths were achieved, considering that the cementitious matrix does not have as good adhesion with PET [23].

The higher the concentration with smaller diameters of PET particles, the lower the probability of an increase in the number of voids and the direct interference in the strength of bricks [7]. In the case of 30% PET samples, Akinyele and Ajede [23] stated that the greater the amount of plastic, the greater the amount of water needed to improve workability;

in this case, it is possible to further state that the lack of water needed for hydration also correlated with lower strength. However, the authors state that this waste can be considered to replace fine aggregates up to 20% in cement-based materials. Additionally, the worse workability of the mixtures with a high amount of PET could be attributed not only to the higher particle size, but also to the rough and irregular shapes of the PET particles, as shown in the SEM images.

Regarding the reference sample (0% PET) and the samples of bricks with different incorporations of PET, they did not reach the average established by [43]. It is also estimated that they may have suffered interference in relation to cement hydration due to the type of cure used (humid chamber), damaging their strength. The incorrect hydration of the cement is generally due to the lack of water or humidity necessary for the cement to react, as shown in [49–52]. As for the ages, there was no difference for most of the treatments due to the cement used being CPV, which has greater strength gain at early ages in view of its better quality due to finer grinding.

3.2.2. Water Absorption

The analysis of water absorption in the bricks in Table 5 showed that there was no significant difference between the mixtures ($p < 0.05$). Thus, it was found that the amount of PET incorporated in all mixtures did not interfere with water absorption.

Table 5. Water absorption of samples at 28 days.

PET (%)	Water Absorption	Standard Deviation
0	16.23 A	1.057
10	15.93 A	0.109
20	15.27 A	0.252
30	15.21 A	0.283

* Values followed by the same letter, uppercase letters in the column do not differ at the 5% level by Tukey's test.

According to the ABNT standard [43], the water absorption values should not present average values higher than 20%, nor individual values higher than 22%. Considering that PET is a material that does not absorb water, it is believed that it contributed to the bricks not presenting high water absorption. This was confirmed by Górak et al. [53] in studies of the effect of incorporating PET waste into cementitious composites, which indicated that the particle size of the waste has a significant effect on water absorption in relation to its porosity [54–57].

4. Conclusions

The soil used in the study, according to its particle distribution and sand characteristics, is considered suitable for use in the manufacturing of soil-cement bricks. Indeed, the soil with 57.1% sand fraction, 24.3% clay fraction and 18.6% silt fraction is classified as sandy clay soil. The PET waste corresponded to 99.9% of sandy particle fraction, with 85.5% coarse particles, 11.5% medium particles and 2.9% fine particles. It was possible to verify that as the addition of PET increased in the mixtures, the higher the content of sandy particles.

As for the optimal moisture content and compaction energy, it was observed that the natural soil (without waste addition) showed optimal moisture value and maximum specific weight satisfactory. For the mixtures, the greater the addition of PET (10%, 20% and 30%), the greater the optimum humidity of compaction, in addition to a decrease in maximum specific weight. This revealed that the higher the addition of water, the lower the workability of the mixture, thus interfering in the mechanical strength.

The evaluated PET waste can be classified as sandy grain size and non-plastic waste. Through its microstructure it was possible to verify the rough and irregular shapes of the PET due to its crushing and processing.

Based on the evaluation of the physical and mechanical tests of soil-cement bricks, it was possible to verify improvement in their properties with the incorporation of PET.

The average compressive strength with the incorporation of 20% PET managed to reach the value of 1.80 MPa, i.e., above 1.50 MPa which is the minimum value established by the Brazilian standard. As for water absorption, the bricks showed satisfactory values and complied with the values established by the standard, i.e., not presenting average values higher than 20% or individual values higher than 22%.

For future studies related to the results of this work, it would be important to consider that although the amount of water in the mixture increased according to the amount of PET, it could compensate with an additional hydration of cement particles, suggesting higher strengths of the bricks.

Thus, the incorporation of PET in soil-cement bricks can be considered an alternative for non-structural applications, such as closing walls in building construction. Moreover, this study verifies that it is possible to reduce the environmental impacts of this type of waste, as demonstrated in the bricks made with 20% PET waste. However, it is necessary to consider further studies regarding the life cycle of this type of material, especially its final cycle and durability, in order to enhance applications and avoid greater environmental impacts.

Author Contributions: Conceptualization, M.T.M., M.G., T.R.d.S.; methodology, T.R.d.S. and J.A.; software, M.T.M.; validation, M.G.; formal analysis, F.C.d.S., S.N.M.; investigation, M.T.M., M.G., T.R.d.S.; resources, A.R.G.d.A.; data curation, F.C.d.S. and F.G.F.; writing—original draft preparation, I.V., T.R.d.S.; writing—review and editing, A.R.G.d.A. and S.N.M.; visualization, I.V. and F.G.F.; supervision, I.V., D.C. and A.R.G.d.A.; project administration, D.C., A.R.G.d.A. and S.N.M.; funding acquisition, S.N.M. All authors have read and agreed to the published version of the manuscript.

Funding: This research received no external funding.

Institutional Review Board Statement: Not applicable.

Informed Consent Statement: Not applicable.

Conflicts of Interest: The authors declare no conflict of interest.

References

1. Araújo, C.K.D.C.; Salvador, R.; Piekarski, C.M.; Sokulski, C.C.; De Francisco, A.C.; Camargo, S.K.D.C.A. Circular Economy Practices on Wood Panels: A bibliographic analysis. *Sustainability* **2019**, *11*, 1075. [CrossRef]
2. Krishna, R.; Mishra, J.; Meher, S.; Das, S.K.; Mustakim, S.; Singh, S.K. Industrial solid waste management through sustainable green technology: Case study insights from steel and mining industry in Keonjhar, India. *Mater. Today Proc.* **2020**, *33*, 5243–5249. [CrossRef]
3. Luiz, N.F.; Cecchin, D.; Azevedo, A.R.G.; Alexandre, J.; Silva, F.C.; Paes, A.L.C.; Pinheiro, V.D.; Do Carmo, D.F.; Ferraz, P.F.P.; Hüther, C.M.; et al. Characterization of materials used in the manufacture of ceramic tile with incorporation of ornamental rock waste. *DuraSpace* **2020**, *18*, 904–914. [CrossRef]
4. Da Silva, T.R.; de Azevedo, A.R.; Cecchin, D.; Marvila, M.T.; Amran, M.; Fediuk, R.; Vatin, N.; Karelina, M.; Klyuev, S.; Szelag, M. Application of Plastic Wastes in Construction Materials: A Review Using the Concept of Life-Cycle Assessment in the Context of Recent Research for Future Perspectives. *Materials* **2021**, *14*, 3549. [CrossRef] [PubMed]
5. Areias, I.O.R.; Vieira, C.M.F.; Colorado, H.A.; Delaqua, G.C.G.; Monteiro, S.N.; Azevedo, A.R.G. Could city sewage sludge be directly used into clay bricks for building construction? A comprehensive case study from brazil. *J. Build. Eng.* **2020**, *31*, 2566. [CrossRef]
6. United Nations. The state of plastics: World Environment Day Outlook 2018. In *UN Environment Programme*; ONU: New York, NY, USA, 2018; p. 11.
7. Spósito, F.A.; Higuti, R.T.; Tashima, M.M.; Akasaki, J.L.; Melges, J.L.P.; Assunção, C.C.; Bortoletto, M.; Silva, R.G.; Fioriti, C.F. Incorporation of PET wastes in rendering mortars based on Portland cement/hydrated lime. *J. Build. Eng.* **2020**, *32*, 101506. [CrossRef]
8. World Wide Fund for Nature (WWF). *Solving Plastic Pollution through Accountability*; WWF Word: New York, NY, USA, 2019; p. 23.
9. De Castro, A.M.; Carniel, A.; Stahelin, D.; Junior, L.S.C.; Honorato, H.D.A.; de Menezes, S.M.C. High-fold improvement of assorted post-consumer poly(ethylene terephthalate) (PET) packages hydrolysis using Humicola insolens cutinase as a single biocatalyst. *Process. Biochem.* **2019**, *81*, 85–91. [CrossRef]
10. Da Costa, A.M.; Lopes, V.R.D.O.; Vidal, L.; Nicaud, J.-M.; de Castro, A.M.; Coelho, M.A.Z. Poly(ethylene terephthalate) (PET) degradation by Yarrowia lipolytica: Investigations on cell growth, enzyme production and monomers consumption. *Process. Biochem.* **2020**, *95*, 81–90. [CrossRef]

11. Almeida, F.C.; Sales, A.; Moretti, J.P.; Mendes, P.C. Sugarcane bagasse ash sand (SBAS): Brazilian agroindustrial by-product for use in mortar. *Constr. Build. Mater.* **2015**, *82*, 31–38. [CrossRef]
12. De Azevedo, A.R.A.R.G.; Marvila, M.T.; Tayeh, B.A.; Cecchin, D.; Pereira, A.C.; Monteiro, S.N.S.N. Technological performance of açaí natural fibre reinforced cement-based mortars. *J. Build. Eng.* **2021**, *33*. [CrossRef]
13. Azevedo, A.; de Matos, P.; Marvila, M.; Sakata, R.; Silvestro, L.; Gleize, P.; Brito, J. Rheology, Hydration, and Microstructure of Portland Cement Pastes Produced with Ground Açaí Fibers. *Appl. Sci.* **2021**, *11*, 3036. [CrossRef]
14. Abdel-Ghani, N.T.; El-Sayed, H.A.; El-Habak, A.A. Utilization of by-pass cement kiln dust and air-cooled blast-furnace steel slag in the production of some "green" cement products. *HBRC J.* **2018**, *14*, 408–414. [CrossRef]
15. Carasek, H.; Girardi, A.C.C.; Araújo, R.C.; Angelim, R.; Cascudo, O. Study and evaluation of construction and demolition waste recycled aggregates for masonry and rendering mortars. *Cerâmica* **2018**, *64*, 288–300. [CrossRef]
16. Rahman, Z.A.; Saleh, N.M.M.; Idris, W.M.R.; Lihan, T. Thermal effect on mechanical characteristics of drinking water sludge brick incorporated with rice husk Ash (Kesan Suhu terhadap Cirian Mekanik Bata Sisa Rawatan Air Campuran Abu Sekam Padi). *Sains Malays.* **2019**, *48*, 2541–2549. [CrossRef]
17. Azevedo, A.; Marvila, M.; Fernandes, W.J.; Alexandre, J.; Xavier, G.; Zanelato, E.; Cerqueira, N.; Pedroti, L.; Mendes, B. Assessing the potential of sludge generated by the pulp and paper industry in assembling locking blocks. *J. Build. Eng.* **2019**, *23*, 334–340. [CrossRef]
18. Contreras, M.; Teixeira, S.R.; Lucas, M.C.; Lima, L.C.N.; Cardoso, D.S.L.; da Silva, G.A.C.; Gregório, G.C.; de Souza, A.E.; dos Santos, A. Recycling of construction and demolition waste for producing new construction material (Brazil case-study). *Constr. Build. Mater.* **2016**, *123*, 594–600. [CrossRef]
19. Jannat, N.; Hussien, A.; Abdullah, B.; Cotgrave, A. Application of agro and non-agro waste materials for unfired earth blocks construction: A review. *Constr. Build. Mater.* **2020**, *254*, 119346. [CrossRef]
20. Pooja, P.; Vaitla, M.; Sravan, G.; Reddy, M.P.; Bhagyawati, M. Study on behavior of concrete with partial replacement of fine aggregate with waste plastics. *Mater. Today Proc.* **2019**, *8*, 182–187. [CrossRef]
21. Alfahdawi, I.H.; Osman, S.A.; Hamid, R.; L-Hadithi, A.I.A. Influence of PET wastes on the environment and high strength concrete properties exposed to high temperatures. *Constr. Build. Mater.* **2019**, *225*, 358–370. [CrossRef]
22. Perera, S.; Arulrajah, A.; Wong, Y.C.; Horpibulsuk, S.; Maghool, F. Utilizing recycled PET blends with demolition wastes as construction materials. *Constr. Build. Mater.* **2019**, *221*, 200–209. [CrossRef]
23. Akinyele, J.O.; Ajede, A. The use of granulated plastic waste in structural concrete. *Afr. J. Sci. Technol. Innov. Dev.* **2018**, *10*, 169–175. [CrossRef]
24. Mohammed, A.A.; Rahim, A.A.F. Experimental behavior and analysis of high strength concrete beams reinforced with PET waste fiber. *Constr. Build. Mater.* **2020**, *244*, 118350. [CrossRef]
25. Akinyele, J.O.; Igba, U.T.; Adigun, B.G. Effect of waste PET on the structural properties of burnt bricks. *Sci. Afr.* **2020**, *7*, e00301. [CrossRef]
26. Paschoalin Filho, J.A.; Storopoli, J.H.; Dias, A.J.G. Evaluation of compressive strength and water absorption of soil-cement bricks manufactured with addition of pet (polyethylene terephthalate) wastes. *Acta Sci. Technol.* **2016**, *38*, 163–171. [CrossRef]
27. Akinwumi, I.I.; Domo-Spiff, A.H.; Salami, A. Marine plastic pollution and affordable housing challenge: Shredded waste plastic stabilized soil for producing compressed earth bricks. *Case Stud. Constr. Mater.* **2019**, *11*, e00241. [CrossRef]
28. França, B.R.; Azevedo, A.; Monteiro, S.N.; Filho, F.D.C.G.; Marvila, M.; Alexandre, J.; Zanelato, E.B. Durability of Soil-Cement Blocks with the Incorporation of Limestone Residues from the Processing of Marble. *Mater. Res.* **2018**, *21*, 1–6. [CrossRef]
29. Campos, A.; Brito, L.T. Structural behavior of soil-cement bricks using different sources of water and healing methods Comportement structurel de sol ciment briques utilisant différentes sources fuentes de agua y métodos de sanación. *Interações (Campo Grande)* **2017**, *20*, 283–296.
30. Al-Fakih, A.; Mohammed, B.S.; Liew, M.S.; Nikbakht, E. Incorporation of waste materials in the manufacture of masonry bricks: An update review. *J. Build. Eng.* **2019**, *21*, 37–54. [CrossRef]
31. Reis, F.M.D.; Ribeiro, R.P.; Reis, M.J. Physical-mechanical properties of soil-cement bricks with the addition of the fine fraction from the quartzite mining tailings (State of Minas Gerais—Brazil). *Bull. Eng. Geol. Environ.* **2020**. [CrossRef]
32. Kongkajun, N.; Laitila, E.A.; Ineure, P.; Prakaypan, W.; Cherdhirunkorn, B.; Chakartnarodom, P. Soil-cement bricks produced from local clay brick waste and soft sludge from fiber cement production. *Case Stud. Constr. Mater.* **2020**, *13*, e00448. [CrossRef]
33. Kouamé, A.N.; Doubi, B.I.H.G.; Konan, L.K.; Tognonvi, M.; Oyetola, S. The effect of Shea butter wastes on Physical Properties of Compressed Earth Bricks (CEB) and Cement Stabilized. *Aust. J. Basic Appl. Sci.* **2020**, *13*, 19–26. [CrossRef]
34. Vilela, A.P.; Eugênio, T.M.C.; De Oliveira, F.F.; Mendes, J.F.; Ribeiro, A.G.C.; Vaz, L.E.V.D.S.B.; Mendes, R.F. Technological properties of soil-cement bricks produced with iron ore mining waste. *Constr. Build. Mater.* **2020**, *262*, 120883. [CrossRef]
35. Brazilian Association of Technical Standards (ABNT). *NBR 10833: Production of Soil-Cement Bricks and Blocks Using a Manual or Hydraulic Press—Procedure*; ABNT: Rio de Janeiro, Brazil, 2013; 3p.
36. Brazilian Association of Technical Standards (ABNT). *NBR 7181: Granulometric Analysis*; ABNT: Rio de Janeiro, Brazil, 2016; 12p.
37. Brazilian Association of Technical Standards (ABNT). *NBR 6459: Soil—Determination of Liquidity Limit*; ABNT: Rio de Janeiro, Brazil, 2017; 5p.
38. Brazilian Association of Technical Standards (ABNT). *NBR 7180: Soil—Determination of the Plasticity Limit*; ABNT: Rio de Janeiro, Brazil, 2016; 3p.

39. Brazilian Association of Technical Standards (ABNT). *NBR 7182: Soil—Compaction Test*; ABNT: Rio de Janeiro, Brazil, 2020; 9p.
40. Brazilian Association of Technical Standards (ABNT). *NBR 12023: Soil-Cement—Compaction Test*; ABNT: Rio de Janeiro, Brazil, 2012; 7p.
41. Azevedo, A.; Cecchin, D.; Carmo, D.; Silva, F.; Campos, C.; Shtrucka, T.; Marvila, M.; Monteiro, S. Analysis of the compactness and properties of the hardened state of mortars with recycling of construction and demolition waste (CDW). *J. Mater. Res. Technol.* **2020**, *9*, 5942–5952. [CrossRef]
42. Brazilian Association of Technical Standards (ABNT). *NBR 8491: Soil-Cement Brick—Requirements*; ABNT: Rio de Janeiro, Brazil, 2012; 5p.
43. Brazilian Association of Technical Standards (ABNT). *NBR 8492: Soil-Cement Brick—Dimensional Analysis, Determination of Compressive Strength and Water Absorption—Test Method*; ABNT: Rio de Janeiro, Brazil, 2012; 4p.
44. Brazilian Association of Technical Standards (ABNT). *NBR 6502: Rocks and Soils—Terminology*; ABNT: Rio de Janeiro, Brazil, 1995; 18p.
45. Souza, M.I.B.; Segantini, A.A.S.; Pereira, J.A. Soil-cement pressed bricks made with concrete wastes. *Rev. Bras. Eng. Agrícola Ambient.* **2008**, *12*, 205–212. [CrossRef]
46. Brazilian Portland Cement Association (ABCP). *Fabricação de Tijolos de Solo-Cimento com a Utilização de Prensas Manuais*, 3rd ed.; ABCP: São Paulo, Brazil, 2000; 16p, ISBN 85-87024-28-0. (In Portuguese)
47. Ferreira, D.; Luso, E.; Cruz, M. Ecological soil-cement blocks with waste incorporation. In Proceedings of the REHABEND 2018: Construction Pathology, Rehabilitation Technology and Heritage Management, Caceres, Spain, 15–18 May 2018; pp. 1368–1376.
48. Murmu, A.L.; Patel, A. Towards sustainable bricks production: An overview. *Constr. Build. Mater.* **2018**, *165*, 112–125. [CrossRef]
49. Siqueira, F.B.; Holanda, J.N.F. Effect of incorporation of grits waste on the densification behavior of soil-cement bricks. *Ceramica* **2015**, *61*, 414–419. [CrossRef]
50. Umasabor, R.I.; Daniel, S.C. The effect of using polyethylene terephthalate as an additive on the flexural and compressive strength of concrete. *Heliyon* **2020**, *6*, e04700. [CrossRef] [PubMed]
51. Brazilian Association of Technical Standards (ABNT). *NBR 15270-1: Ceramic Components—Blocks and Bricks for Masonry—Part 1: Requirements*; ABNT: Rio de Janeiro, Brazil, 2017; 26p.
52. Santos, A.C.D. Investigation of the Effect of Incorporating PET Waste in Cementitious Composites. Ph.D. Thesis, Federal University of São João Del-rei, São João del-Rei, Brazil, 2012; p. 102. (In Portuguese).
53. Górak, P.; Postawa, P.; Trusilewicz, L.N.; Łagosz, A. Lightweight PET based composite aggregates in Portland cement materials—Microstructure and physicochemical performance. *J. Build. Eng.* **2020**, *34*, 101882. [CrossRef]
54. Martins, R.O.G.; Alvarenga, R.D.C.S.S.; Pedroti, L.G.; de Oliveira, A.F.; Mendes, B.C.; de Azevedo, A.R.G. Assessment of the durability of grout submitted to accelerated carbonation test. *Constr. Build. Mater.* **2018**, *159*, 261–268. [CrossRef]
55. De Azevedo, A.R.G.; Marvila, M.T.; Rocha, H.A.; Cruz, L.R.; Vieira, C.M.F. Use of glass polishing waste in the development of ecological ceramic roof tiles by the geopolymerization process. *Int. J. Appl. Ceram. Technol.* **2020**, *17*, 2649–2658. [CrossRef]
56. Carvalho, A.; de Castro Xavier, G.; Alexandre, J.; Pedroti, L.G.; de Azevedo, A.R.G.; Vieira, C.M.F.; Monteiro, S.N. Environmental durability of soil-cement block incorporated with ornamental stone waste. *Mater. Sci. Forum* **2014**, *798–799*, 548–553. [CrossRef]
57. Neuba, L.M.; Pereira Junio, R.F.; Ribeiro, M.P.; Souza, A.T.; Lima, E.S.; Filho, F.C.G.; Monteiro, S.N. Promising mechanical, thermal, and ballistic properties of novel epoxy composites reinforced with cyperus malaccensis sedge fiber. *Polymers* **2020**, *12*, 1776. [CrossRef] [PubMed]

materials

MDPI

Article

Dismantling of Waste Printed Circuit Boards with the Simultaneous Recovery of Copper: Experimental Study and Process Modeling

Szabolcs Fogarasi [1,2], Árpád Imre-Lucaci [1,*] and Florica Imre-Lucaci [2,*]

[1] Department of Chemical Engineering, Faculty of Chemistry and Chemical Engineering, Babeş-Bolyai University, 11 Arany Janos Street, 400028 Cluj-Napoca, Romania; szabolcs.fogarasi@ubbcluj.ro

[2] Interdisciplinary Research Institute on Bio-Nano-Sciences, Babeş-Bolyai University, 42 Treboniu Laurian Street, 400271 Cluj-Napoca, Romania

* Correspondence: arpad.imre@ubbcluj.ro (Á.I.-L.); florica.imre@ubbcluj.ro (F.I.-L.); Tel.: +40-264-593833 (Á.I.-L.)

Abstract: The study was carried out with the aim to demonstrate the applicability of a combined chemical–electrochemical process for the dismantling of waste printed circuit boards (WPCBs) created from different types of electronic equipment. The concept implies a simple and less polluting process that allows the chemical dismantling of WPCBs with the simultaneous recovery of copper from the leaching solution and the regeneration of the leaching agent. In order to assess the performance of the dismantling process, various tests were performed on different types of WPCBs using the 0.3 M $FeCl_3$ in 0.5 M HCl leaching system. The experimental results show that, through the leaching process, the electronic components (EC) together with other fractions can be efficiently dismounted from the surface of WPCBs, with the parallel electrowinning of copper from the copper rich leaching solution. In addition, the process was scaled up for the dismantling of 100 kg/h WPCBs and modeled and simulated using process flow modelling software ChemCAD in order to assess the impact of all steps and equipment on the technical and environmental performance of the overall process. According to the results, the dismantling of 1 kg of WPCBs requires a total energy of 0.48 kWh, and the process can be performed with an overall low environmental impact based on the obtained general environmental indexes (GEIs) values.

Keywords: WPCBs dismantling; metals dissolution; copper recovery; economic assessment; process modeling and simulation

Citation: Fogarasi, S.; Imre-Lucaci, Á.; Imre-Lucaci, F. Dismantling of Waste Printed Circuit Boards with the Simultaneous Recovery of Copper: Experimental Study and Process Modeling. *Materials* **2021**, *14*, 5186. https://doi.org/10.3390/ma14185186

Academic Editors: Rossana Bellopede and Lorena Zichella

Received: 31 July 2021
Accepted: 6 September 2021
Published: 9 September 2021

1. Introduction

According to the literature data, in the period 2016–2030, the quantity of waste electrical and electronic equipment (WEEEs) will increase at world level from 44.7 million tonnes to 110 million tonnes [1–4]. Thus, increasing the quantity of WEEEs requires the urgent development of environmentally friendly and low cost recycling technologies [5]. Waste printed circuit boards (WPCBs) represents an important part of WEEEs (3–5 wt.%) with more than 30% metals (of which ~30% copper) and 70% non-metals [6–8]. For this reason, the recycling of WPCBs can constitute a potential secondary source of raw materials for different industrial sectors [9], contributing to preservation of natural resources [10–12]. With this regard, a lot of research has been done that has established different techniques for WPCBs recycling based on physico-mechanical [13,14], pyrometallurgical [15], pyrolytic [11,16,17], and hydrometallurgical approaches [18,19]. One of the major issues that needs to be solved regarding the recycling of WPCBs is related to its dismantling into different material fractions that can be further processed into value added products [20,21]. Some studies so far conducted to WPCBs dismantling involve the thermal processing of WPCBs, but this operation is not eco-friendly because it occurs with the release of toxic gases [22], high energy consumption, and the burning of components that could be reused [23,24]. On the other hand, hydrometallurgical recovery techniques of valuable materials from WPCBs are achieved by grinding followed by acid or basic leaching of base metals from the obtained powder with the generation of other unusable or toxic residues; at the same time, they are non-selective processes [25–27]. Typically, the approach does not allow the full recovery of metallic copper, which is the main base metal of the EC found on the motherboards. Considering that traditional approaches are highly energy-demanding and environmentally dangerous [28,29], our research group has been carrying out the development of chemical and electrochemical processes for the recovery of metals from WPCBs, which allow the oxidative dissolution of metals with the simultaneous electrochemical regeneration of the oxidant, leading to the minimization of several drawbacks [30–33]. It was found that the metallic parts that hold together the different parts of WPCBs can be dissolved in the combined chemical–electrochemical processes, leading to the disassembly of WPCBs into different material fractions, such as plastics, printed circuit boards without EC, chips and small EC, and sludge [31,33–35]. Considering the success achieved with particular types of WPCBs [31], in this work, it is our intention to prove the applicability of this method for the dismantling of other types of WPCBs simultaneously with the electrochemical regeneration of the leaching agent and the partial electrodeposition of dissolved copper. In addition, to evaluate the contribution of all stages and equipment to the technical–environmental performance of the process, besides the ones used in the experimental studies, the process was extended, modeled, and simulated using ChemCAD process flow modelling software for a higher productivity.

In order to accomplish the abovementioned aims, the following sections are discussed after the materials and methods section: (i) *Theoretical background for the dissolution of metals* to better understand the processes that occur in the case of WPCBs disassembly; (ii) *experimental dismantling process of WPCBs* meant to provide the necessary background for the scale up of the process; (iii) *the scaled-up dismantling process* in order to evaluate the contribution of all steps and equipment to the performance of the overall process and to provide the necessary data for the environmental assessment; (iv) *environmental assessment of the scaled-up dismantling process* to assess the environmental impact of the process in the early phase of development using the Biwer–Heinzle method.

2. Materials and Methods

The WPCB dismantling experiments were performed using four types of motherboards, which differ in terms of technical details (Table 1) and metal content (Table 2) due to the different periods of production. For each case, before inserting the WPCBs into the rotating drum of the chemical reactor (CR), large pieces of aluminum and stainless steel were removed from the surface of the WPCBs, because they can be easier recycled this way. Additionally, it is not justified, due to the high power consumption, to bring the metals from these parts into the solution. Next, the WPCBs were cut into 40–100 cm^2 pieces to fit into the rotating drum.

Table 1. WPCBs samples used in the experimental studies.

Motherboard 1	Motherboard 2	Motherboard 3	Motherboard 4
Type: ATI Radeon Xpress 200 (2004–>) Processor (Socket 478): Pentium 4	Type: MSI P54C TR4 (1996–>) Processor (Socket 7): Pentium; AMD K5; Cyrix 6 × 86	Type: Gigabyte GA-BX2000 (1999–>) Processor (Slot 1): Pentium II; Pentium III	Type: Intel Acorp 6VIA/ZX85 (2000–>) Processor (Socket 370): Pentium III
	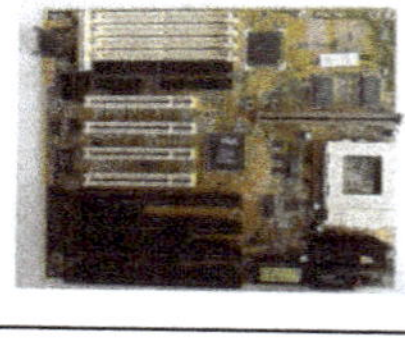		

Table 2. Concentration (wt.%) of the most important metals in the WPCBs samples.

Sample	Mass, g	Cu	Au	Ag	Sn	Pb	Ni	Zn	Fe
Motherboard 1	459.14	18.29	0.011	0.018	3.92	2.31	1.11	1.71	2.81
Motherboard 2	488.45	19.6	0.009	0.015	4.1	2.16	1.27	1.92	3.01
Motherboard 3	444.06	22.1	0.013	0.02	3.51	2.59	1.06	1.53	2.57
Motherboard 4	482.68	23.1	0.01	0.016	3.76	2.64	1.19	1.67	2.93

The experimental setup used for the dismantling of the WPCBs samples consisted of two reactors connected in series, including a 2L CR with perforated rotating drum and a 3L divided electrochemical reactor (ER) by a ceramic separator. In all experiments, 5L of 0.3 M $FeCl_3$ in 0.5 M HCl solution was recirculated between the two reactors using two Medorex TC200 pumps (Medorex, Nörten-Hardenberg, Germany). The solution was evacuated at the bottom of the CR and supplied at the bottom of the cathode compartment of the ER. From the top of the cathode compartment, the solution was transported to the bottom of the anode compartment, and with the help of the second pump, the solution was pumped from the top of the anode compartment back into the CR. In consequence, a cross flow of electrolyte between the two reactors is achieved. The rectangular cathode and anode were made of copper and graphite, respectively, each with an area of 570 cm^2. Two $Ag/AgCl/KCl_{sat}$ reference electrodes were used to measure the cathodic and anodic potentials. The experiments were carried out in the optimal operating conditions identified for the combined chemical-electrochemical processes in previous studies [31,32,36]: drum rotation speed (30 rpm), solid/liquid ratio (1/8), constant current density 4 mA/cm^2, initial electrolyte composition 0.3 M $FeCl_3$ in 0.5 M HCl, and a flow rate of 400 mL/min. An atomic absorption spectrometer was used to determine the metal content and different material fractions of the solutions, while the surface morphology and chemical composition of the cathodic deposits were characterized with a scanning electron microscope (SEM, Thermo Fisher Scientific, Waltham, MA, USA) equipped with an energy dispersive X-ray spectrometer (SEM/EDAX, FEI QUANTA 3D).

3. Results and Discussions

3.1. Theoretical Background for the Dissolution of Metals

To better understand the processes that occur in the case of WPCB disassembly, through the dissolution of metals with Fe^{3+} in the presence of Cl^-, the standard apparent potentials (E') were calculated for the potentially active reactions (PAR) based on the standard normal potentials (E^0) and the thermodynamic equilibrium constants for the formation of complexes [37–39]. In order to calculate the E, it is necessary to identify the chemical species present in the solution, with respect to the equilibriums in which they are involved. In the case of copper, the species involved in the process were considered on the basis of the Pourbaix diagram for the copper–chlorine–water system at 25 °C at a total Cl^- concentration of 1.5 M [40].

In the case of the other metals, the species involved in the chemical process have been identified from the literature based on their stability in the reaction environment [37,38,41,42]. The E' values (Table 3) were obtained (Equation (1)) based on the Nernst equation applied for redox reactions, which involves the complexation of the oxidized form of the redox couple [21].

$$E' = E^0 - \frac{0.059}{n} logK_{f_{ox}} \quad (1)$$

where E^0 is the standard normal potential, n is the number of electrons changed, and $K_{f_{ox}}$ is the thermodynamic equilibrium constant for the formation of complexes with the oxidized form of the redox couple.

Table 3. E' values for the involved potential active reaction.

Potential Active Reaction	E° [V/HNE]	E' [V/HNE]	Thermodynamic Equilibrium Constants $logK_{fox}$
$AuCl_4^- + 3e^- \rightarrow Au + 4Cl^-$	1.52	1.02	25.42
$AuCl_2^- + e^- \rightarrow Au + 2Cl^-$	1.83	1.25	9.83
$AgCl_2^- + e^- \rightarrow Ag + 2Cl^-$	0.79	0.48	5.25
$CuCl_{2aq} + 2e^- \rightarrow Cu + 2Cl^-$	0.34	0.294	1.56
$CuCl_2^- + e^- \rightarrow Cu + 2Cl^-$	0.52	0.21	5.14
$PbCl_4^{2-} + 2e^- \rightarrow Pb + 4Cl^-$	−0.126	−0.17	1.6
$PbCl_6^{2-} + 4e^- \rightarrow Pb + 6Cl^-$	0.762	0.747	1.05
$SnCl_4^{2-} + 2e^- \rightarrow Sn + 4Cl^-$	−0.137	−0.19	1.79
$SnCl_6^{2-} + 4e^- \rightarrow Sn + 6Cl^-$	−0.0085	−0.025	1.11
$NiCl_4^{2-} + 2e^- \rightarrow Ni + 4Cl^-$	−0.26	−0.28	0.7
$FeCl_4^- + 3e^- \rightarrow Fe + 4Cl^-$	−0.036	−0.031	−0.09
$FeCl_4^{2-} + 2e^- \rightarrow Fe + 4Cl^-$	−0.44	−0.402	−1.276
$ZnCl_4^{2-} + 2e^- \rightarrow Zn + 4Cl^-$	−0.76	−0.77	0.2

As was expected, the standard apparent potential decreases when the complexing agent employs the oxidized form, and the potential shift will be greater as the stability of the complexes is higher. Therefore, E' differs the most from E^0 in the case of Au, Ag, and Cu(I), for which the thermodynamic equilibrium constants for the formation of chloro-complexes are the highest. In the case of the other metals, where the complexation equilibrium is less shifted to the formation of chloro-complexes, the redox potential varies insignificantly under the experimental conditions. Additionally, from the E′ values calculated for the PAR, it can be seen that Fe^{3+} is an efficient oxidant in the dissolution of metals, with the exception of gold.

It is also important to note that the dissolution processes lead to chloride complexes of Cu, Sn, Pb, Fe, in which the metals may have different oxidation forms. As a result, the E' values (Table 4) for these redox couples, in which both forms are involved in complexation processes, was calculated by the following equation:

$$E' = E^0 + \frac{0.059}{n} log \frac{K_{fox}}{K_{fred}}. \quad (2)$$

Table 4. E' values for PAR that involve the complexation of both forms of the redox couple.

Potential Active Reaction	E^0 [V/HNE]	E' [V/HNE]
$FeCl_4^- + e^- \rightarrow FeCl_4^{2-}$	0.77	0.7
$CuCl_{2aq} + e^- \rightarrow CuCl_2^-$	0.16	0.37
$PbCl_6^{2-} + 2e^- \rightarrow PbCl_4^{2-} + 2Cl^-$	1.65	1.664
$SnCl_6^{2-} + 2e^- \rightarrow SnCl_4^{2-} + 2Cl^-$	0.154	0.14

As can be seen from Table 4, the existence of significant differences between the E′ values of these redox couples implies a state of un-equilibrium in these working conditions. The evolution of the system towards equilibrium determines the change in the concentration of the electroactive species, especially in the case of metals that may have different oxidation forms. The most important redox reactions leading to the installation of redox equilibrium are those in which Fe^{3+} oxidizes Cu^+ to Cu^{2+} and Sn^{2+} to Sn^{4+}. From the strong positive value of E′ for the couple $PbCl_6^{2-}/PbCl_4^{2-}$, it is concluded that Pb is more stable in the reduced form $PbCl_4^{2-}$.

Therefore, from the oxidation reaction of Pb by Fe^{3+}, only Pb^{2+} is formed without its subsequent oxidation to Pb^{4+}. The above conclusions are in agreement with the values of redox equilibrium constants (K_r) for the dissolution reactions of the metals from the WPCBs samples, calculated on the basis of Equation (4) using the E′ values from Tables 3 and 4.

For a redox reaction in the general form:

$$m\ ox_1 + p\ red_2 \rightleftharpoons m\ red_1 + p\ ox_2 \tag{3}$$

The equilibrium constant is defined as follows [32]:

$$K_r = 10^{\frac{mp(E'_1 - E'_2)}{0.059}} \tag{4}$$

where mp—number of electrons transferred between the redox couples.

The K_r values increase (the equilibrium will be shifted to the right) with the number of electrons transferred between the redox couples and greater the difference between the values of E' for the two systems. The values of the redox equilibrium constants, calculated according to Equation (4), are presented in Table 5.

Table 5. The values of redox equilibrium constants calculated for the leaching reactions.

Redox Reaction	Redox Equilibrium Constant
$3\ FeCl_4^- + Au + 4Cl^- \rightleftharpoons AuCl_4^- + 3\ FeCl_4^{2-}$	5.35×10^{-17}
$FeCl_4^- + Au + 2Cl^- \rightleftharpoons AuCl_2^- + FeCl_4^{2-}$	4.76×10^{-10}
$FeCl_4^- + Ag + 2Cl^- \rightleftharpoons AgCl_2^- + FeCl_4^{2-}$	5.35×10^{3}
$FeCl_4^- + Cu + 2Cl^- \rightleftharpoons CuCl_2^- + FeCl_4^{2-}$	2.01×10^{8}
$FeCl_4^- + CuCl_2^- \rightleftharpoons CuCl_{2aq} + FeCl_4^{2-}$	3.91×10^{5}
$2\ FeCl_4^- + Cu + 2Cl^- \rightleftharpoons CuCl_{2aq} + 2\ FeCl_4^{2-}$	5.79×10^{13}
$2\ FeCl_4^- + Sn + 4Cl^- \rightleftharpoons SnCl_4^{2-} + 2\ FeCl_4^{2-}$	1.47×10^{30}
$2\ FeCl_4^- + SnCl_4^{2-} + 2Cl^- \rightleftharpoons SnCl_6^{2-} + 2\ FeCl_4^{2-}$	9.61×10^{18}
$4\ FeCl_4^- + Sn + 6Cl^- \rightleftharpoons SnCl_6^{2-} + 4\ FeCl_4^{2-}$	1.42×10^{49}
$2\ FeCl_4^- + Pb + 4Cl^- \rightleftharpoons PbCl_4^{2-} + 2\ FeCl_4^{2-}$	3.1×10^{29}
$2\ FeCl_4^- + PbCl_4^{2-} \rightleftharpoons PbCl_6^{2-} + 2\ FeCl_4^{2-}$	2.09×10^{-33}
$4\ FeCl_4^- + Pb + 6Cl^- \rightleftharpoons PbCl_6^{2-} + 4\ FeCl_4^{2-}$	6.5×10^{-4}
$2\ FeCl_4^- + Ni + 4Cl^- \rightleftharpoons NiCl_4^{2-} + 2\ FeCl_4^{2-}$	1.66×10^{33}
$2\ FeCl_4^- + Fe + 4Cl^- \rightleftharpoons 3\ FeCl_4^{2-}$	2.2×10^{37}
$2\ FeCl_4^- + Zn + 4Cl^- \rightleftharpoons ZnCl_4^{2-} + 2\ FeCl_4^{2-}$	3.1×1^{49}

From the values of redox equilibrium constants, Table 5, it is observed that, for all metals except gold, redox reactions are strongly displaced towards their dissolution with the formation of chloro-complexes. Based on the redox equilibrium constant values from Table 5, it can be assumed that the dissolution rate of metals will have following order: Zn > Sn > Fe > Ni > Pb > Cu > Ag. Tin is located after Zn, although from the E′ value from Table 3, Sn should be between Pb and Cu. This can be explained by the fact that the redox equilibrium constant depends both on the potential difference between the redox couples and on the number of electrons transferred. Considering that Sn changes $4e^-$ while the other metals change only $2e^-$, the redox equilibrium constant is much higher (1.42×10^{49}), which is assumed to promote its dissolution.

3.2. Experimental Dismantling Process of WPCBs

The results presented in Table 6 prove that the combined chemical–electrochemical process leads to the complete dismantling of the WPCBs samples, leading to different material fractions. It can be observed that the obtained fractions can be easily separated due to the major differences in physical characteristic, such as size and density.

Table 6. Material fractions resulted from the dismantling process.

Plastics	Boards	Chips and Small EC	Sludge	Solution

The obtained material fractions still have in their composition undissolved metals, considering that, in the dismantling process, only the metals, which were accessible to the leaching solution, from the surface of the WPCBs were dissolved. To determine their metallic composition, they were grounded to a fine powder and treated with aqua regia to determine the metallic composition, which is presented in Table 7. It can be noticed that, for all the samples, the plastic fractions do not contain metals, which means that all the pins were dissolved. Similarly, the boards contain only copper in a relatively high concentration, and this material can be further processed for high purity copper recovery. As for the other two solid fractions, all the metals are present with the remark that they have high Au and Ag concentration, especially in the case of the obtained sludge. Additionally, Table 7 indicates that almost 70% of the metals present in the initial samples were dissolved during the dismantling process, and the major component is Cu, which represents, on average, 50% of the dissolved metals.

Table 7. The metallic composition of the different fractions obtained in the dismantling process.

Sample	Type of Fractions	Total Mass of Fractions	Metallic Composition, %							
			Cu	Au	Ag	Sn	Pb	Ni	Zn	Fe
Motherboard 1	Plastics	147.6	-	-	-	-	-	-	-	-
	Boards	174.46	17.2	-	-	-	-	-	-	-
	Chips and small EC	29.21	8.3	0.05	0.1	0.75	0.15	0.75	0.15	1.64
	Sludge	9.73	3.65	0.37	0.05	28.72	36.7	1.09	0.02	0.4
	Dissolved metals	98.14	52.16	0	0.05	15.27	7.12	4.86	7.95	12.62
Motherboard 2	Plastics	166	-	-	-	-	-	-	-	-
	Boards	172.2	23	-	-	-	-	-	-	-
	Chips and small EC	35.05	8.4	0.06	0.1	0.55	0.15	0.83	0.13	1.29
	Sludge	5.2	3.1	0.44	0.06	29.53	39.1	1.05	0.022	0.51
	Dissolved metals	110	48.20	0	0.03	16.63	7.70	5.33	8.48	12.93
Motherboard 3	Plastics	147.9	-	-	-	-	-	-	-	-
	Boards	170.2	23.5	-	-	-	-	-	-	-
	Chips and small EC	20.8	8.7	0.03	0.1	0.68	0.13	0.91	0.18	1.51
	Sludge	8.1	3.65	0.64	0.05	26.83	34.8	1.02	0.025	0.47
	Dissolved metals	97.06	57.73	0	0.07	13.67	8.92	4.57	6.96	11.40
Motherboard 4	Plastics	145.8	-	-	-	-	-	-	-	-
	Boards	141.55	20.2	-	-	-	-	-	-	-
	Chips and small EC	57.25	8.1	0.05	0.1	0.61	0.16	0.81	0.15	1.37
	Sludge	2.65	3.65	0.74	0.05	27.52	38.1	1.12	0.02	0.44
	Dissolved metals	135.43	57.72	0	0.01	12.60	8.60	3.88	5.89	9.85

This is the main reason why the leaching solution is suitable for high purity copper production simultaneously with the regeneration of the leaching agent and dismantling of the WPCB samples. Considering that the cathodic deposits were obtained at the same current density and flow rate and a similar solution composition, they showed very similar, almost identical, morphological properties and elementary composition. According to the composition analysis, in all the experimental studies, the cathodic deposit contains more than 99.95 wt.% copper, the only impurity being Sn. The elementary analysis results presented in Figure 1 confirm that the developed process leads to the recovery of a high purity copper deposit. It is also important to note that the obtained copper deposits were compact, which is sustained by the characterization made by SEM.

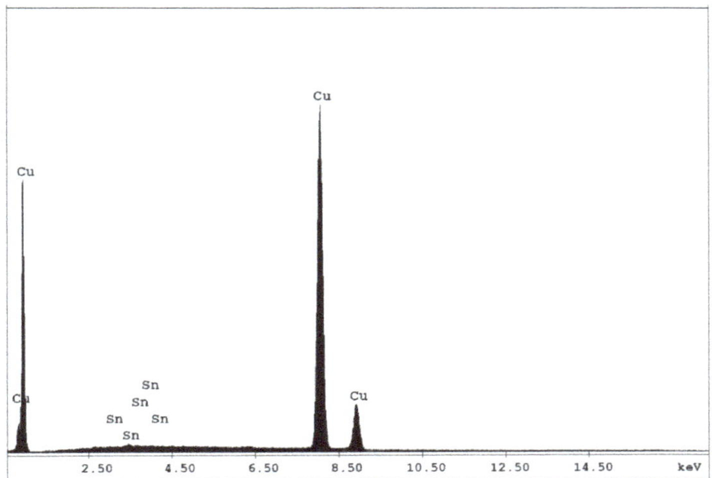

Figure 1. EDAX spectrum of the copper deposit.

In addition, the SEM images from Figure 2 show that the copper deposit presents rougher surfaces with larger nuclei and pyramidal growth, which is characteristic for copper deposition from chloride solutions in accordance with the literature data [43].

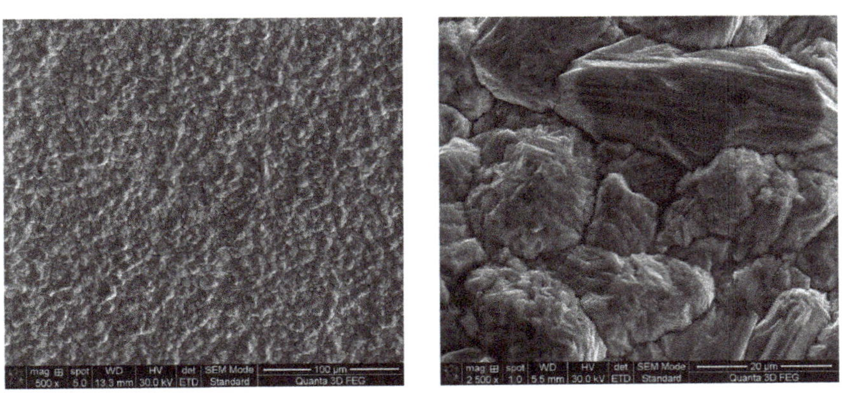

(**a**) (**b**)

Figure 2. SEM images for the copper deposit at magnitude: ×500 (**a**); ×2500 (**b**).

3.3. The Scaled-Up Dismantling Process

The experimental results obtained for the dismantling of the four types of WPCBs were used to design a conceptual recycling plant at higher scale of production for the dismantling of 100 kg/h of WPCBs. It was assumed that the WPCBs fed into the recycling plant have the metallic composition of Table 8, which presents an average composition calculated based on the data from Table 2.

Table 8. Metallic composition (wt.%) of WPCBs feed into the recycling plant.

Cu	Au	Ag	Sn	Pb	Ni	Zn	Fe
20.77	0.011	0.017	3.83	2.42	1.16	1.71	2.84

The dismantling process was modeled and simulated using process flow modeling software ChemCAD 7.1.5 in order to evaluate the contribution of all steps and equipment to the performance of the overall process and to provide the necessary data for the environmental assessment. Chemical and phase equilibrium were assumed based on Gibbs free energy minimization model. Other or complementary input data and assumptions used for modelling and simulation of the dismantling process are presented in Table 9.

Table 9. Model assumptions and input data [31–33,38].

Unit	Parameters
Leaching reactor	Temperature: 25 °C Residence time 24 h 0.3 M $FeCl_3$ in 0.5 M HCl
Electrochemical reactor	$r_e = 90.12\%$ $r_c = 72.69\%$ $W_c = 1.59$ kWh/kg
Processed WPCBs	100 kg/h
Pump efficiency	90%
Heat exchanger $\Delta T_{min.}$	10 °C
Heat exchanger pressure drop	1–3%

Figure 3 shows that the recycling plant designed for the dismantling of WPCBs with the simultaneous recovery of copper includes two main subsystems: Phase 1 for the dissolution of the accessible metallic parts from the surface of WPCBs along with the separation of the obtained material fractions and Phase 2, where copper is electrodeposited and the leaching solution is regenerated. In addition, Phase 1 deals with the processing of the chips, small EC, and boards, considering that these fractions have high copper content and low concentration in other metals. In the model, these material fractions are ground to a fine powder in order to enhance the dissolution reaction; then, the obtained material is contacted in a second chemical reactor with the leaching solution generating two streams: (i) the electrolyte solution used for copper and leaching agent production and (ii) polymers and fiberglass that exit the system as byproduct. According to Figure 3, the solution obtained in the ER is only partially recirculated into the process, due to the fact that it contains other metals besides copper that need to be extracted, after which it can be reused in the dissolution step. This is the reason why, in this model, new reagents are feed into the system, which, in a complete recovery plant, would be significantly, or possibly entirely, reduced.

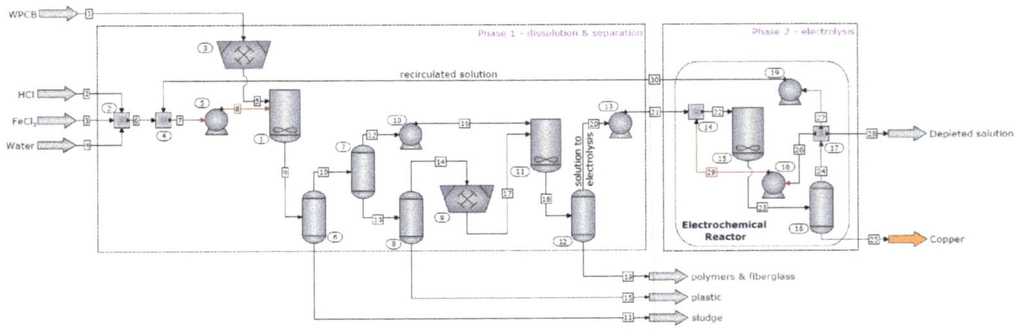

Figure 3. Recycling plant process flow diagram.

The overall mass balance (Table 10) of the scaled-up dismantling process indicates that the WPCBs were completely dismantled and processed into different materials fractions. Based on the initial amount of copper feed into the system with the WPCBs and the amount of obtained copper deposit, it is evident that more 90% of the copper is extracted during the dismantling process. As was expected, the organic matter, consisting of plastic and epoxy resin, represents the most important solid fraction obtained in the dismantling process, followed by fiber glass. It can also be seen that the amount of solution evacuated from the system is slightly higher than at the inlet due to the presents of the dissolved metals.

Table 10. Overall mass balance of the scaled-up dismantling process.

Input		Output	
Component	**kg/h**	**Component**	**kg/h**
WPCBs	100	Copper deposit	18.7
Leaching solution	623.4	Sludge	3.7
		Plastic	32.4
		Epoxy resin	14.5
		Fiber glass	15.1
		Depleted solution	634
TOTAL	**723.4**	**TOTAL**	**723.4**

In accordance with the mass balance data (Table 10), equipment types, and operating conditions, the overall energy balance of the scaled-up dismantling process was established (Table 11). The results indicate that the highest energy consumption is associated with the electrochemical process followed by the separation of different material fractions during the dismantling process. Combining the mass and energy balance data, it was determined that the dismantling of 1 kg of WPCBs requires 0.48 kWh, while the total specific energy consumption for copper production is 2.59 kWh/kg. It is also important to note that the overall energy balance data reveals that the process generates 47% more energy that it consumes. However, the potential usability of the generated energy is discussible, considering that it is low grad heat generated during the dissolution processes in comparison to the consumed electrical power. However, if the energy would be valorized for heating purposes, then the process would be useful for energy production in parallel with the dismantling of WPCBs and recovery of copper.

Table 11. Overall energy balance of the scaled-up dismantling process.

Process Equipment		Energy, (MJ/h)	
Type	**ID**	**Added**	**Extracted**
Crusher	3, 9	0.212	-
Reactor	1	2.08	211.545
Reactor	11	1.07	45.407
Separators	6, 8, 12	25.617	-
Electrochemical Reactor	15	144.17	-
Pumps	5, 10, 13, 18, 19	1.398	-
TOTAL:		174.5470	256.952

3.4. Environmental Assessment of the Scaled-Up Dismantling Process

Considering that the process presented in the current study is in the early phase of development, the Biwer–Heinzle method was applied for its environmental impact assessment. According to the methodology described in the literature [30,44] the environmental factors were determined for six impact groups in the case of the input materials (Figure 4) and 11 impact groups in the case of the output materials (Figure 5) by allocating each of the materials streams to a class of toxicity (A = 1—highly toxic substances, B = 0.3—less toxic substances, C = 0—nontoxic substances). The general effect indices (GEIs) were also calculated by dividing the sum of environmental indices to the total mass indices obtained from the mass balance data (Table 9) of the dismantling process.

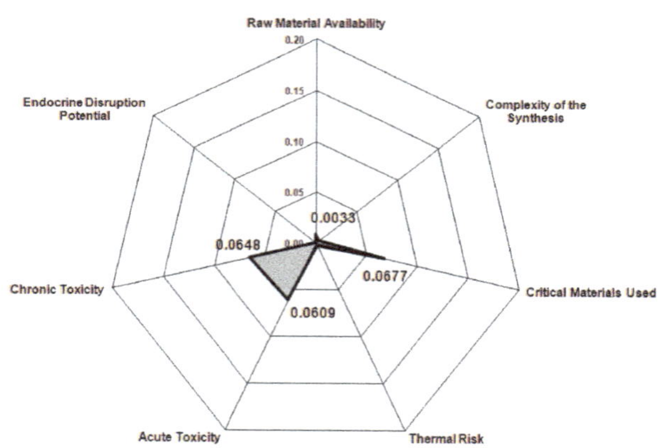

Figure 4. Environmental impact assessment for input streams.

It was found that the category "critical materials used" has the highest value due to the large amount of $FeCl_3$ feed into the process. Additionally, $FeCl_3$ is the main reason for the high chronic and acute toxicity of the input material stream. Among the other input streams, HCl and copper have the most important contributions to the results shown in Figure 4. The reason that copper surpasses other toxic components in the WPCBs, such as lead, is due to the fact that, even if it is less toxic than other components, it has the highest concentration. Similarly, for the output materials, chronic and acute toxicity remain important categories, but they are preceded by the "acidification potential" due to the depleted solution. Nevertheless, the GEIs values calculated for the input materials (0.064) is lower than for the output materials (0.039) indicating that the process diminishes the environmental impact of the WPCBs. Moreover, considering that both values are close to the minimum possible (0), it can be concluded that the dismantling process has an overall low environmental impact.

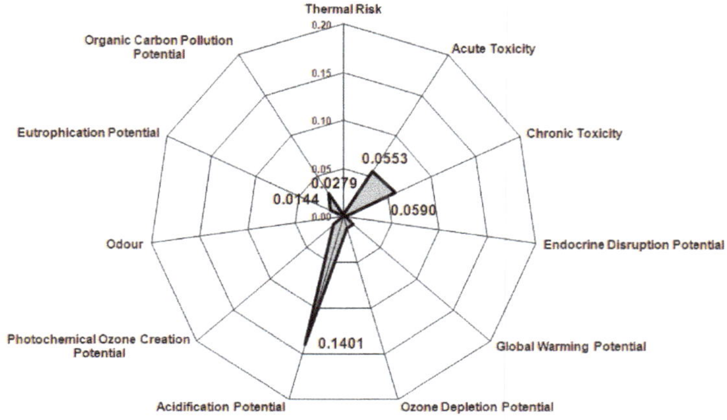

Figure 5. Environmental impact assessment for output streams.

4. Conclusions

The obtained results prove that the combined chemical–electrochemical process can be applied efficiently for the dismantling of different types of WPCBs with the parallel recovery of copper and regeneration of the leaching agent. As it was expected, in accordance with the redox equilibrium constants, all the metals, with the exception of gold, were dissolved during the leaching process. The experimental results demonstrate that the high purity (>99.95%) copper deposit can be produced with high current efficiency (72.69%) and low specific energy consumption (1.59 kWh/kg Cu). It is also important to note that the obtained copper deposits were compact and presented rough surfaces with larger nuclei and pyramidal growth, which is characteristic for copper deposition from chloride solutions. Modeling and simulating the conceptual recycling plant led to the conclusion that more than 90% of the copper can be extracted during the dismantling process by introducing a complementary step for the mechanical pretreatment of the boards, EC, and chips. It was also found that the dismantling of 1 kg of WPCBs requires 0.48 kWh, while the total specific energy consumption for copper production is 2.59 kWh/kg if the energy consumption of all process steps is considered. The overall energy balance revealed the possibility to use the process for a combined recycling–heating purpose, considering that it generates 47% more heat than the electrical power consumed. Based on the results of the environmental impact assessment, it can be concluded that the dismantling process can be performed not just with high technical performance but with low environmental impact as well. Still, further work is recommended to improve the technical performances and to assess the economic potential of the developed conceptual recycling plant.

Author Contributions: Conceptualization, S.F. and Á.I.-L.; methodology, S.F. and F.I.-L.; formal analysis, Á.I.-L.; investigation, F.I.-L., S.F. and Á.I.-L.; writing—original draft preparation, F.I.-L. and S.F.; writing—review and editing, S.F. and Á.I.-L.; supervision, S.F. All authors have read and agreed to the published version of the manuscript.

Funding: This research received no external funding.

Institutional Review Board Statement: Not applicable.

Informed Consent Statement: Not applicable.

Data Availability Statement: Data sharing is not applicable to this article.

Acknowledgments: We are grateful for the administrative and financial support offered by the Babes-Bolyai University.

Conflicts of Interest: The authors declare no conflict of interest.

References

1. Tan, Q.; Liu, L.; Yu, M.; Li, J. An innovative method of recycling metals in printed circuit board (pcb) using solutions from pcb production. *J. Hazard. Mater.* **2020**, *390*, 121892. [CrossRef] [PubMed]
2. Hao, J.; Wang, Y.; Wu, Y.; Guo, F. Metal recovery from waste printed circuit boards: A review for current status and perspectives. *Resour. Conserv. Recycl.* **2020**, *157*, 104787. [CrossRef]
3. Bressanelli, G.; Saccani, N.; Pigosso, D.C.A.; Perona, M. Circular economy in the weee industry: A systematic literature review and a research agenda. *Sustain. Prod. Consum.* **2020**, *23*, 174–188. [CrossRef]
4. Ning, C.; Hadi, P.; Aghdam, E.; Zhu, S.; Hui, D.C.-W.; Lin, C.S.K.; McKay, G. Environmental emission analysis of a waste printed circuit board-derived adsorbent production process. *Chem. Eng. J.* **2017**, *326*, 594–602. [CrossRef]
5. Rieger, T.; Oey, J.C.; Palchyk, V.; Hofmann, A.; Franke, M.; Hornung, A. Chemical recycling of weee plastics—Production of high purity monocyclic aromatic chemicals. *Processes* **2021**, *9*, 530. [CrossRef]
6. Shin, S.-R.; Mai, V.D.; Lee, D.-S. Chemical recycling of used printed circuit board scraps: Recovery and utilization of organic products. *Processes* **2019**, *7*, 22. [CrossRef]
7. Siddiqi, M.M.; Naseer, M.N.; Abdul Wahab, Y.; Hamizi, N.A.; Badruddin, I.A.; Hasan, M.A.; Zaman Chowdhury, Z.; Akbarzadeh, O.; Johan, M.R.; Kamangar, S. Exploring e-waste resources recovery in household solid waste recycling. *Processes* **2020**, *8*, 1047. [CrossRef]
8. Arya, S.; Patel, A.; Kumar, S.; Pau-Loke, S. Urban mining of obsolete computers by manual dismantling and waste printed circuit boards by chemical leaching and toxicity assessment of its waste residues. *Environ. Pollut.* **2021**, *283*, 117033. [CrossRef] [PubMed]
9. Cucchiella, F.; D'Adamo, I.; Rosa, P.; Terzi, S. Automotive printed circuit boards recycling: An economic analysis. *J. Clean. Prod.* **2016**, *121*, 130–141. [CrossRef]
10. Wang, R.; Zhu, Z.; Tan, S.; Guo, J.; Xu, Z. Mechanochemical degradation of brominated flame retardants in waste printed circuit boards by ball milling. *J. Hazard. Mater.* **2020**, *385*, 121509. [CrossRef]
11. Flandinet, L.; Tedjar, F.; Ghetta, V.; Fouletier, J. Metals recovering from waste printed circuit boards (wpcbs) using molten salts. *J. Hazard. Mater.* **2012**, *213-214*, 485–490. [CrossRef] [PubMed]
12. Cucchiella, F.; D'Adamo, I.; Lenny Koh, S.C.; Rosa, P. A profitability assessment of european recycling processes treating printed circuit boards from waste electrical and electronic equipments. *Renew. Sustain. Energy Rev.* **2016**, *64*, 749–760. [CrossRef]
13. Sarvar, M.; Salarirad, M.M.; Shabani, M.A. Characterization and mechanical separation of metals from computer printed circuit boards (pcbs) based on mineral processing methods. *Waste Manag.* **2015**, *45*, 246–257. [CrossRef] [PubMed]
14. Bizzo, W.A.; Figueiredo, R.A.; de Andrade, V.F. Characterization of printed circuit boards for metal and energy recovery after milling and mechanical separation. *Materials* **2014**, *7*, 4555–4566. [CrossRef] [PubMed]
15. Wang, H.; Zhang, S.; Li, B.; Pan, D.A.; Wu, Y.; Zuo, T. Recovery of waste printed circuit boards through pyrometallurgical processing: A review. *Resour. Conserv. Recycl.* **2017**, *126*, 209–218. [CrossRef]
16. Wang, J.; Guo, J.; Xu, Z. An environmentally friendly technology of disassembling electronic components from waste printed circuit boards. *Waste Manag.* **2016**, *53*, 218–224. [CrossRef] [PubMed]
17. Zhao, C.; Zhang, X.; Shi, L. Catalytic pyrolysis characteristics of scrap printed circuit boards by tg-ftir. *Waste Manag.* **2017**, *61*, 354–361. [CrossRef] [PubMed]
18. Birloaga, I.; Coman, V.; Kopacek, B.; Veglio, F. An advanced study on the hydrometallurgical processing of waste computer printed circuit boards to extract their valuable content of metals. *Waste Manag.* **2014**, *34*, 2581–2586. [CrossRef]
19. Szalatkiewicz, J. Metals recovery from artificial ore in case of printed circuit boards, using plasmatron plasma reactor. *Materials* **2016**, *9*, 683. [CrossRef]
20. Bacher, J.; Kaartinen, T. Liberation of printed circuit assembly (pca) and dust generation in relation to mobile phone design in a size reduction process. *Waste Manag.* **2017**, *60*, 609–617. [CrossRef] [PubMed]
21. Lee, J.; Kim, Y.; Lee, J.C. Disassembly and physical separation of electric/electronic components layered in printed circuit boards (pcb). *J. Hazard. Mater.* **2012**, *241*, 387–394. [CrossRef] [PubMed]
22. Khayyam Nekouei, R.; Maroufi, S.; Assefi, M.; Pahlevani, F.; Sahajwalla, V. Thermal isolation of a clean alloy from waste slag and polymeric residue of electronic waste. *Processes* **2020**, *8*, 53. [CrossRef]
23. Zhang, Y.; Ji, G.; Ma, D.; Chen, C.; Wang, Y.; Wang, W.; Li, A. Exergy and energy analysis of pyrolysis of plastic wastes in rotary kiln with heat carrier. *Process Saf. Environ. Prot.* **2020**, *142*, 203–211. [CrossRef]
24. Zhang, Y.; Ji, G.; Chen, C.; Wang, Y.; Wang, W.; Li, A. Liquid oils produced from pyrolysis of plastic wastes with heat carrier in rotary kiln. *Fuel Process. Technol.* **2020**, *206*, 106455. [CrossRef]
25. Kumar, A.; Holuszko, M.; Espinosa, D.C.R. E-waste: An overview on generation, collection, legislation and recycling practices. *Resour. Conserv. Recycl.* **2017**, *122*, 32–42. [CrossRef]
26. Silvas, F.P.; Correa, M.M.; Caldas, M.P.; de Moraes, V.T.; Espinosa, D.C.; Tenorio, J.A. Printed circuit board recycling: Physical processing and copper extraction by selective leaching. *Waste Manag.* **2015**, *46*, 503–510. [CrossRef] [PubMed]
27. Ippolito, N.M.; Medici, F.; Pietrelli, L.; Piga, L. Effect of acid leaching pre-treatment on gold extraction from printed circuit boards of spent mobile phones. *Materials* **2021**, *14*, 362. [CrossRef] [PubMed]
28. Cucchiella, F.; D'Adamo, I.; Lenny Koh, S.C.; Rosa, P. Recycling of weees: An economic assessment of present and future e-waste streams. *Renew. Sustain. Energy Rev.* **2015**, *51*, 263–272. [CrossRef]

29. Torretta, V.; Ragazzi, M.; Istrate, I.A.; Rada, E.C. Management of waste electrical and electronic equipment in two eu countries: A comparison. *Waste Manag.* **2013**, *33*, 117–122. [CrossRef]
30. Fogarasi, S.; Imre-Lucaci, F.; Ilea, P.; Imre-Lucaci, A. The environmental assessment of two new copper recovery processes from waste printed circuit boards. *J. Clean. Prod.* **2013**, *54*, 264–269. [CrossRef]
31. Fogarasi, S.; Imre-Lucaci, F.; Imre-Lucaci, Á.; Ilea, P. Copper recovery and gold enrichment from waste printed circuit boards by mediated electrochemical oxidation. *J. Hazard. Mater.* **2014**, *273C*, 215–221. [CrossRef]
32. Fogarasi, S.; Imre-Lucaci, F.; Egedy, A.; Imre-Lucaci, A.; Ilea, P. Eco-friendly copper recovery process from waste printed circuit boards using fe(3+)/fe(2+) redox system. *Waste Manag.* **2015**, *40*, 136–143. [CrossRef]
33. Fogarasi, S.; Imre-Lucaci, F.; Fogarasi, M.; Imre-Lucaci, A. Technical and environmental assessment of selective recovery of tin and lead from waste solder alloy using direct anodic oxidation. *J. Clean. Prod.* **2019**, *213*, 872–883. [CrossRef]
34. Cocchiara, C.; Dorneanu, S.-A.; Inguanta, R.; Sunseri, C.; Ilea, P. Dismantling and electrochemical copper recovery from waste printed circuit boards in h2so4–cuso4–nacl solutions. *J. Clean. Prod.* **2019**, *230*, 170–179. [CrossRef]
35. Imre-Lucaci, A.; Fogarasi, M.; Imre-Lucaci, F.; Fogarasi, S. Chemical-electrochemical process concept for lead recovery from waste cathode ray tube glass. *Materials* **2021**, *14*, 1546. [CrossRef]
36. Fogarasi, S.; Imre-Lucaci, F.; Varga, T.; Ilea, P. Eco-friendly leaching of base metals from waste printed circuit boards: Experimental study and mathematical modeling. *Studia UBB Chem.* **2012**, *LVII*, 91–100.
37. Hodisan, T.; Cimpoiu, C.; Haiduc, I.; Hodisan, S. *Teorie si Aplicatii in Chimia Analitica*; Risoprint: Cluj-Napoca, Romania, 2004.
38. Imre-Lucaci, F. *Electrochemical Methods for the Recovery of Copper from Waste Water and Solid Waste*; Babes-Bolyai University: Cluj-Napoca, Romania, 2011.
39. Chen, J.-Z.; Cao, H.-Z.; Li, B.; Yuan, H.-J.; Zheng, G.-Q.; Yang, T.-z. Thermodynamic analysis of separating lead and antimony in chloride system. *Trans. Nonferrous Met. Soc. China* **2009**, *19*, 730–734. [CrossRef]
40. Beverskog, B. *Pourbaix Diagrams for the System Copper-Chlorine at 5–100 °C*; SKI Rapport 98:19. Available online: https://inis.iaea.org/search/search.aspx?orig_q=reportnumber:%22SKI-R--98-19%22 (accessed on 9 September 2021).
41. Welham, N.J.; Kelsall, G.H.; Diaz, M.A. Thermodynamics of ag-cl-h2o, ag-br-h2o and ag-i-h2o systems at 298 k. *J. Electroanal. Chem.* **1993**, *361*, 39–47. [CrossRef]
42. Senanayake, G. Gold leaching in non-cyanide lixiviant systems: Critical issues on fundamentals and applications. *Miner. Eng.* **2004**, *17*, 785–801. [CrossRef]
43. Luyima, A.; Cui, W.; Heckman, C.; Moats, M.S. Examination of copper electrowinning smoothing agents. Part iv: Nucleation and growth of copper on stainless steel. *Miner. Metall. Process.* **2016**, *33*, 39–46. [CrossRef]
44. Imre-Lucaci, Á.; Nagy, M.; Imre-Lucaci, F.; Fogarasi, S. Technical and environmental assessment of gold recovery from secondary streams obtained in the processing of waste printed circuit boards. *Chem. Eng. J.* **2017**, *309*, 655–662. [CrossRef]

materials

MDPI

Article

Electronic Waste Low-Temperature Processing: An Alternative Thermochemical Pretreatment to Improve Component Separation

Juliana S. S. Oliveira, Ronald R. Hacha, Felipe S. d'Almeida, Caroline A. Almeida, Francisco J. Moura, Eduardo A. Brocchi and Rodrigo F. M. Souza *

Department of Chemical and Materials Engineering, Pontifical Catholic University of Rio de Janeiro, Rio de Janeiro 22451-900, RJ, Brazil; juliana.sette@yahoo.com.br (J.S.S.O.); rojas@esp.puc-rio.br (R.R.H.); felipe_dalmeida@hotmail.com (F.S.d.); caroline.azvdo@gmail.com (C.A.A.); moura@puc-rio.br (F.J.M.); ebrocchi@puc-rio.br (E.A.B.)

* Correspondence: rsouza@puc-rio.br

Abstract: The production of electronic waste due to technological development, economic growth and increasing population has been rising fast, pushing for solutions before the environmental pressure achieves unprecedented levels. Recently, it was observed that many extractive metallurgy alternatives had been considered to recover value from this type of waste. Regarding pyrometallurgy, little is known about the low-temperature processing applied before fragmentation and subsequent component separation. Therefore, the present manuscript studies such alternative based on scanning electron microscopy characterization. The sample used in the study was supplied by a local recycling center in Rio de Janeiro, Brazil. The mass loss was constant at around 30% for temperatures higher than 300 °C. Based on this fact, the waste material was then submitted to low-temperature processing at 350 °C followed by attrition disassembling, size classification, and magnetic concentration steps. In the end, this first report of the project shows that 15% of the sample was recovered with metallic components with high economic value, such as Cu, Ni, and Au, indicating that such methods could be an interesting alternative to be explored in the future for the development of alternative electronic waste extraction routes.

Keywords: low-temperature processing; WEEE (waste of electric and electronic equipment) recycling; materials separation; process characterization; SEM/EDS

Citation: Oliveira, J.S.S.; Hacha, R.R.; d'Almeida, F.S.; Almeida, C.A.; Moura, F.J.; Brocchi, E.A.; Souza, R.F.M. Electronic Waste Low-Temperature Processing: An Alternative Thermochemical Pretreatment to Improve Component Separation. *Materials* **2021**, *14*, 6228. https://doi.org/10.3390/ma14206228

Academic Editors: Rossana Bellopede and Lorena Zichella

Received: 2 September 2021
Accepted: 11 October 2021
Published: 19 October 2021

1. Introduction and Literature Review

The production of urban waste is becoming a subject of environmental pressure in modern society with impacts of sanitary, social, and financial importance [1–3]. The observed economic growth and population increase of recent decades have pushed for the manufacturing of more products and devices, intensifying issues associated with environmental pollution and depletion of natural resources, and are now driving initiatives related to the treatment of this class of wastes [4–7] In this context, the waste of electric and electronic equipment (WEEE) is considered a critical byproduct of urban lifestyles [8–10]. WEEE is an unconventional waste, typically with high metal content that is challenging to recycle based on traditional metallurgical processes [9].

WEEE has a wide variety of different components and devices, the most common being copper wire, batteries, structural components, LCD screens and printed circuit boards (PCBs). The mass percentage of the components (e.g., metal, polymers, and ceramics) varies a lot with the type of equipment and brand [11,12]. From a resource perspective, this type of waste has higher concentrations of metals than those found in the typical run-of-mines (ROM) [13], making WEEE recycling a possible secondary source of metals [14,15]. From an economical perspective, it was estimated that in 2017, WEEE accumulated a worldwide value of EUR 55 billion in raw material [16,17]. Thus, it is necessary to develop technological routes and managements policies to consolidate the recycling of WEEE to recover valuable materials [18]. High levels of metal recovery from WEEE have been reported and the costs associated with it are becoming more competitive but are still higher than those associated with mining operations [5,19]. In terms of economic potential in Brazil, a previous work from our research group estimated, based on a population survey and mass balance, that the stockpile value of devices in hibernation could be as high as USD 797 million [20].

In parallel to this context, PCBs are present in every electronic device, representing about 3 to 7% of the equipment's mass [21–24]. Computer-based PCBs are composed, essentially, of an epoxy resin or fiberglass coated with a thin layer of copper and are classified according to the composition of the insulator used. Fire-resistant material made of fiberglass and epoxy is the most used today [25]. In addition to Cu, PCBs contain a wide variety of metals, for example, Au, Pd, Ag and Ni [26]. However, hazardous metals such as Cd, Pb and Be may also be present [27]. Therefore, many kinds of research have been carried out to the recover precious metals and to the remove harmful elements/compounds [28–31]. In addition to that, there are also some environmental sustainability issues associated with the already-established routes which need to be faced in the coming years [32].

The combination of chemical and physical methods is a commonly used route in WEEE recycling. However, due to cost associated with chemical inputs and easy operation, most of the PCBs are processed using incineration and acid leaching, producing a substantial amount of hazardous emissions [33–35]. Moreover, poor WEEE disposal and processing could be related to the production of a variety of dangerous compounds, such as dioxins, that could be responsible for serious health issues [16,36,37]. Additionally, there are also concerns associated with persistent pollutants from the polymeric contents of the WEEE [10], such as polycyclic aromatic hydrocarbons [38], polychlorinated biphenyls [39,40], polychlorinated dibenzo-p-dioxins and dibenzofurans [40,41], and brominated and organophosphate compounds [39,42]. In this perspective, low-temperature processing-related processes could be regarded as appealing options to deal with these risks [23,32,43].

Typically, regarding thermal processing, most WEEE recycling proposals start with physical beneficiation, essentially disassembling and grinding the PCBs samples [23,44]. This mimics a typical extractive metallurgy approach, with the same energy-intensive requirements to reduce the particle size. Another disadvantage of this type of physical processing is the fact that it is not possible to obtain a pure material in its present form, and the consequent production of fine comminution powder [45]. These two conditions present a challenge to the cost. According to Quan et al. (2010) [27], excessive fragmentation limits the recovery of fiberglass and can significantly increase metal losses while imposing a high energy consumption for the operation of the fragmentation equipment due to the high hardness of the PCBs. Nevertheless, the high-temperature processing routes generate products that can be recovered and reused [23,27,46,47]. The review by Ambaye et al. (2020) [48] showed that WEEE recycling by means of pyrometallurgy is an energy-intensive alternative, majorly focused on copper recovery. Moreover, it seems that little is known on the effect of pyrometallurgical processes being applied before the initial physical processing, particularly regarding the investigation of low-temperature processing effects on PCB constituent separation. For instance, Ma et al. (2018) [49] assessed this type of processing from a heat transfer perspective while Guo et al. (2014) [50] dealt with calorific capacities of a PCB sample. In praxis, the typical pyrometallurgical processes are furnace smelting and alkali fusion, according to Chauhan et al. (2018) [35] and Ding et al. (2019) [51]. Additionally, thermal processing of WEEE is regarded as a promising alternative for recycling the non-metallic fractions [52].

Therefore, the present work has the motivation of providing a first look into such alternatives, to offer conditions for easy-to-implement physical disassembling without major particle size reduction. The technical support of such proposition is related to previous observations in which the thermal processing was investigated to produce a solid, free of the volatile organic fraction, to hydrometallurgical leaching without excessive fragmentation [53]. The possibility of a processing alternative without the prior comminution, bypassing the initial physical beneficiation step, was also reported for PCBs with the chemical characterization of the oil-based resin [27].

The present study has the interest of exploring some of the material's behavior in a pyrometallurgical process through electron microscopy characterization of the resulting materials, as it seems that the most relevant effort in material characterization has been reported for hydrometallurgical or for pure mineral processing approaches [54–56].

The technological context of this proposition is associated with the present Brazilian context, in which some important initiatives towards WEEE collection, disassembling and parts recycling can be observed but with little advance in material recovery through chemical processing [57–60]. According to Nithya et al. [61] Brazil is among the top five 2019 WEEE producers after China, USA, India and Japan, producing more than 2 million tons per year [62]. Moreover, it is also recognized as a trans-boundary destination of electronic wastes with lack of proper infrastructure related to waste management [63]. Dias et al. [64] present an alarming scenario in which the Brazilian recycling system operates towards valuable constituent concentration and undertakes shipping abroad to further processing of materials.

Under this perspective, the present manuscript's purpose is related to the investigation of the PCBs material's behavior in a low temperature pyrometallurgical processing operation, prior to physical fragmentation based on scanning electron microscopy (SEM/EDS) characterization. The study also covers a thermogravimetric analysis (TGA) to identify the lower temperature in which the process could be carried out to provide material disassembling without major fragmentation.

2. Materials and Methods

Samples of connectors from hard-disk drives (HD) and random-access memories (RAM) were received from a local recycling center in Rio de Janeiro, Brazil, that operates with the collection, disassembly, and parts separation for reuse. The samples were collected from the supplier stockpile of non-recyclable parts and locally processed to concentrate valuable metallic content.

After receiving the sample, a visual classification was carried out and three patterns of connectors that were categorized as Type 1, Type 2 and Type 3 varieties. In terms of the visual distinctions of each variety, it was observed that Type 1 and 2 were presented with a more distinguished goldish yellow with differences in the morphology of the metallic parts. On the other hand, Type 3 presents a pale gold color with similarities in contact shape with Type 2. Figure 1 shows the macroscopic features of the received material. The WEEE samples, as well as all the solid materials produced in this study, were characterized using scanning electron microscopy coupled with energy-dispersive X-ray spectroscopy (SEM/EDS), using a Hitachi TM3000 microscope (Hitachi, Tokyo, Japan) connected with an Oxford Swift ED3000 microanalysis system. The detection mode for SEM is Backscattered Electrons. The thermogravimetric analysis (TGA) was carried out on NETZSCH STA 449 F3 Jupiter equipment (NETZSCH, Selb, Germany) using a 10 K·min^{-1} heating from room temperature until 1000 °C. The WEEE samples had their thermal behavior evaluated under a chemically inactive as well as oxidizing atmosphere. The former was conducted with high-purity nitrogen (99.98%) while the latter was taken into effect with a synthetic air mixture composed of 80% N_2 and 20% O_2. Both gas mixtures were manufactured by Linde company (Dublin, Ireland). The applied flowrate entering the reaction chamber was fixed at 20 L·min^{-1}.

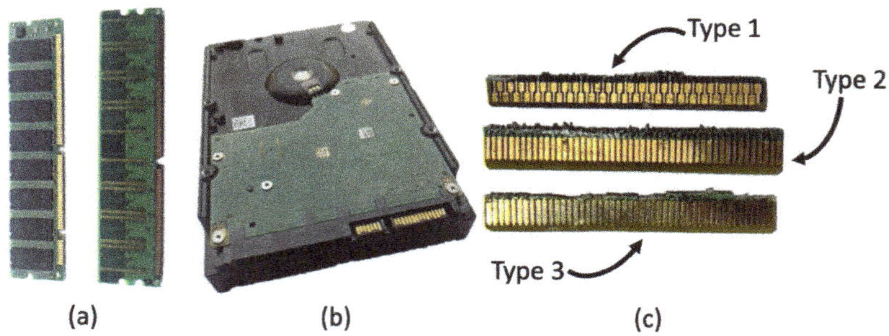

Figure 1. Macroscopic features of the received WEEE sample: (**a**) RAM; (**b**) HD; (**c**) Varieties of contacts from RAM and HD as received.

After the WEEE microscopical and thermal characterizations, samples were submitted to the chemical process of low-temperature processing, physical disassembling with glass bodies, size classification and magnetic separation. Figure 2 presents a schematic representation of the proposed route to process WEEE samples and recover valuable metals. The sequence of unit operation was defined to avoid major shredding and fine powder formation, as the pyrometallurgical process could provide the volatilization of chemicals responsible for the structural integrity of the PCBs.

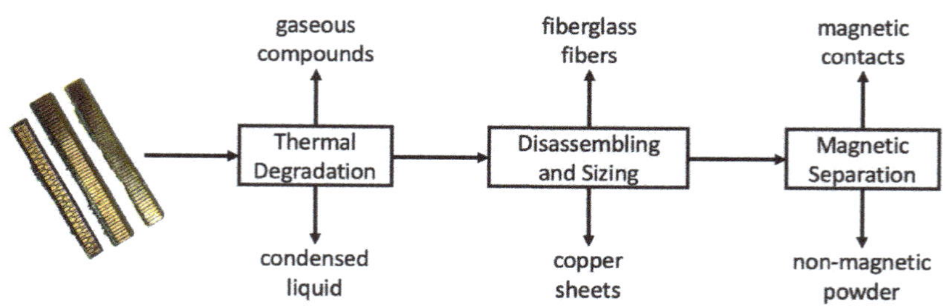

Figure 2. The proposed chemical processing route for the received WEEE samples.

Samples of WEEE were submitted to isothermal pyrometallurgical processing in a tubular furnace. The experiments were conducted in compressed air (incineration) and ultrapure argon (99.998%) atmosphere (inert processing). The latter was also supplied by Linde. It was defined that the WEEE samples would be accommodated at room temperature inside the furnace and then heated until the desired temperature was reached. The reaction time for processing was fixed at 60 min for the defined process temperature. After thermal degradation, the solid products were cooled down until 80 °C and then removed from the furnace.

The disassembling operations were conducted with a hand-operated mill using glass pebbles as friction bodies. This option was taken on purpose as it would exemplify how easily the physical detachment of constituents occurs after high-temperature processing without deleterious effects on components liberation. The produced particulate system was then collected and classified by size, using sieves with openings of 4.75, 2.80, 1.4, 0.71, 0.50, 0.21 and 0.18 mm. Finally, the small-sized fraction was exposed to a magnetic field from a hand-magnet, again because of the simplicity of the operation. The recovered materials were then characterized by SEM/EDS to assess the performance of the proposed route as an alternative to separating some of the constituents.

3. Results and Discussion

3.1. Characterization of WEEE Samples

Samples of the three varieties of WEEE were submitted to SEM/EDS to identify the chemical distribution in the PCB through a semiquantitative approach. The detection mode for this study is related to Backscattered Electrons. In this context, Figure 3 presents such results for a sample of the Type 1 variety. It can be observed that as expected the metallic area of the PCB features a bright shade of grey, associated with a higher average atomic number, while the dark area is associated with the polymeric material. Some clouds of dark structures could also be observed over the bright area, possibly associated with oxidation of the metallic parts, possibly associated with nickel and copper. It is noteworthy to mention that the image and the respective EDS results are mainly associated with the surface area of the sample. In the dark area, a major presence of carbon, oxygen, and bromine can be observed, while in the bright area, gold stands as the major constituent, at least at the sample surface. These are interesting and expected findings as brominated compounds are used in this context as flame retardants, while gold layers on the surface of the metallic contacts are responsible for increasing efficient current transmission.

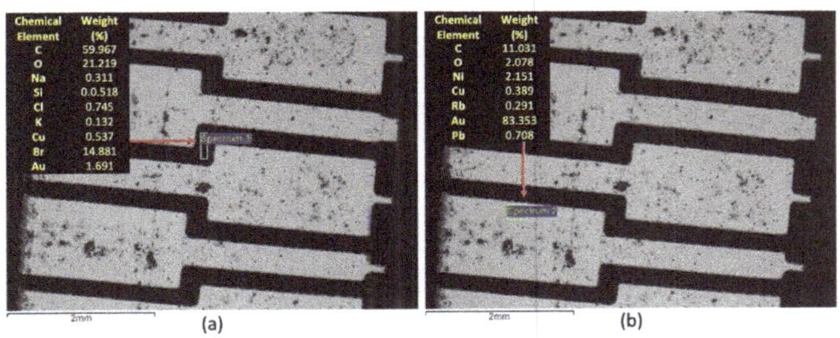

Figure 3. SEM/EDS characterization of Type 1 WEEE sample: (**a**) microanalysis of the dark area; (**b**) microanalysis of the bright area.

However, the amounts of other valuable metals, such as copper (in the inner contact) and aluminum (in the fiberglass) could only be qualitatively assessed by utilizing a chemical mapping that analyses the composition distribution deeper in the sample. Such analysis is presented in Figure 4.

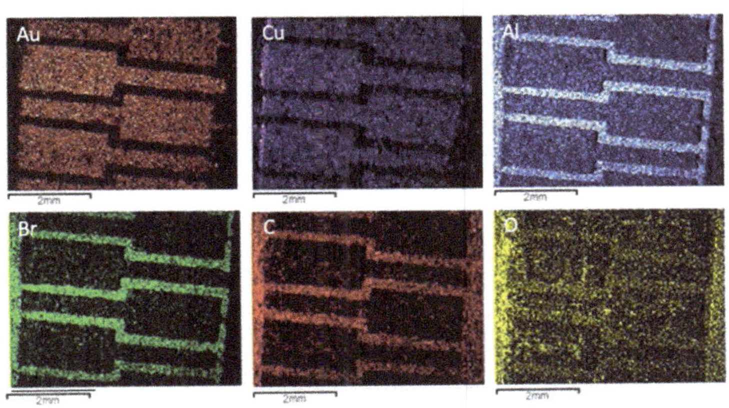

Figure 4. Chemical mapping of the most relevant elements detected in the microanalysis of the dark and bright areas of the Type 1 WEEE sample.

Figure 5 presents the SEM/EDS analysis while Figure 6 shows the qualitative chemical mapping of the Type 2 variety of WEEE in the received sample. It can be observed that Type 2 follows Type 1 in terms of the overall chemical composition of the bright (metallic) and darker (polymeric) areas, with a cleaner surface regarding the presence of clouds potentially associated with metal oxidation. Comparing the two varieties, at a semiquantitative level, the composition in both areas in each sample is relatable.

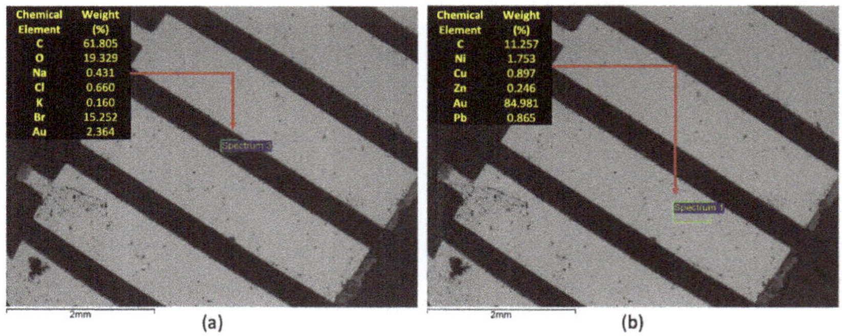

Figure 5. SEM/EDS characterization of Type 2 WEEE sample: (**a**) microanalysis of the dark area; (**b**) microanalysis of the bright area.

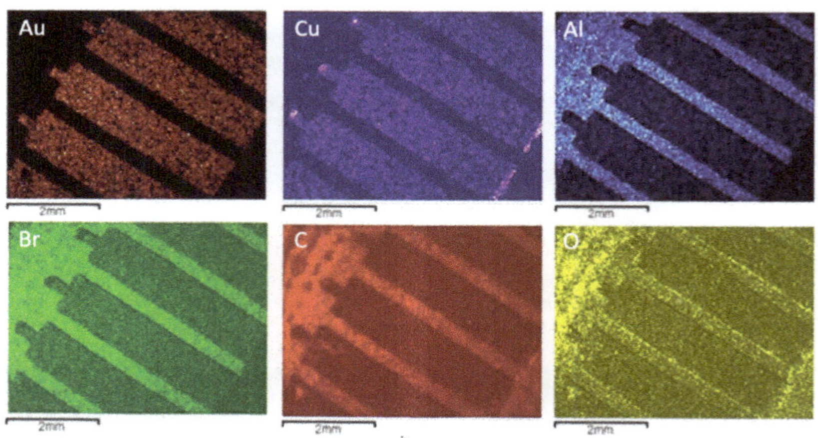

Figure 6. Chemical mapping of the most relevant elements detected in the microanalysis of the dark and bright areas of the Type 2 WEEE sample.

On the other hand, the Type 3 variety, with a paler metallic hue, does not follow the other two in terms of composition and surface integrity, as shown in Figure 7. It was verified that the samples of this variety present a similar overall composition for the dark (polymeric) area as the one observed in the previous cases, but with a larger number of elements in the lower range of relevance. For the bright (metallic) area, the distinction is clear, with nickel as the major metallic constituent and with gold still present at an important level. The dark clouds were most present in this variety, which could be related to lower levels of Au in the surface.

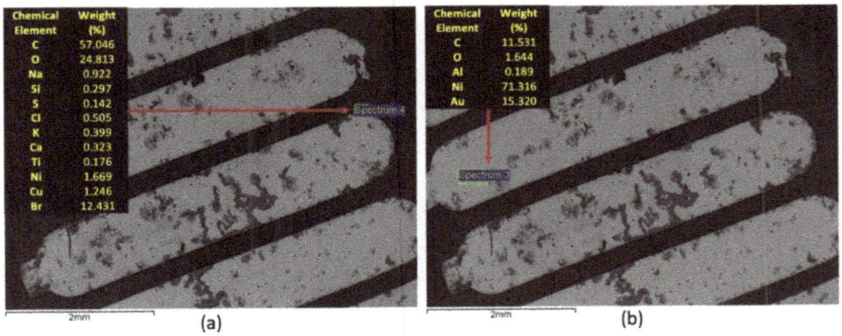

Figure 7. SEM/EDS characterization of Type 3 WEEE sample: (**a**) microanalysis of the dark area; (**b**) microanalysis of the bright area.

Following the same approach, Figure 8 displays the chemical mapping of the Type 3 variety. The major qualitative difference between this variety and the others is related to the lower presence of gold and the higher distribution of oxygen, which corroborates the behavior associated with the more distinguished dark clouds observed in Figure 7.

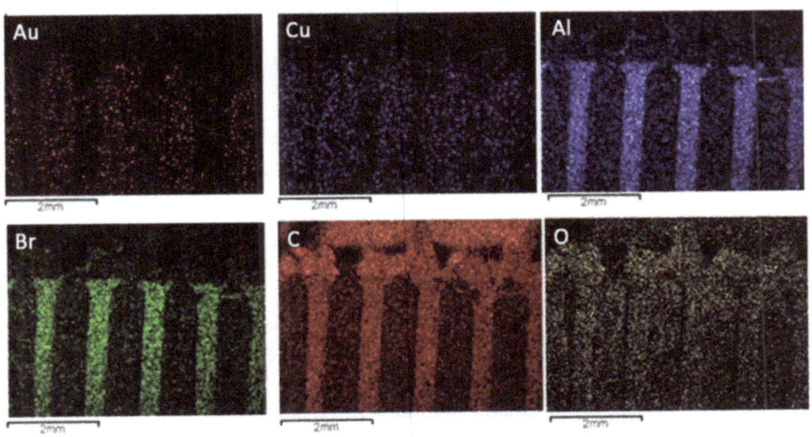

Figure 8. Chemical mapping of the most relevant elements detected in the microanalysis of the dark and bright areas of the Type 3 WEEE sample.

In parallel to the SEM/EDS characterization, samples of each of one of the three varieties of WEEE were also submitted to non-isothermal TGA in oxidizing and chemical inert conditions, as presented in Figure 9. In praxis, the same thermal behavior as a function of temperature regardless of PCB type and reaction atmosphere was verified. The total weight variation indicates a mass loss of 30%. Regarding the chemical environment, at N_2 atmosphere, carbon and hydrogen were volatilized as organic compounds, possibly carrying flame-retardant components, whereas in the oxidative experiment these elements were possibly being transported to the gas phase as oxidized compounds such as water, monoxide, and carbon dioxide [65]. The observed degradation temperature is in accordance with the presence of thermoplastic materials in the sample, at about 250 °C [27].

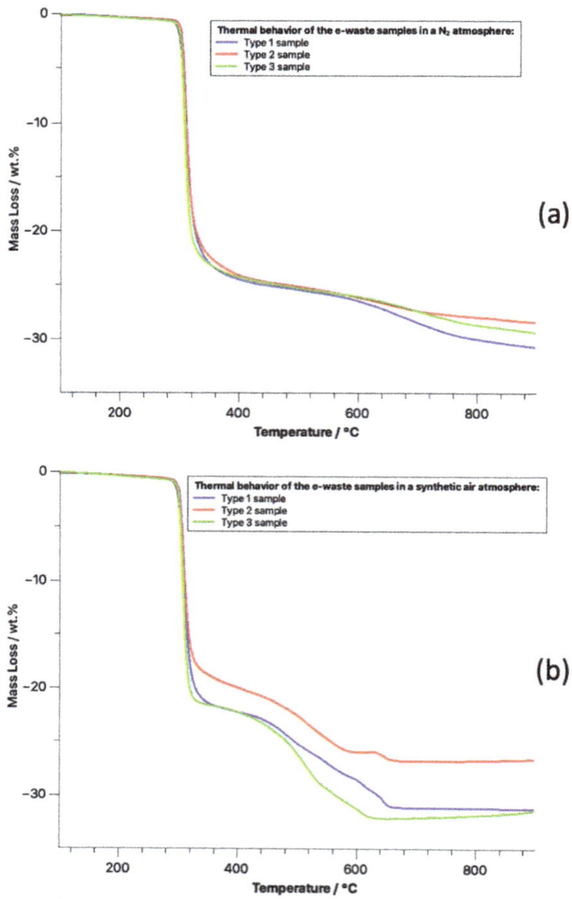

Figure 9. Thermal behavior characterization of WEEE sample: (**a**) in an inert atmosphere; (**b**) in an oxidizing atmosphere.

Both materials' characterization results indicate similarities between the three varieties, particularly regarding the polymeric fraction, and therefore to establish the chemical process more simply, the thermal processing of the received sample was considered without the variety distinction, to remove at least a fraction of the organic phase and to liberate constituents. Consequently, the thermochemical processing of the WEEE samples was carried-out for the material in the same condition as it was received from the local recycling center, without any classification, in a tubular furnace above 300 °C.

3.2. *Thermal Processing of WEEE Samples*

Table 1 shows the experimental results associated with the pyrometallurgical processing of 2.5 g of PCB connector at 300 and 400 °C, using compressed air (incineration) and argon (inert processing). As expected, the weight loss was again around 30%. However, at the inert gas atmosphere, the formation of some droplets of a black liquid at the far end of the tubular furnace were observed. This was interesting, and an indicative of the volatilization and condensation of the organic fraction, as reported previously by other authors [27,53].

Table 1. Observed WEEE samples' mass loss after thermal processing in a tubular furnace as a function of the temperature and the atmospheric chemical composition.

Temperature above Sample in the Furnace (°C)	Mass Loss in a Compressed-Air Atmosphere (wt.%)	Mass Loss in an Ultrapure Argon Atmosphere (wt.%)
300	31.5	28.5
400	32.0	30.1

To produce more liquid and to generate more solid material, another thermal degradation experiment at 350 °C was carried out in argon using 54.5 g of the WEEE sample. The reaction time of 60 min was also applied to this test. A weight loss of 30.1% was observed, resulting in a solid weight of 38.1 g. In this context, Figure 10 presents the macroscopic aspect variation before and after thermal degradation. It is noteworthy that some substance has been removed from PCBs, exposing the metallic compounds, some copper sheets as well as the inner glass fibers, now covered with some dark material, possibly carbon black.

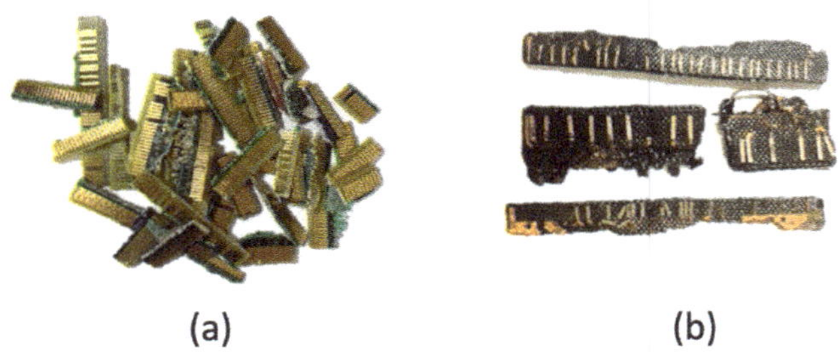

Figure 10. Macroscopic aspects of the WEEE samples: (**a**) before thermal processing; (**b**) after thermal processing.

The disassembling and sizing unit operation was applied to the solid product, and, in this context, Table 2 presents the size classification of the obtained material after these operations.

Table 2. Size classification of the thermal processing solid product after grinding operation using glass pebbles as friction agents for materials disassembling.

Sieve Opening (mm)	Retained Mass (g)	Retained Mass (wt.%)
4.75	30.67	80.71
2.80	0.77	2.03
1.40	0.67	1.76
0.71	4.20	11.05
0.50	0.75	1.97
0.21	0.41	1.08
0.18	0.29	0.76
Bottom	0.24	0.63
Total	38.00	100.00

It can be observed that most of the size classification is associated with large particulate material. A total of 80.71% is associated with glass fibers (17.4 g), copper sheets (1.3 g) and non-liberated material (12.0 g). The fact that 31.58% of the sample is associated with non-liberated material indicates that the disassembling unit operation could be the subject of future developments, to optimize larger recovery of metals. Moreover, the non-liberated material could also be submitted to hydrometallurgical processes as most of the organic phase has been removed from it, in a roast–leach type of route. Material below 2.80 mm was submitted to magnetic separation. It was observed that 4.7 g was susceptible to the effects of the magnetic field and recovered easily. Therefore, it can be said that 12.4% of the solid product is composed of magnetic-metal-containing materials. Figure 11 illustrates the macroscopic aspect of the most relevant materials recovered.

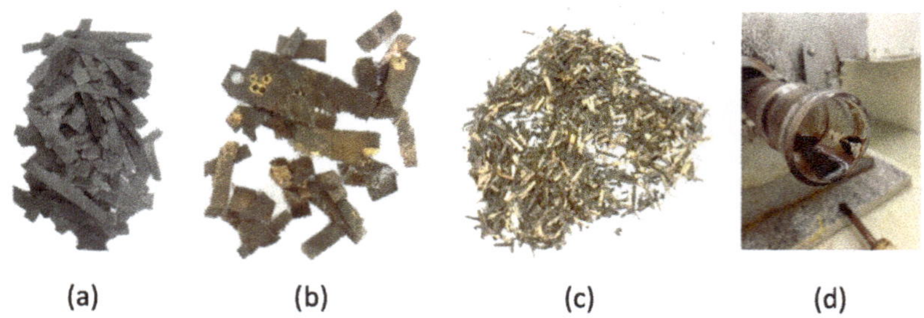

Figure 11. Macroscopic aspects of some relevant materials recovered in the process: (**a**) fiberglass; (**b**) copper sheets; (**c**) magnetic contacts; (**d**) condensed liquid.

3.3. *Characterization of the Thermal Processing Products*

The recovered glass fiber was characterized utilizing SEM/EDS and its results are presented in Figure 12. It is possible to note that Si and Ca are the major metals in the fiberglass while oxygen is the major overall component. The preeminent levels of C also followed the carbon-black-deposition expectations. Some minor contents of Al, Ti and Cu were also detected. Some minor, bright, particulate material within the knitting pattern can be observed. Since such elements are heavier than C, and combined with the Backscattered Electrons detection mode, it is possible to affirm that some metallic powder is also being transported with the fiberglass.

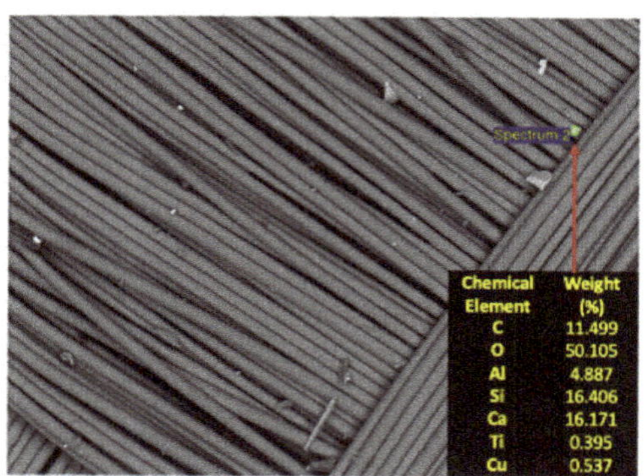

Figure 12. SEM/EDS characterization of the fiberglass recovered after the size separation.

Figure 13 is associated with the copper sheets recovered in this study, presenting the SEM/EDS analysis as well as the chemical mapping for the same area. The morphological aspect of the image indicates some deposition of organic matter (dark scales) over the copper sheet (bright area). This is indicative that some fraction of the polymeric material is not being transferred to the gas phase in the thermal degradation. The qualitative chemical mapping clarifies that suggestion, as copper is the major component of this material while the scales have carbon and bromine in their composition. This context indicates that temperature and reaction time could also be optimized to separate copper from non-metals and flame-retardant elements.

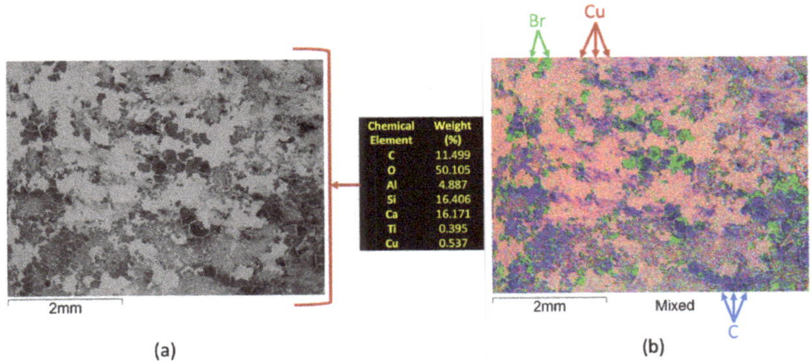

Figure 13. SEM/EDS characterization of the copper sheets recovered after the size separation: (**a**) Backscattered Electrons Image with microanalysis; (**b**) chemical mapping.

Figure 14 presents the SEM/EDS analysis of the recovered magnetic connector. It was verified, through the EDS microanalysis, that the Cu, Ni and Au are being recovered in this material (Figure 14a). Without size reduction to powder, some non-magnetic metals remain linked to nickel providing its concentration being subjected to a magnetic field. These elements are the main constituents of the bright area of the connector that covers the inner layer of copper (Figure 14b). The presence of carbon is also perceptible, both in the EDS microanalysis as well as through the dark matter over the metallic phase.

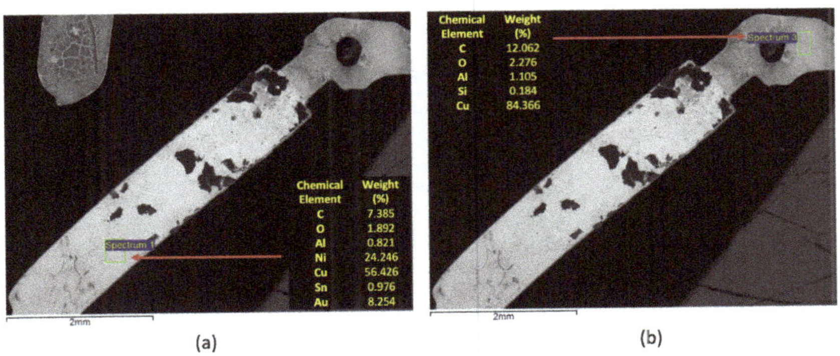

Figure 14. SEM/EDS characterization of the metallic contacts recovered after the size separation and magnetic separation: (**a**) microanalysis of the surface layer; (**b**) microanalysis of the inner contact.

4. Final Remarks

In short, the proposed route was established as an alternative for processing PCB samples and to valorize them, concentrating metals, separating these from some pf the ceramic content and organic substances. The process produced a black liquid which condensates from the gas phase, therefore diminishing atmospheric emissions and possibly providing a source of materials in some technological applications.

The SEM/EDS analyses showed that the support for the alloy connectors was composed of carbon and oxygen, which indicates, as expected, a typical polymeric material. The presence of bromine, related to flame-retardant additives, was also observed, as were small fractions of metals. The chemical composition for this part of the PCBs samples does not change between varieties. Analyses also showed that the metal connectors have some compositional distinctions between the three types. It was observed that all three samples contain gold and nickel in the surface. The main difference is the relative amounts, with Au as the main metallic constituent for Types 1 and 2 while it is Ni for Type 3. The chemical mapping indicated a clear presence of copper in the inner layers of the connectors. The distribution of chemical elements is also clear following the SEM/EDS approach.

The TGA analyses showed a mass loss of 30%. The observed degradation temperature of about 250 °C is related to the presence of thermoplastic materials. This behavior indicates that all three samples could be processed simultaneously to volatilize the organic fraction. The tubular reactor processing also showed a mass loss of around 30% in weight. The formation of black liquid under inert processing condition was also observed. The characterization of the produced fluid will be assessed in future developments.

It is interesting to observe that combing all metallic contacts in a single fraction in the magnetic separation step is a significant contribution and a simple alternative for metals separation. It should be noted that these characterization and processing experiments could contribute to future recycling route development, lowering environmental impacts associated with this type of WEEE.

Author Contributions: Conceptualization, F.J.M. and R.F.M.S.; methodology, J.S.S.O., R.R.H., F.J.M. and R.F.M.S.; formal analysis, J.S.S.O., R.R.H., F.S.d. and C.A.A.; investigation, J.S.S.O., F.S.d. and C.A.A.; resources, F.J.M., E.A.B. and R.F.M.S.; data curation, J.S.S.O.; writing—original draft preparation, F.J.M. and R.F.M.S.; writing—review and editing, F.S.d., C.A.A. and R.F.M.S.; supervision, F.J.M. and R.F.M.S.; project administration, F.J.M. and R.F.M.S.; funding acquisition, E.A.B. and R.F.M.S.; All authors have read and agreed to the published version of the manuscript.

Funding: The APC was funded by Fundação Carlos Chagas Filho de Amparo à Pesquisa do Estado do Rio de Janeiro (FAPERJ)–Project "Apoio para publicações científicas em revistas qualificadas e de open access, priorizando a participação de alunos" (Edital 05/2020).

Institutional Review Board Statement: Not applicable.

Informed Consent Statement: Not applicable.

Data Availability Statement: Data will be made available based on request to the authors.

Acknowledgments: This study was financed in part by the Coordenação de Aperfeiçoamento de Pessoal de Nível Superior—Brasil (CAPES)—Financial Code 001. The authors are also grateful for the support throughout the research to VRAc/PUC-Rio (Vice-Reitoria para Assuntos Acadêmicos) and CNPq (Conselho Nacional de Desenvolvimento Científico e Tecnológico).

Conflicts of Interest: The authors declare that they have no known competing financial interests or personal relationships that could have appeared to influence the work reported in this paper.

References

1. Margallo, M.; Ziegler, K.; Vázquez-Rowe, I.; Aldaco, R.; Irabien, A.; Kahhat, R. Enhancing waste management strategies in Latin America under a holistic environmental assessment perspective: A review for policy support. *Sci. Total Environ.* **2019**, *689*, 1255–1275. [CrossRef]
2. Zorpas, A.A. Strategy development in the framework of waste management. *Sci. Total Environ.* **2020**, *716*, 137088. [CrossRef]
3. Van Fan, Y.; Klemeš, J.J.; Walmsley, T.; Bertók, B. Implementing circular economy in municipal solid waste treatment system using P-graph. *Sci. Total Environ.* **2020**, *701*, 134652. [CrossRef]
4. Dutta, T.; Kim, K.-H.; Uchimiya, M.; Kwon, E.E.; Jeon, B.-H.; Deep, A.; Yun, S.-T. Global demand for rare earth resources and strategies for green mining. *Environ. Res.* **2016**, *150*, 182–190. [CrossRef]
5. Tansel, B. From electronic consumer products to e-wastes: Global outlook, waste quantities, recycling challenges. *Environ. Int.* **2017**, *98*, 35–45. [CrossRef]
6. Xiao, S.; Dong, H.; Geng, Y.; Brander, M. An overview of China's recyclable waste recycling and recommendations for integrated solutions. *Resour. Conserv. Recycl.* **2018**, *134*, 112–120. [CrossRef]
7. Breukelman, H.; Krikke, H.; Löhr, A. Failing services on urban waste management in developing countries: A review on symptoms, diagnoses, and interventions. *Sustainability* **2019**, *11*, 6977. [CrossRef]
8. Widmer, R.; Oswald-Krapf, H.; Sinha-Khetriwal, D.; Schnellmann, M.; Böni, H. Global perspectives on e-waste. *Environ. Impact Assess. Rev.* **2005**, *25*, 436–458. [CrossRef]
9. Kaya, M. Recovery of metals and nonmetals from electronic waste by physical and chemical recycling processes. *Waste Manag.* **2016**, *57*, 64–90. [CrossRef] [PubMed]
10. Sahajwalla, V.; Gaikwad, V. The present and future of e-waste plastics recycling. *Curr. Opin. Green Sustain. Chem.* **2018**, *13*, 102–107. [CrossRef]

11. Abu Bakar, M.S.; Rahimifard, S. Ecological and economical assessment of end-of-life waste recycling in the electrical and electronic recovery sector. *Int. J. Sustain. Eng.* **2008**, *1*, 261–277. [CrossRef]
12. Parajuly, K.; Fitzpatrick, C. Understanding the impacts of transboundary waste shipment policies: The case of plastic and electronic waste. *Sustainability* **2020**, *12*, 2412. [CrossRef]
13. Hsu, E.; Barmak, K.; West, A.C.; Park, A.-H.A. Advancements in the treatment and processing of electronic waste with sustainability: A review of metal extraction and recovery technologies. *Green Chem.* **2019**, *21*, 919–936. [CrossRef]
14. Zhang, L.; Xu, Z. A review of current progress of recycling technologies for metals from waste electrical and electronic equipment. *J. Clean. Prod.* **2016**, *127*, 19–36. [CrossRef]
15. Isildar, A.; Rene, E.R.; van Hullebusch, E.D.; Lens, P.N.L. Electronic waste as a secondary source of critical metals: Management and recovery technologies. *Resour. Conserv. Recycl.* **2018**, *135*, 296–312. [CrossRef]
16. Bakhiyi, B.; Gravel, S.; Ceballos, D.; Flynn, M.A.; Zayed, J. Has the question of e-waste opened a Pandora's box? An overview of unpredictable issues and challenges. *Environ. Int.* **2018**, *110*, 173–192. [CrossRef]
17. Kuehr, R. Waste electrical and electronic equipment. In *Waste—A Handbook for Management*, 2nd ed.; Letcher, T.M., Vallero, D.A., Eds.; Elsevier: Bonn, Germany, 2019; pp. 477–487.
18. Cocchiara, C.; Dorneanu, S.-A.; Inguanta, R.; Sunseri, C.; Ilea, P. Dismantling and electrochemical copper recovery from Waste Printed Circuit Boards in H_2SO_4–$CuSO_4$–NaCl solutions. *J. Clean. Prod.* **2019**, *230*, 170–179. [CrossRef]
19. Zeng, X.; Mathews, J.A.; Li, J. Urban mining of e-waste is becoming more cost-effective than virgin mining. *Environ. Sci. Technol.* **2018**, *52*, 4835–4841. [CrossRef]
20. D'Almeida, F.S.; de Carvalho, R.B.; dos Santos, F.S.; de Souza, R.F.M. On the hibernating electronic waste in Rio de Janeiro higher education community: An assessment of population behavior analysis and economic potential. *Sustainability* **2021**, *13*, 9181. [CrossRef]
21. Bernardes, A.; Bohlinger, I.; Rodriguez, D.; Milbrandt, H.; Wuth, W. Recycling of printed circuit boards by melting with oxidising/reducing top blowing process. In Proceedings of the Extraction and Processing Division Symposium, TMS Annual Meeting, Orlando, FL, USA, 9–13 February 1997; Volume 126, pp. 363–375.
22. Hadi, P.; Gao, P.; Barford, J.; McKay, G. Novel application of the nonmetallic fraction of the recycled printed circuit boards as a toxic heavy metal adsorbent. *J. Hazard. Mater.* **2013**, *252–253*, 166–170. [CrossRef]
23. Wang, H.; Zhang, S.; Li, B.; Pan, D.; Wu, Y.; Zuo, T. Recovery of waste printed circuit boards through pyrometallurgical processing: A review. *Resour. Conserv. Recycl.* **2017**, *126*, 209–218. [CrossRef]
24. Vučinić, A.A.; Bedeković, G.; Šarc, R.; Premur, V. Determining metal content in waste printed circuit boards and their electronic components. *J. Sustain. Dev. Energy Water Environ. Syst.* **2020**, *8*, 590–602. [CrossRef]
25. Urey, H.; Holmstrom, S.; Yalcinkaya, A.D. Electromagnetically actuated FR4 scanners. *IEEE Photonics Technol. Lett.* **2007**, *20*, 30–32. [CrossRef]
26. Rao, M.D.; Singh, K.K.; Morrison, C.A.; Love, J.B. Challenges and opportunities in the recovery of gold from electronic waste. *RSC Adv.* **2020**, *10*, 4300–4309. [CrossRef]
27. Quan, C.; Li, A.; Gao, N.; Dan, Z. Characterization of products recycling from PCB waste pyrolysis. *J. Anal. Appl. Pyrolysis* **2010**, *89*, 102–106. [CrossRef]
28. Hao, J.; Wang, Y.; Wu, Y.; Guo, F. Metal recovery from waste printed circuit boards: A review for current status and perspectives. *Resour. Conserv. Recycl.* **2020**, *157*, 104787. [CrossRef]
29. Zhu, J.; Chen, X.; Zhao, N.; Wang, W.; Du, J.; Ruan, J.; Xu, Z. Bromine removal from resin particles of crushed waste printed circuit boards by vacuum low-temperature heating. *J. Clean. Prod.* **2020**, *262*, 121390. [CrossRef]
30. Baniasadi, M.; Graves, J.; Ray, D.A.; De Silva, A.L.; Renshaw, D.; Farnaud, S. Closed-loop recycling of copper from waste printed circuit boards using bioleaching and electrowinning processes. *Waste Biomass Valoriz.* **2021**, *12*, 3125–3136. [CrossRef]
31. Charitopoulou, M.A.; Kalogiannis, K.G.; Lappas, A.A.; Achilias, D.S. Novel trends in the thermo-chemical recycling of plastics from WEEE containing brominated flame retardants. *Environ. Sci. Pollut. Res.* **2020**, 1–24. [CrossRef]
32. Debnath, B.; Chowdhury, R.; Ghosh, S.K. Sustainability of metal recovery from E-waste. *Front. Environ. Sci. Eng.* **2018**, *12*, 2. [CrossRef]
33. Li, J.; Lu, H.; Guo, J.; Xu, Z.; Zhou, Y. Recycle technology for recovering resources and products from waste printed circuit boards. *Environ. Sci. Technol.* **2007**, *41*, 1995–2000. [CrossRef]
34. Priya, A.; Hait, S. Comparative assessment of metallurgical recovery of metals from electronic waste with special emphasis on bioleaching. *Environ. Sci. Pollut. Res.* **2017**, *24*, 6989–7008. [CrossRef] [PubMed]
35. Chauhan, G.; Jadhao, P.R.; Pant, K.; Nigam, K. Novel technologies and conventional processes for recovery of metals from waste electrical and electronic equipment: Challenges & opportunities—A review. *J. Environ. Chem. Eng.* **2018**, *6*, 1288–1304. [CrossRef]
36. Zhang, Y.; Xu, X.; Chen, A.; Davuljigari, C.B.; Zheng, X.; Kim, S.; Dietrich, K.N.; Ho, S.-M.; Reponen, T.; Huo, X. Maternal urinary cadmium levels during pregnancy associated with risk of sex-dependent birth outcomes from an e-waste pollution site in China. *Reprod. Toxicol.* **2018**, *75*, 49–55. [CrossRef] [PubMed]
37. Singh, N.; Ogunseitan, O.A.; Tang, Y. Systematic review of pregnancy and neonatal health outcomes associated with exposure to e-waste disposal. *Crit. Rev. Environ. Sci. Technol.* **2021**, *51*, 2424–2448. [CrossRef]
38. Soler, A.; Conesa, J.A.; Iñiguez-Cantos, M.E.; Ortuño, N. Pollutant formation in the pyrolysis and combustion of materials combining biomass and e-waste. *Sci. Total Environ.* **2018**, *622–623*, 1258–1264. [CrossRef] [PubMed]

39. Li, T.-Y.; Zhou, J.-F.; Wu, C.-C.; Bao, L.-J.; Shi, L.; Zeng, E.Y. Characteristics of polybrominated diphenyl ethers released from thermal treatment and open burning of E-Waste. *Environ. Sci. Technol.* **2018**, *52*, 4650–4657. [CrossRef]
40. Hong, W.-J.; Jia, H.; Ding, Y.; Li, W.-L.; Li, Y.-F. Polychlorinated biphenyls (PCBs) and halogenated flame retardants (HFRs) in multi-matrices from an electronic waste (e-waste) recycling site in Northern China. *J. Mater. Cycles Waste Manag.* **2018**, *20*, 80–90. [CrossRef]
41. Chakraborty, P.; Selvaraj, S.; Nakamura, M.; Prithiviraj, B.; Cincinelli, A.; Bang, J.J. PCBs and PCDD/Fs in soil from informal e-waste recycling sites and open dumpsites in India: Levels, congener profiles and health risk assessment. *Sci. Total Environ.* **2018**, *621*, 930–938. [CrossRef]
42. Li, H.; La Guardia, M.; Liu, H.; Hale, R.C.; Mainor, T.M.; Harvey, E.; Sheng, G.; Fu, J.; Peng, P. Brominated and organophosphate flame retardants along a sediment transect encompassing the Guiyu, China e-waste recycling zone. *Sci. Total Environ.* **2019**, *646*, 58–67. [CrossRef]
43. Tange, L.; Drohmann, D. Waste electrical and electronic equipment plastics with brominated flame retardants—from legislation to separate treatment—Thermal processes. *Polym. Degrad. Stab.* **2005**, *88*, 35–40. [CrossRef]
44. Abdelbasir, S.M.; Hassan, S.S.M.; Kamel, A.H.; El-Nasr, R.S. Status of electronic waste recycling techniques: A review. *Environ. Sci. Pollut. Res.* **2018**, *25*, 16533–16547. [CrossRef]
45. Veit, H.M. Reciclagem de Cobre de Sucatas de Placas de Circuito Impresso. Doctoral Thesis, Universidade Federal do Rio Grande do Sul, Porto Alegre, Brazil, 2005.
46. Hense, P.; Reh, K.; Franke, M.; Aigner, J.; Hornung, A.; Contin, A. Pyrolysis of waste electrical and electronic equipment (WEEE) for recovering metals and energy: Previous achievements and current approaches. *Environ. Eng. Manag. J.* **2015**, *14*, 1637–1647. [CrossRef]
47. Chandrasekaran, S.R.; Avasarala, S.; Murali, D.; Rajagopalan, N.; Sharma, B.K. Materials and energy recovery from E-Waste Plastics. *ACS Sustain. Chem. Eng.* **2018**, *6*, 4594–4602. [CrossRef]
48. Ambaye, T.G.; Vaccari, M.; Castro, F.; Prasad, S.; Rtimi, S. Emerging technologies for the recovery of rare earth elements (REEs) from the end-of-life electronic wastes: A review on progress, challenges, and perspectives. *Environ. Sci. Pollut. Res.* **2020**, *27*, 36052–36074. [CrossRef] [PubMed]
49. Ma, H.; Du, N.; Lin, X.; Li, C.; Lai, J.; Li, Z. Experimental study on the heat transfer characteristics of waste printed circuit boards pyrolysis. *Sci. Total Environ.* **2018**, *633*, 264–270. [CrossRef] [PubMed]
50. Guo, X.; Qin, F.G.; Yang, X.; Jiang, R. Study on low-temperature pyrolysis of large-size printed circuit boards. *J. Anal. Appl. Pyrolysis* **2014**, *105*, 151–156. [CrossRef]
51. Ding, Y.; Zhang, S.; Liu, B.; Zheng, H.; Chang, C.-C.; Ekberg, C. Recovery of precious metals from electronic waste and spent catalysts: A review. *Resour. Conserv. Recycl.* **2019**, *141*, 284–298. [CrossRef]
52. Wang, R.; Xu, Z. Recycling of non-metallic fractions from waste electrical and electronic equipment (WEEE): A review. *Waste Manag.* **2014**, *34*, 1455–1469. [CrossRef]
53. Gurgul, A.; Szczepaniak, W.; Zabłocka-Malicka, M. Incineration and pyrolysis vs. steam gasification of electronic waste. *Sci. Total Environ.* **2018**, *624*, 1119–1124. [CrossRef]
54. Dutta, T.; Kim, K.-H.; Deep, A.; Szulejko, J.; Vellingiri, K.; Kumar, S.; Kwon, E.E.; Yun, S.-T. Recovery of nanomaterials from battery and electronic wastes: A new paradigm of environmental waste management. *Renew. Sustain. Energy Rev.* **2018**, *82*, 3694–3704. [CrossRef]
55. Xavier, L.H.; Giese, E.C.; Duthie, A.C.R.; Lins, F.A.F. Sustainability and the circular economy: A theoretical approach focused on e-waste urban mining. *Resour. Policy* **2019**, 101467. [CrossRef]
56. Barnwal, A.; Dhawan, N. Recycling of discarded mobile printed circuit boards for extraction of gold and copper. *Sustain. Mater. Technol.* **2020**, *25*, e00164. [CrossRef]
57. Neto, J.F.O.; Silva, M.M.; Santos, S.M. A mini-review of e-waste management in Brazil: Perspectives and challenges. *Clean Soil Air Water* **2019**, *47*, 1–10. [CrossRef]
58. Abbondanza, M.; Souza, R. Estimating the generation of household e-waste in municipalities using primary data from surveys: A case study of Sao Jose dos Campos, Brazil. *Waste Manag.* **2019**, *85*, 374–384. [CrossRef] [PubMed]
59. Ottoni, M.; Dias, P.; Xavier, L.H. A circular approach to the e-waste valorization through urban mining in Rio de Janeiro, Brazil. *J. Clean. Prod.* **2020**, *261*, 120990. [CrossRef]
60. Rodrigues, A.C.; Boscov, M.E.; Günther, W.M. Domestic flow of e-waste in São Paulo, Brazil: Characterization to support public policies. *Waste Manag.* **2020**, *102*, 474–485. [CrossRef]
61. Nithya, R.; Sivasankari, C.; Thirunavukkarasu, A. Electronic waste generation, regulation and metal recovery: A review. *Environ. Chem. Lett.* **2021**, *19*, 1347–1368. [CrossRef]
62. Adrian, C.S.; Drisse, M.B.; Cheng, Y.; Devia, L.; Deubzer, O. *The Global E-Waste Monitor 2020*; UNU-United Nations University: Geneva, Switzerland, 2020; ISBN 9789280891140.
63. Ghosh, S.K.; Debnath, B.; Baidya, R.; De, D.; Li, J.; Ghosh, S.K.; Zheng, L.; Awasthi, A.K.; Liubarskaia, M.; Ogola, J.S.; et al. Waste electrical and electronic equipment management and Basel Convention compliance in Brazil, Russia, India, China and South Africa (BRICS) nations. *Waste Manag. Res.* **2016**, *34*, 693–707. [CrossRef]

64. Dias, P.; Machado, A.; Huda, N.; Bernardes, A.M. Waste electric and electronic equipment (WEEE) management: A study on the Brazilian recycling routes. *J. Clean. Prod.* **2018**, *174*, 7–16. [CrossRef]
65. Evangelopoulos, P.; Kantarelis, E.; Yang, W. Investigation of the thermal decomposition of printed circuit boards (PCBs) via thermogravimetric analysis (TGA) and analytical pyrolysis (Py–GC/MS). *J. Anal. Appl. Pyrolysis* **2015**, *115*, 337–343. [CrossRef]

materials

MDPI

Article

Carbonized Solid Fuel Production from Polylactic Acid and Paper Waste Due to Torrefaction

Kacper Świechowski [1,*], Christian Zafiu [2] and Andrzej Białowiec [1]

[1] Department of Applied Bioeconomy, Wrocław University of Environmental and Life Sciences, 37a Chełmońskiego Str., 51-630 Wrocław, Poland; andrzej.bialowiec@upwr.edu.pl

[2] Department of Water, Atmosphere and Environment, Institute of Waste Management, University of Natural Resources and Life Sciences, Muthgasse 107, 1190 Wien, Austria; christian.zafiu@boku.ac.at

* Correspondence: kacper.swiechowski@upwr.edu.pl

Abstract: The quantity of biodegradable plastics is increasing steadily and taking a larger share in the residual waste stream. As the calorific value of biodegradable plastic is almost two-fold lower than that of conventional ones, its increasing quantity decreases the overall calorific value of municipal solid waste and refuse-derived fuel which is used as feedstock for cement and incineration plants. For that reason, in this work, the torrefaction of biodegradable waste, polylactic acid (PLA), and paper was performed for carbonized solid fuel (CSF) production. In this work, we determined the process yields, fuel properties, process kinetics, theoretical energy, and mass balance. We show that the calorific value of PLA cannot be improved by torrefaction, and that the process cannot be self-sufficient, while the calorific value of paper can be improved up to 10% by the same process. Moreover, the thermogravimetric analysis revealed that PLA decomposes in one stage at ~290–400 °C with a maximum peak at 367 °C, following a 0.42 reaction order with the activation energy of 160.05 kJ·(mol·K)$^{-1}$.

Keywords: torrefaction; solid fuel; waste to carbon; circular economy; biodegradable materials; calorific value

Citation: Świechowski, K.; Zafiu, C.; Białowiec, A. Carbonized Solid Fuel Production from Polylactic Acid and Paper Waste Due to Torrefaction. *Materials* **2021**, *14*, 7051. https://doi.org/10.3390/ma14227051

Academic Editors: Rossana Bellopede and Lorena Zichella

Received: 23 October 2021
Accepted: 16 November 2021
Published: 20 November 2021

1. Introduction

1.1. Background of Current Situation

The negative impact of plastic waste accumulated in the environment (in oceans, soils, and air), including the form of microplastics, is undeniable. Most of the commonly used polymers are based on fossil resources and resistant to biodegradation, which means that once released to the environment, they will persist for a long time. Currently, there is a risk of the release of chemicals from all plastic that is unproperly landfilled into the soil and groundwater. Plastic waste that has leaked into oceans is a cause of death of marine life and is a source of microplastic that pollutes the air we breathe and water we drink [1–4].

Geyer et al. [5] estimated in 2017 that, since the 1950s, over 8300 Mt of plastics were ever produced globally, out of which 56% (ca. 4700 Mt) of the ever-produced plastics were landfilled or ended up in the environment [5]. In 2019 alone, 368 Mt of plastic were produced [6], and it is estimated that that annual production will increase by four times in 2050 [7]. To date, to cover plastic production, around 4% of the total extracted fossil fuels (e.g., natural gas, oil, and coal), are needed annually, and by 2050 this number could increase to 20% [8]. Currently, the largest plastic producers are China (31%), North America (19%), and the European Union (EU) (16%) [6]. According to "Global Plastic Flow 2018" that was prepared by Conversio Market & Strategy GmbH [9], the global plastic consumption was 385 Mt which consisted of 172 Mt of packaging waste and 213 Mt of non-packaging waste. At the same time, 250 Mt of plastic waste was generated, of which only 175 Mt was collected, and hence 75 Mt was improperly disposed or released to the environment. Only

~28.5% of the collected plastic waste was recycled; a similar amount was incinerated, and 43% was landfilled [9].

A large share of plastic materials (almost 40% in the EU) is used for packaging, which has the shortest life cycle. Other sectors that consume large amounts of plastic are building and construction (~20%) and automotive (~10%) [6]. These shares are most probably similar for the rest of the world. According to the Ellen Macarthur Foundation [10], only 14% of produced packaging plastic globally was collected for recycling purposes, wherein 4% was lost during recycling processes, 8% was recycled in cascaded recycling (waste plastic was converted into other, lower-value products), and only 2% of produced plastic had a closed-recycling loop (wasted plastic was converted into the same or similar quality products) [10].

At the first glance, the presented data show that the most abundant type of plastic waste (packaging) is hard to recycle, or its recycling is not economical yet. The reasons for this are the low quality of the recycled plastics in comparison to the virgin material, cost-intensive recycling processes, and lack of proper infrastructure [1]. Some plastic materials, such as high-density polyethylene (HD-PE), or polyethylene terephthalate (PET) can be recycled economically, due to their high market value, whereas low-density polyethylene (LD-PE), and other foil materials are used for refuse-derived fuel (RDF) production [1,11]

1.2. The Problem of Bioplastic Solution

With increasing awareness of citizens about ecology and sustainability, an increasing number of producers replace conventional packing plastic with biobased and biodegradable plastics. In 2020, around 47% of all produced bioplastic was used in the packaging sector. According to the European Bioplastics organization, bioplastic (biodegradable and non-biodegradable) represents about 1% of all produced plastic. The organization also estimates that, due to the rising demand, the bioplastic market will increase by ~40% up to 2025 [11,12].

Bio-based plastics are a potential solution for problems related to fossil-based plastic. In theory, bioplastics open new end-of-life scenarios, such as composting or anaerobic digestion, and lead to a reduction in conventional plastics pollution. In practice, however, there are problems with proper management [11]. Different biodegradable plastics need different environmental conditions to be biodegraded, e.g., biodegradable PLA-based biowaste bags need relatively high temperatures for overcoming the glass transition temperature (~70 °C) and initiating biodegradability. Such temperatures can be achieved in industrial composting plants, but not in home composters. In practice, biodegradable plastic is not usually decomposed during anaerobic digestion [13]. As a result, some countries, and some municipalities in the EU, allow the use of biodegradable bags for kitchen waste collection, while others do not [13]. At the same time, the bioplastic products also increase their share in the municipal solid waste (MSW) stream as, to date, no strategies exist for the collection and processing of bioplastic wastes. The reason for this is that these plastics are still a minority in the waste stream, are difficult to detect, and require sophisticated methods for proper separation [14]. Therefore, most of the bioplastic waste goes to residual fraction of municipal solid waste or is collected with conventional plastic. In both cases, biodegradable plastics are used for RDF production or are landfilled, if the local regulations allow it. As a result, biodegradable plastics do not lead to a decrease in plastic pollutions and additionally decrease the calorific value of RDF made from waste. The calorific value of the most abundant plastic (PE-LD) used for RDF production is ca. 40 $MJ \cdot kg^{-1}$ [15], while the most common biodegradable plastic used to replace it, is PLA with ~19 $MJ \cdot kg^{-1}$. A simple simulation in Figure A1 shows that when biodegradable plastic share increases, the high heating value of RDF decreases from 28 to 18 $MJ \cdot kg^{-1}$.

1.3. The RDF Quality Importance

Refused-derived fuel (RDF), also known as solid recovered fuel (SRF), is mainly made from MSW. The RDF can also be made from other waste such as used tires, sewage sludges, textiles, wood, and others. The main properties of RDF decisive of its quality are calorific value and ash content. The higher the calorific value and lower ash content, the better quality of RDF. The calorific value of RDF depends on the share of RDF components and can differ from 11 MJ·kg^{-1} [16] to 36 MJ·kg^{-1} [17]. From a calorific point of view, the most valuable materials are plastics such as PP and PE ~46 MJ·kg^{-1}, PS ~41 MJ·kg^{-1}, and PET ~26 MJ·kg^{-1} [18], whereas organic waste, paper, and fabrics lead to a decrease in RDF energetic potential [19]; however, this increases the renewable energy availability. Organic waste such as kitchen and food wastes are also the main source of moisture that further decreases the energetic potential of RDF [19]. Similarly, the ash content of RDF depends on materials share, and the ash amount in plastic wastes is much lower than in other waste.

The high-quality RDF is needed for specialized incineration plants and for cement plants where RDF replaces coal and provides cleaner and partly renewable energy. In particular, cement plants need high calorific value RDF to keep the cement production process stable and safe for the environment. During waste incineration (also applies to RDF), there is a need to keep the temperature of exhaust gases above 850 °C for at least 2 s to eliminate the formation of harmful compounds. In the case of a cement plant, the waste needs to generate higher temperatures for clinker burning, and when the RDF calorific value is not high enough, the required temperature will not be obtained [20].

The RDF is usually produced in the mechanical-biological treatment plant (MBT), where MSW are valorized by various mechanical and biological methods. Mechanical methods include material separation, screening, and grinding [21]. These methods are applied to increase calorific value, increase homogeneity, and decrease ash and other pollutants (Hg, Cl) content. On the other hand, a biological method such as bio-drying is used to remove water from MSW. If RDF, produced in the MBT plant, does not meet the required quality, it can be upgraded in the future by mixing other more energetic industrial materials, by the densification process (pelletization), or by thermal processing such as torrefaction or carbonization in low temperatures [21,22]. Thermal processing in conventional pyrolysis temperatures is not applicable, as most plastics are concerting into oil and gas instead of solid carbonized fuel; as result, the calorific value of solid carbonized fuel starts to decrease [23]. Furthermore, mixing, densification, and thermal processing can be combined to maximize the quality of RDF. Here, it is important to note that each of the mentioned processes requires energy, and the legitimacy of the use of these methods depends on a specific situation.

While conventional plastics PP, PE, and PET are usually subjected to mechanical recycling after separate collection, biodegradable plastics recycling has not been developed yet. Therefore, the decreasing share of conventional plastics and increasing share of bioplastics in RDF induces a need for research on the torrefaction of these biodegradable materials, as a perspective for CSF production from MSW in the future.

1.4. Study Aim

In this work, PLA wastes, PLA-made cups, and paper-made cups with the addition of PLA were subjected to thermal processing-torrefaction. The main aim was to check the legitimacy of low-temperature processing of PLA wastes for fuel parameters improvement. PLA wastes were processed at 200–300 °C to check the possibilities of thermal upgrading. As torrefaction and low-temperature pyrolysis of mixed waste turned out to increase the calorific value of RDF [23], we assumed that similar results will be obtained for PLA wastes. As result, a decreasing calorific value of MSW and RDF with an increasing biodegradable plastic share will be overcome. For this reason, the fuel properties of torrefied PLA wastes, torrefaction kinetics, and theoretical energy required for torrefaction were determined.

1.5. Methods of Thermal Processes Analysis

There are many various methods and techniques for thermal study performance and thermal process analysis. The most common are studies using small, lab-scale reactors made for specific situations, or by adopting other equipment such as muffle furnaces or autoclaves. These types of equipment allow performing thermal conversion of materials to produce enough carbonized material used for other analyses such as proximate analysis, elemental analysis, etc. Such small reactors are in favor of testing new and non-standard materials as they provide a lot of information about process efficiency and product quality [24]. On the other hand, these reactors have limited potential for thermal process reaction analysis. Most of them work similar to a black-box and only the beginning and the final product is measured, without intermediates. For that reason, thermal analysis is also performed using thermogravimetric equipment, allowing us to measure changes in materials mass, occurred reactions, quality, and chemical compositions of intermediate products. The basic thermogravimetric analysis is TGA that provides information about mass losses during a time at a defined temperature, and differential scanning calorimetry (DSC) provides information about energy flow through sample. Additionally, TGA/DSC equipment can be coupled with other instruments that identify released gasses and their chemical composition. As result, emissions and evolved pollution during the process can be quantified and managed [25–27].

2. Materials and Methods

2.1. Materials

The samples of biodegradable materials for the experiment were prepared from commercially available one-use cups. Paper (PAP) served as reference material and was obtained from cups that were made of 99% of paper, and 1% of PLA. The PLA material was obtained from cups made of 100% PLA plastic. The paper cups were ground using a laboratory knife mill (Testchem, model LMN-100, Pszów, Poland), through a 3 mm sieve, while the PLA cups were cut manually into pieces of ~1 cm^2 as the PLA was melting and blocked the mill. Then, the crumbled material was subjected to a torrefaction process. Samples of raw and torrefied materials were stored in plastic containers at room temperature (~20 °C).

2.2. Methods

Before the experiment, raw, crumbled materials were dried at 105 °C using a laboratory dryer (WAMED, KBC-65W, Warsaw, Poland) until a constant mass was obtained. These dry materials were used for CSF production. After that, the materials and produced CSFs were subjected to proximate analysis and higher heating value (HHV) determination analysis. Next, dry raw samples of raw materials were subjected to thermogravimetric analysis (TGA) for kinetic parameters determination and differential scanning calorimetry analysis (DSC) for determination of endo and exothermal reaction presence. Next, data from the CSF production process and proximate analysis were used to build regression models that show and describe quantitatively the effect of process temperature and time on CSF properties.

2.2.1. Torrefaction Process—CSF Production

The CSF was produced at different temperatures of 200–300 °C in intervals of 20 °C and kept for 20, 40, and 60 minutes each. For the torrefaction procedure, 10 g of dry samples were placed in ceramic crucibles. These crucibles were placed into the chamber of the muffle furnace (Snol 8.1/1100, Utena, Lithuania), which was purged with CO_2 gas to create an inert atmosphere before the samples were heated to the setpoint temperature. During the torrefaction process, CO_2 gas was continuously supplied to the chamber to prevent sample ignition. The CO_2 flow was shut off after the treatment period and when the temperature of the chamber declined to <150 °C. The mass of samples before and after the process was used to calculate mass yield following Equation (1). Then, using the results of HHV, the energy densification ratio was calculated (Equation (2)), and then the energy yield of CSF was determined according to Equation (3).

$$MY = \frac{m_b}{m_r} \cdot 100 \tag{1}$$

where $MY.$ is the mass yield, %; m_b is the mass of material after torrefaction, g (CSF); and m_r is the mass of material before torrefaction, g.

$$EDr = \frac{HHV_b}{HHV_r} \quad (2)$$

Where $EDr.$ is the energy densification ratio; $HHV_b.$ is the high heating value of material after torrefaction (J·g^{-1}) (CSF); and m_r is the high heating value of material before torrefaction (J·g^{-1}).

$$EY = MY{\cdot}EDr \quad (3)$$

where EY is the energy yield, %; $MY.$ is the mass yield, %; and EDr is the energy densification ratio.

2.2.2. Proximate Analysis and HHV Determination

For all samples, the proximate analysis was performed. The moisture content (MC) was determined by the drying method at 105 °C using a laboratory dryer (WAMED, KBC-65W, Warsaw, Poland) according to PN-EN 14346:2011 standard [28]. The volatile matter (VM) was measured by a thermogravimetric method using a tubular furnace (Czylok, RST 40 × 200/100, Jastrzębie-Zdrój, Poland), according to [29]. The ash content (AC) was measured by sample incineration in a muffle furnace (Snol 8.1/1100, Utena, Lithuania) according to PN-Z-15008-04:1993 standard [30], and fixed carbon was measured by difference. Additionally, samples were tested for volatile solids content (VS) and combustible part content (CP) using the muffle furnace (Snol 8.1/1100, Utena, Lithuania) according to PN-EN 15169:2011 [31] and PN-Z-15008-04:1993 [30] standards, respectively. All samples were tested for high heating value using a calorimeter (IKA, C200, Staufen, Germany), according to PN EN ISO 18125:2017-07 [32]. To ensure repeatability, each experiment was triplicated.

2.2.3. Statistical Analyses

Results of CSF production and proximate analysis were subjected to regression analyses to provide empirical equations. These equations are used to describe the following properties of CSF: MY, EDr, EY, VM, AC, FC, VS, CP, and HHV depending on process temperature and time. The regression was performed according to previous work [19]. In brief, experimental data were subjected to four regression models: (I) linear equation, (II) second-order polynomial equation, (III) factorial regression equation, and (IV) response surface regression equation. Then, determination coefficient (R^2) and Akaike value (AIC) were calculated for each model. Next, models with the greatest R^2 and the lowest AIC value were chosen as the best fit to experimental data; the other models were rejected. In the case chosen model had some insignificant regression coefficients (a_n), they were removed, and regression analysis was performed again.

To check if process conditions have an impact on fuel properties, ANOVA was performed, with a post hoc Tukey test to test the pairwise significance ($p < 0.05$).

2.2.4. Thermal Analysis

The dry samples were subjected to TG/DTG/DSC thermal analysis using a simultaneous thermal analyzer (Netzsch, 449 F1 Jupiter, Selb, Germany). Term TG/DTG/DSC stands for thermogravimetry/difference thermogravimetry/differential scanning calorimetry. TG/DTG results present how material decomposes in the function of temperature, while the DSC results show transformations and reactions occurring at a particular temperature.

The sample was placed into a corundum crucible. The mixture of nitrogen and argon 4:1 was used as an inert gas. The sample was heated 10 °C·min^{-1} from 30–800 °C. As a reference, an empty crucible was used. TGA/DTG/DSC analyzer automatically recalculated DSC data to mW·mg^{-1} and determined DTG from TG.

The TG data was used to determine kinetic parameters according to the Coats–Redfern (CR) method. The CR's kinetic triplet is activation energy (Ea), pre-exponential factor (A), and order of reaction (n). The methodology of CR determination was presented elsewhere [24].

2.2.5. Theoretical Mass and Energy Balance of the Torrefaction Process

Using part of the data from analyses that have been mentioned in the earlier paragraphs, theoretical energy balance for the torrefaction of PLA and paper waste was calculated. The calculations refer to the production of 1 g of CSF and include the determination of the:

- Mass of substrate used to produce 1 g of CSF;
- Energy contained in the raw material used to produce 1 g of CSF;
- External energy provided to the reactor to heat the proper amount of substrate to setup temperature, to produce 1 g of CSF;
- Energy contained in 1 g of CSF;
- Mass of gas generated during the production of 1 g of CSF;
- Energy contained in gas after production of 1 g of CSF.

For calculations, data of MY, HHV, and DSC results were used. The scheme of energy balance determination is shown in Figure 1. The green squares represent the order of calculations, the grey squares represent experimental/calculated data used for energy balance determination, and the blue squares stand for input and output data results.

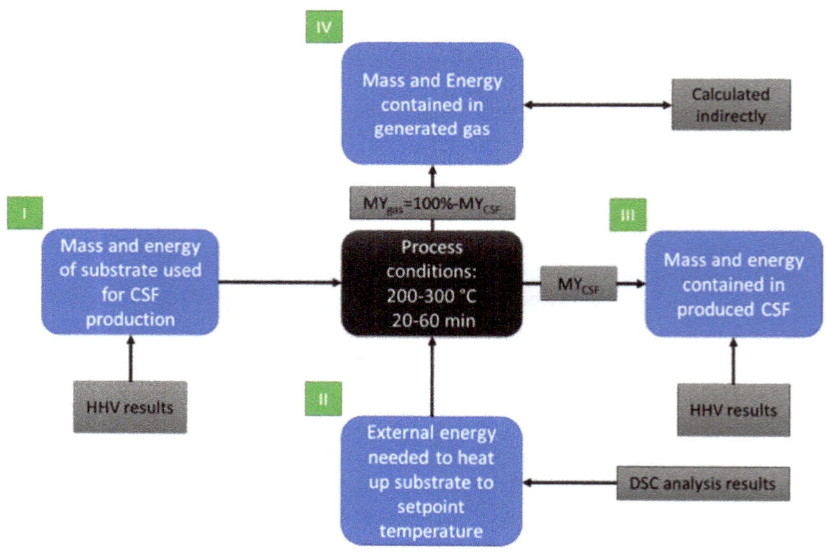

Figure 1. Scheme of mass and energy balance determination.

In step I, the mass yield of CSF production was used to determine the mass of substrate to produce 1 g of CSF by Equation (4), which allowed us to calculate the energy contained in the substrate used to produce 1 g of CSF by Equation (5).

$$M_s = \frac{Mr_{CSF}}{MY_{CSF}} \tag{4}$$

where: M_s—mass of substrate used to produce the required amount of CSF, (here 1 g), g; Mr_{CSF}—required mass of CSF, (here 1 g), g; and MY_{CSF}—mass yield of CSF production, % (Equation (1)).

$$E_s = M_s \cdot HHV_s \tag{5}$$

where: E_s—energy contained in the substrate used to produce CSF, J; M_s—mass of substrate used to produce CSF, g; and HHV_s—high heating value of substrate, J·g^{-1}.

For step II, the results from DSC were used as input in the form of a power flow by the sample during heating. The DSC was converted from mW·mg^{-1} to J mg^{-1} by the multiplication by time in seconds, providing information about the energy in J used to increase the temperature for 1 g of substrate. The energy demand to heat to setpoint temperature and mass of substrate demand produce CSF per g were used to calculate the demand of external energy to produce 1 g of CSF.

For step III, it is assumed, that the energy contained in 1 g of CSF equals the HHV, which was determined by the experiment.

In step IV, the energy contained in the gas was calculated indirectly. The energy in the gas is assumed to be a sum of external energy from step II, and the difference between energy contained in substrate and energy contained in CSF obtained from torrefaction, following Equation (6).

$$E_{gas} = E_{external} + E_{substrate} - E_{CSF} \tag{6}$$

where: E_{gas}—energy contained in the gas, J; $E_{external}$—external energy provided to the-reactor to heat the substrate to setup temperature, J; $E_{substrate}$—energy contained in the substrate used to produce CSF, J; and E_{CSF}—energy contained in produced CSF, J.

To keep calculations as simple as possible, the calculations were performed following assumptions:

- Moisture content in substrate = 0%;
- External energy is used to provide heat for the process;
- No heat losses of the reactor;
- The energy contained in the gas is a sum of chemical energy related to the chemical composition of gas and heat; here it was assumed that CSF is cooled down after the process, and all heat goes to gas.

3. Results and Discussion

3.1. Torrefaction Process—CSF Production

In Figures 2–4, process temperature and time effect on mass yield, energy densification ratio, and energy yield of carbonized solid fuel made from PLA and PAP were presented. The equations for these models were summarized in Table A1.

The mass yield of CSF made from PLA was almost not affected by process conditions. Small weight loss was observed in CSF produced at 300 °C in 60 minutes, where the MY decrease to 92%. For comparison, MY of CSF started to decrease from the lowest temperatures, at 200 °C and 20 min, the MY had around 80%, which decreased to 40% at 300 °C and 60 min (Figure 2). The reason for the very high MY of CSF made from PLA is the PLA decomposition resistances in the torrefaction temperatures range. It has been confirmed later in this work by TG/DTG results, that PLA decomposition began around 290 °C, and peaked at 367 °C (Figure 5a). For comparison, the PAP's main decomposition started already around 240 °C and peaked at 326 °C (Figure 5a). Although TG/DTG results are useful to investigate the thermochemical characteristics of a material, such as the temperature of decomposition, it is insufficient to determine the mass yield in certain temperature regimes or reaction times for different reactors due to different geometries, sample sizes, or thermal properties. Depending on the temperature regime, which has the main effect on decomposition, the time can result in less or more significant mass losses, especially in temperature regimes that include the main decomposition reactions and long residence time [33]. Therefore, empirical models for MY of PLA and PAP samples were developed (Table A1) to correct the challenges of the experiments.

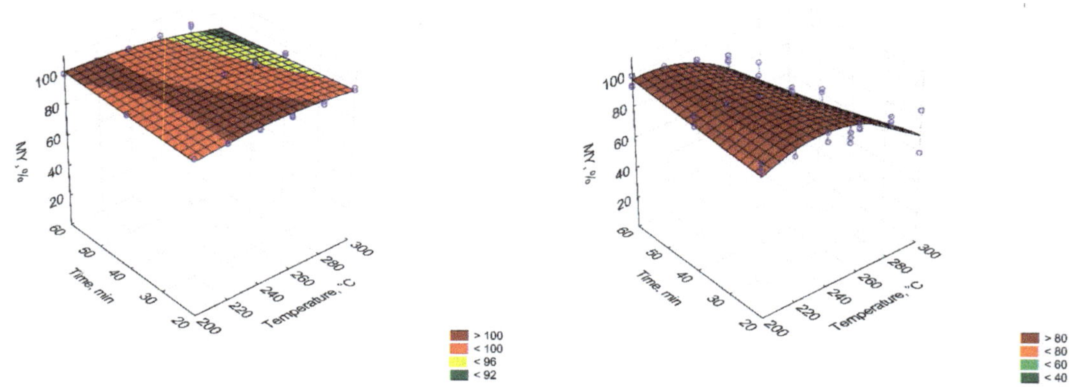

(a) (b)

Figure 2. Temperature and time effect on the mass yield (MY) of carbonized solid fuel made from (**a**) PLA, (**b**) PAP.

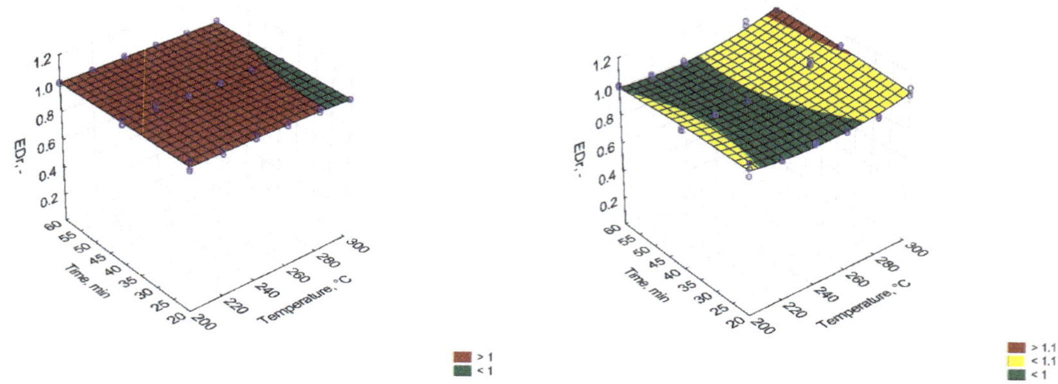

(a) (b)

Figure 3. Temperature and time effect on the energy densification ratio (EDr) of carbonized solid fuel made from (**a**) PLA and (**b**) PAP.

Figure 2 shows the process temperature and time effects on the energy densification ratio (EDr). The EDr shows how much more energy is contained in the CSF in comparison to unprocessed material. When EDr is equal to 1, no effect of a process for energy improvement is observed. When EDr is lower than 1, it means that there is less energy in CSF than it was initially in a substrate, and when EDr is higher than 1, it means that there is more energy in CSF than it was in a substrate. In this study, no statistically significant ($p > 0.05$) effect of torrefaction on EDr of PLA could be observed. However, a small effect of CSF made from paper could be observed. Here, EDr increased at a statistically significant level ($p < 0.05$) at setpoint temperatures higher than 280 °C.

The studied material was characterized by low enhancement in EDr. Typically, processed biomass is characterized by EDr from 1.2 to 1.4 [34]. The EDr increase was a result of the increase in HHV. The calorific value increase was probably a result of higher deoxygenation in comparison to the less intense decarbonization of material. When torrefaction temperature increases, the relative oxygen content decreases, in favor of relative carbon content which leads to an increase in HHV of CSF [35]. In the case of PLA, the process was below decomposed temperature so proper deoxygenation could not take place, while the PAP probably did not release enough oxygen compared to carbon to significantly increase HHV.

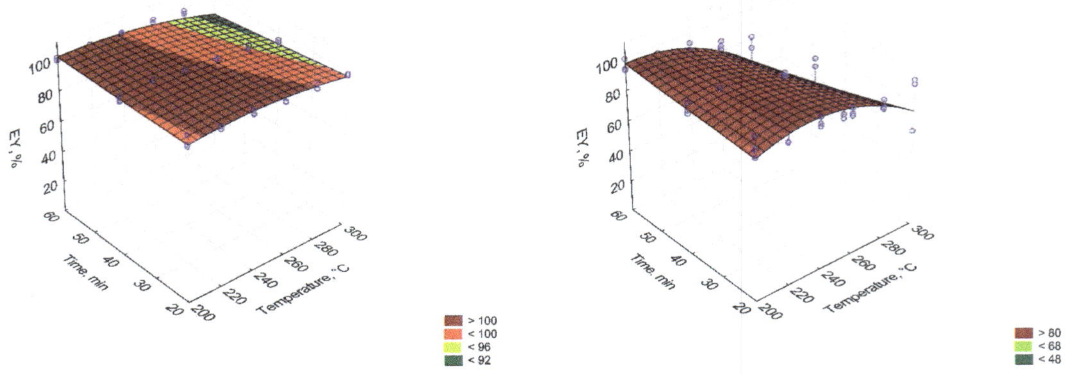

(a) (b)

Figure 4. Temperature and time effect on the energy yield (EY) of carbonized solid fuel made from (**a**) PLA and (**b**) PAP.

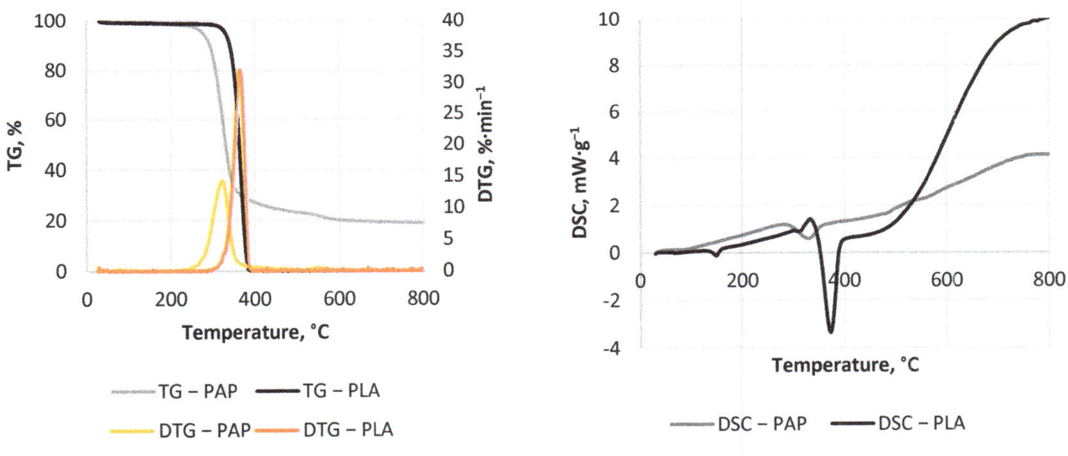

(a) (b)

Figure 5. Thermal analysis results, (**a**) TG/DTG, (**b**) DSC.

The energy yield (EY) shows how much energy that is contained in the substrate remains in the CSF after the process. With the increasing process temperature and time, the solid mass of substrate decreases as more gases and later also liquids are formed. Each of the products needs some chemical energy for its formation, which results in a decrease in the EY of CSF. Therefore, carbon and oxygen migration is an important factor during torrefaction [35]. The EY of CSF made from PLA was not affected by the process conditions for experimental conditions that were lower than 300 °C and 40 min (Figure 4). Under these conditions, MY remained constant and at lower temperatures, no significant changes in the HHV of torrefied PLA could be found. Therefore, the trend for EY was similar to MY. In the case of PAP, an EY decrease at temperatures higher than 280 °C was found, which resulted in a carbon migration to gas and liquid products [35,36].

3.2. Proximate Analysis and HHV Results

The samples of materials used to produce CSF were also analyzed for volatile matter (VM), ash content (AC), fixed carbon (FC), volatile solids (VS), combustibles parts (CP), and high heating value (HHV). The PLA materials had 100%, 0%, 0%, 100%, and 100% of VM, AC, FC, VS, and CP, respectively, while the PAP material had 88.2%, 3.6%, 8.2%, 96.3%, and 96.4% of VM, AC, FC, VS, and CP, respectively. The HHV of PLA and PAP were 19,420 and 17,525 $J \cdot g^{-1}$, respectively (Table 1).

Table 1. Results of proximate analysis and calorific value of CSF, as dry basis.

Material	Temp., °C	Time, min	VM, %	FC, %	AC, %	VS, %	CP, %	HHV, $J \cdot g^{-1}$
PLA	-	-	100.0	0.0	0.0	100.0	100.0	19,420
	200	20	100.0	0.0	0.0	100.0	100.0	19,675
		40	100.0	0.0	0.0	100.0	100.0	19,598
		60	100.0	0.0	0.0	100.0	100.0	19,512
	220	20	100.0	0.0	0.0	100.0	100.0	19,631
		40	100.0	0.0	0.0	100.0	100.0	19,799
		60	100.0	0.0	0.0	100.0	100.0	19,613
	240	20	100.0	0.0	0.0	100.0	100.0	19,703
		40	100.0	0.0	0.0	100.0	100.0	19,654
		60	100.0	0.0	0.0	100.0	100.0	19,682
	260	20	100.0	0.0	0.0	100.0	100.0	19,399
		40	100.0	0.0	0.0	100.0	100.0	19,372
		60	100.0	0.0	0.0	100.0	100.0	19,592
	280	20	100.0	0.0	0.0	100.0	100.0	19,529
		40	100.0	0.0	0.0	100.0	100.0	19,510
		60	100.0	0.0	0.0	100.0	100.0	19,410
	300	20	100.0	0.0	0.0	100.0	100.0	19,346
		40	100.0	0.0	0.0	100.0	100.0	19,294
		60	100.0	0.0	0.0	100.0	100.0	19,571
PAP	-	-	88.2	8.2	3.6	96.3	96.4	17,525
	200	20	86.6	9.9	3.4	96.6	96.6	17,889
		40	86.2	10.1	3.6	96.3	96.4	17,283
		60	86.7	9.8	3.5	96.5	96.5	17,653
	220	20	88.0	8.6	3.4	96.5	96.6	17,185
		40	86.7	10.0	3.3	96.4	96.7	17,504
		60	86.4	10.1	3.5	96.4	96.5	17,368
	240	20	85.5	10.9	3.5	96.3	96.5	17,446
		40	84.7	11.8	3.6	96.2	96.4	17,366
		60	84.8	11.7	3.5	96.2	96.5	17,434
	260	20	86.2	10.2	3.6	96.1	96.4	17,163
		40	84.0	12.4	3.6	96.0	96.4	17,389
		60	81.9	14.1	4.0	95.7	96.0	17,220
	280	20	83.6	12.7	3.7	96.3	96.3	17,352
		40	67.9	26.0	6.1	93.7	93.9	19,048
		60	66.9	26.2	7.0	92.8	93.0	19,146
	300	20	69.3	24.8	5.9	93.9	94.1	18,758
		40	60.8	31.5	7.7	91.8	92.3	19,520
		60	55.7	34.6	9.7	89.9	90.3	19,346

For PLA samples, an unexpected result was found for FC and AC, 0%, while VM, VS, and CP were 100%. The same results were obtained for all CSF made from PLA (Table 1). Moreover, the Tukey test shows that there were no significant changes between the HHV of CSF made from PLA. Therefore, it can be stated that torrefaction does not affect PLA fuel properties. These unexpected results can be explained in two ways: (I) The amount of ash (minerals) in PLA was too small to be detected by equipment that was used, or (II) there were no minerals in the PLA material at all. In case of a lack of minerals (case II), the results would be correct, as all organic matter was incinerated/devoltalized during experiments. In the other case (I), a correction for the undetected mass would have to be performed. However, the error of the undetected mass is ±0.1 mg at an input of 1 g and therefore negligible.

In the literature, both cases can be found for PLA. In favor of assumption (II) were results from Camacho-Muñoz et al. [37] that showed 100% of vs. in a PLA sample. However, Jing et al. [38] showed that PLA is a type of thermally degradable material that burns at a relatively rapid heat release rate with negligible chars, suggesting that at least some FC should remain.

For CSF made of PAP, a decrease in VM with increasing temperature and time was observed. With increasing process temperature and time from 200 °C and 20 min to 300 °C and 60 min, the VM decreases from 86.6% to 55.7%, while FC and AC increase from 9.9% to 34.6%, and from 3.4% to 9.7%, respectively (Table 1). The observed decrease in VM is related to the devoltalization of materials. On a molecular level, large cellulose molecules in PAP are broken into smaller ones until they are small enough to be removed by convection [39]. Depending on the chemical composition, more or fewer of such small molecules are released and, as a result, different values of VM can be observed. Unlike VM, the AC and FC content increase mainly as a result of the loss in VM. Unlike AC, which is related to the mineral present in the sample, additional FC can be produced during secondary reactions [40]. Nevertheless, for biomass, the presence of components such as hemicellulose and cellulose is the main contributor of VM production, while lignin is the same for FC production [41].

Both tested materials were characterized by a relatively high level of VM, and low and zero content of FC (PAP and PLA, respectively). For comparison, wood biomass has 86% of VM, 15% of FC, and 0.4% of AC [42], torrefied wood at 300 °C in 30 min has 71% of VM, 29% of FC, and 0.4% of AC [43], while high-rank bituminous (coal) has 27.6% of VM, 65% of FC, and 7.4% of AC [44]. It is clear that fuel properties of torrefied paper and biodegradable plastic are not close to conventional solid fuels. Nevertheless, the positive aspect of PLA material is its zero-ash content, which decreases the costs for managing the ash.

The high heating value of 19.4 MJ·kg^{-1} for PLA is more than twice lower than that of conventional plastics such as polyethylene [45]. Moreover, torrefaction does not increase the HHV of PLA (Table 1). On the other hand, torrefaction was found to be suitable for PAP. The HHV of PAP increased from 17.5 MJ·kg^{-1} to 19.5 MJ·kg^{-1} in CSF produced at 300 °C; 40 min. Though these values seem to suffice when they are compared to energetic biomasses (HHV ~ 18 MJ·kg^{-1}) [46], they are still small in comparison with coals 30 MJ·kg^{-1} [47] or conventional plastics 40 MJ·kg^{-1} [45].

3.3. Thermal Analysis Results

Figure 5a shows the TG/DTG results. The PLA mass was almost constant up to around 290 °C, where thermal decomposition started. The PLA decomposed totally in one step at temperatures of ~300–400 °C, with the maximum peak at 367 °C. Backes et al. [48] show that PLA composition (additive presence) affects thermal degradation, and some components reduce the activation energy of initiation of thermo-degradation reactions. As a result, the decomposition onset temperature and maximum peak can differ up to 40 °C depending on the processed PLA [48]. Additionally, maximum decomposition peaks occur at 353–385 °C [48]. The DSC analysis results are shown in Figure 5b. The analysis shows that during PLA pyrolysis several reactions related to polymer phase transition occurred. The first phase transition at 64 °C is the glass transition of PLA. At 149 °C, the endothermal melting transformation was observed and finally, at 372 °C, the main endothermal decomposition peak was found. These findings agreed well with the result of Sousa et al. [49]. The results show that, for some reason, the DSC decomposition peak was shifted in comparison to DTG at about 5 °C (Figure 5a,b). Nevertheless, these findings explain that torrefaction could not significantly change the properties of PLA, as the temperature was too low for efficient devolatilization.

For PAP, three peaks were observed by DTG. First at 80 °C, second at 326 °C, and third at 550 °C with 1.3%, 74.6%, and 5.4% mass change (Figure 5a, grey curve), respectively. The first and third peaks are almost not visible in Figure 5a. The first peak is related to residual water evaporation, while the second peak is probably related to cellulose decomposition. This is due to the fact that white paper is made mainly from cellulose, (85–99%) with the addition of lignin of 0–15% [50]. Nevertheless, reprocessed paper (e.g., newspaper) has less cellulose (40–55%), more lignin (18–30%), and comparable content hemicellulose (25–40%) in comparison to white paper [50]. Additionally, the previously mentioned substances could affect the PAP sample decomposition. Typically, the hemicellulose, cellulose, and lignin decompose at 225–325 °C, 305–375 °C, and 250–500 °C, respectively [51]. According to Porshnov et al. [52], the temperature range of 250–300 °C is a characteristic interval for hemicellulose decomposition, 300–350 °C for cellulose decomposition, while above 400 °C the residue of lignocellulosic substances decomposed at a very slow rate. Lignin decomposition reactions were reported to occur at up to 900 °C [52]. Therefore, it is highly probable that PAP's third peak is related to lignin decomposition. The DSC results showed that, during PAP pyrolysis, four endothermal transformations occurred. The first transformation at 91.4 °C was probably related to residual moisture removal [53], and the following transformations were related to the decomposition of elements of the PAP sample. Similar results were obtained by Yang et al. [53], who tested clean cellulose and found the main endothermal peak related to decomposition at 355 °C. In this study, this peak was found at 329.6 °C (Figure 5b) and, similarly to the PLA, the DSC peak of PAP was shifted in comparison to DTG at about 3.6 °C.

The kinetic parameters were determined at $\beta = 10\ °C \cdot min^{-1}$ using the Coats–Redfern method. The kinetic triplets were determined for the whole process (30–800 °C) and the main peaks observed at TG/DTG plots (Figure 5a). The whole decomposition process for PAP and PLA were described by a reaction order of 1.56 and 2.02, respectively, and relative low activation energy of 33.11 kJ·(mol·K)$^{-1}$, and 46.24 kJ·(mol·K)$^{-1}$, respectively (Table 2). Here, it is worth noting that, for PLA, the determination coefficient was low, at 0.66, which was a result of the one-stage decomposition process, which occurred at 290–400 °C. Additionally, other kinetic triplets were determined with high determination coefficients (Table 2). The main PLA decomposition reaction was described by a reaction order of 0.42 and 160.05 kJ·(mol·K)$^{-1}$ activation energy, while PAP exhibited an order of 2.12, and 122.55 kJ·(mol·K)$^{-1}$ (Table 2). The first peak for PAP was omitted, as it was only residual water evaporation. It is worth noting that the suspected lignin decomposition at the third peak of the PAP sample had the highest activation energy of 173.05 kJ·(mol·K^{-1}), which was about 51 kJ·(mol·K)$^{-1}$ larger than the main decomposition of cellulose. This finding is contrary to Noszczyk et al. [54] who studied several types of biomass materials and noticed that the cellulose content had a significant impact on the E_a, and the highest E_a was observed at the second stage of reaction, which was related to the cellulose decomposition [54].

Table 2. Kinetic triplets determined at $\beta = 10\ °C \cdot min^{-1}$ using Coats-Redfern method.

Material	Note	Temperature, °C	n	Ea, kJ·(mol·K)$^{-1}$	A, s^{-1}	R^2
PLA	Whole process	30–800	2.02	46.24	2.91×10	0.66
	Main decomposition peak	290–400	0.42	160.05	2.37×10^{10}	0.96
PAP	Whole process	30–800	1.56	33.11	5.88×10^{-1}	0.89
	Main decomposition peak	240–400	2.12	122.55	1.74×10^{8}	0.96
	Third decomposition peak	668–760	3.00	173.05	4.90×10^{10}	0.91

3.4. *Theoretical Mass and Energy Balance of the Torrefaction Process*

Table 3 summarizes the theoretical mass and energy balance to produce 1 g CSF s given. The table compares the temperature and time. The third and fourth headings present the input mass needed to produce 1 g of CSF, and the chemical energy contained in this material. The fifth heading presents external heat provided to the torrefaction process. The sixth heading shows energy contained in 1 g of CSF. The seventh heading present a mass of gas released from the substrate during torrefaction, and the last heading show energy contained in this gas. The energy in gas was calculated as a sum of external energy provided to conduct a process and energy of substrate that was not converted into CSF.

Table 3. Torrefaction mass and energy balance for production of 1 g of CSF from PLA and Paper wastes.

Temp., °C	Time, min	Mass of Substrate Used to Produce 1 g of CSF, g		Energy Contained in the Raw Material Used to Produce 1 g of CSF, J		External Energy Needed to Produce 1 g of CSF, J *		Energy Contained in 1 g of CSF, J **		Mass of Gas Generated during the Production of 1 g of CSF, g		Energy Contained in Gas after Production of 1 g of CSF, J ***	
		PLA	PAP	PLA	PAP	PLA	PAP	PLA	PAP	PLA	PAP	PLA	PAP
200	20	1.004	1.054	19,500	18,475	86	328	19,675	17,889	0.004	0.054	−89	914
	40	1.006	1.048	19,540	18,367	86	328	19,598	17,283	0.006	0.048	27	1412
	60	1.006	1.055	19,538	18,482	86	328	19,512	17,653	0.006	0.055	112	1157
220	20	1.003	1.074	19,483	18,817	133	425	19,631	17,185	0.003	0.074	−15	2056
	40	1.004	1.053	19,505	18,459	133	425	19,799	17,504	0.004	0.053	−161	1380
	60	1.007	1.060	19,552	18,582	133	425	19,613	17,368	0.007	0.060	72	1639
240	20	1.005	1.053	19,512	18,454	194	536	19,703	17,446	0.005	0.053	3	1543
	40	1.007	1.078	19,562	18,886	194	536	19,654	17,366	0.007	0.078	101	2056
	60	1.013	1.096	19,676	19,207	194	536	19,682	17,434	0.013	0.096	188	2309
260	20	1.010	1.066	19,608	18,683	267	663	19,399	17,163	0.010	0.066	477	2184
	40	1.011	1.102	19,642	19,308	267	663	19,372	17,389	0.011	0.102	537	2583
	60	1.007	1.170	19,562	20,499	267	663	19,592	17,220	0.007	0.170	237	3942
280	20	1.014	1.131	19,685	19,822	355	803	19,529	17,352	0.014	0.131	510	3273
	40	1.025	1.357	19,909	23,778	355	803	19,510	19,048	0.025	0.357	754	5534
	60	1.022	1.550	19,839	27,163	355	803	19,410	19,146	0.022	0.550	784	8820
300	20	1.012	1.288	19,646	22,571	458	940	19,346	18,758	0.012	0.288	758	4753
	40	1.043	2.357	20,247	41,303	458	940	19,294	19,520	0.043	1.357	1,410	22,722
	60	1.227	2.485	23,833	43,551	458	940	19,571	19,346	0.227	1.485	4,719	25,144

* value determined using DSC analysis result. ** value determined using calorimetric analysis result (HHV). *** value is the sum of chemical energy contained in gas and heat from external energy, assuming that no external energy stays in CSF.

The result shows that more PAP than PLA substrate is needed to produce 1 g of CSF. In the case of 300 °C at 60 min, the double mass of PAP is needed compared to PLA (Table 3). The reason for this large input substrate demand originates from the low mass yield of PAP torrefaction (Figure 2b). As a result, much more chemical energy contained in PAP is put into the process to produce 1 g of CSF (23,833 J for PLA vs. 43,551 J for PAP). Additionally, the DSC results showed that more energy was needed to heat PAP than PLA to 300 °C, (458 $J \cdot g^{-1}{}_{CSF}$ vs. 940 $J \cdot g^{-1}{}_{CSF}$) (Table 3). This is caused probably by the mostly higher specific heat value (Sp) of PAP in comparison to PLA. Depending on chemical composition, Sp of PAP varies from 1150 to 1650 $J \cdot (g \cdot K)^{-1}$ [55] while, for PLA, the value varies from 1180 to 1210 $J \cdot (g \cdot K)^{-1}$ [56]. On the other hand, PLA has a higher thermal conductivity, 0.12–0.15 $W \cdot (m \cdot K)^{-1}$ than PAP 0.08–0.11 $W \cdot (m \cdot K)^{-1}$ [55,56].

During torrefaction, torrgas are produced. The analysis showed that a small mass of torrgas is produced from PLA and, depending on process conditions, these vary from 0.004 $g \cdot g^{-1}{}_{CSF}$ to 0.227 $g \cdot g^{-1}{}_{CSF}$. As the production of 1 g of CSF from PAP needs far more substrate, much more torrgas is produced and varies from 0.054 $g \cdot g^{-1}{}_{CSF}$ to 1.485 $g \cdot g^{-1}{}_{CSF}$ (Table 3). As a result, during torrefaction at 300 °C, for each gram of produced CSF, around 1.5 g of torrgas is generated, and these torrgas contain more energy than produced CSF, while for PLA it is only 0.23 g of torrgas with around four times less energy than produced CSF (Table 3).

When energy contained in torrgas is higher than the external energy needed to produce CSF, it theoretically can be assumed that the process is self-sufficient. This is true when torrgas are incinerated to provide heat for a substrate. With that assumption can be stated that PLA torrefaction can be self-sufficient at process temperatures higher than 300 °C and 40 min, while PAP is similar from 200 °C and 20 min (Table 3). Nevertheless, these results do not include heat losses, process efficiency, and energy needed for water evaporation that is in the real feedstock. Due to many different approaches to reactors design, it is hard to assume any heat losses and process efficiency. However, the contribution of water can be calculated and added to the results obtained in this study. To remove 1% of the water from solid fuel, at least 22.57 J (2257 $J \cdot g^{-1}{}_{H2O}$ is the latent heat of water evaporation at 100 °C) is needed, as well as the energy needed to heat this water to 100 °C [57]. For this reason, the herein presented calculations serve as a starting point that has to be adapted for a particular reactor system and different feedstocks.

4. Summary

The results of this study showed that PLA's fuel properties cannot be improved by torrefaction, as no calorific values increase were observed with increasing process temperature and time. The reason is that PLA hardly decomposes, with negligible charring effects at torrefaction temperatures. On the other hand, PAP's fuel properties can be improved up to 10% by applying temperatures higher than 280 °C, which is probably caused by a partial cellulose decomposition. Additionally, the kinetic analysis revealed that PLA is decomposed in a one-stage process, that takes place at ~290–400 °C, with Ea of 160.05 kJ·(mol·K)$^{-1}$, while PAP is decomposed in a two-stage process, at ~240–400 °C, and ~668–760 °C, with Ea of 122.55 kJ·(mol·K)$^{-1}$ and 173.05 kJ·(mol·K)$^{-1}$, respectively. Moreover, the calculations showed that PLA torrefaction cannot be self-sufficient for CSF production and external energy is required, while CSF production from PAP proves to be self-sufficient under assumptions of no heat loss.

These results provide the first step towards an understanding of the PLA torrefaction process, but further research is needed to investigate higher temperatures of thermal PLA processing embracing gaseous and liquid products rather than solids, as PLA decomposes entirely into volatile components. Moreover, future studies should focus on PLA co-pyrolysis with conventional plastic, as a separation in waste management facilities is currently not possible from the MSW stream. Such a separation may be possible for separately collected and clean plastic wastes, but will fail in the case of plastics with organic adhesions, which are typical for plastic in MSW.

Regarding waste management scenarios, our study showed that the thermal properties of PLA qualify this material neither as a fuel surrogate in waste incinerators nor for an improvement by torrefaction process when we compare PLA with conventional high energy plastics. Therefore, a successive substitution of high caloric plastics by PLA may be reasonable when the end-of-life-scenario for the material is composting, but will raise the demand of conventional fuel when its thermally treated.

Supplementary Materials: The following are available online at https://www.mdpi.com/article/10.3390/ma14227051/s1, Table S1: CSF Production, Table S2: Proximate Analysis, Table S3: TG-DTG-DSC.

Author Contributions: Conceptualization, K.Ś; methodology, K.Ś.; software, K.Ś.; validation, K.Ś. and A.B, formal analysis, K.Ś.; investigation, K.Ś.; resources, K.Ś and A.B.; data curation, K.Ś.; writing—original draft preparation, K.Ś; writing—review and editing, K.Ś., C.Z., and A.B; visualization, K.Ś.; supervision, C.Z. and A.B.; project administration, K.Ś.; and funding acquisition, A.B. All authors have read and agreed to the published version of the manuscript.

Funding: This research received no external funding.

Institutional Review Board Statement: Not applicable.

Informed Consent Statement: Not applicable.

Data Availability Statement: All data derived during the experiments are given in the paper or the Supplementary Materials.

Acknowledgments: The presented article results were obtained as part of the activity of the leading research team—Waste and Biomass Valorization Group (WBVG). Paper has been prepared during research scholarship of Kacper Świechowski at BOKU financed under the Leading Research Groups support project from the subsidy increased for the period 2020–2025 in the amount of 2% of the subsidy referred to Art. 387 (3) of the Law of 20 July 2018 on Higher Education and Science, obtained in 2019—project number N040/0012/20.

Conflicts of Interest: The authors declare no conflict of interest.

Glossary

PLA	polylactic acid
PAP	paper
CSF	carbonized solid fuel
EU	European Union
HD-PE	high-density polyethylene
PET	polyethylene terephthalate
LD-PE	low-density polyethylene
RDF	refuse-derived fuel
MSW	municipal solid waste
SRF	solid recovered fuel
MBT	mechanical-biological treatment plant for waste
PP	Polypropylene
PE	Polyethylene
PS	Polystyrene
HHV	higher heating value
TGA	thermogravimetric analysis
DSC	differential scanning calorimetry analysis
MY	mass yield
EDr	energy densification ratio
EY	energy yield
MC	moisture content
VM	volatile matter
AC	ash content
VS	volatile solids content
CP	combustible part content
R^2	determination coefficient
AIC	Akaike value
a_n	regression coefficients,
DTG	differential thermogravimetry
CR	Coats–Redfern method
Ea	activation energy
A	pre-exponential factor
n	order of reaction
Sp	specific heat value

Appendix A

In Figure A1, the predictions of RDFs HHV depending on biodegradable plastic share in conventional plastic was presented. The calculation was completed using data from previous work [58]. Predictions were made based on the following assumptions: RDF components (component name, component share, components HHV): carton, 10%, 14.6 MJ·kg^{-1}; fabric 10%, 17.6 MJ·kg^{-1}; kitchen waste 5%, 16,4 MJ·kg^{-1}; paper 10%, 13.8 MJ·kg^{-1}; plastic 50%, 38.5 MJ·kg^{-1}; rubber 5%, 31 MJ·kg^{-1}; tetrapack 5%, 21.6 MJ·kg^{-1}; and wood 5%, 18.7 MJ·kg^{-1} [58]. The HHV of biodegradable plastic 19 MJ·kg^{-1} [this study].

In Table A1, torrefaction temperature and time effect on process yields and fuel properties are summarized. Presented equations are valid for CSF properties determination at T = 200–300 °C and t = 20–60 min. The R^2 stands for the determination coefficient. The higher = better.

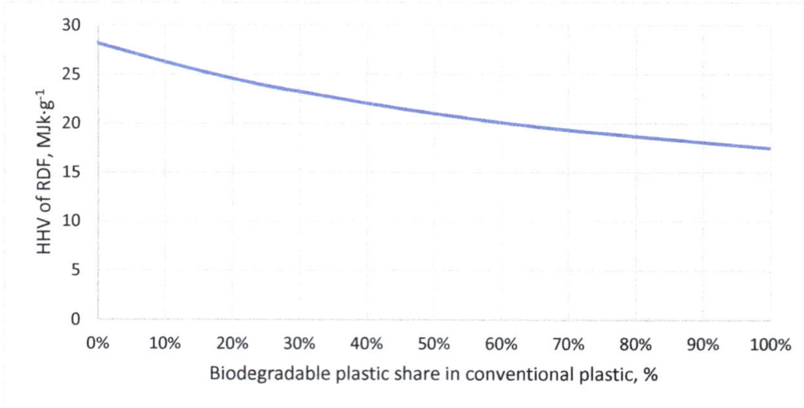

Figure A1. Effect of biodegradable plastic share in conventional plastic on RDF quality.

Table A1. Empirical equations for determination of torrefaction process and fuel properties of CSF produced from PLA and PAP; the equations boundary, T = 200–300 °C, t = 20–60 min.

Material	Equation	R^2
PLA	MY(T,t), % = 0.759 × T − 0.00139 × T^2 + 0.678 × t − 0.00303 × T × t	0.55
	EDr(T,t), % = 0.975 + 0.000545 × T − 0.00000160 × T^2 − 0.00101 × t + 0.00000220 × t^2 + 0.00000340 × T × t	0.13
	EY(T,t), % = 0.792 × T − 0.00151 × T^2 + 0.596 × t − 0.00270 × T × t	0.58
	VM, % = 100	1.00
	FC, % = 0	1.00
	AC, % = 0	1.00
	VS, % = 100	1.00
	CP, % = 100	1.00
	HHV, J·g^{-1} = 19549 ± 140	1.00
PAP	MY(T,t), % = − 340.901 + 3.558 × T − 0.00712 × T^2 + 2.079 × t − 0.00952 × T × t	0.86
	EDr(T,t), % = 2.404 − 0.0119 × T + 0.0000243 × T^2 − 0.00189 × t − 0.0000268 × t^2 + 0.0000184 × T x t	0.77
	EY(T,t), % = − 260.469 + 2.876 × T − 0.00570 × T^2 + 1.946 × t − 0.00889 × T × t	0.78
	VM(T,t), % = -153.308 + 2.021 × T − 0.00418 × T^2 +0.899 × t − 0.00421 × T × t	0.92
	FC(T,t), % = 184.153 − 1.583 × T + 0.00336 × T^2 − 0.00609 × t^2 + 0.00245 × T × t	0.90
	AC(T,t), % = 53.879 − 0.409 × T + 0.000815 × T^2 − 0.232 × t + 0.00105 × T × t	0.94
	VS, % = 47.732 + 0.396 × T − 0.000790 × T^2 + 0.239 × t − 0.00109 × T × t	0.94
	CP(T,t), % = 46.120 + 0.409 × T − 0.000815 × T^2 + 0.232 × t − 0.00105 × T × t	0.94
	HHV(T,t), J·g^{-1} = 39,926.103 − 198.210 × T + 0.425 × T^2 + 0.0447 × T × t	0.77

References

1. Narancic, T.; O'Connor, K.E. Plastic waste as a global challenge: Are biodegradable plastics the answer to the plastic waste problem? *Microbiology* **2019**, *165*, 129–137. [CrossRef] [PubMed]
2. North, E.J.; Halden, R.U. Plastics and environmental health: The road ahead. *Rev. Environ. Health* **2013**, *28*, 1–8. [CrossRef]
3. Jambeck, J.R.; Geyer, R.; Wilcox, C.; Siegler, T.R.; Perryman, M.; Andrady, A.; Narayan, R.; Lavender Law, K. Plastic waste inputs from land into the ocean. *Science* **2015**, *347*, 768–771. [CrossRef]
4. Katyal, D.; Kong, E.; Villanueva, J. Microplastics in the environment: Impact on human health and future mitigation strategies. *Environ. Health Rev.* **2020**, *63*, 27–31. [CrossRef]
5. Geyer, R.; Jambeck, J.R.; Law, K.L. Production, use, and fate of all plastics ever made. *Sci. Adv.* **2017**, *3*, e1700782. [CrossRef]
6. PlascticsEurope. Plastics—The Facts 2020 An analysis of European Plastics Production, Demand and Waste Data. 2020. Available online: https://plasticseurope.org/knowledge-hub/plastics-the-facts-2020/ (accessed on 15 November 2021).
7. Qureshi, M.S.; Oasmaa, A.; Pihkola, H.; Deviatkin, I.; Tenhunen, A.; Mannila, J.; Minkkinena, H.; Pohjakallioc, M.; Laine-Ylijokid, J. Pyrolysis of plastic waste: Opportunities and challenges. *J. Anal. Appl. Pyrolysis* **2020**, *152*, 104804. [CrossRef]
8. Rujnić-Sokele, M.; Pilipović, A. Challenges and opportunities of biodegradable plastics: A mini review. *Waste Manag. Res.* **2017**, *35*, 132–140. [CrossRef] [PubMed]

9. Conversio Market & Strategy. Global Plastic Flow 2018. 2020. Available online: https://www.carboliq.com/pdf/19_conversio_global_plastics_flow_2018_summary.pdf (accessed on 15 November 2021).
10. Ellen Macarthur Foundation. The New Plastics Economy: Rethinking the Future of Plastics & Catalysing Action. 2017. Available online: https://ellenmacarthurfoundation.org/the-new-plastics-economy-rethinking-the-future-of-plastics-and-catalysing (accessed on 15 November 2021).
11. Dilkes-Hoffman, L.S.; Pratt, S.; Lant, P.A.; Laycock, B. The role of biodegradable plastic in solving plastic solid waste accumulation. In *Plastics to Energy: Fuel, Chemicals, and Sustainability Implications*; Elsevier: Amsterdam, The Netherlands, 2018; pp. 469–505. [CrossRef]
12. European Bioplastic Conference. Bioplastics Market Development 2020. 2020. Available online: https://docs.european-bioplastics.org/conference/Report_Bioplastics_Market_Data_2020_short_version.pdf (accessed on 15 November 2021).
13. European Environment Agency. *Bio-Waste in Europe-Turning Challenges into Opportunities*; Van der Linden, A., Almut, R., Eds.; European Environment Agency: Copenhagen, Denmark, 2020. [CrossRef]
14. Singh, N.; Hui, D.; Singh, R.; Ahuja, I.P.S.; Feo, L.; Fraternali, F. Recycling of plastic solid waste: A state of art review and future applications. *Compos. Part B Eng.* **2017**, *115*, 409–422. [CrossRef]
15. Eriksson, O.; Finnveden, G. Plastic waste as a fuel—CO_2-neutral or not? *Energy Environ. Sci.* **2009**, *2*, 907–914. [CrossRef]
16. Boumanchar, I.; Chhiti, Y.; M'hamdi Alaoui, F.E.; Sahibed-dine, A.; Bentiss, F.; Jama, C.; Bensitel, M. Municipal solid waste higher heating value prediction from ultimate analysis using multiple regression and genetic programming techniques. *Waste Manag. Res.* **2019**, *37*, 578–589. [CrossRef]
17. Dianda, P.; Mahidin, M.; Munawar, E. Production and characterization refuse derived fuel (RDF) from high organic and moisture contents of municipal solid waste (MSW). In *IOP Conference Series: Materials Science and Engineering*; Institute of Physics Publishing: Banda Aceh, Indonesia, 2018; Volume 334. [CrossRef]
18. Gug, J.I.; Cacciola, D.; Sobkowicz, M.J. Processing and properties of a solid energy fuel from municipal solid waste (MSW) and recycled plastics. *Waste Manag.* **2015**, *35*, 283–292. [CrossRef] [PubMed]
19. Świechowski, K.; Stępień, P.; Syguła, E.; Koziel, J.; Białowiec, A. Lab-Scale Study of Temperature and Duration Effects on Carbonized Solid Fuels Properties Produced from Municipal Solid Waste Components. *Materials* **2021**, *14*, 1191. [CrossRef] [PubMed]
20. Wielgosiński, G. Thermal Waste Conversion. Racibórz: Nowa Energia. 2020. Available online: https://nowa-energia.com.pl/wydawnictwa-ksiazkowe/ (accessed on 15 November 2021).
21. Caputo, A.C.; Pelagagge, P.M. RDF production plants: I Design and costs. *Appl. Therm. Eng.* **2002**, *22*, 423–437. [CrossRef]
22. Nobre, C.; Vilarinho, C.; Alves, O.; Mendes, B.; Gonçalves, M. Upgrading of refuse derived fuel through torrefaction and carbonization: Evaluation of RDF char fuel properties. *Energy* **2019**, *181*, 66–76. [CrossRef]
23. Syguła, E.; Świechowski, K.; Stępień, P.; Koziel, J.A.; Białowiec, A. The Prediction of Calorific Value of Carbonized Solid Fuel Produced from Refuse-Derived Fuel in the Low-Temperature Pyrolysis in CO_2. *Materials* **2020**, *14*, 49. [CrossRef]
24. Syguła, E.; Świechowski, K.; Hejna, M.; Kunaszyk, I.; Białowiec, A. Municipal Solid Waste Thermal Analysis—Pyrolysis Kinetics and Decomposition Reactions. *Energies* **2021**, *14*, 4510. [CrossRef]
25. Miranda, T.; Nogales, S.; Román, S.; Montero, I.; Arranz, J.I.; Sepúlveda, F.J. Control of several emissions during olive pomace thermal degradation. *Int. J. Mol. Sci.* **2014**, *15*, 18349–18361. [CrossRef]
26. Abd-Elghany, M.; Klapötke, T.M. A review on differential scanning calorimetry technique and its importance in the field of energetic materials. *Phys. Sci. Rev.* **2018**, *3*, 103. [CrossRef]
27. Mansa, R.; Zou, S. Thermogravimetric analysis of microplastics: A mini review. *Environ. Adv.* **2021**, *5*, 100117. [CrossRef]
28. PN-EN 14346:2011 Standard. Waste Characteristics. Calculation of Dry Mass on the Basis of Dry Residue or Water Content. Available online: https://sklep.pkn.pl/pn-en-14346-2011p.html (accessed on 15 November 2021).
29. Torquato, L.D.M.; Crnkovic, P.M.; Ribeiro, C.A.; Crespi, M.S. New approach for proximate analysis by thermogravimetry using CO_2 atmosphere: Validation and application to different biomasses. *J. Therm. Anal. Calorim.* **2017**, *128*, 1–14. [CrossRef]
30. PN-Z-15008-04:1993 Standard. Municipal Solid Waste. Analysis of Combustible and Non-Combustible Content. Available online: https://sklep.pkn.pl/pn-z-15008-04-1993p.html (accessed on 15 November 2021).
31. PN-EN 15169:2011 Standard. Waste Characteristics. Determination of Organic Matter Content for Waste, Slurry and Sludge. Available online: https://sklep.pkn.pl/pn-en-15169-2011p.html (accessed on 15 November 2021).
32. PN EN ISO 18125:2017-07 Solid Biofuels—Determination of Calorific Value. Available online: https://sklep.pkn.pl/pn-en-iso-18125-2017-07p.html (accessed on 15 November 2021).
33. Poudel, J.; Karki, S.; Oh, S.C. Valorization of waste wood as a solid fuel by torrefaction. *Energies* **2018**, *11*, 1641. [CrossRef]
34. Dyjakon, A.; Noszczyk, T. Alternative fuels from forestry biomass residue: Torrefaction process of horse chestnuts, oak acorns, and spruce cones. *Energies* **2020**, *13*, 2468. [CrossRef]
35. Chen, D.; Gao, A.; Cen, K.; Zhang, J.; Cao, X.; Ma, Z. Investigation of biomass torrefaction based on three major components: Hemicellulose, cellulose, and lignin. *Energy Convers. Manag.* **2018**, *169*, 228–237. [CrossRef]
36. Tumuluru, J.S.; Sokhansanj, S.; Hess, J.R.; Wright, T.C.; Boardman, R.D. A review on biomass torrefaction process and product properties for energy applications. *Ind. Biotechnol.* **2011**, *7*, 384–401. [CrossRef]
37. Camacho-Muñoz, R.; Villada-Castillo, H.S.; Solanilla-Duque, J.F. Anaerobic biodegradation under slurry thermophilic conditions of poly(lactic acid)/starch blend compatibilized by maleic anhydride. *Int. J. Biol. Macromol.* **2020**, *163*, 1859–1865. [CrossRef]

38. Jing, J.; Zhang, Y.; Tang, X.; Fang, Z. Synthesis of a highly efficient phosphorus-containing flame retardant utilizing plant-derived diphenolic acids and its application in polylactic acid. *RSC Adv.* **2016**, *6*, 49019–49027. [CrossRef]
39. Caillat, S.; Vakkilainen, E. Large-scale biomass combustion plants: An overview. In *Biomass Combustion Science, Technology and Engineering*; Woodhead Publishing Limited: Sawston, UK, 2013; pp. 189–224. [CrossRef]
40. Elyounssi, K.; Blin, J.; Halim, M. High-yield charcoal production by two-step pyrolysis. *J. Anal. Appl. Pyrolysis* **2010**, *87*, 138–143. [CrossRef]
41. Narzari, R.; Borkotoki, B. Chapter 2-Biochar: An Overview on its Production, Properties and Potential Benefits. In *Biology, Biotechnology and Sustainable Development*; Research India Publications: Delhi, India, 2015.
42. Tong, S.; Xiao, L.; Li, X.; Zhu, X.; Liu, H.; Luo, G.; Worasuwannarak, N.; Kerdsuwan, S.; Fungtammasan, B.; Yao, H. A gas-pressurized torrefaction method for biomass wastes. *Energy Convers. Manag.* **2018**, *173*, 29–36. [CrossRef]
43. Ramos-Carmona, S.; Pérez, J.F.; Pelaez-Samaniego, M.R.; Barrera, R.; Garcia-Perez, M. Effect of torrefaction temperature on properties of patula pine. *Maderas: Cienc. Y Tecnol.* **2017**, *19*, 39–50. [CrossRef]
44. Yi, L.; Feng, J.; Qin, Y.H.; Li, W.Y. Prediction of elemental composition of coal using proximate analysis. *Fuel* **2017**, *193*, 315–321. [CrossRef]
45. Zhou, H.; Meng, A.; Long, Y.; Li, Q.; Zhang, Y. An overview of characteristics of municipal solid waste fuel in China: Physical, chemical composition and heating value. *Renew. Sustain. Energy Rev.* **2014**, *36*, 107–122. [CrossRef]
46. Słupska, M.; Dyjakon, A.; Stopa, R. Determination of strength properties of energy plants on the example of miscanthus-giganteus, rosa multiflora and salix viminalis. *Energies* **2019**, *12*, 3660. [CrossRef]
47. Parikh, J.; Channiwala, S.A.; Ghosal, G.K. A correlation for calculating HHV from proximate analysis of solid fuels. *Fuel* **2005**, *84*, 487–494. [CrossRef]
48. Backes, E.H.; Pires, L.d.N.; Costa, L.C.; Passador, F.R.; Pessan, L.A. Analysis of the degradation during melt processing of pla/biosilicate® Composites. *J. Compos. Sci.* **2019**, *3*, 52. [CrossRef]
49. Sousa, S.; Costa, A.; Silva, A.; Simões, R. Poly(lactic acid)/Cellulose films produced from composite spheres prepared by emulsion-solvent evaporation method. *Polymers* **2019**, *11*, 66. [CrossRef] [PubMed]
50. Byadgi, S.A.; Kalburgi, P.B. Production of Bioethanol from Waste Newspaper. *Procedia Environ. Sci.* **2016**, *35*, 555–562. [CrossRef]
51. Van der Stelt, M.J.C. *Chemistry and Reaction Kinetics of Biowaste Torrefaction*; Technische Universiteit Eindhoven: Eindhoven, The Netherlands, 2011.
52. Porshnov, D.; Ozols, V.; Ansone-Bertina, L.; Burlakovs, J.; Klavins, M. Thermal decomposition study of major refuse derived fuel components. *Energy Procedia* **2018**, *147*, 48–53. [CrossRef]
53. Yang, H.; Yan, R.; Chen, H.; Lee, D.H.; Zheng, C. Characteristics of hemicellulose, cellulose and lignin pyrolysis. *Fuel* **2007**, *86*, 1781–1788. [CrossRef]
54. Noszczyk, T.; Dyjakon, A.; Koziel, J.A. Kinetic parameters of nut shells pyrolysis. *Energies* **2021**, *14*, 682. [CrossRef]
55. Lavrykov, S.A.; Ramarao, B.V. Thermal Properties of Copy Paper Sheets. *Dry. Technol.* **2012**, *30*, 297–311. [CrossRef]
56. Zmeskal, O.; Marackova, L.; Lapcikova, T.; Mencik, P.; Prikryl, R. Thermal properties of samples prepared from polylactic acid by 3D printing. *AIP Conf. Proc.* **2020**, *2305*, 020022. [CrossRef]
57. Water-Heat of Vaporization Online Water Heat of Vaporization Calculator. Available online: https://www.engineeringtoolbox.com/water-properties-d_1573.html (accessed on 15 November 2021).
58. Świechowski, K.; Syguła, E.; Koziel, J.A.; Stępień, P.; Kugler, S.; Manczarski, P.; Białowiec, A. Low-Temperature Pyrolysis of Municipal Solid Waste Components and Refuse-Derived Fuel—Process Efficiency and Fuel Properties of Carbonized Solid Fuel. *Data* **2020**, *5*, 48. [CrossRef]

materials

MDPI

Article

The Influence of Low-Temperature Food Waste Biochars on Anaerobic Digestion of Food Waste

Kacper Świechowski *, Bartosz Matyjewicz, Paweł Telega and Andrzej Białowiec

Department of Applied Bioeconomy, Wrocław University of Environmental and Life Sciences, 37a Chełmońskiego Str., 51-630 Wrocław, Poland; 110829@student.upwr.edu.pl (B.M.); pawel.telega@upwr.edu.pl (P.T.); andrzej.bialowiec@upwr.edu.pl (A.B.)

* Correspondence: kacper.swiechowski@upwr.edu.pl

Abstract: The proof-of-the-concept of application of low-temperature food waste biochars for the anaerobic digestion (AD) of food waste (the same substrate) was tested. The concept assumes that residual heat from biogas utilization may be reused for biochar production. Four low-temperature biochars produced under two pyrolytic temperatures 300 °C and 400 °C and under atmospheric and 15 bars pressure with 60 min retention time were used. Additionally, the biochar produced during hydrothermal carbonization (HTC) was tested. The work studied the effect of a low biochar dose (0.05 $g_{BC} \times g_{TSsubstrate}^{-1}$, or 0.65 $g_{BC} \times L^{-1}$) on AD batch reactors' performance. The biochemical methane potential test took 21 days, and the process kinetics using the first-order model were determined. The results showed that biochars obtained under 400 °C with atmospheric pressure and under HTC conditions improve methane yield by 3.6%. It has been revealed that thermochemical pressure influences the electrical conductivity of biochars. The biomethane was produced with a rate (k) of 0.24 d^{-1}, and the most effective biochars increased the biodegradability of food waste (FW) to 81% compared to variants without biochars (75%).

Keywords: methane fermentation; biogas; biomethane; biochar; pyrolysis; hydrothermal carbonization; biochemical methane potential; biogas production kinetics

Citation: Świechowski, K.; Matyjewicz, B.; Telega, P.; Białowiec, A. The Influence of Low-Temperature Food Waste Biochars on Anaerobic Digestion of Food Waste. *Materials* **2022**, *15*, 945. https://doi.org/10.3390/ma15030945

Academic Editors: Rossana Bellopede and Lorena Zichella

Received: 3 December 2021
Accepted: 22 January 2022
Published: 26 January 2022

Publisher's Note: MDPI stays neutral with regard to jurisdictional claims in published maps and institutional affiliations.

1. Introduction

1.1. Background

The implementation of a circular economy induces the new approaches of closing the loops of material and energy flows within the systems, including the new solutions for food waste management. The high biodegradability and high biogas potential of food waste may be utilized for both biogas and organic fertilizer production. The biogas yield may be enhanced, and the fertilizer quality may be improved by the addition of biochars derived from food waste. It may bring added value to food waste, a component of municipal solid waste (MSW) sustainable management. Progressing economic development is conducive to an increase in waste production. MSW causes environmental problems such as water, air, and soil pollution, loss of biodiversity, and resource depletion, and over-use of land [1]. To counteract the negative waste effects and to counteract resource depletion, the European Union (EU) goes to a circular economy, where waste becomes a new resource. According to the directive 2008/98/EC on waste [2], EU states should move towards a circular economy by achieving targets for preparing, reusing, and recycling MSW. These targets were set to a minimum of 55%, 60%, and 65% (by weight) by 2025, 2030, and 2035, respectively [2]. To meet the directive targets and goals of the circular economy, the Council of the European Union adopted a rule for the collection of bio-waste. By 2023, all EU states must collect bio-waste separately or recycle it at the source (home composting) [3].

The bio-waste term refers to biodegradable garden and park waste, food and kitchen waste from households, restaurants, caterers, and retailers, and comparable waste from food processing plants [2]. Bio-waste accounts for about 30% of the MSW stream and about 60% of bio-waste is made from food waste (FW) [4].

1.2. Bio-Waste Processing Methods

Currently, in the EU, MSW (containing bio-waste from households) are mainly processed in the mechanical-biological treatment plants (MBT). In the MBT, in the first step, waste is treated mechanically by screening to separate fractions' streams. An undersize fraction constitutes mainly minerals and wet organic waste, while an oversize fraction consists of plastics and other flammable materials. The screening process is not perfect, and, therefore, part of plastics and other flammable materials go to the undersize fraction, while some organic waste stays in the oversize fraction. As a result, an undersize fraction is unfit for organic recycling, and plastics quality is lower in comparison to plastics collected separately at the source. After screening, the undersize fraction is processed by composting or anaerobic digestion to stabilize, where waste is converted into a low-grade compost-like output (CLO), which must be landfilled [5,6].

On the other hand, when MSW are collected separately, the recycling rate of materials increases, and organics recycling of bio-waste is possible. Waste streams collected separately have higher quality than mixed [5,6], and bio-waste can be converted by composting or anaerobic digestion to fertilizer. In both processes, microorganisms are used to break down organic matter. Compositing is the process under controlled conditions in the presence of oxygen, at an appropriate temperature and humidity of ~60%. Depending on composting technology, it may be done in pits, by piling and heaping [7], or in closed reactors with forced aeration also known as in-vessel systems [8]. During composting, organic matter can heat itself to 70 °C at the thermophilic phase, ensuring the destruction of pathogens [7]. The process also leads to a decrease in the mass and volume because of water evaporation and organic matter decomposition. Besides composting advantages like low-cost technology and easy process control, several drawbacks exist. The process requires external energy for heaps turning and/or aeration, and when out of a vessel system technology is used, gaseous and liquid emissions occur. Processing of green waste results in CO, CO_2, CH_4, H_2, NH_3, N_2O, CH_4, and volatile organic compounds (VOCs) emissions which cannot be avoided [9]. Therefore, if composting does not follow in closed reactors, a better option for biowaste processing is anaerobic digestion (AD).

Methane fermentation is a decomposition of organic matter under an oxygen-free atmosphere by anaerobic microorganisms at 37 °C or 55 °C. The main process product is flammable biogas consisting of CO_2 and CH_4 about 1:1, and digestate residues that can be used as fertilizer or solid fuel as well. Similarly, to composting, a lot of different technologies exist. For an organic fraction of MSW, (i) solid-state anaerobic digestion, (ii) continuous digestion with thermophilic conditions, and (iii) plug flow and continuous stirring tank rectors [6] are the most suitable. Though investment costs are much higher for anaerobic processes compared to aerobic ones, surplus energy production, comparable quality fertilizer, and almost zero emissions are plays in favor of AD [10]. Therefore, biogas plants will gradually replace composting ones.

1.3. Problems with AD of Bio-Waste

Due to a variable of bio-waste composition, conducting the AD process entails certain difficulties. To maintain biogas production at a stable level, many monitoring parameters need to be taken into count (feedstock size, total solids, volatile solids, pH value, ammonium nitrogen, volatile fatty acids (VFA), redox potential, alkalinity ratio, biogas composition (CH_4, CO_2, H_2 and H_2S), temperature, trace elements concentration, organic loading rate (OLR), and hydraulic retention time (HRT)). As a result, trained workers with laboratory equipment are needed [11]. Lack of concise process control and optimization of bio-waste composition lead to harmful intermediate compounds' production and process instability. It is due to organic waste nature. Most FW has acidic pH which consumes digested feedstock alkalinity and is quickly decomposed during the hydrolysis phase.

Quick decomposition with a combination of high protein and lipids content leads to rapid generation and accumulation of ammonia (NH_3), and VFAs over inhibitory levels [12]. Though high VFA concentration does not have to inhibit the process since VFAs are essential nutrients for bacteria growth, pH value needs to be kept at an optimal level to balance the inhibitory effects of VFAs and NH_3 [13]. As a result of difficulties, AD of bio-waste (especially FW) is often performed at a low OLR of 2–3 $g_{COD} \times (L \times d)^{-1}$ [12]. For that reason, different substances improving process stability and performance are added [13]. One such substance getting attention recently is biochar.

Biochar is considered as the material improving the methane fermentation process [14]. Biochar can absorb compounds such as H_2S and CO_2, and it also has the potential to mitigate the inhibition of ammonia and acids. It also creates an optimal environment for the growth of microorganisms, which results in faster colony development and higher biogas yield. The effect of biochar addition (positive or negative) depends on the specific situation like reactor type (batch, continuous) substrate type, type of fermentation, type of the biochar, and others [14].

The biochar is produced from organic materials during thermal processing at temperatures above 300 °C in a free oxygen atmosphere. Depending on conditions, the process is called torrefaction (200–320 °C), pyrolysis (>300 °C) [15], or hydrothermal carbonization (180–320 °C) [16]. Besides temperature, other parameters specify these processes, inter alia residence time, pressure, and initial moisture. Torrefaction and pyrolysis are performed at atmospheric pressure for pre-dried materials, while hydrothermal carbonization is performed at overpressure for wet materials. Each process has pros and cons and is used for different materials and purposes. The amount and quality (desired properties) of carbonaceous material obtained from thermal processing depends on feedstock type and process conditions. In general, the higher the process temperature, the more energy-consuming the thermal processing, and the lower amount of biochar is produced in favor of the yield of other products (liquid and gases) [15–17]. Therefore, low-temperature biochars produced with lower energy demand than under high-temperature pyrolysis may be considered as a sustainable source of structural additive for FW AD. The scientific question on its influence on AD performance may be derived.

1.4. Study Aim

All the advantages of the AD process improvement by biochar addition have not been fully explored because biochar can be produced from various substrates, under different conditions, and various substrates can be processed by AD. Additionally, the application of biochar produced from the same materials as being processed under AD has been rarely studied [18]. In this work, five low-temperature biochars that potentially could be made using residual heat from biogas combined heat and power units (300–400 °C) were produced and used to enhance the AD of FW. Moreover, biochars were produced from the substrate (here food waste) under torrefaction, low-temperature pyrolysis, and hydrothermal carbonization conditions.

2. Materials and Methods

2.1. Materials

2.1.1. Inoculum Preparation

As inoculum for biochemical methane potential tests, digestate from the 1 MW_{el} commercial agricultural biogas plant (Bio-Wat Sp. z o.o., Świdnica, Poland) was used. The biogas plant is operating on wet (dry mas < 10%) and mesophilic conditions (37 °C). The digestate was collected to plastic canisters and was taken to the laboratory where it was stored at room temperature for ~24 h. The next day, the digestate was filtered through gauze to separate liquid from solid particles: unprocessed substrate, plastics, etc. Then, the liquid digestate was stored in the climate chamber (Pollab, model 140/40, Wilkowice, Poland) at 4 °C before the biochemical methane potential test.

2.1.2. Food Waste Preparation

The food waste mixture for biochemical methane potential tests was prepared from food purchased in the grocery store. The mixture consists of 3.67% of orange, 8.67% of banana, 7.33% of apple, 1.33% of lemon, 24.33% of potatoes, 4.67% of onion, 3.33% of salad, 3.33% of cabbage, 2.33% of tomatoes, 6% of rice, 6% of pasta, 3% of bread, 3% of meat, 12% of fish meat, and 11% of cheese by fresh mass. The fresh food waste mixture had 64.2% of moisture content (MC), while volatile solids (VS) constituted 95.8% of dry mass. The ash content (AC) of the mixture was 4.2%. The FW composition was based on the work of Valta et al. [19]. The properties of moisture content, total solids (TS), volatile solids (organic matter content), and ash content, of used food materials, and mixture composition per fresh, dry, and volatile solids percentage share bases are presented in Table 1.

Table 1. Food waste properties and its share in food waste mixtures.

Material	Basic Properties				Share in Mixture		
	MC, % *	TS, % *	VS, % **	AC, % **	By Fresh Mass, %	by Dry Mass, %	by VS, %
Mixture	64.2	35.8	95.8	4.2	-	-	-
Orange	86.2	13.8	95.3	4.7	3.67	1.42	1.43
Banana	81.4	18.6	87.8	12.2	8.67	4.51	4.19
Apple	87.4	12.6	95.4	4.6	7.33	2.58	2.60
Lemon	85.4	14.6	93.5	6.5	1.33	0.55	0.54
Potatoes	61.6	38.4	93.1	6.9	24.33	26.11	25.73
Onion	89.2	10.8	93.4	6.6	4.67	1.41	1.40
Salad	94.9	5.1	85.7	14.3	3.33	0.48	0.43
Cabbage	92.2	7.8	91.6	8.4	3.33	0.72	0.70
Tomatoes	95.1	4.9	82.1	17.9	2.33	0.32	0.32
Rice	13.2	86.8	99.4	0.6	6.00	14.55	15.31
Pasta	11.6	88.4	95.5	4.5	6.00	14.84	15.00
Bread	22.5	77.5	95.2	4.8	3.00	6.50	6.54
Meat	69.8	30.2	96.0	4.0	3.00	2.53	2.57
Fish meat	81.7	18.3	95.5	4.5	12.00	6.12	6.19
Cheese	43.5	56.5	92.8	7.2	11.00	17.37	17.06

* as received base. ** as dry base.

FW components were dried in the laboratory dryer (WAMED, model KBC-65W, Warsaw, Poland) at 105 °C and shredded. Drying time differed depending on the food type. Then, dry food was ground through a 1 mm screen using a laboratory knife mill (Testchem, model LMN-100, Pszów, Poland). Ground FW samples were stored in plastic string bags, at room temperature. The mixture for AD was prepared from ground dry food materials according to data presented in Table 1. To ensure mixture homogeneity, one portion of 1 kg was prepared before the biochemical methane potential test. In addition, all tests were done using this mixture.

2.1.3. Low-Temperature Biochar Preparation and Analyses

The low-temperature biochars, low-temperature and low-pressure biochars, and low-pressure hydro-char were produced using a prototype batch laboratory reactor (WUELS, RBMT2020-1.1, Wrocław, Poland) presented in Figure 1. A full reactor design description is available elsewhere [20]. In short, the reactor is steel-made, an air-tight vessel of 22.3 dm^3, wrapped in a 3 kW heating jacket and insulations (4). The process gas can be released by the upper (6) or lower valve (8). In this study, gas was released by the upper valve and went through a cooler that kept its temperature below 200 °C (to protect the manometer) (1).

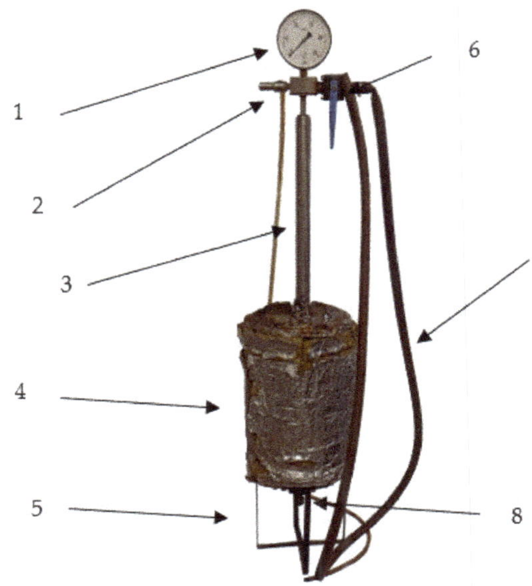

Figure 1. Reactor RBMT2020-1.1 used for biochar production, 1—manometer, 2—safety valve, 3—gas cooler, 4—reactor chamber wrapped by heating jacket and insulation, 5—stand, 6—upper valve, 7—exhaust gas pipe, 8—lower valve.

The biochars were produced from a dry FW mixture at 300 °C and 400 °C in 60 min, at atmospheric pressure, and overpressure of 15 bars. For each process, the residence time of 60 min was counted since the setpoint temperature inside the reactor was reached. For the process at overpressure, when the pressure in the reactor increased over 15 bars, it was released manually up to 14 bars. An exemplary biochar production parameters' diagram is presented in Figure A1. The outer reactor wall temperature was around 150 °C higher than the setpoint temperature (inside the reactor). For low-pressure hydrothermal carbonization (15 bars), a dry FW mixture was mixed with water to obtain 64.2% moisture content (to simulate the initial moisture of FW). The setpoint temperature for hydrothermal carbonization was 280 °C.

For each process, a total sample mass of 250 g was used. Each sample was divided into five smaller samples of ~50 g that were placed into aluminum trays that next were covered with aluminum foil. Then, the five trays were placed evenly inside the reactor. The reason for sample dividing was to place it in a different part of the reactor to assure better heat transfer from the reactor's walls to samples. The reason for covering trays with aluminum foil was to avoid sample incineration at the initial stage where some air could have been present in the reactor.

After 60 min, since the setpoint temperature inside the reactor was reached, the heating jacket was turned off. Additionally, in the case of overpressure processes, the upper valve has been opened to release pressure. Then, the reactor was left to cool down. After cooling down to room temperature, samples were removed. The difference between the initial and end mass of solids was used to calculate the mass yield of the biochar production following Equation (1):

$$MY = \frac{m_b}{m_r} \times 100 \quad (1)$$

where:

MY—mass yield, %;
m_b—dry mass of biochar after the process, g,
m_r—dry mass of material before process, g.

Produced biochars were analyzed for specific surface area (BET), total pore volume <50 nm (Vt), and average pore size <50 nm (L) by adsorption analyzer (Micromeritics, ASAP 2020, Norcross, GA, USA).

2.2. Methods

2.2.1. Biochemical Methane Potential Test

Biochemical methane potential (BMP) tests were performed using an automatic methane potential test system (BPC Instruments AB, AMPTS® II, Lund, Sweden) presented in Figure 2. The system consists of 15 reactors (500 mL) with agitation (2) placed in water batch (1), gas volume meters (4) as well as a built-in data acquisition system that can be displayed on PC (5). Due to the presence of CO_2 absorption units filled with NaOH solution (4), only CH_4 volume was measured.

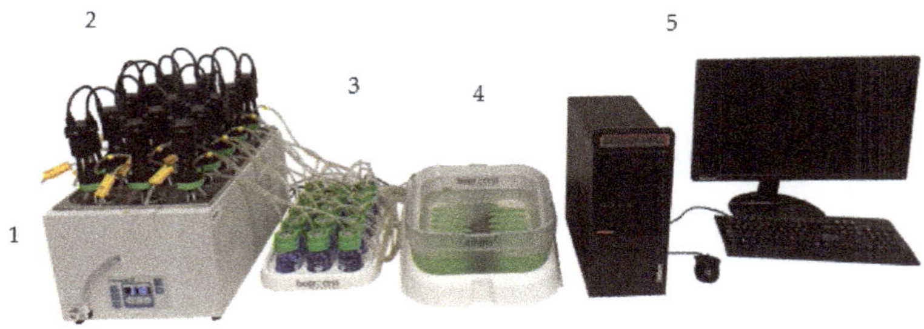

Figure 2. Biochemical methane potential test equipment AMPTS II, 1—water bath, 2—reactors with agitation, 3—CO_2 absorption units, 4—gas volume meters, 5—computer.

A biomethane potential test took 21 days and was performed twice. Each replication consists of two reactors filled with digestate; two reactors filled with digestate and FW, and two reactors filled with digestate, FW, and biochar according to the matrix presented in Table 2. For each reactor, 300 g of liquid digestate was used. For each reactor (excluding the first two), 3.96 g of dry FW mixture was added, and for reactors with BC, 0.1982 g of dry biochar was added. As a result, the substrate to inoculum ratio (SIR) was 0.4 by VS (or 0.25 by TS), the total solids in the reactors were 6.53–6.59%, and biochar share in FW was 5% (by total solids). At the beginning and end of the test, ph and electrical conductivity (EC) was measured using a ph/EC meter (Elmetron, CPC-411, Zabrze, Poland).

Table 2. Anaerobic digestion experiment matrix, D—digestate, FW—food waste, BC_—specific biochar derived under the following conditions: temperature, °C/residence time, min./pressure, bar.

Sample	Digestate	Food Waste Mixture	Biochar
D	+	-	-
D	+	-	-
D + FW	+	+	-
D + FW	+	+	-
D + FW + BC_300/60/0	+	+	+
D + FW + BC_300/60/0	+	+	+
D + FW + BC_300/60/15	+	+	+
D + FW + BC_300/60/15	+	+	+
D + FW + BC_400/60/0	+	+	+
D + FW + BC_400/60/0	+	+	+
D + FW + BC_400/60/15	+	+	+
D + FW + BC_400/60/15	+	+	+
D + FW + BC_ HTC280	+	+	+
D + FW + BC_ HTC280	+	+	+

D—digestate; FW—food waste mixture, BC_300/60/0—biochar produced at 300 °C in 60 min and atmospheric pressure, BC_300/60/15—biochar produced at 300 °C in 60 min and overpressure pressure of 15 bars, BC_400/60/0—biochar produced at 300 °C in 60 min and atmospheric pressure, BC_400/60/150—biochar produced at 300 °C in 60 min and overpressure pressure of 15 bars, HTC280—biochar/hydrochar produced in hydrothermal carbonization process at 280 °C in 60 min at a pressure of up to 15 bars.

The SIR of 0.4 was chosen due to works of [21,22], which show that, for FW, the optimal SIR varies from 0.33 to 0.5, while a 5% BC share in food waste by TS was chosen due to our previous work [18]. In addition, a 5% share of biochar addition considered in the current study is equal to biochar addition of 0.05 $g_{BC} \times g_{TSsubstrate}{}^{-1}$, or 0.65 $g_{BC} \times L^{-1}$.

2.2.2. Materials and Process Residue Analysis

All material used in the study was subjected to moisture content, total solids, volatile solids, and ash content determination [23]. The moisture content and total solids were determined using the laboratory dryer (WAMED, model KBC-65W, Warsaw, Poland), according to the PN-EN 14346:2011 standard [24], while volatile solids and ash content were determined using the muffle furnace (SNOL, 8.1/1100, Utena, Lithuania) according to the PN-EN 15169:2011 standard [25]. Additionally, biochars were analyzed for pH and EC. The measurements were performed in measured in solution: 1 g of dry mass to 10 mL of deionized water, after 30 min since being mixed [26].

FW mixture was additionally subjected to ultimate analysis for determination of the elemental composition (C, H, N, S, O). The ultimate analysis was performed using a CHNS analyzer (PerkinElmer, 2400 CHNS/O Series II, Waltham, MA, USA) according to 12902:2007 [27]. The oxygen content was calculated by the difference according to Equation (2):

$$O = 100 - C - H - N - S - AC \tag{2}$$

where:

O—oxygen % share in dry mass, %;
C—carbon % share in dry mass, %;
H—hydrohen % share in dry mass, %; S—sulfur % share in dry mass, %;
AC—ash % share in dry mass, %.

The elemental composition was used for the calculation of theoretical biogas composition and the theoretical biochemical methane potential (TBMP). Calculations were done according to Boyle modification of Buswell and Mueller stoichiometric formulas, Equation (3) [28]:

$$\begin{aligned} C_aH_bO_cN_dS_e + &\left(a - \tfrac{b}{4} - \tfrac{c}{2} + \tfrac{3d}{4} + \tfrac{e}{2}\right)H_2O \\ &\to \left(\tfrac{a}{2} + \tfrac{b}{8} - \tfrac{c}{4} - \tfrac{3d}{8} - \tfrac{e}{4}\right)CH_4 + \left(\tfrac{a}{2} - \tfrac{b}{8} + \tfrac{c}{4} + \tfrac{3d}{8} + \tfrac{e}{4}\right)CO_2 + dNH_3 + eH_2S \end{aligned} \tag{3}$$

where:

$C_aH_bO_cN_dS_e$—elemental composition of the substrate, C—carbon, H—hydrogen, O—oxygen, N—nitrogen, S—sulphury, and a, b, c, d, e stands for molar % share of specific elements of the volatile solids of biomass [29].
H_2O—water needed for substrate decomposition, mol;
CH_4—methane, mol;
CO_2—carbon dioxide, mol;
NH_3—ammonia, mol;
H_2S—hydrogen sulfide, mol.

The mols of biogas products (CH_4, CO_2, NH_3, H_2S) were recalculated for volume in standard conditions (p = 1013.25 hPa, T = 273.15 K) by multiplication obtained mols by 22.415 obeying Avogadro's law. Knowing the elemental composition of substrates and the molar mass of each element, the mass of 1 mol of the substrate was calculated. Next, the volume of each gas component was divided by the mass of 1 mole of substrate used for its production, providing a result in dm^3 per gram of dry substrate. Then, knowing the volatile solids of a substrate, results were recalculated to dm^3 of gas per gram of volatile solids of a substrate.

Additionally, the FW biodegradability was calculated using data of cumulative methane production and theoretical maximum methane production following Equation (4) [30], and CH_4 production effect, Equation (5):

$$BD = \frac{EBMP}{TBMP} \times 100 \quad (4)$$

where:

BD—biodegradability of FW obtained in the methane fermentation process, %;
$EBMP$—experimental biochemical methane potential, ml × $g_{VS}{}^{-1}$;
$TBMP$—theoretical biochemical methane potential, ml × $g_{VS}{}^{-1}$;

$$CH_4 production\ effect = \frac{CH4_{with\ BC} - CH4_{without\ BC}}{CH4_{without\ BC}} \times 100 \quad (5)$$

where:

$CH_4 production\ effect$—change of CH_4 produced after biochar addition to the process, %;
$CH4_{with\ BC}$—CH_4 produced from a sample without biochar added, ml;
$CH4_{without\ BC}$—CH_4 produced from a sample with biochar added, ml.

2.2.3. Methane Production Kinetics

The results of the BMP test were subjected to kinetics determination. The first-order equation (Equation (6)) was used to provide information about the rate of methane production and the estimated value of maximum methane production potential with the application of Statistica 13.0 software (TIBCO Software Inc., Palo Alto, CA, USA). Afterward, the methane production rate was calculated (Equation (7)) [18]:

$$BMP = EBMP_e \times \left(1 - e^{(-k \times t)}\right) \quad (6)$$

$$r = k \times EBMP_e \quad (7)$$

where:

BMP—the cumulative methane production obtained from a substrate after time t, $ml_{CH4} \times g_{VS}{}^{-1}$;
$EBMP_e$—the estimated value of experimental maximum methane production obtains from a substrate, $ml_{CH4} \times g_{VS}{}^{-1}$;
k—constant reaction rate, d^{-1};
t—process time, d;
r—methane production rate, $ml_{CH4} \times (g_{VS} \times d)^{-1}$.

2.2.4. Statistical Analysis of Biochar Effect

To check if biochar addition had a statistically significant effect (positive or negative) on the methane fermentation, the one-way analysis of variance with post-hoc Tukey tests was performed at the level of $\alpha = 0.05$, with the application of Statistica 13.0 software (TIBCO Software Inc., Palo Alto, CA, USA).

3. Results and Discussion

3.1. Substrate and Biochar Properties

The liquid digestate used for BMP had 7.86 of pH, 68.8 µS × cm^{-1} of EC, 94.7% of MC, 5.3% of TS, 59.3% of VS, and 40.7% of AC, while the FW mixture (substrate) used for BMP tests had 5.6% of MC, 94.4% of TS, 95.8 of VS, and 4.2% of AC (Table 1). The elemental analysis showed that FW mixture was characterized by 44–47.8%, 5.7–6.2%, 39.9–44.4%, 1.45–1.58%, 0.24–0.26% of C, H, O, N, S, respectively (by dry mass base). In addition, the FW mixture was characterized by a pH of 5.62 and EC of 3.6 mS × cm^{-1}.

The five types of biochars were used depending on the production conditions as follows: temperature/time/pressure; however, the HTC280 means a hydrothermal carbonization process at 280 °C in 60 min. The biochars were characterized by MY ranging from 34.3% to 56.4% for 400/60/15 and HTC280, respectively (Table 3). The highest MY was noted in the case of HTC and the 300/60/15 process (Table 3). As result, for biochars with high MY, less substrate and energy are needed for their production in comparison to biochars with low MY. Nevertheless, in such a scenario, the substrate is less converted, and biochar may not have the desired properties [31]. Produced biochars had a relatively low volatile solid content compared to FW used for biochar production. On the other hand, biochars had a much higher ash content than the FW mixture. The ash content in biochar varied from 10.4% to 39.1%, while the FW mixture had only 4.2% of ash. The produced biochar was also analyzed for specific surfaces area (SSA) according to BET theory, total pore volume <50 nm (V_t), and average pore size <50 nm (L). Moreover, produced biochars had a value of SSA ranging from 0.26 to 0.64 g $\times$ m^{-2}, and pore size ranging from 5.2 to 7.1 nm (Table 3). The total pore volume ranged from 3.3 $\times$ 10^{-4} cm^3 $\times$ g^{-1} to 8.2 $\times$ 10^{-4} cm^3 $\times$ g^{-1}, excluding 400/60/15 biochar that had V_t of 11.3 $\times$ 10^{-4} cm^3 $\times$ g^{-1} (Table 3). The pyrolysis results in biochars' pH increase from 5.62 to 8.61–10.75, except HTC280, for which pH decreased to 5.59. Except for biochar produced at 300 °C, all biochars had higher EC in comparison to the FW (Table 3).

Table 3. Low-temperature biochar properties.

Material	MY, % **	MC, % *	TS, % *	VS, % **	AC, % **	SSA, m^2 $\times$ g^{-1}	V_t, cm^3 $\times$ g^{-1}	L, nm	pH ***	EC, mS $\times$ cm^{-1} ***
300/60/0	42.6	4.5	95.5	79.5	20.5	0.62	8.2 $\times$ 10^{-4}	5.2	8.61	3.04
300/60/15	45.9	3.3	96.7	89.6	10.4	0.26	3.3 $\times$ 10^{-4}	5.0	8.04	3.57
400/60/0	37.4	4.4	95.6	77.3	22.7	0.61	7.6 $\times$ 10^{-4}	5.0	10.19	4.53
400/60/15	34.3	4.0	96.0	60.9	39.1	0.64	11.3 $\times$ 10^{-4}	7.1	10.75	7.69
HTC280	56.4	18.4	81.6	88.1	11.9	0.38	5.6 $\times$ 10^{-4}	5.9	5.59	4.71

* as-received base, ** dry base, *** measured in solution: 1 g $_{BC}$ to 10 mL $_{deionized\ water}$, after 30 min.

The pore volume, pore size, specific surface area, pH, elemental composition, surface functional groups, electrical conductivity (EC), and cation exchange capacity (CEC) are considered as key biochar physicochemical properties which affect the AD and biogas production [32]. Porosity is considered a key factor to recognize the plausible relations with microbes in AD. The porosity is characterized in terms of the average diameter [33] and is described by three main pore type: micropores (<2 nm), mesopores (2–50 nm), and macropores (>50 nm). For activated carbon, a specific surface area of micropores may constitute up to 95% of the total SSA of activated carbon. As result, micropores decide about the adsorption capacity. On the other hand, mesopores significantly contribute to the adsorption of larger particles, such as dye or humic acids [34]. Generally, pores with a radius over 25 nm are considered transport pores, while pores smaller than 25 nm are considered adsorbing ones [35]. Besides absorption, pores provide a microorganism habitat for proliferating since the typical size of bacteria is 0.3 μm to 13 μm. The higher the SSA, the more effective biochar is in the interaction with the surrounding species [33]. The SSA of biochar varied significantly depending on substrate and process conditions. The SSA in activated carbons varies from 419 to 3102 m^2 $\times$ g^{-1} [36], while for low-temperatures and not activated biochar (350–500 °C), it varies from 0.36 to 5.31 g $\times$ m^{-2}. Moreover, pore volume and average pore size in such biochars vary from 10 $\times$ 10^{-4} to 80 10^{-4} cm^3 $\times$ g^{-1}, and 2.39 to 14.60 nm, respectively [37]. It means that biochars produced in the current study do not differ significantly in comparison with other biochars produced at similar temperatures but have incomparably smaller SSA in comparison to activated carbon.

Since electrically conductive materials (i.e., mineral particles, carbon materials) added to AD show a reduction in lag phase and increased methane production rates, electrically conductive materials found more attention. Conductive materials (i.e., biochar, graphite, activated carbon) added to AD can promote direct interspecies electron transfer (DIET) between syntrophic partners [38]. The DIET is an alternative to interspecies H_2/formate transfer for syntrophic electron exchange between microbial species. In AD, some methanogens can receive electrons from other microorganisms by molecular electric connections or by conductive materials [39]. For that reason, materials with good electrical conductivity properties are assumed to help enhance methane fermentation. The biochar electrical conductivity can be measured in solid-state [40], as powder [41], or in water solution, like soil EC is measured [42]. The EC varies depending on the method, and therefore caution is needed when data are compared between studies. Nevertheless, results from the same method show that an increase in pyrolysis temperature increases EC value. In addition, this is due to higher carbonization and an increase in ash content [41]. Biochar EC values may vary from 0.04 mS × cm^{-1} to 54.2 mS × cm^{-1}, and besides pyrolysis temperature, the feedstock affects EC as well [42]. These show that biochar produced in this study had relatively low EC (3.04–7.69 mS × cm^{-1}) in comparison to biochars found in the literature.

The pH is an important factor affecting the BMP test results and will be described in more detail later. It is worth noting here that all biochars except HTC280 were alkaline, and their pH increased with process temperature, while HTC280 become more acidic. In addition, it is worth noting that pH did not change when pressure was applied, while EC increased, 3.04 vs. 3.57 mS × cm^{-1} for biochars made at 300 °C, and 4.53 vs. 7.69 mS × cm^{-1} for biochars made at 400 °C. This suggests that pressure may potentially be a parameter that can be used to modify EC. This finding should be further investigated.

3.2. Biochemical Methane Potential—Theoretical and Experimental

The effect of low-temperature biochar addition on the cumulative biomethane production process for 21 days was investigated (Figure 3). The result shows that the highest methane production was obtained for biochar from hydrothermal carbonization (HTC280) and biochar produced at 400/60/0. The control reactors obtained 347.9 $ml_{CH4} \times {g_{VS}}^{-1}$, while reactors with biochars 400/60/0 and HTC280 had 360.1 $ml_{CH4} \times {g_{VS}}^{-1}$ and 365.2 $ml_{CH4} \times {g_{VS}}^{-1}$, respectively (Figure 3). The lowest value of BMP was obtained for reactors where biochar 400/60/15 was added (331.7 $ml_{CH4} \times {g_{VS}}^{-1}$).

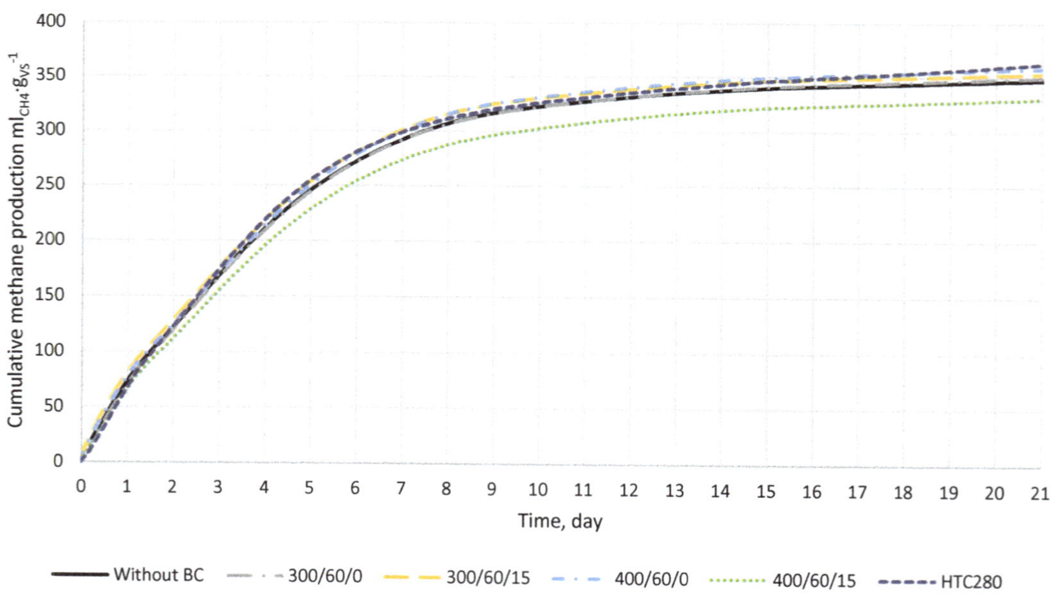

Figure 3. The biomethane production from food waste (n = 4). The results show CH_4 production in ml per gram of food waste volatile solids, and the CH_4 produced by inoculum (digestate) was subtracted.

The theoretical biochemical methane potential of the food waste mixture was 460 $ml_{CH4} \times g_{VS}^{-1}$ (Equation (3)). In addition, theoretical calculations showed that, for complete substrate conversion into biogas, 437 $ml_{CO2} \times g_{VS}^{-1}$, 25 $ml_{NH3} \times g_{VS}^{-1}$, and 2 $ml_{H2S} \times g_{VS}^{-1}$ will be produced. The experimental BMP test for control samples after 21 days obtained 347.9 $ml_{CH4} \times g_{VS}^{-1}$ (Figure 3) reaching 75.5% substrate biodegradation.

Experimental BMP values obtained in this study are lower than the BMP value for source-separated domestic FW collected in the EU, for which BMP ranges from 420 to 470 $ml_{CH4} \times g_{VS}^{-1}$ [43]. Nevertheless, the theoretical potential is in this range, and most reactors reached BD over 75%, which suggests that BMP was done properly, especially since the processing time was only 21 days.

The CH_4 production effect shows a difference between the value obtained from the control (D + FW) and the reactor with biochar (Table 4). When the value is greater than 0, biochar increased the methane production, while when the value is lower than 0, biochar decreased methane production in comparison to control. The biochar addition had a positive effect on methane production from FW. Only biochar 400/60/15 showed a decrease in methane production. For this biochar, all reactors produced less methane than control. For other biochars, mean value from the repetitions was generally positive, and more methane was produced than by control. Nevertheless, biochars produced at 300 °C led to a decrease in methane production in some repetitions. The highest methane production was obtained from reactors where 400/60/0 and HTC280 were added, 3.5%, and 3.6% respectively (Table 4). Among literature, various effects of biochar addition on methane production effect can be found. Results differ from total process inhibition to a several-fold increase in methane production. The effect is highly dependent on factors such as initial conditions of the batch test, used inoculum and substrate, the substrate to inoculum ratio, biochar dose, biochar type, and conditions of its production) [22,44–47]. Kaur et al. [47] added biochars produced at 550 °C and 700 °C from wood, oilseed rape, and wheat straw at a dose of 10 $g_{BC} \cdot L^{-1}$ to co-fermentation of food waste and sewage sludge under a high SIR level of 11.5 by VS. As a result, cumulative methane production increased from 4.5% to 24%. In addition, the highest increase was observed for biochar made from wheat straw at 550 °C, and the lowest for oilseed rape produced at 700 °C [47]. On the other hand, Sunyoto et al. [22] added biochar made from pine sawdust at 650 °C to anaerobic digestion of food waste. Biochar doses of 8.3, 16.6, 25.1, and 33.3 $gBC \cdot L^{-1}$ were studied, and results showed that only a dose of 8.3 increased methane production by 6.2%, while others decreased methane production up to 12.9%. It is also worth noting that biochar doses that increased methane potential did not do it significantly, while biochar doses higher than 25.1 $g_{BC} \cdot L^{-1}$ significantly decreased methane production (at the p-value of 0.002) [22]. Furthermore, the results of Zhang et al. [45] that conducted methane fermentation of FW at thermophilic conditions showed that the lowest of tested biochar doses (6 $g_{BC} \cdot L^{-1}$) gave the highest cumulative methane production [45]. Because, in the current study, only one dosage of 0.65 $g_{BC} \cdot L^{-1}$ was tested, and other research proved that a biochar dose of up to 10 $g_{BC} \cdot L^{-1}$ can improve methane production, higher doses of 400/60/0 and HTC280 should be tested in the future.

The initial pH in all reactors with FW and biochar differed from 7.62 to 7.91, while EC differed from 56.1 to 67.9 $\mu S \times cm^{-1}$. After 21 days of the process, pH differed from 7.92 to 8.03, and EC differed from 68.7 to 77.7 $\mu S \times cm^{-1}$ (Table 4). For comparison, digestate alone had an initial pH and EC of 7.86, and 66.8 $\mu S \times cm^{-1}$, respectively, while, after 21 days, these parameters were 8 and 71.8 $\mu S \times cm^{-1}$, respectively (Table 4). The initial pH is an important parameter affecting methane yield in batch experiments, but no one value would show the correctness of the process [48]. The initial pH and then its changes during the process affect product yield, as optimal pH was reported value from 6.8 to 7.4 [49]. Anaerobic digestion is a four-stage process consisting of hydrolysis, acidogenesis, acetogenesis, and methanogenesis. The pH is crucial in each stage, and each of them required a different value. A positive correlation was found between the hydrolysis rate and pH [49]. The optimal pH for acidogenesis is 5.5–6.5 [50], while methanogenesis is effective when pH is around 6.5–8.2 (with optimum pH of 7.0) [51]. Even though methanogenesis is effective at 6.5, the methanogens' growth rate is reduced significantly at a pH lower than 6.6 [52]. Therefore, the best result of AD can be obtained by a division process into two-stage hydrolysis with acidogenesis, and acetogenesis with methanogenesis [49]. The pH also affects the decomposition of total solids, and volatile solids in the reactor, as well as volatile fatty acid composition [53,54]. Nevertheless, in this study, biochar addition did not significantly change pH ($p < 0.05$), and as result, all reactors had similar conditions. Here it is worth noting that, for some reason, biochars with completely different pH, 10.19 vs. 5.59 for 400/60/0 and HTC280, respectively, showed the best methane production enhancement. The reason for that may be some other biochar properties that were not considered in this study. Maybe these biochars enhanced buffer capacity in the highest way despite different pH, and, as a result, provided better conditions for microorganism growth.

Table 4. The biochar addition effect on the process residues and methane production, after 21 days.

Biochar	No.	Initial		End		Process Residues' Properties				Mass Reduction, %	BD, %	CH_4 Production Effect, %
		pH	EC, μS × cm^{-1}	pH	EC, μS × cm^{-1}	MC, %	TS, %	VS, %	AC, %			
D + FW	1	7.91	61.4	7.92	76.1	95.8	4.2	61.0	39.0	3.6	79.6	-
	2	7.85	63.6	7.92	72.7	95.6	4.4	58.9	41.1	3.7	78.2	-
	3	7.69	65.7	8.02	73.5	95.8	4.2	59.6	40.4	2.1	73.3	-
	4	7.68	65.1	7.99	73.5	95.8	4.2	60.3	39.7	2.5	71.7	-
	Mean	7.78	64.0	7.96	74.0	95.7	4.3	59.9	40.1	3.0	75.5	-
300/60/0	1	7.82	56.1	7.97	75.6	95.6	4.4	60.7	39.3	3.0	78.7	−0.2
	2	7.85	58.6	7.92	74.4	95.6	4.4	60.5	39.5	3.0	78.4	−0.7
	3	7.62	66.1	7.96	74.1	95.7	4.3	63.1	36.9	2.5	75.5	4.1
	4	7.67	66.3	7.96	74.7	95.6	4.4	61.0	39.0	2.3	72.3	−0.3
	Mean	7.74	61.8	7.95	74.7	95.7	4.3	61.3	38.7	2.7	76.2	0.7
300/60/15	1	7.85	66.1	7.93	74.1	95.6	4.4	59.5	40.5	3.1	81.9	3.8
	2	7.84	63.8	7.93	73.5	95.6	4.4	61.5	38.5	3.1	81.2	2.9
	3	7.67	66.6	8.02	75.6	95.7	4.3	59.6	40.4	2.2	72.4	−0.2
	4	7.65	65.1	8.02	74.4	95.6	4.4	62.2	37.8	2.4	72.7	0.3
	Mean	7.75	65.4	7.98	74.4	95.6	4.4	60.7	39.3	2.7	77.0	1.7
400/60/0	1	7.86	57.9	7.92	75.1	95.6	4.4	58.5	41.5	3.1	82.6	4.7
	2	7.84	65.1	7.92	75.9	95.6	4.4	59.6	40.4	3.1	81.6	3.4
	3	7.65	65.7	7.95	74.3	95.7	4.3	61.1	38.9	2.2	75.4	3.9
	4	7.64	64.5	8.01	74.5	95.5	4.5	61.0	39.0	2.3	73.9	1.9
	Mean	7.75	63.3	7.95	75.0	95.6	4.4	60.0	40.0	2.7	78.4	3.5
400/60/15	1	7.83	65.2	7.93	77.7	95.7	4.3	60.1	39.9	2.7	72.4	−8.2
	2	7.85	65.8	7.92	76.5	95.7	4.3	59.9	40.1	3.4	72.0	−0.7
	3	7.68	67.9	8.03	73.9	95.8	4.2	64.4	35.6	2.4	-	-
	4	7.67	61.6	8.00	72.5	95.6	4.4	61.8	38.2	2.3	-	-
	Mean	7.76	65.1	7.97	75.2	95.7	4.3	61.6	38.4	2.7	72.7	−4.5
HTC280	1	7.78	64.7	7.95	75.9	95.7	4.3	61.0	39.0	3.0	81.6	3.4
	2	7.82	63.2	7.93	76.8	95.6	4.4	60.3	39.7	3.9	81.5	3.3
	3	7.64	66.0	7.99	72.0	95.7	4.3	69.6	30.4	2.3	75.4	4.0
	4	7.64	67.4	8.02	68.7	95.7	4.3	61.4	38.6	3.0	-	-
	Mean	7.72	65.3	7.97	73.4	95.7	4.3	63.1	36.9	3.1	79.5	3.6

The EC shows the number of dissolved salts in solutions and is proportional to the quantity of these salts. The solutions with higher salt concentration have a greater ability to conduct an electrical current [42]. In the methane fermentation process, this parameter alone is rather useless. Nevertheless, EC can be used in online monitoring of biogas plants for prediction in advanced methane production of up to two days [55], or alkalinity [56]. As mentioned previously, conductive materials can enhance methane production by DIET. Nevertheless, in this study, biochar addition did not change the electrical conductivity of the solution significantly ($p < 0.05$); therefore, it is highly probable that DIET had no effect here.

Generally, biochar addition did not lead to significant ($p < 0.05$) changes in pH, and EC obtained biodegradability, substrate mass reduction, and amount of produced CH_4. However, even though no statistically significant differences were found, results of biochar made at 400/60/0 and HTC280 showed to always have higher methane production than control, on average by 3.5% (Table 4). At first sight, it looks small; however, when the 1 MWe FW biogas plant working for 8000 h per year is considered, after the addition of BC, the additional 280 MWh of electricity may be produced. It is worth noting that usually biogas plants have problems with the utilization of heat, which in this case may be used for BC production.

3.3. Biomethane Production Kinetics

The mean kinetic parameters evaluated by the model for control (D + FW) were $k = 0.240\ d^{-1}$, $EBMP_e = 351.4\ ml_{CH4} \times g_{VS}^{-1}$ and $r = 84.43\ ml_{CH4} \times (g_{VS} \times d)^{-1}$ (Table 5). All determined kinetics had a high determination coefficient (R > 0.99) (Table 5), which suggests that the used model fits the experimental data well. In general, the first-order model is used for quickly and abruptly stopping degradation substrates [57]. Furthermore, there was no need to use more sophisticated models like the modified Gompertz equation (good fitting when a lag phase is present), the mondo model (good fitting when gas production slowly declining at the end of the process), or two first-order equations (good fitting when two separate degradation profiles occur) [57] since here no such situation took place and biodegradation of over 75% was obtained in 21 days (Table 4).

Table 5. Kinetic of CH4 production for all experiments.

Variant	No.	k, d^{-1}	$EBMP_e$, $ml_{CH_4} \times g_{VS}^{-1}$	r, $ml_{CH4} \times (g_{VS} \times d)^{-1}$	R^2, -
Control	1	0.265	362.13	95.89	0.997
	2	0.270	354.13	95.48	0.996
	3	0.217	348.40	75.46	0.993
	4	0.208	340.94	70.88	0.992
	Mean	0.240	351.40	84.43	0.995
300/60/0	1	0.266	357.43	95.25	0.996
	2	0.264	357.42	94.29	0.996
	3	0.205	357.16	73.31	0.995
	4	0.202	343.32	69.23	0.993
	Mean	0.234	353.83	83.02	0.995
300/60/15	1	0.281	371.93	104.62	0.997
	2	0.273	371.08	101.45	0.997
	3	0.212	342.20	72.62	0.993
	4	0.217	344.88	74.90	0.993
	Mean	0.246	357.52	88.40	0.995
400/60/0	1	0.249	377.05	93.77	0.996
	2	0.268	368.88	98.99	0.996
	3	0.200	356.88	71.20	0.994
	4	0.222	347.68	77.29	0.994
	Mean	0.235	362.62	85.31	0.995
400/60/15	1	0.250	326.62	81.75	0.995
	2	0.208	341.82	70.96	0.994
	3	-	-	-	-
	4	-	-	-	-
	Mean	0.229	334.22	76.36	0.995
HTC280	1	0.254	361.80	91.93	0.992
	2	0.238	364.77	86.82	0.992
	3	0.210	356.53	74.69	0.995
	4	-	-	-	-
	Mean	0.234	361.04	84.48	0.993

The biochar addition changed the values of kinetic parameters slightly, but these changes were not statistically significant ($p < 0.05$). The highest constant production rate of biomethane was observed for 300/60/15 ($k = 0.246\ d^{-1}$), while the lowest for 400/60/15 ($k = 0.229\ d^{-1}$). Overall, 400/60/15 addition resulted in the worst kinetics, and the $EBMP_e$ and r were 334.22 $ml_{CH4} \times g_{VS}^{-1}$ and 76.36 $ml_{CH4} \times (g_{VS} \times d)^{-1}$, respectively. On the other hand, the best kinetics were obtained for 300/60/0 and 400/60/0 (Table 5). These results are a little confusing since the experiment showed that the highest methane production was for 400/60/0 and HTC280; nevertheless, this is probably due to a simplification of the model, which was not able to consider the increase in CH_4 production after 17 days visible for HTC280 (Figure 3).

Overall, the results of methane production kinetics were determined accurately. The maximum methane potential and process kinetics are highly dependent on substrate, inoculum, equipment, and process conditions such as TS and pH. Deepanraj et al. [58] analyzed the kinetic of biogas production from kitchen waste at different TS concentrations (5–15%) and pH (5–9). The results of Deepanraj et al. [58] showed that first-order model kinetics (Gompertz model name by author) fit well to experimental data and had a determination coefficient >0.994. Moreover, results showed that the highest biogas production was obtained for TS = 7.5% and pH of 7 [58]. Is worth noting that these values are close to the ones used in this study (TS varied from 6.53 to 6.59%, and pH varied from 7.62 to 7.91). This suggests that those are important parameters for food waste anaerobic digesting and should be always considered when a BMP test of FW is prepared. There are pieces of evidence in the literature for which biochar addition can improve anaerobic digestion of food waste e.g., by improving process stability, decreasing lag phase, increasing methane yield, etc. Some theories described a process, how biochar enhances AD. Nevertheless, the abundance of food waste and used equipment/procedures lead to different AD enhancement results among studies—bearing in mind that biochar production consumes energy, and biochar transport to biogas plants costs as well. Different low-temperature biochars that potentially could be made using residual heat from biogas combined heat and power unit (CHP) (300–400 °C) were tested. It must be noted that biochars were made from a substrate used in a biogas plant and added to reactors at only one low dose ($0.05\ g_{BC} \times g_{TSsubstrate}{}^{-1}$, or $0.65\ g_{BC} \times L^{-1}$). The application of different BC doses might influence biomethane production more significantly. It should be further investigated.

4. Conclusions

Executed experiments, on the application of biochar produced from the same substrate as used for the anaerobic digestion (food waste) under different low-temperature and pressure conditions, indicated that:

- not all low-temperature biochars at the presented dose can improve biomethane production yield;
- the biomethane yield changes are visible for extreme cases. The worst biochar led to an average 4.5% CH_4 decrease, while two of the best biochars increased CH_4 production on average by 3.5%;
- biomethane production was improved on average by 3.5% by biochar made at 400 °C in 60 min at atmospheric pressure, and by low-pressure hydrochar produced at 280 °C, while the biodegradability of FW was higher than 81% in those variants;
- the theoretical CH_4 potential of food waste was $460\ ml_{CH4} \times g_{VS}{}^{-1}$, while the first-order constant reaction rate was $k = 0.24\ d^{-1}$;
- the FW thermal treatment pressure may influence the EC of biochar.

Further research is needed at low-temperature biochars since this study did not clearly reveal the dependence between low-temperature biochars addition and methane production yield. More trials with different biochar production pressure variants, biochar doses, and at different food waste concentrations should be performed for the validity of the low-temperature biochar application in AD.

Author Contributions: Conceptualization, K.Ś., B.M., A.B.; methodology, K.Ś.; software, K.Ś.; validation, K.Ś. formal analysis, B.M, P.T., K.Ś.; investigation, B.M.; resources, B.M.; data curation, K.Ś.; writing—original draft preparation, K.Ś. and B.M.; writing—review and editing, K.Ś., B.M., A.B.; visualization, K.Ś.; supervision, A.B.; project administration, K.Ś. and A.B.; funding acquisition, B.M. and A.B. All authors have read and agreed to the published version of the manuscript.

Funding: The research was financed under the individual student research project, Młode umysły—Young Minds Project" from the subsidy increased for the period 2020–2025 in the amount of 2% of the subsidy referred to Art. 387 (3) of the Law of 20 July 2018 on Higher Education and Science, obtained in 2019. The APC is financed by Wroclaw University of Environmental and Life Sciences.

Institutional Review Board Statement: Not applicable.

Informed Consent Statement: Not applicable.

Data Availability Statement: All data derived during the experiments are given in the paper.

Conflicts of Interest: The authors declare no conflict of interest and declare that the funders had no role in the design of the study; in the collection, analyses, or interpretation of data; in the writing of the manuscript, or in the decision to publish the results.

Appendix A

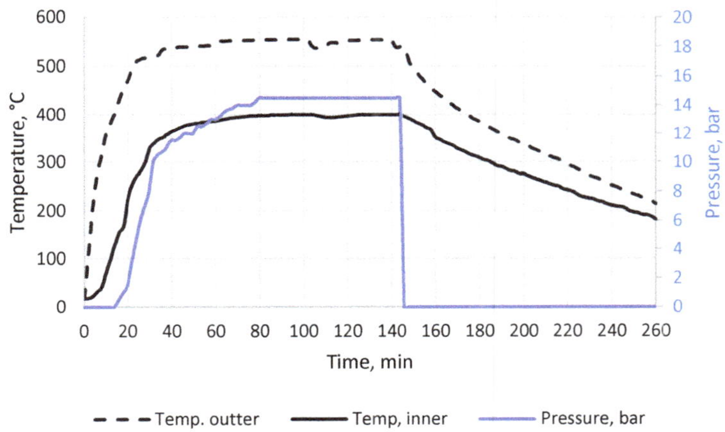

Figure A1. The biochar 400/60/15 production process, process parameters.

References

1. Rockström, J.; Steffen, W.; Noone, K.; Persson, Å.; Chapin, F.S.; Lambin, E.; Lenton, T.M.; Scheffer, M.; Folke, C.; Schellnhuber, H.; et al. Planetary boundaries: Exploring the safe operating space for humanity. *Ecol. Soc.* **2009**, *14*, 1–32. Available online: https://www.ecologyandsociety.org/vol14/iss2/art32/ (accessed on 29 November 2021). [CrossRef]
2. Directive 2008/98/EC of the European Parliament and of the Council of 19 November 2008 on Waste and Repealing Certain Directives. Available online: https://eur-lex.europa.eu/legal-content/EN/TXT/?uri=celex%3A32008L0098 (accessed on 29 November 2021).
3. Store, J. Waste Management and Recycling: Council Adopts New Rules. Available online: https://www.consilium.europa.eu/en/press/press-releases/2018/05/22/waste-management-and-recycling-council-adopts-new-rules/# (accessed on 29 November 2021).
4. European Environment Agency Bio-Waste in Europe-Turning Challenges into Opportunities. Available online: https://www.eea.europa.eu/publications/bio-waste-in-europe (accessed on 29 November 2021).
5. Połomka, J.; Jedrczak, A. Potential of mineral fraction in compost-like-output, methods of its obtaining and the possibility of using it in the context of circular economy. *Materials* **2020**, *13*, 3023. [CrossRef] [PubMed]
6. Zamri, M.F.M.A.; Hasmady, S.; Akhiar, A.; Ideris, F.; Shamsuddin, A.H.; Mofijur, M.; Fattah, I.M.R.; Mahlia, T.M.I. A comprehensive review on anaerobic digestion of organic fraction of municipal solid waste. *Renew. Sustain. Energy Rev.* **2021**, *137*, 110637. [CrossRef]
7. Sarkar, S.; Pal, S.; Chanda, S. Optimization of a vegetable waste composting process with a significant thermophilic phase. *Procedia Environ. Sci.* **2016**, *35*, 435–440. [CrossRef]
8. Sobieraj, K.; Stegenta-Dąbrowska, S.; Koziel, J.A.; Białowiec, A. Modeling of Co accumulation in the headspace of the bioreactor during organic waste composting. *Energies* **2021**, *14*, 1367. [CrossRef]
9. Stegenta-Dabrowska, S.; Sobieraj, K.; Koziel, J.A.; Bieniek, J.; Bialowiec, A. Kinetics of biotic and abiotic CO production during the initial phase of biowaste composting. *Energies* **2020**, *13*, 5451. [CrossRef]

10. Bouallagui, H.; Cheikh, R.B.; Marouani, L.; Hamdi, M. Mesophilic biogas production from fruit and vegetable waste in a tubular digester. *Bioresour. Technol.* **2003**, *86*, 85–89. [CrossRef]
11. Drosg, B. Process Monitoring in Biogas Plants; IEA Bioenergy. 2013. Available online: https://www.ieabioenergy.com/blog/publications/process-monitoring-in-biogas-plants/ (accessed on 29 November 2021).
12. Xu, F.; Li, Y.; Ge, X.; Yang, L.; Li, Y. Anaerobic digestion of food Waste–Challenges and opportunities. *Bioresour. Technol.* **2018**, *247*, 1047–1058. [CrossRef]
13. Issah, A.A.; Kabera, T.; Kemausuor, F. Biogas optimisation processes and effluent quality: A review. *Biomass Bioenergy* **2020**, *133*. [CrossRef]
14. Qiu, L.; Deng, Y.F.; Wang, F.; Davaritouchaee, M.; Yao, Y.Q. A Review on biochar-mediated anaerobic digestion with enhanced methane recovery. *Renew. Sustain. Energy Rev.* **2019**, *115*, 109373. [CrossRef]
15. Chen, W.H.; Wang, C.W.; Ong, H.C.; Show, P.L.; Hsieh, T.H. Torrefaction, pyrolysis and two-stage thermodegradation of hemicellulose, cellulose and lignin. *Fuel* **2019**, *258*, 116168. [CrossRef]
16. Nizamuddin, S.; Baloch, H.A.; Griffin, G.J.; Mubarak, N.M.; Bhutto, A.W.; Abro, R.; Mazari, S.A.; Ali, B.S. An overview of effect of process parameters on hydrothermal carbonization of biomass. *Renew. Sustain. Energy Rev.* **2017**, *73*, 1289–1299. [CrossRef]
17. Morales, V.L.; Pérez-Reche, F.J.; Hapca, S.M.; Hanley, K.L.; Lehmann, J.; Zhang, W. Reverse engineering of biochar. *Bioresour. Technol.* **2015**, *183*, 163–174. [CrossRef] [PubMed]
18. Dudek, M.; Świechowski, K.; Manczarski, P.; Koziel, J.A.; Białowiec, A. The effect of biochar addition on the biogas production kinetics from the anaerobic digestion of brewers' spent grain. *Energies* **2019**, *12*, 1518. [CrossRef]
19. Valta, K.; Sotiropoulos, A.; Malamis, D.; Kosanovic, T.; Antonopoulou, G.; Alexandropoulou, M.; Jonuzay, S.; Lyberatos, G.; Loizidou, M. Assessment of the effect of drying temperature and composition on the biochemical methane potential of in-house dried household food waste. *Waste Manag. Res.* **2019**, *37*, 461–468. [CrossRef]
20. Matyjewicz, B.; Świechowski, K.; Koziel, J.A.; Białowiec, A. Proof-of-Concept of high-pressure torrefaction for improvement of pelletized biomass fuel properties and process cost reduction. *Energies* **2020**, *13*, 4790. [CrossRef]
21. Cai, J.; He, P.; Wang, Y.; Shao, L.; Lü, F. Effects and Optimization of the Use of Biochar in Anaerobic Digestion of Food Wastes. *Waste Manag. Res. J. A Sustain. Circ. Econ.* **2016**, *34*, 409–416. [CrossRef]
22. Sunyoto, N.M.S.; Zhu, M.; Zhang, Z.; Zhang, D. Effect of biochar addition on hydrogen and methane production in two-phase anaerobic digestion of aqueous carbohydrates food waste. *Bioresour. Technol.* **2016**, *219*, 29–36. [CrossRef]
23. de Blasio, C. *Fundamentals of Biofuels Engineering and Technology*; Green Energy and Technology; Springer International Publishing: Cham, Switzerland, 2019; ISBN 978-3-030-11598-2.
24. PN-EN 14346:2011 Standard. Waste Characteristics. Calculation of Dry Mass on the Basis of Dry Residue or Water Content. Available online: https://sklep.pkn.pl/pn-en-14346-2011p.html?options=cart (accessed on 29 November 2021).
25. PN-EN 15169:2011 Standard. Waste Characteristics. Determination of Organic Matter Content for Waste, Slurry and Sludge. Available online: https://sklep.pkn.pl/pn-en-15169-2011p.html (accessed on 29 November 2021).
26. Al-Wabel, M.I.; Al-Omran, A.; El-Naggar, A.H.; Nadeem, M.; Usman, A.R.A. Pyrolysis temperature induced changes in characteristics and chemical composition of biochar produced from conocarpus wastes. *Bioresour. Technol.* **2013**, *131*, 374–379. [CrossRef]
27. PKN ISO/TS 12902:2007 Solid Mineral Fuel-Determination of Total Carbon, Hydrogen and Nitorgen-Instrumental Methods. Available online: https://sklep.pkn.pl/pkn-iso-ts-12902-2007p.html?options=cart (accessed on 29 November 2021).
28. Achinas, S.; Euverink, G.J.W. Theoretical analysis of biogas potential prediction from agricultural waste. *Resour.-Effic. Technol.* **2016**, *2*, 143–147. [CrossRef]
29. Zhang, Y. Anaerobic Digestion Fundamentals II Thermodynamics; Jyväskylä. 2013. Available online: http://www.valorgas.soton.ac.uk/Pub_docs/JyU%20SS%202013/VALORGAS_JyU_2013_Lecture%202.pdf (accessed on 29 November 2021).
30. Orangun, A.; Kaur, H.; Kommalapati, R.R. Batch anaerobic co-digestion and biochemical methane potential analysis of goat manure and food waste. *Energies* **2021**, *14*, 1952. [CrossRef]
31. Stępień, P.; Serownik, M.; Koziel, J.A.; Bialowiec, A. Waste to carbon: Estimating the energy demand for production of carbonized refuse-derived fuel. *Sustainability* **2019**, *11*, 5685. [CrossRef]
32. Kumar, M.; Dutta, S.; You, S.; Luo, G.; Zhang, S.; Show, P.L.; Sawarkar, A.D.; Singh, L.; Tsang, D.C.W. A Critical Review on Biochar for Enhancing Biogas Production from Anaerobic Digestion of Food Waste and Sludge. *J. Clean. Prod.* **2021**, *305*, 127143. [CrossRef]
33. Codignole Luz, F.; Cordiner, S.; Manni, A.; Mulone, V.; Rocco, V. Biochar characteristics and early applications in anaerobic digestion-a review. *J. Environ. Chem. Eng.* **2018**, *6*, 2892–2909. [CrossRef]
34. Zatorska, J. Badania Porowatych Materiałów Węglowych Otrzymywanych Poprzez Karbonizację Poli (Tereftalanu Etylenu) w Mieszaninie z Wybranymi Związkami Magnezu, PhD Thesis, Szczecin. 2013. Available online: https://zbc.ksiaznica.szczecin.pl/dlibra/publication/31080/edition/29484?language=en (accessed on 29 November 2021).
35. Zdravkov, B.D.; Čermák, J.J.; Šefara, M.; Janků, J. Pore Classification in the characterization of porous materials: A perspective. *Cent. Eur. J. Chem.* **2007**, *5*, 385–395. [CrossRef]
36. Xu, M.; Li, D.; Yan, Y.; Guo, T.; Pang, H.; Xue, H. Porous high specific surface area-activated carbon with co-doping N, S and P for high-performance supercapacitors. *RSC Adv.* **2017**, *7*, 43780–43788. [CrossRef]

37. Gaffar, S.; Dattamudi, S.; Baboukani, A.R.; Chanda, S.; Novak, J.M.; Watts, D.W.; Wang, C.; Jayachandran, K. Physiochemical characterization of biochars from six feedstocks and their effects on the sorption of atrazine in an organic soil. *Agronomy* **2021**, *11*, 716. [CrossRef]
38. Cruz Viggi, C.; Simonetti, S.; Palma, E.; Pagliaccia, P.; Braguglia, C.; Fazi, S.; Baronti, S.; Navarra, M.A.; Pettiti, I.; Koch, C.; et al. Enhancing methane production from food waste Fermentate using biochar: The added value of electrochemical testing in pre-selecting the most effective type of biochar. *Biotechnol. Biofuels* **2017**, *10*, 303. [CrossRef]
39. Chen, S.; Rotaru, A.-E.; Liu, F.; Philips, J.; Woodard, T.L.; Nevin, K.P.; Lovley, D.R. Carbon cloth stimulates direct interspecies electron transfer in syntrophic Co-Cultures. *Bioresour. Technol.* **2014**, *173*, 82–86. [CrossRef]
40. Gabhi, R.S.; Kirk, D.W.; Jia, C.Q. Preliminary investigation of electrical conductivity of monolithic biochar. *Carbon* **2017**, *116*, 435–442. [CrossRef]
41. Hoffmann, V.; Rodriguez Correa, C.; Sautter, D.; Maringolo, E.; Kruse, A. Study of the electrical conductivity of biobased carbonaceous powder materials under moderate pressure for the application as electrode materials in energy storage technologies. *GCB Bioenergy* **2019**, *11*, 230–248. [CrossRef]
42. Singh, B.; Dolk, M.M.; Shen, Q.; Arbestain, M.C. Biochar PH, Electrical Conductivity and Liming Potential. In *Biochar: A Guide to Analytical Methods*; Singh, B., Camps-Arbestain, M., Lehmann, J., Eds.; CRC Press: Boca Raton, FL, USA, 2017; pp. 23–38. Available online: https://www.researchgate.net/publication/319206365 (accessed on 29 November 2021).
43. Banks, C.; Heaven, S.; Zhang, Y.; Baier, U. *Food Waste Digestion: Anaerobic Digestion of Food Waste for a Circular Economy*; IEA Bioenergy: Göteborg, Sweden, 2018; ISBN 9781910154588. Available online: https://www.ieabioenergy.com/blog/publications/food-waste-digestion-anaerobic-digestion-of-food-waste-for-a-circular-economy/ (accessed on 29 November 2021).
44. Pan, X.; Lv, N.; Cai, G.; Zhou, M.; Wang, R.; Li, C.; Ning, J.; Li, J.; Li, Y.; Ye, Z.; et al. Carbon- and metal-based mediators modulate anaerobic methanogenesis and phenol removal: Focusing on stimulatory and inhibitory mechanism. *J. Hazard. Mater.* **2021**, *420*, 126615. [CrossRef] [PubMed]
45. Zhang, L.; Lim, E.Y.; Loh, K.C.; Ok, Y.S.; Lee, J.T.E.; Shen, Y.; Wang, C.H.; Dai, Y.; Tong, Y.W. Biochar enhanced thermophilic anaerobic digestion of food waste: Focusing on biochar particle size, microbial community analysis and pilot-scale application. *Energy Convers. Manag.* **2020**, *209*, 112654. [CrossRef]
46. Zhao, W.; Yang, H.; He, S.; Zhao, Q.; Wei, L. A review of biochar in anaerobic digestion to improve biogas production: Performances, mechanisms and economic assessments. *Bioresour. Technol.* **2021**, *341*, 125797. [CrossRef]
47. Kaur, G.; Johnravindar, D.; Wong, J.W.C. Enhanced volatile fatty acid degradation and methane production efficiency by biochar addition in food waste-sludge Co-Digestion: A step towards increased organic loading efficiency in Co-Digestion. *Bioresour. Technol.* **2020**, *308*, 123250. [CrossRef]
48. Chen, T.-H.; Hashimoto, A.G. Effects of PH and substrate: Inoculum ratio on batch methane fermentation. *Bioresour. Technol.* **1996**, *56*, 179–186. [CrossRef]
49. Mao, C.; Feng, Y.; Wang, X.; Ren, G. Review on research achievements of biogas from anaerobic digestion. *Renew. Sustain. Energy Rev.* **2015**, *45*, 540–555. [CrossRef]
50. Kim, J.; Park, C.; Kim, T.-H.; Lee, M.; Kim, S.; Kim, S.-W.; Lee, J. Effects of various pretreatments for enhanced anaerobic digestion with waste activated sludge. *J. Biosci. Bioeng.* **2003**, *95*, 271–275. [CrossRef]
51. Lee, D.H.; Behera, S.K.; Kim, J.W.; Park, H.-S. Methane production potential of leachate generated from korean food waste recycling facilities: A lab-scale study. *Waste Manag.* **2009**, *29*, 876–882. [CrossRef]
52. Zhang, P.; Chen, Y.; Zhou, Q. Waste activated sludge hydrolysis and short-chain fatty acids accumulation under mesophilic and thermophilic conditions: Effect of PH. *Water Res.* **2009**, *43*, 3735–3742. [CrossRef]
53. Dinamarca, S.; Aroca, G.; Chamy, R.; Guerrero, L. The influence of PH in the hydrolytic stage of anaerobic digestion of the organic fraction of urban solid waste. *Water Sci. Technol.* **2003**, *48*, 249–254. [CrossRef] [PubMed]
54. Fang, H.H.P.; Liu, H. Effect of PH on hydrogen production from glucose by a mixed culture. *Bioresour. Technol.* **2002**, *82*, 87–93. [CrossRef]
55. Klein, R.; Slaný, V.; Krčálová, E. Conductivity measurement for control of a biogas plant. *Acta Univ. Agric. Et Silvic. Mendel. Brun.* **2018**, *66*, 1151–1156. [CrossRef]
56. Xue, B.; Zifu, L.; Xuemei, W.; Xi, H.; Shikun, C.; Xiaofeng, B.; Ruiling, G. Online measurement of alkalinity in anaerobic Co-Digestion using linear regression method. *Int. J. Agric. Biol. Eng.* **2017**, *10*, 176–183. [CrossRef]
57. Filer, J.; Ding, H.H.; Chang, S. Biochemical Methane Potential (BMP) assay method for anaerobic digestion research. *Water* **2019**, *11*, 921. [CrossRef]
58. Deepanraj, B.; Sivasubramanian, V.; Jayaraj, S. Experimental and kinetic study on anaerobic digestion of food waste: The effect of total solids and PH. *J. Renew. Sustain. Energy* **2015**, *7*, 063104. [CrossRef]

materials

MDPI

Article

Use of Arc Furnace Slag and Ceramic Sludge for the Production of Lightweight and Highly Porous Ceramic Materials

Gamal A. Khater [1,*], Bassem S. Nabawy [2], Amany A. El-Kheshen [1], Manal Abdel-Baki Abdel Latif [1] and Mohammad M. Farag [1]

1 Glass Research Department, National Research Centre, Cairo 12622, Egypt; aelkheshen1@yahoo.com (A.A.E.-K.); manalbaki@yahoo.com (M.A.-B.A.L.); mmfaragnrc@gmail.com (M.M.F.)
2 Geophysical Sciences Department, National Research Centre, Cairo 12622, Egypt; bsnabawy@yahoo.co.uk
* Correspondence: j.khater@yahoo.com

Abstract: The utility of recycling some intensive industries' waste materials for producing cellular porous ceramic is the leading aim of this study. To achieve this purpose, ceramic samples were prepared utilizing both arc furnace slag (AFS) and ceramic sludge, without any addition of pure chemicals, at 1100 °C. A series of nine samples was prepared via increasing AFS percentage over sludge percentage by 10 wt.% intervals, reaching 10 wt.% sludge and 90 wt.% AFS contents in the ninth and last batch. The oxide constituents of waste materials were analyzed using XRF. All synthesized samples were investigated using XRD to detect the precipitated minerals. The developed phases were β-wollastonite, quartz, gehlenite, parawollastonite and fayalite. The formed crystalline phases were changed depending on the CaO/SiO_2 ratio in the batch composition. Sample morphology was investigated via scanning electron microscope to identify the porosity of the prepared ceramics. Porosity, density and electrical properties were measured; it was found that all these properties were dependent on the composition of starting materials and formed phases. When increasing CaO and Al_2O_3 contents, porosity values increased, while increases in MgO and Fe_2O_3 caused a decrease in porosity and increases in dielectric constant and electric conductivity. Sintering of selected samples at different temperatures caused formation of two polymorphic structures of wollastonite, either β-wollastonite (unstable) or parawollastonite (stable). β-wollastonite transformed into parawollastonite at elevated temperatures. When increasing the sintering temperature to 1150 °C, a small amount of fayalite phase (Fe_2SiO_4) was formed. It was noticed that the dielectric measurements of the selected sintered samples at 1100 °C were lower than those recorded when sintering temperatures were 1050 °C or 1150 °C.

Keywords: industrial wastes; arc furnace slag; ceramic sludge; β-wollastonite; parawollastonite; porous ceramics

Citation: Khater, G.A.; Nabawy, B.S.; El-Kheshen, A.A.; Abdel Latif, M.A.-B.; Farag, M.M. Use of Arc Furnace Slag and Ceramic Sludge for the Production of Lightweight and Highly Porous Ceramic Materials. *Materials* **2022**, *15*, 1112. https://doi.org/10.3390/ma15031112

Academic Editor: A. Javier Sanchez-Herencia

Received: 17 December 2021
Accepted: 21 January 2022
Published: 31 January 2022

1. Introduction

The growth of industrial waste materials has become a significant public and ecological problem due to enlarged populations and rising industrialization. Solid waste handling is one of the most critical environmental issues in many developing countries. Solid waste is the byproduct of human activities such as the construction, utilization, and sharing of various resources in the world. Many studies have considered various aspects such as technology, innovation, recycling of solid waste management in growing and urbanized countries [1–4].

These wastes present many severe problems related to storage, shipping, and atmosphere or environmental pollution. The use of industrial waste in the fabrication of concrete, building materials, and cement is essential for hindering environmental pollution, lowering production expenses, and saving energy consumption. Thus, developing many technologies for recycling such waste is of great importance [5].

Among all industries, metallurgical industries are significantly costly to operate. Their byproducts are also the most abundant industrial wastes. The many kinds of waste products, such as junk, slags, and refractory wastes, make them a major waste-producing industry. Research by Mukhamedzhanova et al. [6] showed that the addition of accurate quantities of modifiers to metallurgical wastes could allow for the fabrication of many kinds of building materials.

Ferrous slag is a main waste byproduct obtained from iron and steel making. Various investigations have considered the use of iron slag as an economical raw material [7–9]. Many countries are interested in manufacturing construction materials via iron slag recycling. In Italy, 25–40 wt.% iron slag was added to clay to synthesize wall tiles. In South Africa, iron slag was used (up to 75 wt.%) in batches to obtain floor tiles with a water absorption ability of approximately 0.2 wt.% and bending strength of 43.8 MPa [10–12].

Arc furnace slag (AFS) is a byproduct of the steelmaking procedure, resulting from the use of electric arc furnaces, which account for more than 40% of the global steel production. AFS is created after melting and preliminary acid refining of liquid steel. It is a rocky and non-rugged material that can be crushed for use as aggregate in concrete batches [13–15].

A basic assumption is that AFS has an excellent prospect for use as a raw material in ceramic production on account of its aluminosilicate content and the matching of alkali and alkaline earth ingredients via their fluxing function in a prepared frit outline. These qualities can assist in the technical procedures of established processing. Ceramic wall and floor tiles are good candidates for producing controlled porosity bodies. Therefore, it is greatly successful to manufacture such ceramics through recycling of industrial byproducts [16].

Earlier researchers found that blast furnace slag was primarily anticipated to be an additive in Portland cement [17]. This was because of its pozzolanic and cementitious properties. By comparison, AFS does not have these features, and instead has higher iron content [18]. Pioro et al. [19] recommended that metallurgical slag be used to generate construction materials such as ceramic tiles. A complete combination of clays and AFS results in a balanced composition where the most wanted superior properties of tiles may be obtained by forming anorthite and wollastonite phases [20]. Wollastonite, also known as calcium silicate ($CaSiO_3$), has been broadly studied due to its enormous applications in ceramics, dental inserts, structural design, and building, where it is used as a floor material as an alternative to granite and natural marble [21,22]. The most interesting character of this material is its ability to produce both curved and flat sheets [23].

Many efforts were made in earlier studies to build ceramic tiles employing 30–40 wt.% of AFS [18]. Teo et al. [16] established that the use of 40 wt.% AFS was capable of producing excellent mechanical properties in ceramic tiles [24].

The most widespread waste material produced from ceramic manufacturing is ceramic sludge. Several familiar artificial ceramics include wall tiles, floor tiles, sanitary ware, domestic ceramics, and traditional pottery. They are frequently created utilizing natural materials including clay and minerals. Ceramic wastes are classified into two groups according to the raw materials used [25]. The first group is wastes created by structural ceramic industries employing only red pastes for creating blocks, bricks, and roof tiles. The second category is fired ceramic wastes, which are formed in production of stoneware ceramic (floor and wall tiles and sanitary ware). Studies have revealed that during ceramic manufacturing, approximately 30% of the matter is wasted [26,27], and presently these wastes are not constructively reused, which causes much pollution to the environment. This indicates the need for research into pioneering methods of reusing ceramic wastes [28,29]. Ceramic sludge typically has high contents of SiO_2 and Al_2O_3 oxides [30].

The recycling of ceramic sludge and electric arc furnace slag to produce ceramic tile material depends on the reaction between the SiO_2 and Al_2O_3 of the ceramic sludge and the CaO of AFS, enhancing the properties of ceramic tile by developing calcium-aluminate-silicate crystals, such as anorthite and wollastonite. Therefore, it is a prominent way to cancel out the negative effect of iron oxide content in AFS on the densification process.

This recycling process has many advantages in terms of cleaning the environment, storing energy, and using sustainable virgin materials [31].

The main goal of this study is to advance the use of two industrial wastes, arc furnace slag and ceramic sludge, for production of lightweight and porous ceramic materials that can be used for environmentally friendly construction. To fulfill this goal, ceramic samples containing parawollastonite, β-wollastonite and gehlenite minerals were prepared at different sintering temperatures. Microstructural, morphological, electrical, and some physical properties of those samples were studied.

2. Experimental Techniques

2.1. Batch Calculation and Samples Preparation

Nine ceramic samples were prepared from arc furnace slag (Al-Ezz-Dekheila Iron Steel Company, Alexandria, Egypt) and ceramic sludge (Ceramica Venezia, Cairo, Egypt). The arc furnace slag content in their composition varied from 10 wt.% to 90 wt.% at 10 wt.% intervals, and accordingly, the ceramic sludge content varied from 90 wt.% to 10 wt.%. The ceramic batches were prepared by calculating the appropriate proportions of arc furnace slag and ceramic sludge. The required silica and calcium oxide for all compositions were obtained from silica sand and limestone, respectively. The other minor elements present in the industrial wastes were considered during batch calculation. X-ray fluorescence (XRF) was used for chemical analysis of both waste materials (Table 1). The ceramic sample compositions were calculated based on ceramic sludge percentage with each successive increase in arc furnace slag percentage. These samples were designated AFS1, AFS2, AFS3, AFS4, AFS5, AFS6, AFS7, AFS8, and AFS9, as listed in Table 2.

Approximately 10 kg, as representative samples, of arc furnace slag, ceramic sludge, limestone, and silica sand were collected and crushed into −100 mesh powders. An adequate representative quantity for all stages of the laboratory investigation was obtained using the quartering technique. The prepared ceramic batches, after being accurately weighed to yield approximately 4 g for each sample, were closely dry-mixed in a ball mill for approximately 40 min until they became utterly homogeneous. Approximately 5% water was added to each sample as a binder, and then the powder samples were shaped into cylinders of 40 mm diameter and 4 mm thickness via uniaxial pressing at 20 MPa. The prepared ceramic batches were dried in an oven at a temperature of 100 °C for 24 h and then sintered at 1100 °C for one hour; this is shown in Figure 1. Figure 2 exhibits the visual appearance of the prepared samples.

Table 1. Chemical analysis of used raw materials.

Oxide (wt.%)	Arc Furnace Slag	Ceramic Sludge	Limestone	Silica Sand
SiO_2	16.09	61.89	0.15	99.2
CaO	38.59	6.58	55.7	0.1
Al_2O_3	5.24	17.28	0.22	0.28
MgO	14.62	0.9	0.1	trace
Na_2O	0.62	1.83	trace	trace
BaO	0	0.64	nil	trace
MnO	0.74	0.02	nil	trace
Fe_2O_3	10.29	0.99	0.03	0.03
TiO_2	0.16	0.74	nil	trace
K_2O	0.13	1.24	trace	trace
L.O.I.	11.15	5.86	44.02	0.4

Table 2. Chemical compositions and corresponding batch percentages of investigated samples.

Batch No.	Nominal Composition (wt.%)		Calculated Oxide Constituents (wt.%)					CaO/SiO_2	Batch Ingredients (wt.%)			
	AF	S	SiO_2	CaO	Al_2O_3	Fe_2O_3	MgO		Arc Furnace Slag	Sludge	Limestone	Silica Sand
AFS1	10	90	47.86	32.73	13.42	1.6	1.9	0.68	6.52	58.64	34.84	0
AFS2	20	80	47.8	30.22	13.48	2.58	3.3	0.63	14.6	58.39	27.08	0
AFS3	30	70	47.65	27.35	13.53	3.71	4.97	0.57	24.84	57.95	17.21	0
AFS4	40	60	47.52	23.79	13.61	5.14	6.97	0.5	38.32	57.48	4.2	0
AFS5	50	50	52.58	21.46	10.7	5.37	7.37	0.41	42.91	42.91	0	14.17
AFS6	60	40	48.54	24.52	9.56	6.25	8.69	0.51	51.37	34.25	0	14.38
AFS7	70	30	48.62	25.55	7.81	6.61	9.25	0.52	55.74	23.88	0	20.38
AFS8	80	20	49.44	26.02	6.19	6.82	9.6	0.53	58.71	14.68	0	26.61
AFS9	90	10	49.8	26.62	4.85	7.04	9.97	0.53	61.64	6.85	0	31.51

AF = arc furnace slag, S = ceramic sludge.

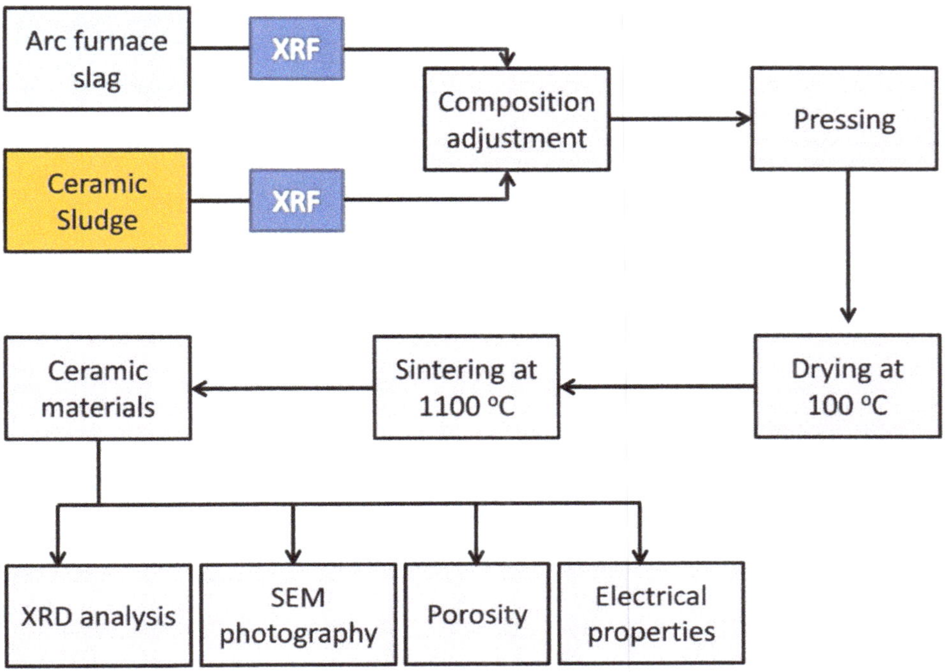

Figure 1. Schematic presentation of ceramic production from industrial wastes.

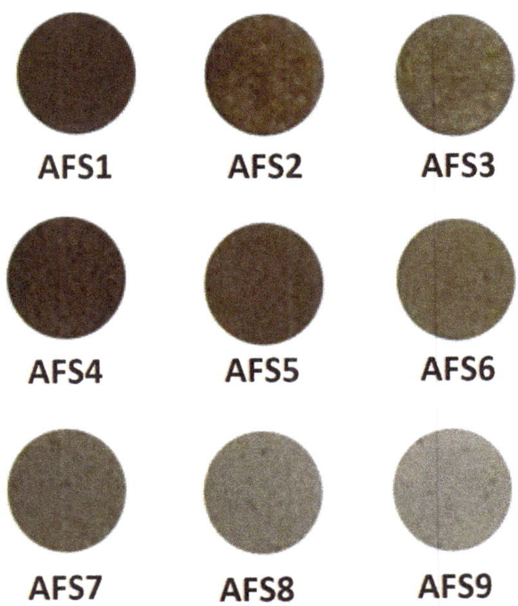

Figure 2. The surface appearance of ceramic samples prepared from industrial wastes after being treated at 1100 °C for 1 h.

Sample AFS5 was chosen, representing the average composition of all other samples, to study the effect of different sintering temperatures on the developed phases and microstructures.

2.2. X-ray Diffraction

The precipitated phases were identified with X-ray diffraction using a Bruker AXA diffractometer (D8—ADVANCE, Bruker, Germany) with Cu–Ka radiation, operating at 40 Kv, 40 mA, and a scanning rate of 10°/min.

2.3. Microstructure Photographs

The microstructure of the prepared ceramic samples was investigated utilizing a scanning electron microscope (SEM: JEOL, XL30, Philips, Amsterdam, Netherlands), which operated at an acceleration voltage of 20 kV. Each freshly broken sample was coated with a gold layer on the fracture surface (to reduce any charging effect) in order to observe the internal microstructure.

2.4. Porosity and Density Measurements

The dimensions, bulk volume (v_b), and dry weight (w_d) of the prepared samples were measured directly. For measuring weight and dimensions, a digital balance with 0.1 mg precision and a highly precise caliper sensitive to 0.01 mm were used, respectively. Then, the bulk density (ρ_b) was estimated.

The grain density (ρ_g) of the studied samples was measured using a helium pycnometer at the ambient conditions. Then, the helium porosity ($\varnothing_{He}$) was estimated considering the bulk and grain densities, as follows [32–34]:

$$\varnothing_{He}\,(\%) = 100 \times (\rho_g - \rho_b)/\rho_g \tag{1}$$

Density measurements were taken five times for each sample, and the average value was calculated.

2.5. Dielectric and Electric Measurements

For the present samples, the capacitance effect (*Cp*), electric loss tangent (*D*), and electric resistance (*Rp*) were measured in a parallel series using a computerized LCR Hitester impedance bridge (IM3536) in the range of 50 Hz–8 MHz at 350 AC frequency points. Then, the dielectric constant (ε'), dielectric loss (ε''), imaginary part (M'') of the complex electric modulus, and electric conductivity (σ) were estimated as follows:

$$\sigma = \frac{L}{A \times R_P} \tag{2}$$

$$\varepsilon' = \frac{C_P \times L}{A \times \varepsilon_0} \tag{3}$$

$$\varepsilon'' = \varepsilon' \times D \tag{4}$$

$$M'' = \frac{\varepsilon''}{\varepsilon'^2 + \varepsilon''^2} \tag{5}$$

where ε' is the dielectric constant, Cp is the electric capacitance in parallel, L is the sample thickness, A is the contact area of the measuring non-polarized electrode, ε'' is the dielectric loss, ε_o is the dielectric constant of the air (8.85×10^{-12} F/m), D is the measured electric loss tangent (tan δ), R_P is the measured electric resistance in parallel, σ is the electric conductivity, and M'' is the imaginary electric modulus.

More details on the applied techniques are discussed by many authors [35–37].

For determining the implications of different sintering temperatures on the electric and dielectric properties of the studied samples, three samples of the same composition

(50% arc furnace ceramic slag + 50% sludge; AFS5), were sintered at 1050 °C, 1100 °C, and 1150 °C. Then, their density, porosity, and electric and dielectric parameters were measured.

3. Results and Discussions

3.1. X-ray Diffraction Patterns

X-ray diffraction graphs (Figure 3) of investigated ceramic samples AFS1–AFS9, after sintering at 1100 °C for one hour, showed all precipitated phases of the samples. The major crystalline phases formed were β-wollastonite ($CaSiO_3$) (JCPDS No.29-372), parawollastonite ($CaSiO_3$) (JCPDS No.27-88), gehlenite ($Ca_2Al_2SiO_7$) (JSPDS No.35-755), and low quartz (SiO_2) (JCPDS No.5-0490). Sample AFS1 contained β-wollastonite ($CaSiO_3$) as the dominant phase with lines at 3.33, 2.96, 3.16, and 3.06 Å. Low quartz (SiO_2), which has distinctive lines 4.23, 3.33, 2.29, and 1.82 Å, represented the second phase. Gehlenite ($CaAl_2SiO_7$) was the third phase with lines at 3.72, 2.85, 2.43, 2.04, and 1.75 Å. Samples AFS2–AFS4 showed a significant decrease in quartz phase content. In addition, it is noticed that there was a transformation of β-wollastonite into parawollastonite, as seen through reduction in line heights at 3.33 and 2.51 Å corresponding to β-wollastonite and increases in intensity for the main lines of parawollastonite, which are 2.96, 3.16, and 3.83 Å; thus, parawollastonite was the main phase, and gehlenite was considered the second phase with small amounts of low quartz. In sample AFS5, β-wollastonite and low quartz phases reappeared with lines at 3.33, 2.96, 3.16, and 3.06 Å for β-wollastonite and lines at 4.23 3.33, 2.29, and 1.82 Å for low quartz. Sample AFS6 showed the development of gehlenite phase, which became the main phase with increasing intensities of lines at 2.85, 3.72, 2.43, 2.04, and 1.75 Å. AFS6's pattern indicated the formation of β-wollastonite and quartz, while samples AFS7, AFS8, and AFS9 showed development of both low quartz and β-wollastonite phases and a noticeable decrease in the intensity of gehlenite lines at 3.72, 2.85, 2.43, 2.04, and 1.75 Å.

Through a deep glance at Table 2, it can be found that sample AFS1 contained a CaO/SiO_2 ratio of approximately 0.68, which confirms the X-ray diffraction results. That accounted for β-wollastonite as a primary phase with low quartz and gehlenite as secondary phases. For samples AFS2–AFS4, the CaO/SiO_2 ratio decreased from 0.63 to 0.50, corresponding with the transformation of β-wollastonite into parawollastonite. A decrease in the percentage of low quartz phase and an increase in the percentage of gehlenite were also noticed in samples AFS2, AFS3, and AFS4. The CaO/SiO_2 ratio decreased to 0.41 in sample AFS5, which helped form β-wollastonite and gehlenite as the major phases and low quartz as secondary phase. For sample AFS6, the CaO/SiO_2 ratio increased to 0.51; this led to the formation of gehlenite as the main phase with β-wollastonite and low quartz as secondary phases. Samples AFS7–AFS9 had nearly the same CaO/SiO_2 ratio, approximately 0.52–0.53, which helped form β-wollastonite as the primary phase with low quartz as the second phase and gehlenite as the third phase. All the above results indicate that β-wollastonite (unstable phase) was formed more often than low quartz, with a small amount of gehlenite. The β-wollastonite phase turned into parawollastonite (stable phase) when the CaO/SiO_2 ratio decreased. The parawollastonite phase was formed with greater amounts of gehlenite and smaller amounts of low quartz.

The obtained results indicate that a low CaO/SiO_2 ratio enhances the formation of β-wollastonite and low quartz phases but hinders the development of gehlenite phases (samples AFS7–AFS9). At the same time, increasing the CaO/SiO_2 ratio promotes the formation of parawollastonite and gehlenite phases but hinders the development of the low quartz phase (samples AFS2–AFS4).

Fan et al. [38] explained that by increasing the CaO/SiO_2 ratio, the basicity increases, and consequently, the crystallization of the CaO-Al_2O_3-MgO-SiO_2 system is improved [39,40]. Tabit et al. [41] found that parawollastonite and gehlenite were formed in samples containing higher calcium oxide percentages.

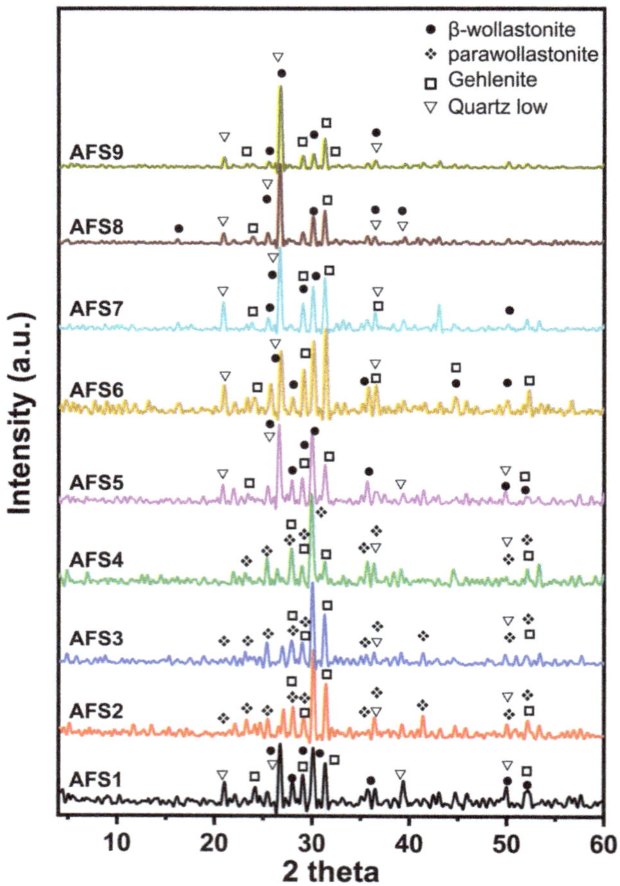

Figure 3. The major phases resulting from industrial wastes, determined via XRD after treatment at 1100 °C for one hour.

Effect of Sintering Temperatures on Crystalline Phases

AFS5 was chosen to study the effect of sintering temperatures on precipitated phases and resulting physical properties, because it represents the mean composition of all the examined samples. The studied sintering temperatures were 1050, 1100, and 1150 °C, all for one hour. Figure 4 shows that, after sintering at 1050 °C, β-wollastonite ($CaSiO_3$) became a primary phase with its characteristic lines at 3.33, 2.967, and 2.51 Å. Low quartz (SiO_2) was the second phase, with its distinctive lines at 4.24, 3.33, and 2.46 Å, and gehlenite ($Ca_2Al_2SiO_7$) was the third phase, with its characteristic lines at 3.74, 3.17, and 2.84 Å. When the temperature was raised to 1100 °C, the gehlenite lines decreased, evident in the mainline 2.84 Å. In addition, β-wollastonite began to convert into parawollastonite. This transformation is demonstrated by the increasing intensity of the line at 2.96 Å and the decreasing intensity of the β-wollastonite line at 3.33 Å. Upon raising the temperature to 1150 °C, β-wollastonite transformed into parawollastonite, which is evident from the decreased intensity of its line at 3.33 Å and increased intensity of the parawollastonite line at approximately 2.97 Å. Gehlenite wholly disappeared, as its distinctive line at 2.84 Å vanished. This increase in sintering temperatures promoted the formation of the fayalite (Fe_2SiO_4) phase. These results agree with Edrees et al. [42], who showed that β-wollastonite is an unstable phase at lower temperatures and turns into parawollastonite (stable phase) at higher temperatures. Ismail et al. [43] found that wollastonite has the property of a polymorphic structure, whether as β-wollastonite (unstable) or parawollastonite (stable). β-wollastonite turns into an independent parawollastonite at high temperatures. Qin et al. [44] also showed that gehlenite was formed with wollastonite and disappeared at high temperatures (1125 °C). The effect of the CaO/SiO_2 ratio on the phase transformation of anorthite, based on steel slag and fly ash, was studied by Tabit et al. [41]. They deduced that the gehlenite phase was precipitated at 1100 °C.

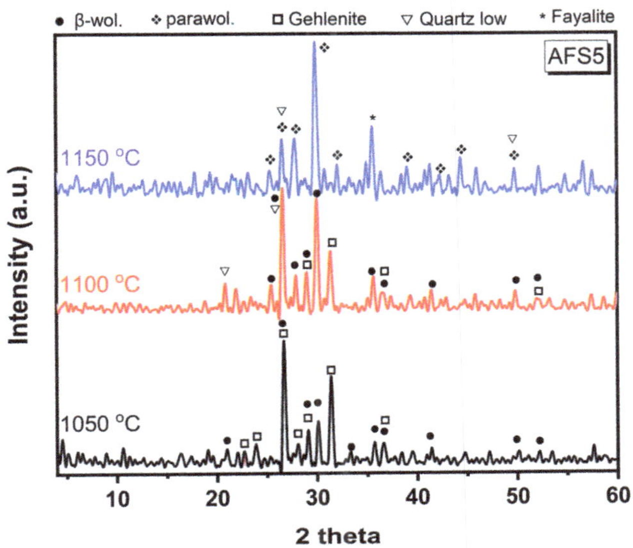

Figure 4. X-ray diffraction patterns of AFS5 samples after sintering at different temperatures for one hour.

3.2. Morphology Studies

Figure 5 presents SEM photographs at different magnifications of the fractured samples AFS1–AFS8 after heat treatment at 1100 °C. It can be observed from the figure that all ceramic samples possessed porous structures, and there were wide variations in the pore size distribution (the pore diameter ranged from 4 to 80 µm) and shape. Circular and irregular pore outlines were noticed in all ceramic samples. Moreover, closed and interconnected pores are apparent in the photos. Porosity decreased and density increased with increases in arc furnace slag (AFS) weight, as is confirmed from the porosity percentage results (decreasing from 41% for AFS1 to 28% for AFS8), and the increasing density of wall pores (Figure 5). This can be attributed to the decrease in CaO and increase in MgO. As MgO molecules are smaller than CaO molecules, this caused a decrease in liquidus temperature, as reported in a previous study [45]. In addition, decreasing Al_2O_3 content caused a decrease in the liquidus temperature of the ceramic particles. When Al_2O_3 content increases, the amount of nonbridging oxygens (NBOs) decreases and consequently, Al–O–Si bonds form and raise viscosity values [46]. This decrease in liquidus temperature leads to more condensed ceramic samples with lower porosity. Moreover, columnar and tabular crystals, which are the characteristic structures of wollastonite crystals, were noticed.

Figure 5. SEM images at different magnifications of microstructures of the investigated samples after treatment at 1100 °C for 1 h.

Effect of Sintering Temperatures on Ceramic Morphology

The applied sintering temperatures for ceramic materials affect properties such as porosity, density, and microstructure. Accordingly, different sintering temperatures were applied on AFS5 to determine the optimum conditions for producing such ceramic. The sample was heat-treated at 1050 °C, 1100 °C, and 1150 °C at a constant time (Figure 6). As shown from the figure, an increase in sintering temperature from 1050 °C to 1100 °C led to a denser and more compact structure. This could be due to the formation of well-developed, dense, and long wollastonite crystals. Upon raising the sintering temperature to 1150 °C, a slight change in microstructure was noticed. This increase in temperature caused the appearance of new rod-shaped crystals with a minor increase in the porous structure and a loss of tiny pores. This might be because of the disappearance of the gehlenite phase and the formation of the fayalite phase, as well as due to the transformation of the β-wollastonite (triclinic) structure into the parawollastonite (monoclinic) structure.

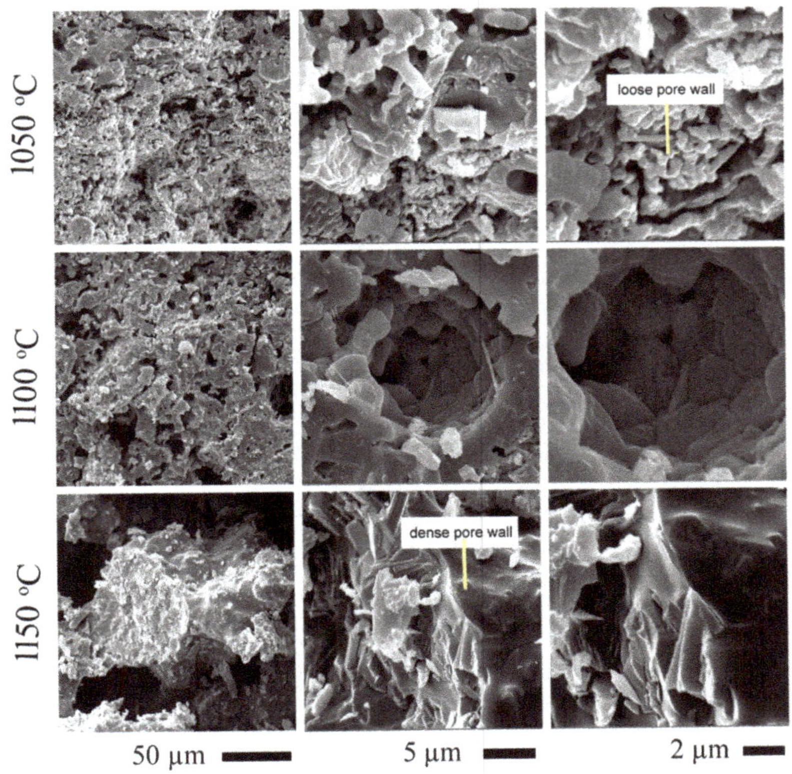

Figure 6. SEM images at different magnifications of microstructures of AFS5 ceramic sample after treatment at 1050 °C, 1100 °C, and 1150 °C.

3.3. *Porosity and Density Studies*

The prepared samples are considered highly porous materials with very high helium porosity values ($\varnothing_{He}$) and shallow bulk density values (ρ_b) (Table 3 and Figure 7). The lowest density values were primarily assigned to the first four samples AFS1–AFS4 and the ninth sample AFS9, which consisted primarily of monoclinic wollastonite, $CaSiO_3$, ρ_g = 2.86–3.09 g/cm^3, [47] and gehlenite, $Ca_2Al_2SiO_7$, ρ_g = 3.04 g/cm^3 [48], with some quartz content, SiO_2, ρ_g = 2.62–2.65 g/cm^3 [49] (Table 3).

Table 3. Main resulting phases and values of porosity and density for the investigated ceramic samples after being treated at 1100 °C for 1 h.

Sample No.	Density g/cm^3	Porosity%	Phases Developed
AFS 1	1.896	41.12	β-woll., Geh., QZ-low
AFS 2	1.906	40.14	Parawoll., Geh., QZ-low
AFS 3	1.875	38.87	Parawoll., Geh., QZ- low
AFS 4	1.824	41.66	Parawoll., Geh., QZ-low
AFS5	2.01	31.16	β-woll., Geh., QZ-low
AFS 6	2.211	29.4	Geh., β-woll., QZ-low
AFS 7	2.203	29	β-woll, Geh., QZ-low
AFS 8	2.106	28.41	β-woll., Geh., QZ-low
AFS 9	1.798	47.26	β-woll., Geh., QZ-low

Parawoll. = parawollastonite, β-woll. = β-wollastonite, Geh. = gehlenite, Qz-low = low quartz.

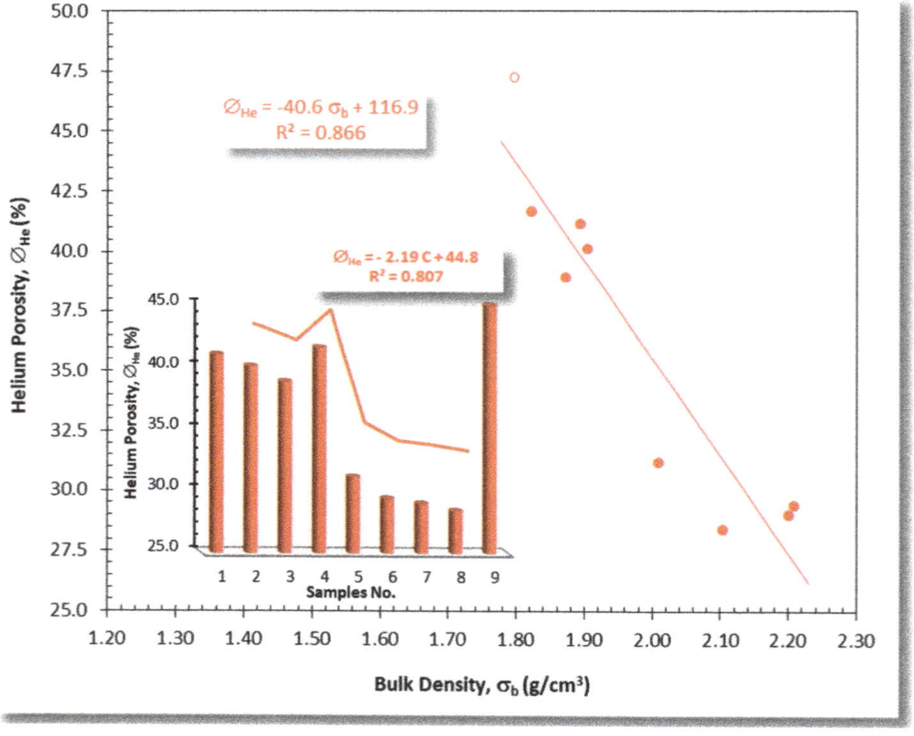

Figure 7. Plot showing the inverse proportional relationship between the bulk density (ρ_b) and the helium porosity ($\varnothing_{He}$) of the AFS samples, and the frequency bars and broken line distribution. Open circle represents sample AFS9, which was removed from processing due to its abnormal porosity value.

The inverse proportional relationship between bulk density and porosity (Figure 7) is due to the high dependence of pore volume and grain density on the principal components of the material. In general, the porosity values decreased from AFS1 to AFS8 (with an exception for sample AFS4). Sample AFS9 showed an increase in porosity value. Therefore, porosity decreased as arc furnace slag content increased from 10% to 80%, in samples AFS1 and AFS8, respectively (Table 2, Figure 7). $\varnothing_{He}$ can be estimated as a function of the bulk density and the arc furnace slag content, based on two mathematical models, as shown in Figure 7.

Effect of Sintering Temperatures on Porosity and Density Values

The implication of sintering temperatures on porosity was checked for sample AFS5, characterized by a median ratio of arc furnace slag to ceramic sludge (50% for each). Porosity decreased with increasing sintering temperature, which was due to the increasing crystallization grade and crystal size with increasing temperatures (Table 4).

Table 4. Effect of sintering temperatures on density, porosity and phases developed for sample AFS5.

Sintering Temperatures	Density g/cm^3	Porosity (%)	Phases Developed
1050 °C	1.691	43.98	Parawoll.,Geh., QZ-low
1100 °C	2.010	31.16	β-woll., Geh., QZ-low
1150 °C	1.990	33.36	Parawoll, Fay., QZ-low

Parawoll. = parawollastonite, β-woll. = β-wollastonite, Geh. = gehlenite, Fay. = Faylite, Qz-low = low quartz.

Table 4 exhibits the porosity and density results of sample AFS5 after sintering at different temperatures, along with the precipitated phases at each temperature. This table shows that porosity decreased upon increasing the sintering temperature from 1050 °C to 1100 °C. This was because of the increase in compactness of the well-sintered sample. However, when sintering temperature was increased to 1150 °C, a slight increase in porosity and decrease in density was found, which can be explained by the transformation of β-wollastonite into parawollastonite as well as due to the complete vanishing of gehlenite. Although the fayalite phase with its high density value, 4.39 [50], was precipitated, the density of AFS5 sample decreased after sintering at 1150 °C. This may be due to its low content in the sample.

3.4. Electric and Dielectric Properties

The dielectric constant values (ε′) of the AFS samples fluctuated between 6.324 for sample AFS2 at 8 MHz and 11.988 for sample AFS8 at 50 Hz (Figure 8), while the electric conductivity values (σ) varied from 0.0105 μS/cm for sample AFS1 at 50 Hz to 43.57 μS/cm for sample AFS8 at 8 MHz (Figure 9). Based on the ε′ values, the AFS samples can be grouped in descending order into two groups: group 1 (AFS8, AFS9, AFS7, AFS5, AFS6 samples) and group 2 (AFS3, AFS4, AFS1, AFS2 samples), as can be seen in Figure 8. The dielectric constant decreased with increasing applied AC frequency, while electric conductivity increased dramatically with increasing AC frequency, a common behavior recorded by many authors [35,36,51]. Following Nabawy and Rochette [38], the increase in conductivity with increasing frequency passes through three stages: (A) a steady stage with negligible increase in σ, (B) a transitional stage, and (C) a last stage with a dramatic increase in σ (Figure 9).

The imaginary electric modulus for samples AFS4, AFS2, and AFS1 was relatively high at 50 Hz (0.1388–0.0145), whereas it was lower for the other samples with different slopes, causing an inflection point in the frequency range of 2.5–3.0 MHz (Figure 10). These different dipolar relaxations were attributed to different relaxations for ε′ and ε″ at higher frequencies and to the crystal surfaces, which become more active at higher frequencies [52–55].

Based on the σ classification of Khater et al. [35,36], differentiation of σ values is difficult at low frequency values (σ = 0.01–0.181 μS/cm, f < 1500 kHz, Figure 9) and refers to poor semiconductors (0.01–1.0 μS/cm, Khater et al. [35,36]), whereas σ values are higher at high frequency values (σ = 0.181–43.57 μS/cm, $f \geq$ 1500 kHz) referring to poor (0.01–1.0 μS/cm) to fair semiconductors (1.0–1000 μS/cm, Khater et al. [35,36]). The measured electric and dielectric parameters are presented graphically as a function of the AC frequency in Figures 8–10.

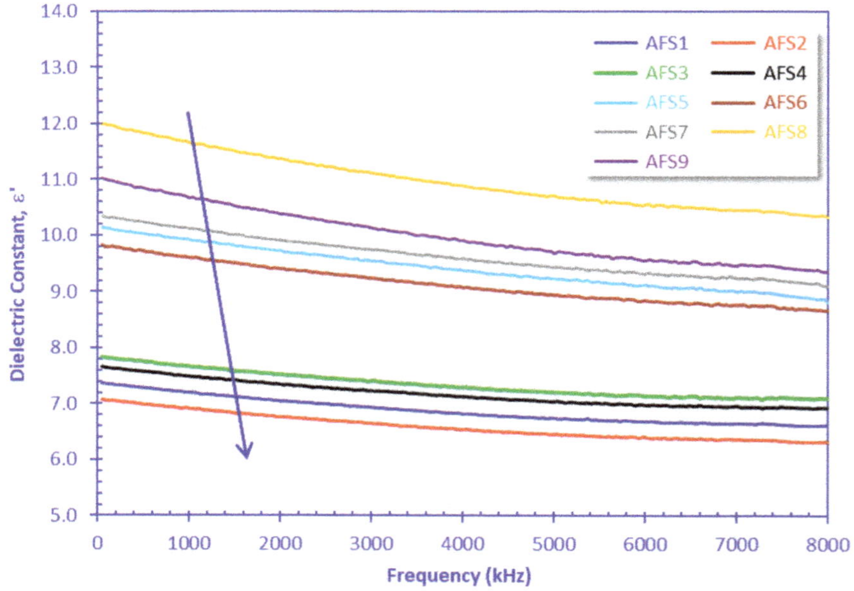

Figure 8. Plotting the dielectric constant of the samples versus the applied frequency.

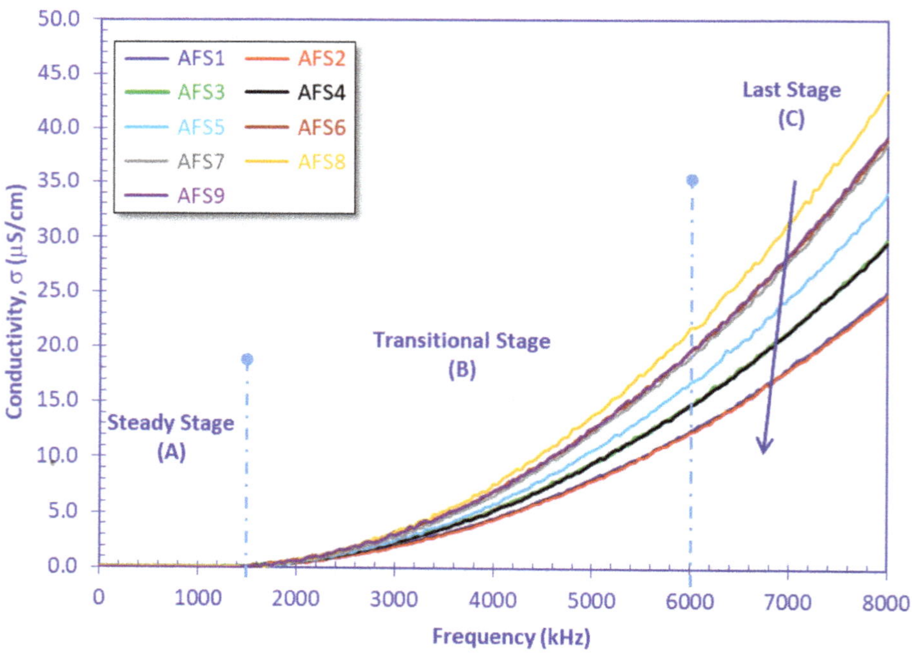

Figure 9. Plotting the electric conductivity of the samples versus the applied frequency.

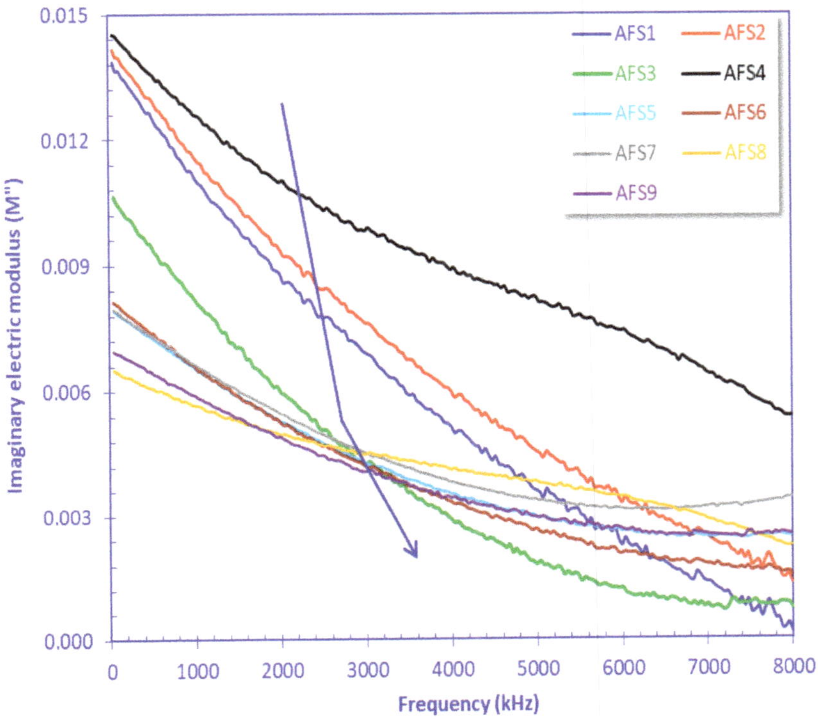

Figure 10. Plotting the imaginary part of the complex electric modulus of the samples versus the applied frequency.

Effect of Sintering Temperatures on Electric Properties

Increases in the crystallization grade were the primary variable affecting the electric and dielectric properties of sample AFS5, as shown in Figure 11. It is indicated that the crystal size was smaller for AFS5 sintered at 1050 °C. At 1100 °C, the crystal size was well-developed with reduced pore volume, which caused a decrease in the electric and dielectric parameters. Sintering at this temperature enhanced dielectric properties, such that it could be considered a more appropriate semiconductor material than the samples sintered at 1050 °C and 1150 °C (Figure 11).

3.5. Implication of Waste Composition on Pore Volume

The incremental increase in CaO and Al_2O_3 fractions was accompanied by increasing porosity values, while MgO and Fe_2O_3 contents were inversely related to the porosity values. The impact of increasing SiO_2 content acted somewhat inversely upon the porosity values (Table 3, Figure 12). Consequently, the decreasing sludge content from AFS1 to AFS9 was responsible for the decreasing pore volume of the studied samples (Tables 2 and 3).

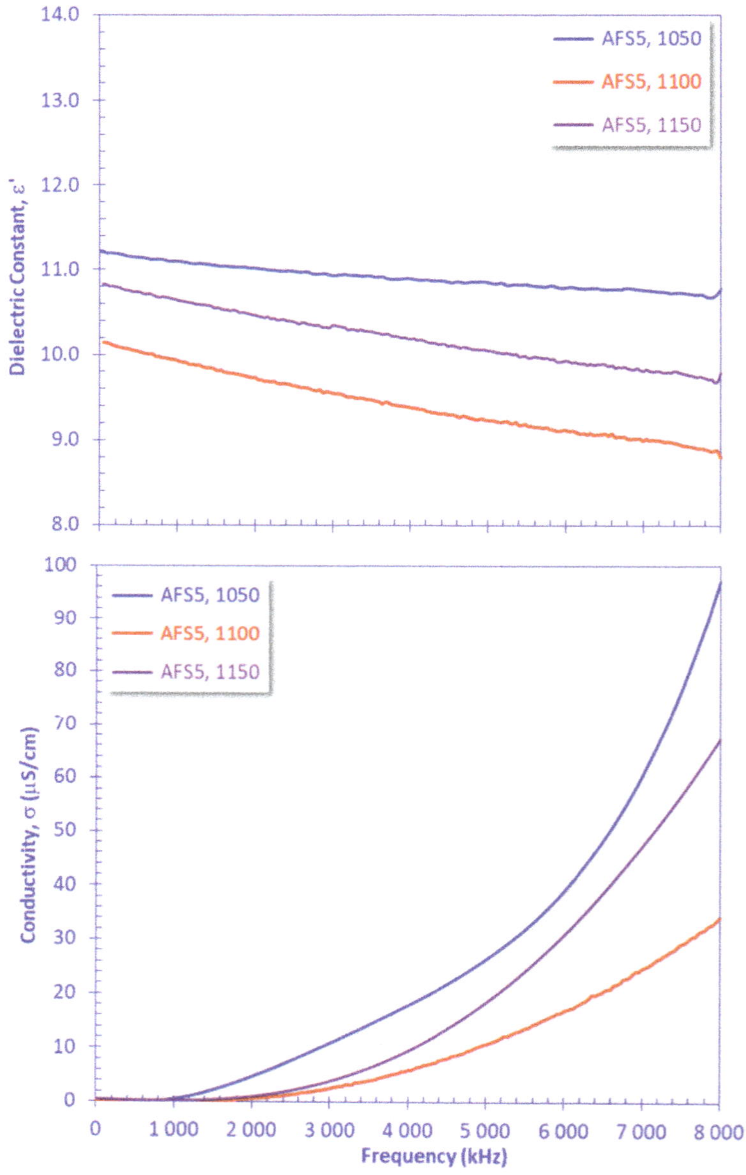

Figure 11. Effect of sintering temperature (1050–1150 °C) of AFS5 on the measured dielectric constant and electric conductivity plotted versus the applied frequency.

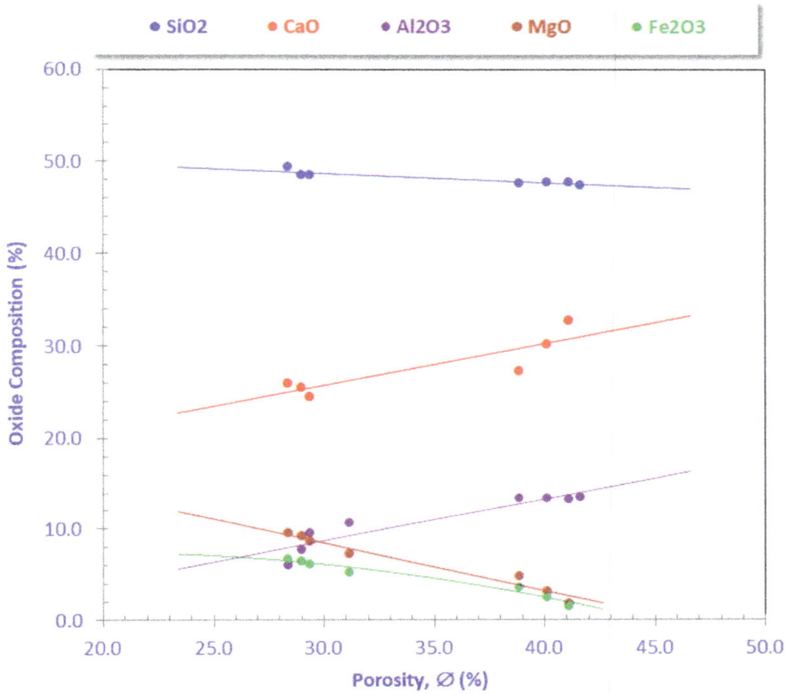

Figure 12. Plotting the main oxide composition versus the incremental percentages of porosity.

3.6. Implication of Waste Composition on Electric Properties

The waste composition of prepared ceramic samples impacted their electric activities. Electric conductivity and capacitance were primarily affected by the arc furnace slag content in an incremental increase from the AFS1 to the AFS9 samples (10–90%, respectively). Based on their dielectric and electric parameters, the AFS samples could be clustered into two groups, AFS1–AFS4 and AFS5–AFS9. These groups may be differentiated by the higher porosity values for the first group (41.12–41.66%) compared to the second group (31.16–47.26%, Table 3). This could be attributed to higher contents of CaO (32.73–23.79%) and Al_2O_3 (13.42–13.61%) in the first group (AFS1–AFS4) than the second one (Table 2).

Likewise, based on the dielectric constant values, the arc furnace slag–sludge mixtures could be clustered into two sample groups; the first one (AFS5–AFS9) was characterized by higher ε' and σ and lower M'' than the second (AFS1–AFS4) (Figures 8–10). This could be attributed to the relatively high contents of Fe_2O_3 and MgO in samples AFS5–AFS9 and the relatively low contents of CaO and Al_2O_3 in these samples (Table 2).

3.7. Implications of Crystal Size and Pore Volume on Electric and Dielectric Properties

The crystal size had a direct effect on the electrical activity of the charged materials. Larger crystal sizes result in reduced surface area and, therefore, less polarized charged surfaces. For samples AFS5–AFS9, the crystals seemed to be well developed, as shown in Figure 5, while for samples AFS1–AFS4, the crystals had a lower surface area.

By contrast, porosity (∅ in%) is a complementary component of the crystal's volume (equal to 100–∅ in%), i.e., it has indirect effects on the studied samples [34,52,53]. Porosity is considered an additional parameter that controls the electrical polarization and conductivity of samples. High porosity means lower grain volume and less polarized surfaces, and vice versa. Samples AFS5-AFS8 had porosity values lower than the remaining samples. In addition, they were characterized by higher ε′ and σ (measured at 4.0 MHz as a midway frequency value), which are inversely related to the porosity values (Figure 13).

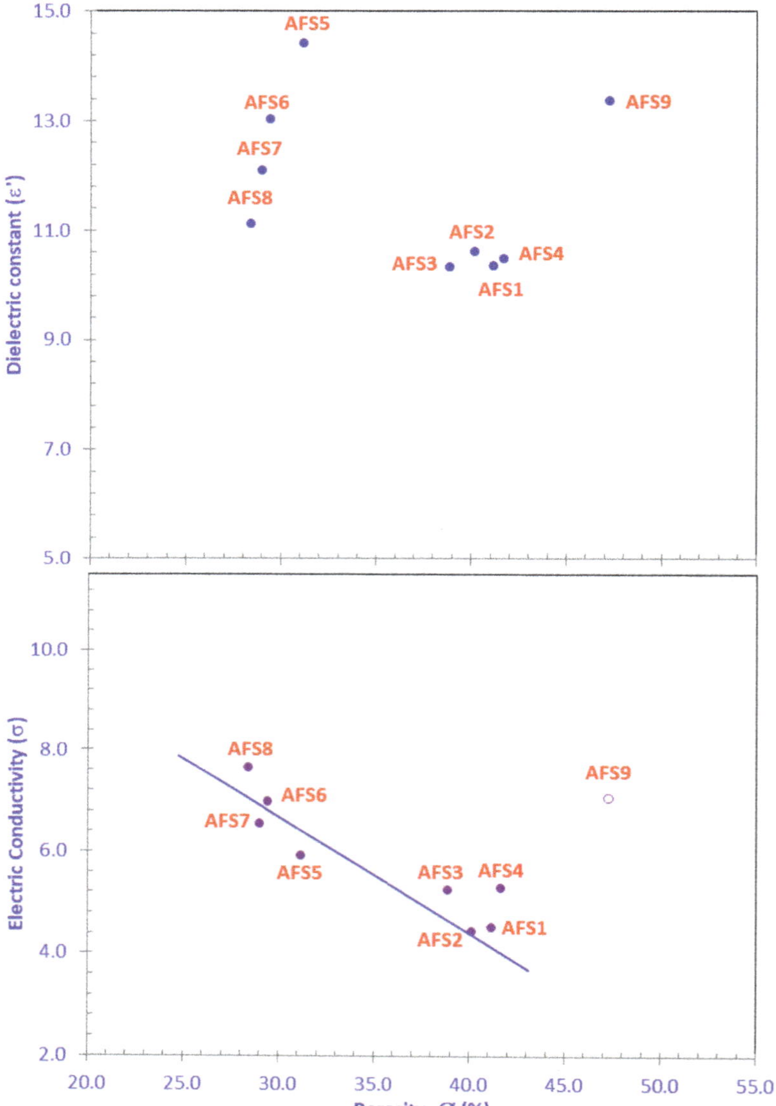

Figure 13. Dielectric constant (ε′) and electric conductivity (σ) versus helium porosity ($\varnothing_{He}$). Sample AFS9 (open circle) was removed from statistical processing.

3.8. Industrial Applications

The highly porous nature of the studied ceramic samples, as produced from both arc furnace slag and sludge waste mixtures, suits them for creation of building materials such as wall tiles. The relatively low electric conductivity, imaginary electric modulus, and dielectric constant values of the studied samples indicate their applications as poor to fair semiconductors (at 50 Hz and higher frequencies). The increasing iron oxide content with increasing arc furnace slag percentages (from 1.6% for AFS1 to 7.04% for AFS9) reduces the possibility of this application by increasing electric activity. In addition, the presence of certain impurities in these samples increases their ability to conduct electric charges on the order of μS (Khater et al., 2019, 2020, Dimitrijev, 2011) [35,36,56].

From an economic point of view, lightweight porous ceramic prepared from recycling hazardous waste materials tends to cost less than porcelain and is much lighter. It is often used for wall and ceiling installations. However, there are some significant restrictions with this material: it is not as strong as porcelain, so it does not make the best walking surface; it can be freezing underfoot in the winter; and heavy tile can be challenging to install. Therefore, this kind of ceramic is a good candidate for wall tile applications.

Porous ceramic wall tile bodies, whose microstructure and composition differ entirely from those of porcelain bodies (higher porosity and smaller glassy phase content), can be incredibly affordable and widely used worldwide. All these requirements are highly fulfilled in this work through the recycling of both electric arc furnace slag and ceramic sludge waste.

In this work, cleaning the environment from industrial waste byproducts was our first goal. The production of lightweight ceramic materials via a low energy consuming process to improve our environment and reduce the headache caused by global warming was our second target.

4. Conclusions

1. Cleaning the environment via waste management of nonrenewable products is very important for our daily life. In the present work, lightweight porous ceramic materials composed of wollastonite (β-wollastonite or parawollastonite), gehlenite and low quartz phases were successfully prepared through recycling two industrial wastes, i.e., arc furnace slag and ceramic sludge. The recycling process consumed little energy, and therefore can be easily applied on pilot and industrial scales.
2. The formed phases depended on the CaO/SiO_2 ratio in both waste materials. Lower CaO/SiO_2 content led to the formation of β-wollastonite and quartz phases with small amounts of the gehlenite phase. A higher CaO/SiO_2 ratio led to the formation of parawollastonite and gehlenite phases but hindered the development of the low quartz phase.
3. By applying different sintering temperatures to a selected sample (AFS5), it was found that both polymorphic structures of wollastonite were formed, either β-wollastonite (unstable) or parawollastonite (stable). β-wollastonite was converted into parawollastonite at high temperatures. By increasing the sintering temperature to 1150 °C, traces of the fayalite phase (Fe_2SiO_4) developed.
4. The porosity of the prepared materials was affected by sintering temperatures. The higher the temperature, the lower the porosity, due to the formation of a compacted microstructure.
5. The porosity, density and electrical properties of prepared materials depended on the composition of starting materials and formed phases. Increases in CaO and Al_2O_3 was accompanied by increasing porosity, while increases in MgO and Fe_2O_3 led to decreasing porosity and increasing dielectric constant and electric conductivity.
6. The selected sample sintered at 1100 °C exhibited lower dielectric parameters than those sintered at 1050 °C and 1150 °C.

Author Contributions: Conceptualization, G.A.K. and A.A.E.-K.; methodology, M.M.F.; software, B.S.N.; validation, G.A.K., A.A.E.-K. and M.A.-B.A.L.; formal analysis, B.S.N.; investigation, M.M.F.; resources, A.A.E.-K.; data curation, M.A.-B.A.L.; writing—original draft preparation, A.A.E.-K.; writing—review and editing, G.A.K.; visualization, M.M.F.; supervision, G.A.K.; project administration, G.A.K.; funding acquisition, G.A.K. All authors have read and agreed to the published version of the manuscript.

Funding: This paper is based upon work supported by the Science, Technology and Innovation Funding Authority ((STDF) under grant No. 41555).

Institutional Review Board Statement: Not applicable.

Informed Consent Statement: Not applicable.

Data Availability Statement: The data of this work is available for first author after publication.

Conflicts of Interest: The authors declare that they have no known competing financial interests or personal relationships that could have influenced the work reported in this paper.

References

1. Aleluia, J.; Ferrão, P. Assessing the costs of municipal solid waste treatment technologies in developing Asian countries. *Waste Manag.* **2017**, *69*, 592–608. [CrossRef] [PubMed]
2. Atkinson, G.; Mourato, S. *Cost-Benefit Analysis and the Environment*; OECD: Paris, France, 2015.
3. Balasubramanian, M. Economics of Solid Waste Management: A Review. In *Strategies of Sustainable Solid Waste Management*; Saleh, H.M., Ed.; IntechOpen: London, UK, 2021; ISBN 978-1-83962-559-6.
4. Rajendran, K.; Varma, V.S.; Mahapatra, D.M.; Kondusamy, D. *Economics of Solid Waste Management. Waste to Wealth*; Springer: Singapore, 2018; pp. 259–275.
5. Bayer Ozturk, Z.; Gultekin, E.E. Preparation of ceramic wall tiling derived from blast furnace slag. *Ceram. Int.* **2015**, *41*, 12020–12026. [CrossRef]
6. Mukhamedzhanova, M.T.; Irkakhodzhaeva, A.P. Ceramic tiles utilizing waste material from a copper-enriching plant. *Glass Ceram.* **1994**, *51*, 328–329. [CrossRef]
7. Khater, G.A. The use of Saudi slag for the production of glass-ceramic materials. *Ceram. Int.* **2002**, *28*, 59–67. [CrossRef]
8. Schilling, P.J.; Butler, L.G.; Roy, A.; Eaton, H.C. 29Si and 27Al MAS-NMR of NaOH-Activated Blast-Furnace Slag. *J. Am. Ceram. Soc.* **1994**, *77*, 2363–2368. [CrossRef]
9. Dana, K.; Das, S.K. High strength ceramic floor tile compositions containing Indian metallurgical slags. *J. Mater. Sci. Lett.* **2003**, *22*, 387–389. [CrossRef]
10. Vincenzini, P. Ceramic Powders: Preparation, Consolidation and Sintering. In Proceedings of the 5th International Meeting on Modern Ceramics Technologies (5th CIMTEC), Lignano Sabbiadoro, Italy, 14–19 June 1982.
11. Semanda, J.; Mutuku, R.N.; Kaluli, J.W. The effect of cement-waste mixes on the physical and strength properties of floor tiles. *Eur. Int. J. Sci. Technol.* **2014**, *3*, 32–43.
12. Badiee, H.; Maghsoudipour, A.; Dehkordi, B.R. Use of Iranian steel slag for production of ceramic floor tiles. *Adv. Appl. Ceram.* **2008**, *107*, 111–115. [CrossRef]
13. Abu-Eishah, S.I.; El-Dieb, A.S.; Bedir, M.S. Performance of concrete mixtures made with electric arc furnace (EAF) steel slag aggregate produced in the Arabian Gulf region. *Constr. Build. Mater.* **2012**, *34*, 249–256. [CrossRef]
14. Zhang, K.; Li, B.; Wu, W. Structural and magnetic properties of steel slag based glass–Ceramics. *Adv. Appl. Ceram.* **2014**, *114*, 121–125. [CrossRef]
15. Liu, Z.; Zong, Y.; Hou, J. Preparation of slag glass ceramic from electric arc furnace slag, quartz sand and talc under various MgO/Al_2O_3ratios. *Adv. Appl. Ceram.* **2015**, *115*, 1–8. [CrossRef]
16. Teo, P.T.; Anasyida, A.S.; Nurulakmal, M.S. Characterization of Ceramic Tiles Added with EAF Slag Waste. *Adv. Mater. Res.* **2014**, *1024*, 211–214. [CrossRef]
17. Fruías Rojas, M.; de Rojas, M.I.S. Chemical assessment of the electric arc furnace slag as construction material: Expansive compounds. *Cem. Concr. Res.* **2004**, *34*, 1881–1888. [CrossRef]
18. Sarkar, R.; Singh, N.; Kumar, S.D. Utilization of steel melting electric arc furnace slag for development of vitreous ceramic tiles. *Bull. Mater. Sci.* **2010**, *33*, 293–298. [CrossRef]
19. Pioro, L.S.; Pioro, I.L. Reprocessing of metallurgical slag into materials for the building industry. *Waste Manag.* **2004**, *24*, 371–379. [CrossRef]
20. Shil'tsina, A.; Vereshchagin, V.I. Ceramic tiles containing diopside and clay raw materials from Khakassia. *Glass Ceram.* **2000**, *57*, 87–90. [CrossRef]
21. Saadaldin, S.A.; Rizkalla, A.S. Synthesis and characterization of wollastonite glass–Ceramics for dental implant applications. *Dent. Mater.* **2014**, *30*, 364–371. [CrossRef]
22. Volochko, A.T.; Podbolotov, K.B.; Zhukova, A.A. Preparation of porous ceramic materials based on wollastonite using silicon-containing components. *Refract. Ind. Ceram.* **2011**, *52*, 186–190. [CrossRef]
23. Pérez, J.M.; Teixeira, S.R.; Rincón, J.M.; Romero, M. Understanding the Crystallization Mechanism of a Wollastonite Base Glass Using Isoconversional, IKP Methods and Master Plots. *J. Am. Ceram. Soc.* **2012**, *95*, 3441–3447. [CrossRef]
24. Hasmaliza, M.; Seman, A.A.; Gan, W.L. Effect of Feldspar Content on Electric Arc Furnace Slag (EAFS) Added Tiles. *Adv. Mater. Res.* **2014**, *1024*, 235–238. [CrossRef]
25. Pacheco-Torgal, F.; Jalali, S. Reusing ceramic wastes in concrete. *Constr. Build. Mater.* **2010**, *24*, 832–838. [CrossRef]

26. Medina, C.; Frías, M.; de Rojas, M.I.S. Microstructure and properties of recycled concretes using ceramic sanitary ware industry waste as coarse aggregate. *Constr. Build. Mater.* **2012**, *31*, 112–118. [CrossRef]
27. Zimbili, O.; Salim, W.; Ndambuki, M. A review on the usage of ceramic wastes in concrete production. *Int. J. Civ. Environ. Struct. Constr. Archit. Eng.* **2014**, *8*, 91–95.
28. Awoyera, P.O.; Ndambuki, J.M.; Akinmusuru, J.O.; Omole, D. Characterization of ceramic waste aggregate concrete. *HBRC J.* **2018**, *14*, 282–287. [CrossRef]
29. Khater, G.A.; Nabawy, B.S.; El-Kheshen, A.A.; Abdel-Baki, M.; Farag, M.; Elsatar, A.A. Preparation and characterization of low-cost wollastonite and gehlenite ceramics based on industrial wastes. *Constr. Build. Mater.* **2021**, *310*, 125214. [CrossRef]
30. McKenna, F.; Archbold, P. Ceramic waste sludge as a partial cement replacement. In Proceedings of the Civil Engineering Research in Ireland Conference, Belfast, UK, 28–29 August 2014.
31. Kuak, L.M. Carbon Dioxide Adsorption on Eaf Steel Slag Derived Adsorbent. Master's Thesis, Universiti Sains Malaysia, Gelugor, Malaysia, 2017.
32. David, C.; Bertauld, D.; Dautriat, J.; Sarout, J.; Menéndez, B.; Nabawy, B.S. Detection of moving capillary front in porous rocks using X-ray and ultrasonic methods. *Front. Phys.* **2015**, *3*, 3. [CrossRef]
33. Nabawy, B.S.; Elgendy, N.T.H.; Gazia, M.T. Mineralogic and Diagenetic Controls on Reservoir Quality of Paleozoic Sandstones, Gebel El-Zeit, North Eastern Desert, Egypt. *Nat. Resour. Res.* **2019**, *29*, 1215–1238. [CrossRef]
34. Elgendy, N.T.H.; Abuamarah, B.A.; Nabawy, B.S.; Ghrefat, H.; Kassem, O.M.K. Pore Fabric Anisotropy of the Cambrian–Ordovician Nubia Sandstone in the Onshore Gulf of Suez, Egypt: A Surface Outcrop Analog. *Nat. Resour. Res.* **2019**, *29*, 1307–1328. [CrossRef]
35. Khater, G.A.; Nabawy, B.; Kang, J.; Mahmoud, M.A. Dielectric Properties of Basaltic Glass and Glass-Ceramics: Modeling and Applications as Insulators and Semiconductors. *Silicon* **2018**, *11*, 579–592. [CrossRef]
36. Khater, G.A.; Nabawy, B.; Kang, J.; Yue, Y.; Mahmoud, M.A. Magnetic and Electrical Properties of Glass and Glass-Ceramics Based on Weathered Basalt. *Silicon* **2020**, *12*, 2921–2940. [CrossRef]
37. Nabawy, B.S.; Rochette, P. A Preliminary Study on the Electrical Signatures of Some Iron and Stony Meteorites and Their Dependence On Nickel Content. *Acta Geophys.* **2016**, *64*, 1942–1969. [CrossRef]
38. Fan, W.-D.; Yang, Q.-W.; Guo, B.; Liu, B.; Zhang, S.-G. Crystallization mechanism of glass-ceramics prepared from stainless steel slag. *Rare Met.* **2018**, *37*, 413–420. [CrossRef]
39. Lu, Z.; Lu, J.; Li, X.; Shao, G. Effect of MgO addition on sinterability, crystallization kinetics, and flexural strength of glass–Ceramics from waste materials. *Ceram. Int.* **2016**, *42*, 3452–3459. [CrossRef]
40. Casasola, R.; Pérez, J.M.; Romero, M. Crystal Growth of F-Phlogopite from Glasses of the SiO_2-Al_2O_3-MgO-K_2O-F System. *J. Am. Ceram. Soc.* **2015**, *99*, 484–491. [CrossRef]
41. Tabit, K.; Hajjou, H.; Waqif, M.; Saâdi, L. Effect of CaO/SiO_2 ratio on phase transformation and properties of anorthite-based ceramics from coal fly ash and steel slag. *Ceram. Int.* **2020**, *46*, 7550–7558. [CrossRef]
42. Edrees, S.J.; Shukur, M.M.; Obeid, M.M. First-principle analysis of the structural, mechanical, optical and electronic properties of wollastonite monoclinic polymorph. *Comput. Condens. Matter* **2018**, *14*, 20–26. [CrossRef]
43. Ismail, H.; Shamsudin, R.; Hamid, M.A.; Awang, R. Characteristics of β-wollastonite derived from rice straw ash and limestone. *J. Aust. Ceram. Soc.* **2016**, *52*, 163–174.
44. Qin, J.; Cui, C.; Cui, X.; Hussain, A.; Yang, C.; Yang, S. Recycling of lime mud and fly ash for fabrication of anorthite ceramic at low sintering temperature. *Ceram. Int.* **2015**, *41*, 5648–5655. [CrossRef]
45. Veit, U.; Rüssel, C. Viscosity and liquidus temperature of quaternary glasses close to an eutectic composition in the CaO–MgO–Al_2O_3–SiO_2 system. *J. Mater. Sci.* **2017**, *52*, 8280–8292. [CrossRef]
46. Leko, V.; Mazurin, O. Analysis of regularities in composition dependence of the viscosity for glass-forming oxide melts: II. Viscosity of ternary alkali aluminosilicate melts. *Glass Phys. Chem.* **2003**, *29*, 16–27. [CrossRef]
47. Almasri, K.A.; Sidek, H.A.A.; Matori, K.A.; Zaid, M.H.M. Effect of sintering temperature on physical, structural and optical properties of wollastonite based glass-ceramic derived from waste soda lime silica glasses. *Results Phys.* **2017**, *7*, 2242–2247. [CrossRef]
48. Marincea, S.; Dumitras, D.-G.; Ghinet, C.; Fransolet, A.-M.; Hatert, F.; Rondeaux, M. Gehlenite from three occurrences of high-temperature skarns, romania: New mineralogical data. *Can. Mineral.* **2011**, *49*, 1001–1014. [CrossRef]
49. Yan, X.; Wen, J. Study on factors influencing the performance of reactive powder concrete. *IOP Conf. Ser. Earth Environ. Sci.* **2020**, *510*, 052075. [CrossRef]
50. Chen, G.Q.; Ahrens, T.J.; Stolper, E.M. Shock-wave equation of state of molten and solid fayalite. *Phys. Earth Planet. Inter.* **2002**, *134*, 35–52. [CrossRef]
51. Abd-El-Messieh, S.L.; Abd-El-Nour, K.N. Effect of curing time and sulfur content on the dielectric relaxation of styrene butadiene rubber. *J. Appl. Polym. Sci.* **2003**, *88*, 1613–1621. [CrossRef]
52. Nabawy, B.S.; Wassif, N.A. Effect of the mineralogical composition on the petrophysical behavior of the amygdaloidal and vesicular basalt of Wadi Wizr, Eastern Desert, Egypt. *J. Afr. Earth Sci.* **2017**, *134*, 613–625. [CrossRef]
53. Nabawy, B.S. Impacts of fossil anisotropy on the electric and permeability anisotropy of highly fossiliferous limestone: A case study. *Mar. Geophys. Res.* **2017**, *39*, 537–550. [CrossRef]

54. Wagner, H.; Richert, R. Measurement and analysis of time-domain electric field relaxation: The vitreous ionic conductor $0.4\,Ca(NO_3)_2$–$0.6\,KNO_3$. *J. Appl. Phys.* **1999**, *85*, 1750–1755. [CrossRef]
55. Turky, G.M.; Fayad, A.M.; El-Bassyouni, G.T.; Abdel-Baki, M. Dielectric and electrical properties of MoO_3-doped borophosphate glass: Dielectric spectroscopy investigations. *J. Mater. Sci. Mater. Electron.* **2021**, *32*, 22417–22428. [CrossRef]
56. Dimitrijev, S. *Principles of Semiconductor Devices (The Oxford Series in Electrical and Computer Engineering)*; Oxford University Press: Oxford, UK, 2011.

materials

MDPI

Review

Sustainable Use of Apple Pomace (AP) in Different Industrial Sectors

Ewelina Gołębiewska [1,*], Monika Kalinowska [1,*] and Güray Yildiz [2]

1 Department of Chemistry, Biology and Biotechnology, Faculty of Civil Engineering and Environmental Science, Institute of Civil Engineering and Energetics, Bialystok University of Technology, Wiejska 45E Street, 15-351 Bialystok, Poland
2 Department of Energy Systems Engineering, Faculty of Engineering, Izmir Institute of Technology, Urla, Izmir 35430, Turkey; gurayyildiz@iyte.edu.tr
* Correspondence: e.golebiewska@pb.edu.pl (E.G.); m.kalinowska@pb.edu.pl (M.K.)

Abstract: In many countries, apple pomace (AP) is one of the most produced types of agri-food waste (globally, it is produced at a rate of ~4 million tons/year). If not managed properly, such bio-organic waste can cause serious pollution of the natural environment and public health hazards, mainly due to the risk of microbial contamination. This review shows that AP can be successfully reused in different industrial sectors—for example, as a source of energy and bio-materials—according to the idea of sustainable development. The recovered active compounds from AP can be applied as preservatives, antioxidants, anti-corrosion agents, wood protectors or biopolymers. Raw or processed forms of AP can also be considered as feedstocks for various bioenergy applications such as the production of intermediate bioenergy carriers (e.g., biogas and pyrolysis oil), and materials (e.g., biochar and activated carbon). In the future, AP and its active ingredients can be of great use due to their non-toxicity, biodegradability and biocompatibility. Given the increasing mass of produced AP, the commercial applications of AP could have a huge economic impact in the future.

Keywords: waste management; sustainability development; apple pomace; active compounds; extraction; renewable energy; anticorrosion agents; biopolymers

Citation: Gołębiewska, E.; Kalinowska, M.; Yildiz, G. Sustainable Use of Apple Pomace (AP) in Different Industrial Sectors. *Materials* **2022**, *15*, 1788. https://doi.org/10.3390/ma15051788

Academic Editors: Rossana Bellopede and Lorena Zichella

Received: 13 December 2021
Accepted: 25 February 2022
Published: 27 February 2022

1. Introduction

Effective waste management is one of the greatest environmental challenges the world is facing today. Technological advancement, economic development, urbanization, population growth and consumer habits have significantly contributed to a rapid increase in waste generation. Moreover, there are no signs of deceleration of this trend in the near future [1,2]. The 2018 edition of *What a Waste 2.0: A Global Snapshot of Solid Waste Management to 2050* estimated that the global municipal solid waste production will be 2.01 billion tons per year, and is projected to grow to 3.4 billion tons per year by 2050. The largest amount of waste (approximately 44% of the total amount of waste generated in the world) is food and green waste [3]. Among them, large groups are by-products of the fruit processing industries, such as the apple juice industry, the disposal of the major by-products of which (i.e., apple pomace—AP) can pose serious environmental problems or even public health hazards if incinerated and/or dumped [4]. The global production of AP is estimated at an average of 4 million/year and is expected to have an increase in the future. Unfortunately, the recovery rate of AP is quite low and insufficient. The most commonly applied disposal method for AP is to discard it directly to the soil in a landfill. It may cause serious soil and water pollution because AP is rich in water (>70%), sugars and organic acids, which are susceptible to fast microbial fermentation. The growth in microbial flora may decrease available nitrogen in the soil and affect the C/N ratio. Moreover, some authors mention the potential toxicity of AP because apple seeds contain a cyanogenic glycoside, amygdalin. However, it is unlikely to cause acute cyanide poisoning in humans because this would require the consumption of 800 g of AP [5].

Therefore, safe and efficient treatment and utilization of AP is required to reduce the possible environmental and health problems. Taking into account the huge amount of produced AP, the commercial applications of AP can create, in the future, great economic impacts. So far, different extraction methods that recover active substances from the AP have been applied. These active substances can later be applied as preservatives, antioxidants or anticorrosion agents used in, e.g., construction, civil engineering, environmental engineering and many other sectors (Figure 1). One of the basic techniques used to isolate the active compounds from materials of plant origin is classical extraction. In recent years, considerable attention has been given to the development of extraction methods that will be both efficient and environmentally friendly, e.g., limiting the amounts of solvent used or favouring energy efficiency [6]. Raw AP or the solid extraction by-products can also be utilised as a feedstock for the production of various types of intermediate bioenergy carriers in liquid, solid or gaseous forms, i.e., bioethanol, biodiesel, biomethane [7], biogas [8], biochar [9] or raw material for batteries [10]. Replacing traditional fossil fuels (FF) with AP-originated biofuels may reduce some undesirable aspects relegated to the production and use of FF, including emissions of greenhouse gases (GHG) (e.g., carbon dioxide (CO_2) and nitrous oxide (N_2O)), which contribute to serious environmental and health problems, and exhaustible resource depletion [11]. AP can also be a potential source of substances for the production of non-toxic and environmentally friendly biopolymers. The literature showed that AP-derived biopolymers were used for the production of biodegradable films, packaging materials, cups, plates and 3D objects [12,13].

Figure 1. Application of apple pomace in production of green, non-toxic and biodegradable products with applications in construction and building.

Currently, due to poor waste management and a lack of environmental awareness in many countries, a large amount of apple pomace is treated as waste with no economic value. Besides, there are some technical limitations affiliated with the effective utilization of apple waste, such as the requirement of immediate treatment after obtaining it (e.g., by drying); this is important to prevent the excessive growth of microorganisms (microbiological contamination), and hence, the loss of overall economic value [14]. To foster sustainability, AP should be treated as a valuable raw material that can be reused or processed. In our work, the sustainable management of AP is based on the recovery and utilization of apple waste, which creates a possibility to reuse it and put it back into the supply chain. Green extraction techniques allow the obtaining of AP extracts that are rich in active compounds in an eco-friendly manner. The solid residues generated during extraction can be stabilised and further transformed into, e.g., alternative energy sources or biopolymers with zero waste. Therefore, sustainable AP management gives the opportunities for reducing environmental pollution and increasing integration into a circular economy [15]. This review presents the scale of AP production, and possible means of its utilization as a source of active compounds and biopolymers, a feedstock for the production of biofuels, and as a source of raw materials that can be utilised in different industrial sectors including the construction and building industry sector, the energy sector, and food or material industries.

Apple and Apple Pomace Production in Poland and the World

According to the statistics, apples took third place (after bananas and watermelons), in 2019, in the terms of the popularity of fresh fruits in the world [16]. In that specific year, about 87.24 million metric tons of apples were produced worldwide (Figure 2). Poland is one of the largest producers of apples in the world (next to China, the United States and Turkey). From 2010 to 2018, the annual production of apples in Poland ranged from 1.877 to 3.9 million tons [16,17]. It is estimated that about 50% of all apples produced in Poland are processed for the production of apple juice concentrate [18]. Literature data show that 25–30 wt.% of the fresh apple used in the production of juice is a fruit by-product (i.e., apple pomace), which is considered a post-industrial organic waste [19]. Based on these facts, it can be calculated that about 0.5 million tons of AP were generated in Poland in 2018. For comparison, a neighbouring country, Germany, produces half as much AP (0.25 million tons/year) [20]. The literature shows that the largest apple producer, China, generates more than one million tons of apple pomace annually [21]. Countries such as New Zealand, Spain and Brazil are characterised by small amounts of generated apple juice by-products (from about 20,000 to 13,750 tons per year) [20]. Global apple production has reached over 87 million tons/year, resulting in 3.915–4.698 million AP (in 2019). Taking into account the fact that the production of 1 litre of apple juice requires about 1.6 kg of apples, 0.40–0.48 kg of AP is produced depending on the apple variety and processing type [22].

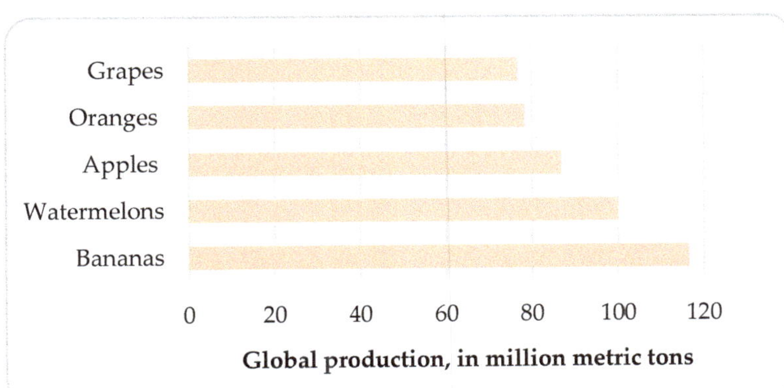

Figure 2. Global-scale production of the most popular fruits in 2019 [16].

A report published by the Food and Agriculture Organization of the United Nations (FAO) indicates that approximately 33.3% (1.6 billion tons) of the food produced worldwide for human consumption is wasted each year [23]. High standards of product quality are required to attract consumers, and hence, the exclusion of the foods lower product quality is one of the major reasons. This problem particularly affects developed countries, including China, Japan, EU countries, Canada and the USA [24]. Such large losses pose a serious threat to the status of food security, the natural environment and the economy.

Fruits, including apples, that deviate from the imposed standards, for example in terms of size and visual elements (colour intensity, discolouration, skin elasticity, defects: bruises, rots) are often discarded by producers, consumers and retailers [25]. It is estimated that approximately 3.7 trillion apples end up in landfills each year [26]. Furthermore, such a waste of food and improper AP waste management causes huge losses of water, land, fertilizer, energy, labour and capital, which are economically unprofitable [24,27]. The amount of generated AP will increase each year due to the growing production of apples in orchards and the increased demand for processed products (juices, concentrates, jams and purees). To avoid the abovementioned problems, it is necessary to introduce economically viable pathways for waste apples and AP, so that they could be further processed into valuable products. Indeed, the cost-effectiveness of such pathways depends on the amount of waste generated/collected, the need for additional storage space and appropriate equipment (wet apple pomace require immediate processing due to high humidity) and related costs for transportation [28].

In this paper, we suggest a number of strategies to reuse the waste originating from apples. Indeed, fresh apples can also be treated as waste and used as renewable feedstocks in many industries, but this review focuses only on the use of AP produced by apple processing processes. In our opinion, AP should be reused according to the principles of the circular economy.

2. The Recovery from Apple Pomace Dried and Powder

2.1. The Pretreatment of Apple Pomace

2.1.1. Biofuels

Due to serious concerns affiliated with climate change, the shift from fossil-based energy production options to various renewable-based alternatives has been a trending topic in the world. The most common renewable energy sources are wind, solar, hydropower, tidal, geothermal and biomass [29]. The latter involves organic materials such as agricultural residues, forestry residues, agri-food industry by-products (e.g., apple pomace), animal wastes, etc. Through various mechanical, biochemical or thermochemical processes, biomass can be transformed into fuels and/or fuel intermediates in solid (e.g., biochar), liquid (e.g., bioethanol, pyrolysis oil, etc.) and gaseous (e.g., biogas/biomethane, etc.) forms. Up to 80% of the organic matter in the pomace can be transformed into biofuel with an energy value of 10–30 $W \cdot m^{-3}$ [30]. Generally, production residues from the food industry are characterised by a low concentration of heavy metals and are good raw materials for biofuel production [31]. AP is rich in fermentable sugars (for example, it has contents of 19.2% fructose and 1.0% sucrose [32]) and is characterised by a low concentration of heavy metals; therefore it could be applied alone as a raw material for biofuel production. However, in practice, installations that use a mixture of several substrates are the most often used. Diversification of substrates favours the obtaining of better biofuel parameters and increases the safety of raw material supplies. Co-fermentation increases the efficiency of the process while reducing the costs incurred by the biogas plant for the purchase of raw material. Batches for energy production should be selected on account of, e.g., the maximization of energy yields, the stability of the fermentation process and the possibility of using the post-fermentation mass. In the study of Olech et al., the results of the analysis of the fermentation medium made of corn silage and apple pomace (in a proportion of 50/50%) showed that the highest methane generation was 61% [8]. The sample efficiency achieved a value of 4460 Nml. It was also demonstrated that olive and apple pomace are good co-substrate in the fermentation of cow slime (excrement) [33].

Bioethanol

Bioethanol (ethyl alcohol) is used as an eco-friendly fuel, most often as an additive to gasoline. It is obtained through the anaerobic fermentation of carbohydrates [34]. This process requires a proper pretreatment of the biomass feedstock (e.g., fruits) allowing the release of simple sugars (glucose, xylose, galactose, etc.) contained in cellulose and hemicellulose. This way, a hydrolysate rich in hexose and pentose sugars can be obtained [7]. In the case of anaerobic fermentation of the apple wastes, it is essential to choose an adequate bacterial strain to deal with a wide variety of sugars that are initially present in lignocellulosic hydrolysates. In addition to the commonly used yeast, *Saccharomyces cerevisiae*, there are references in the literature concerning the use of alternative microorganisms such as *Zymomonas mobilis*, *Kluyveromyces marxianus*, *Kluyveromyces lactis* or *Lachancea thermotolerans* for lignocellulosic biomasses such as apple pomace [7,35]. For example, the potential of apple pomace as a feedstock for bioethanol production was demonstrated in a recent study by Molinuevo-Salces et al. [7]. In their research, scientists assessed the effectiveness of selected bacterial strains on the amount of bioethanol produced from the hydrolysate of dry apple pomace obtained after juice extraction. The results showed that the highest bioethanol concentrations were obtained by testing *Kluyveromyces* sp. and *Lachancea* sp. (between 49.9 and 51.5 g L^{-1}). Total sugar consumption was in the range of 74.5 and 80.0, with bioethanol yields from 0.402 to 0.444 g g^{-1} [7]. In the work of Demiray et al. [36], the influence of a cheap additive—soluble soy protein (at different concentrations: 20, 40, 80, 160 mg/g cellulose)—on enzymatic hydrolysis of AP was investigated. The results showed that the addition of 80 mg/g cellulose soluble soy protein to AP medium hydrolysed with 60 FPU (Filter Paper Units; enzyme concentration) increased the sugar (by 24.8%) and bioethanol concentration (by 8.28% in the case of *Saccharomyces cerevisiae*, and by 20.9% for *Kluyveromyces marxianus*), which makes the bioethanol production from AP process more efficient and still economical [36]. Kut et al. [37], for the first time, conducted the enzymatic hydrolysis of the liquid fraction of AP for the production of bioethanol using a pentose fermenter yeast, namely *Pichia stipitis*. The results of their research indicate that with a properly optimised process (10% (*w*/*v*) AP loading), about 84.1% of the theoretical ethanol yield can be obtained [37].

Biogas

Another type of alternative biofuel in gaseous form, biogas, is produced via a sequence of low temperature (ca. between 30 and 60 °C) processes by which certain microorganisms break down degradable biomaterials in the absence of oxygen. The produced gas mixture consists predominantly of methane (CH_4) and carbon dioxide (CO_2) with volumes of about 40–75% and 15–60%, respectively [38]. The production of biogas is not completely free of GHG, but research is underway to reduce the concentration of CO_2 in biogas [39]; this increases the energy content of the final gas mixture and would result in increased levels of biomethane. Moreover, photosynthetic plants can absorb the CO_2 released via biogas combustion, which results in the emission of less total atmospheric carbon than the classical combustion of coal [38]. In the last decades, the extensive amounts of globally generated apple wastes have received interest in terms of the exploration of their potential as a co-substrate to valorise biofuel production. For example, Olech et al. [8] showed that the anaerobic digestion medium made of apple pomace and corn silage (in the organic mass proportion of 50 to 50%) achieved a satisfactory level of methane yield (about 40%) on the third day of fermentation. The highest daily biogas yield was obtained on the ninth day of measurement and amounted to 4460 Nml [8]. In addition, after processing the fruit residues (with negligible heavy metal content) in biogas production, nutrient-rich organic fertiliser can be obtained [11]. In the study of Claes et al. [40], the influence of biochar and graphene (carbon-based conductive materials) and trace metals supplementation on biogas production from AP was investigated. The results of their study showed that this supplementation significantly improved the biogas production from the AP. At a COD (chemical oxygen demand) concentration of 6000 mg/mL, the addition of (a) trace metals,

(b) biochar and (c) trace metals and biochar, increased the production of biogas by 7.2%, 13.3% and 22.7%, respectively, compared to the control (without supplementation). At a COD concentration of 12 mg/mL, the greatest changes in biogas production were observed for graphene supplementation (increase by 27.8% in comparison with the control). Moreover, in most cases, supplementation also improved the methane yield. In this study, the highest obtained CH_4 yield (increased by 23.0% in comparison with control) was observed in a case of the reactor supplemented with biochar and trace metals (COD = 6000 mg/mL) (468.0 $\pm$ 3.6 mL CH_4/g vs. and 286.0 $\pm$ 6.2 mL CH_4/g COD) [40].

Biochar

Pyrolysis is a thermochemical process in which the organic material is decomposed into liquid (pyrolysis oil), solid (char) and gaseous (CO_2, H_2, CO, CH_4) products. The process is carried out at moderate-to-high temperatures ranging from 300 °C to 650 °C and in the complete absence of oxygen. The type and particle size of biomass used, the heating rate and the residence times of, for instance, the feedstock and the generated primary pyrolysis vapours, are the major parameters that affect the performance and outcome of the pyrolysis process. The successful optimization of such process parameters determines the quantity, composition and quality of the products of the process [41,42]. There are several reports in the literature on AP pyrolysis (Table 1) [9,43–45]. In the work of Kosakowski et al. [43], the rapid pyrolysis of agricultural waste biomass (including AP) resulted in obtaining biochar characterised by higher combustion heat and calorific values than the biomass used [43]. Guerrero et al. [44] investigated the optimal conditions for the slow pyrolysis of AP for the production of gaseous products that can be used as a feedstock for the production of H_2 [44]. Zhang et al. [45] used the biochar obtained in the AP pyrolysis process to create magnetic biochar that could effectively enrich the low concentration of Ag(I) ions in effluents [45]. Xu et al. [9] investigated the effect of the temperature and the type of biomass used on the production of biochar. The results showed that grape residues produced the highest biochar yield, while AP produced the least biochar [9]. According to Table 1, the biochar resulting from the pyrolysis/carbonization of AP had a net caloric value of between 25 and 31 MJ/kg. For comparison, a net value of good-quality milled coal, e.g., eco peat coal, is in the range of 24–26 MJ/kg [45].

Table 1. Apple pomace pyrolysis.

Product	Pyrolysis Type	Pyrolysis Parameters	Products Obtained	Net Caloric Values	Applications	Ref.
Apple pomace	Rapid	The temperature in the reactor was gradually increased to 850 °C over 30 min. Pyrolysis was continued for 60 min at a constant temperature. The experiments were carried out in triplicate with seven gas collection points (450, 515, 585, 650, 715, 785 and 850 °C).	Gas fraction: 47.5% Oil fraction: 14.1% Water: 12.6% Biochar: 25.9%	30.948 ± 168 kJ/kg (biochar) 19.775,2 ± 125,8 kJ/kg (biomass)	Biochar and gaseous products with sufficiently high combustion heat and net calorific values. Can be applied as additives in other fuels	[43]
Flavoured spirits production waste (FSW) (lime, grapefruit and lemon)			Gas fraction: 42.0% Oil fraction: 21.7% Water: 9.7% Biochar: 26.7%	26.598 ± 75 kJ/kg (biochar) 14.904,3 ± 553,8 kJ/kg (biomass)		
Beetroot pulp			Gas fraction: 32.9% Oil fraction: 24.6% Water: 12.3% Biochar: 30.2%	25.572 ± 139 kJ/kg (biochar) 15.169,2 ± 25,8 kJ/kg (biomass)		
Apple pomace	Slow	Temperature: 300–450 °C; heating rate: 5–20 °C/min; residence time: 60 min	CO, CO_2, CH_4	7.639,18 kJ/kg (AP)	Solid product yield is maximum in slow pyrolysis	[44]
Apple pomace	Not defined	Temperature: 600 °C; (pyrolysis followed by immersion ageing in Fe(II)/Fe(III) aqueous solution for obtaining AP-based magnetic biochar)	Magnetic AP biochar	Not defined	Magnetic AP biochar that can be used for enriching Ag(I) in effluents	[45]
Apple pomace	Not defined	Pyrolysis was carried out in a pilot bubbling fluidised bed pyrolyser operating under a range of temperature from 300 to 600 °C and vapour residence times ranging from 2 to 5 s.	Major gases: H_2, CO, CO_2, CH_4; biochar; bio-oil	≈4–6 kJ/g (biomass)	A promising material for biochar production	[9]
Grape residues (GS—grape skins; GSS—grape skins and seeds)				≈0.1–4.1 kJ/g (biomass) ≈0.2–3.5 kJ/g (biomass)		

2.1.2. Sodium-Ion Batteries

Recently, there have been several reports published in the literature concerning the use of hard carbon (HC), obtained from fruit wastes and fruit peels, as an abundant and low-cost material for the production of sodium-ion batteries (SIBs) [10,46–51]. SIBs are cheap and environmentally friendly energy storage tools that are alternatives to the frequently used lithium-ion batteries (LIBs). Moreover, the vast abundance of sodium resources (the sixth most abundant element in the world) compared to the limited abundance of lithium and other elements commonly used in batteries, e.g., copper or nickel, also contribute to an increasing amount of research work being published [52,53]. In LIBs, graphite is used as an anode (negative electrode) material, while in SIBs, graphite is thermodynamically unstable with sodium ions [54]. A work published by Stevens and Dahn in the year 2000 [55] started the interest in hard carbon materials as potential anode materials for SIBs. In their work, the scientists demonstrated that this type of anode delivered a reversible capacity of 300 mAh·g^{-1}, close to that obtained for graphite in LIBs (372 mAh·g^{-1}) [55]. Hard carbon is usually prepared by the pyrolysis of organic precursors (most often from vegetable biomass, coal or petroleum) at temperatures between 1000 °C and 1500 °C, depending on the type of feedstock [56]. There are also some recent literature reports on the use of fruit waste as a source of hard carbon in SIBs. For example, in the study of Wu et al. [10], the electrochemical properties of apple waste-derived hard carbon electrodes were reported. Material for the electrodes (hard carbon) was obtained by a two-step dehydration process of wild apples followed by heat treatment (at 1100 °C) under an argon atmosphere. Then, the hard carbon electrodes (with a final composition of 80 wt.% HC) were prepared. The obtained electrodes demonstrated a very stable capacity of around 245 mAh·g^{-1} (at current rates of 0.1C) with full retention after 80 cycles, and good long-term cycling stability (1000 cycles at 5C). Moreover, the HC electrodes showed a promising rate capability with 112 mAh·g^{-1} at 5C [10]. Another study, conducted by Dou et al. [46], showed that pectin-free apple pomace waste-derived HC have a good overall performance during the galvanostatic long-term cycling at 0.1C (the capacity was around 285 mAh·g^{-1} after 230 cycles). Moreover, the specific capacities at 1.0–0.12 V (slope) and 0.12–0.02 V (plateau) (at 0.1C) during the fifth discharge were recorded. The obtained HC in the slope-like region showed a capacity of 110 mAh·g^{-1}, while in the plateau region, a capacity of 175 mAh·g^{-1} was observed [46]. Interestingly, these results were quite different from those obtained for HC derived from apple pomace containing pectin in other works of the same author. HC from apple waste containing pectin delivered much lower capacity within the plateau as compared to HC from pectin-free apple waste (108 and 175 mAh·g^{-1}, respectively). In the slope-like region, very similar capacities were recorded (112 and 110 mAh·g^{-1}). This indicates the differences in the sodium storage mechanism of HCs [46,57].

2.1.3. Biopolymers

Biopolymers are natural, biocompatible, highly biodegradable and environmentally friendly ("green") alternatives to widespread synthetic plastics. Biopolymers can be obtained/extracted by the following means: *i.* from natural sources (e.g., agricultural waste); *ii.* via direct biosynthesis by microorganisms; and *iii.* through chemical synthesis [58]. AP is a promising raw material for the production of biopolymers due to its high sugar content. It is estimated that the dry mass of AP contains 7–44% cellulose, 14–17% starch, 15–20% lignin and 4–14% pectin, which can be used for the production of biopolymers [19].

There are several recently published literature reports regarding the production of sustainable biomaterials from AP [12,13,59–64]. In the study of Gustafsson et al. [13], AP was used for the production of 3D objects (fibreboards) and biofilms. Solution casting was used to form fibreboards, while film casting was used to produce biofilms. The obtained structures were tested for tensile strength (TS) and elongation at max (EAM). The results showed that the highest value of TS (5.79 MPa) and EAM (1.54%) was reached by the biopolymer made from AP with 30% (*w*/*w*) glycerol. For comparison, the biopolymer produced only from AP showed significantly lower values of TS = 3.71% and EAM = 1.56%. The biopolymer prepared from AP with 7% (*w*/*w*) glycerol had three-times-lower flexibility (EAM = 10.77%) and a four-times-higher TS value (TS = 16.49 MPa) than biofilm prepared from AP without glycerol using values of EAM = 37.39% and TS = 4.20 MPa [13]. In another work [59], AP-derived bioplastic was used in the production of cups. The mechanical properties of bioplastic were measured, and the results showed that the highest values of TS and EAM were also reached using a mixture of washed AP with 30% (*w*/*w*) glycerol content. However, other important parameters, such as water resistance, exposure to environmental factors (e.g., light), or biodegradability, were not investigated in this study. AP-derived biopolymer could be an environmentally friendly replacement for synthetic plastic tableware or additives for the production of structural or building elements (e.g., bricks) [59].

In the work of Liu et al., AP was characterised as a potential source of biopolymers—PHAs (poly-hydroxyalkanoates) [12]. PHAs are biosynthesised by a wide range of Gram-positive and Gram-negative bacteria (e.g., *Azotobacter*, *Clostridium*, *Alcaligenes latus* and *Cupriavidus necator*) and serve as an energy and carbon storage source [60]. Generally, the production of PHAs (2.4 and 5.5 US$/kg) generates much higher costs as opposed to conventional synthetic plastics (1.2 US$/kg) [61]. However, by changing the carbon source used in the production of PHAs to inexpensive agricultural waste (including AP), the production costs can be significantly reduced (up to 50%), which has a great impact on PHAs' applicability in many industries [12,62].

In the work of Pereira [63], *Pseudomonas chlororaphis* sub-sp. *Aurantiaca* was used to produce medium-chain-length PHAs (mcl-PHAs) from apple waste. The obtained mcl-PHAs consisted of, i.e., 3-hydroxydecanoate (42.7 $\pm$ 0.1 mol%), 3-hydroxyoctanoate (17.9 $\pm$ 1.0 mol%), 3-hydroxybutyrate (14.5 $\pm$ 1.1 mol%) and 3-hydroxytetradecanoate (11.1 $\pm$ 0.6 mol%) with a yield of 49.25 $\pm$ 4.08%. The obtained mcl-PHAs biofilms showed attractive mechanical properties (TS = 5.21 $\pm$ 1.09 MPa, EAM = 400.05 $\pm$ 55.8%) [63]. Rebocho et al. [64], in their study, used apple waste as a feedstock for the production of mcl-PHAs using *Pseudomonas citronellolis*. The major components of the obtained biopolymer were 3-hydroxydecanoate (68% mol) and 3-hydroxyoctanoate (22% mol) with a total yield of 1.2 $\pm$ 0.05 (g/L). *P. citronellolis* mcl-PHA films showed high tensile strength (TS = 4.9 $\pm$ 0.68 MPa) and thermal stability [64]. The above research confirms the potential possibilities of using apple waste for the production of biopolymers that could be used as, e.g., packaging materials in many industries [63,64].

3. The Recovery from Apple Pomace Extraction for the Building and Construction Sectors

The considerable quantities of apple pomace produced in the world have been forcing researchers to develop novel and modern methods for their effective use. It is commonly known that apple peel (the main component of the pomace) contains a much higher content of active substances—phenolic antioxidants—than the pulp of the fruit. The relatively low price (compared to the price of raw apples) and widespread availability of AP make them a raw material with great potential [65,66]. However, due to their high water and sugar contents, AP are easily perishable (biologically unstable) and require immediate processing, such as dehydration (drying), as a pretreatment, which is associated with high energy consumption and, hence, additional OPEX. On the other hand, as an adverse side effect, drying can cause the degradation of temperature-sensitive valuable phenolic antioxidants [67]. The extraction of active compounds from AP can be an attractive method of their reuse. In addition, the solid waste generated during the process can be further used in accordance with the ideas of sustainability, e.g., as substrates for energy production.

3.1. Green Extraction Techniques

Green extraction techniques are methods for isolating active phenolic antioxidants from plant-based materials in an environmentally friendly manner. They rely on the utilization of alternative (green) solvents, eliminating the amount of synthetic and/or petroleum-based chemicals, and reducing energy costs and waste generation to obtain high-quality plant extracts [68]. Among green solvents, bioethanol is the most often used one due to its high biodegradability and low price [69]. Water, which is known to be the most natural solvent on the Earth, is effective only for the extraction of polar compounds [70]. Nowadays, various green extraction techniques including ultrasound-assisted, microwave-assisted, enzyme-assisted, pulsed electric field extraction, supercritical fluid extraction or pressurised liquid extraction have been explored [71]. Among them, supercritical fluid extraction (SFE), pressurised hot water extraction (PHWE), ultrasound-assisted extraction (UAE) or a combination of assisted extraction techniques are widely used. Recovered in a green way, phytochemicals from AP can be further used in many industries, including, e.g., construction and building, as anticorrosion agents, wood protectors, preservatives, antioxidants and biopolymers.

The SFE is a relatively new extraction method that is performed in the presence of supercritical fluids (most often liquid carbon dioxide—CO_2). The process is carried out in specialised high-pressure equipment where CO_2 is compressed under high pressure. The simultaneous increase in temperature and the pressure of the system leads CO_2 to reach a supercritical state; in that phase, CO_2 behaves similarly to both a liquid and a gas and mass transfer limitations that slow down the liquid transport are overcome [72]. After extraction, there is no need for additional purification of the extract or CO_2 removal, because gas expands and evaporates at normal temperature and pressure (25 °C and 1 atm). Moreover, the SFE technique does not require air access, which protects the substances contained in the extracted material against oxidation. Another advantage is that the carbon dioxide (extraction solvent) used is non-toxic, odourless, colourless, non-flammable, cheap and reaches a supercritical state at relatively low temperatures (above 31 °C) (Figure 3). Due to such a relatively low temperature, it is possible to obtain plant extracts without losing their properties (degradation of active compounds), which often takes place at higher temperatures [73,74]. However, the SFE method has some disadvantages, with the main ones being the high cost of the aperture and the limited range of substances that can be extracted with CO_2 as the sole solvent due to its non-polar nature [75]. The application of the SFE technique for the extraction of antioxidants from apple pomace was investigated in the work of Giovanna et al. [76]. In that study, fresh, freeze-dried and oven-dried apple pomace was treated with (a) subcritical CO_2 and (b) subcritical CO_2 with ethanol (5%) as a co-solvent, at pressures of 20 and 30 MPa and temperatures of 45 °C and 55 °C. For the comparison, a conventional extraction technology, i.e., Soxhlet extraction with ethanol and boiling water maceration, was also performed. The results of their research showed that the freeze-dried apple pomace extract obtained using the SFE method (at 55 °C, 30 MPa) with the use of ethanol (5%) as a co-solvent had the highest total phenolic antioxidant content (TPC) measured by the Folin–Ciocalteu method (8.87 ± 0.10 mg GAE (gallic acid equivalent)/g of extract) (Table 2). In addition, this extract was also found to have the highest antioxidant activity measured by DPPH$^\bullet$ assay (5.99 ± 0.11 mg TEA (Trolox equivalent antioxidant)/g of extract). For the extract obtained from SFE, carried out on freeze-dried apple pomace at the same conditions (55 °C, 30 MPa), but only with the subcritical CO_2 as a solvent, the TPC was equal to 6.41 ± 0.19 mg GAE/g of extract. Much lower TPC was obtained for the freeze-dried extract obtained using the Soxhlet and boiling water maceration methods, with 4.13 ± 0.90 and 2.37 ± 0.01 mg GAE/g of extract obtained, respectively [76]. The optimal conditions for the SFE apple pomace extraction process were also investigated in the work of De la Peña Armada et al. [77]. The results of their research indicated that the optimal conditions for the SFE process could be established at a temperature of 46 °C and a pressure of 425 bar. Under these conditions, the obtained extracts were characterised by the highest concentration of triterpenic acids (betulinic acid, oleanolic acid, ursolic acid, uvaol, erythrodiol and lupeol) and the highest antioxidant activity tested by means of the ORAC (Oxygen Radical Absorbance Capacity) assay (609.17 ± 96.11 µmol TE (Trolox equivalent)/g extract). For comparison, the extract obtained using the Soxhlet method showed a lower antioxidant capacity (ORAC: 565.95 ± 60.66 μmol TE/g extract) [77].

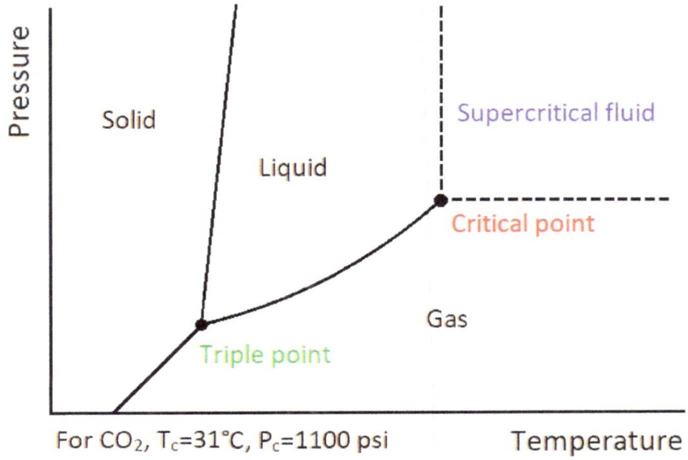

Figure 3. The phase diagram for CO_2.

PHWE is a similar technique to SFE, but in this case, water is used as a solvent. By increasing the temperature and pressure, water obtains similar properties to ethanol; this causes an increase in the solubility of many medium-polar compounds in water and ensures the extraction efficiency. The appropriate temperature of the extractant (water) should be above the atmospheric boiling point (100 °C, 0.1 MPa), but below its critical point (374 °C, 22.1 MPa) [78]. The main advantages of this method are its low cost and environmental friendliness; PHWE limits the use of organic solvents. Besides, the process water can be disposed of without causing any major environmental problems [79]. The applicability of the PHWE technique for the extraction of antioxidants from apple pomace was investigated in the work of Plaza et al. [79]. Using a response surface methodology (RSM), scientists optimised the PHWE parameters by maximising the yield of phenolic antioxidants from AP while minimising the possible formation of undesirable substances (e.g., melanoidins—the final Maillard reaction products). They reported that the highest amount of phenolic compounds (1.8 μmol/g dry AP) was obtained at a temperature of 170 °C and 3 min of extraction time [80].

The UAE generates high-frequency pulses that increase the mass transfer of the extracted biocompounds to the used solvent. This is due to the presence of cavitation bubbles created by ultrasonic waves passing through the solvent. The rupture of cavitation bubbles on the analyte surface causes damage at the impact site and increases the rate of mass transfer of the extracted material to the solvent. UAE can be carried out using two types of device: ultrasonic (US) bath or probe-generating ultrasound (Figure 4). Both of them are equipped with one (US probe) or more (US bath) ultrasound generators called transducers. There is also a temperature control in the ultrasonic bath. Moreover, unlike extraction with a probe, several samples can be extracted simultaneously in an ultrasonic bath. Ultrasonic baths usually operate at frequencies from 37 to 45 kHz, while the ultrasound probes operate at a lower frequency of ca. 20 kHz. Lower frequencies lead to the formation of larger cavitation bubbles. The main disadvantage of this method is the possibility of partial degradation of the analyte compounds. In addition, after UAE, the obtained extract must be filtered to separate it from the extraction residues that sometimes require significant amounts of solvents and can lead to oxygen degradation of the extract. The UAE technique is often combined with other extraction methods, e.g., the Sono–Soxhlet approach involves the combination of UAE with Soxhlet extraction; other approaches include UAE being combined with microwave-assisted extraction, and the combination of UAE and SFE [81,82]. There are several reports in the literature from recent years on the use of the UAE technique to recover active substances from apple pomace [83–86]. For example, in the work of Pollini et al. [86], the effect of the solvent on the TPC of apple pomace extract was investigated. In their research, the extract obtained through UAE, using the mixture of ethanol and water (50:50, vv) as a solvent, had the highest TPC value (1062.9 ± 59.80 μg GAE/g of fresh AP) compared to other solvents used (ethanol:water, 70:30 and 30:70, vv) [86]. Malinowska et al. [84], compared the effect of the solvent used (water and ethanol) and the source of AP on the efficiency of the UAE process. The results showed that AP water extract (from conventional crops) had a two-times-lower TPC value than the AP ethanolic extract and the AP water extract (from ecological crops) [84]. The temperature of the UAE process, time of extraction and ultrasound power (e.g., power intensity) also plays an important role. Overly high temperatures (e.g., much higher than room temperature), power intensities (a wide range of ultrasonic frequencies of 20–100 Hz are applied in the literature) and expanded extraction times (time longer than 30 min) can lead to the deconstruction of valuable compounds [87,88]. The influence of the mentioned extraction conditions (temperature in the range of 10 °C to 40 °C, and ultrasound intensity in the range of 0.764 W/cm^2 to 0.335 W/cm^2) was studied in the work of Pingret et al. [88]. The results of their research indicate that the optimal conditions for the water-extraction of phenolic antioxidants from apple pomace using the UAE method are 40 °C, 40 min and 0.764 W/cm^2 (Table 2) [89].

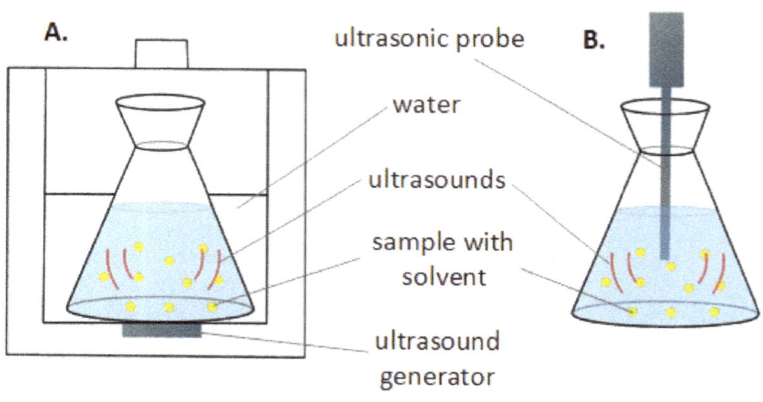

Figure 4. Ultrasonically assisted solvent extraction: (**A**) in an ultrasonic bath and (**B**) with a probe-generating ultrasound.

Table 2. The effects of different extraction techniques (conventional and unconventional) used for the recovery of biocompounds on the apple pomace extract composition.

Material for Research: Apple Variety, Sample Preparation	Extraction Method/Parameters	Analytical/Identification Method	Extract Composition/Identified Compounds	Antioxidant Activity	Ref.
Apple pomace separated from seeds and stems; a variety of apples not defined: (a) Fresh, only enzymatically stabilised; (b) Enzymatically stabilised, freeze-dried (−18 °C for 4h; 35 °C, 1.01 mbar for 20 h; 40 °C, 0.05 mbar for 6 h); (c) Enzymatically stabilised, oven-dried (50 °C for 4 days).	SFE; solvent: CO_2; temp.: 45 and 55 °C; pressure: 20 and 30 MPa; extraction time: 120 min	Total phenolic content (Folin–Ciocalteu)	For 55 °C, 30 MPa: (a) 3.91 ± 0.27 mg GAE/g of extract; (b) 6.41 ± 0.19 mg GAE/g of extract; (c) 5.65 ± 0.14 mg GAE/g of extract	DPPH: (a) 1.93 ± 0.12 mg TEA/g of extract; (b) 3.24 ± 0.11 mg TEA/g of extract; (c) 2.72 ± 0.19 mg TEA/g of extract	[76]
	SFE; solvent: CO_2 and ethanol (5%); temp.: 45 and 55 °C; pressure: 20 and 30 MPa; extraction time: 120 min		For 55 °C, 30 MPa: (a) 6.13 ± 0.16 mg GAE/g of extract; (b) 8.87 ± 0.17 mg GAE/g of extract; (c) 7.31 ± 0.18 mg GAE/g of extract	DPPH: (a) 2.67 ± 0.14 mg TEA/g of extract; (b) 5.99 ± 0.11 mg TEA/g of extract; (c) 4.73 ± 0.11 mg TEA/g of extract	[76]
	Soxhlet; solvent: ethanol; temp.: boiling temp. of ethanol; extraction time: 6 h		(a) 4.01 ± 0.06 mg GAE/g of extract; (b) 4.13 ± 0.90 mg GAE/g of extract; (c) 3.31 ± 0.12 mg GAE/g of extract	DPPH: (a) 1.96 ± 0.10 mg TEA/g of extract; (b) 2.05 ± 0.21 mg TEA/g of extract; (c) 1.38 ± 0.29 mg TEA/g of extract	[76]
	Boiling water maceration; solvent: water; temp.: 100 °C; extraction time: 37 min; 0.01 g/mL (solid-to-solvent ratio)		(a) 2.41 ± 0.01 mg GAE/g of extract; (b) 2.37 ± 0.01 mg GAE/g of extract; (c) 1.08 ± 0.11 mg GAE/g of extract	DPPH: (a) 1.17 ± 0.01 mg TEA/g of extract; (b) 1.14 ± 0.01 mg TEA/g of extract; (c) 0.92 ± 0.01 mg TEA/g of extract	[76]
Apple pomace constituted by seed, stalks, peel and a small proportion of pulp; "Golden Delicious" variety; apple pomace was freeze-dried at −45 °C and then milled to a fine powder	SFE; solvent: CO_2; temp.: 37, 46 and 55 °C; pressure: 300, 425 and 550 bar; extraction time: 100 min	UHPLC (Ultrahigh-performance liquid chromatography)	Main compounds: betulinic acid, oleanolic acid, ursolic acid, uvaol, erythrodiol, lupeol	ORAC: 609.17 ± 96.11 µmol; TE/g extract (46 °C, 425 bar); HORAC: 104.83 ± 8.82 µmol; CAE/g extract (46 °C, 425 bar)	[76]
	Soxhlet; solvent: n-hexane; temp.: 70 °C; extraction time: 6 h		Main compounds: betulinic acid, oleanolic acid, ursolic acid, uvaol, erythrodiol, lupeol	ORAC: 565.95 ± 60.66 µmol; TE/g extract; HORAC: 193.20 ± 17.49 µmol; CAE/g extract	[76]
Apple pomace composed of seeds, cores, stems, skin and parenchyma; obtained from Kiviks Musteri in Kivik, Sweden	PHWE; solvent: n-hexane; temp.: 25, 50, 112, 175 and 200 °C; extraction time: 3, 5, 10, 15 and 17 min; extractions were performed in 11 mL extraction cells, containing 5 g of fresh sample	Total phenols concentration calculated by RSM	1.8 µmol/g of dry AP (170 °C, 3 min)	n.t.	[80]
Apple pomace; "Champion" variety; (a) conventional and (b) ecological crops; fresh apple pomace was stored at 4 °C for 24 h	UAE; solvent: water; temp.: 20 °C; extraction time: 30 min; solid/liquid ratio of 1:20 (g/mL); US bath (50 Hz, 300 W)	Total phenolic content (Folin–Ciocalteu)	(a) 14.33 ± 0.26 mg/l; (b) 31.28 ± 0.29 mg/l	n.t.	[84]
	UAE; solvent: ethanol; temp.: 20 °C; extraction time: 30 min; solid/liquid ratio of 1:20 (g/mL); US bath (50 Hz, 300 W)		(a) 28.46 ± 0.28 mg/l; (b) 44.34 ± 0.44 mg/l	n.t.	[84]
Apple pomace separated from seeds and petioles; "Red Delicious" variety; blended	UAE; solvent: ethanol and water in different ratios ((a) 50:50, (b) 70:30, and (c) 30:70, *v*/*v*); temp.: 60 °C; extraction time: 60 min; solid/liquid ratio of 1:10 (g/mL)	Total phenolic content (Folin–Ciocalteu)	(a) 1062.9 ± 59.80 µg GAE/g of fresh AP; (b) ≈ 900 µg GAE/g of fresh AP; (c) ≈ 800 µg GAE/g of fresh AP	n.t.	[86]
Apple pomace obtained from Val-de-Vire Bioactives (Conde-sur-Vire, France); kept in the dark	UAE; solvent: water; temp.: 40 °C; extraction time: 40 min; solid/liquid ratio 150 g/mL; US bath (25 kHz, 150 W)	Total phenolic content (Folin–Ciocalteu)	Predicted/calculated value: 555 mg of catechin equivalent/100 g of dry AP	n.t.	[89]

Regarding the evaluation of the above-mentioned extraction processes for the recovery of active compounds from apple waste, the efficiency and costs have to be evaluated. In the case of the SFE and PHWE techniques, the costs of extractions are relatively expensive, due to the high cost of specialistic equipment (Figure 5). However, the advantage of these techniques is that they use environmentally friendly solvents, such as CO_2 and H_2O. On the other hand, the cost of the ultrasonic bath or probe-generating ultrasound used in the UAE technique is relatively low, but this method requires larger amounts of solvents than in SFE and PHWE. However, taking into account the production costs of some synthetic compounds, obtaining compounds from the extracts may be a cheaper solution. Natural compounds are also more desirable than synthetic ones [90,91].

EXTRACTION METHOD	ADVANTAGES	DISADVANTAGES
Supercritical fluid extraction (SFE)	- Eco-friendly and inexpensive solvent used (CO_2) - Elimination of organic solvents - No additional purification of the extract required - Air access not required (protection against oxidation) - Low operating temperatures (protection against thermal degradation)	- High equipment cost - High power consumption - Limited range of substances that can be extracted with CO_2 as the sole solvent due to its non-polar nature
Pressurized hot water extraction (PHWE)	- Eco-friendly and inexpensive solvent used (H_2O) - Limits the use of organic solvents	- High equipment cost - Not suitable for heat sensitive compounds
Ultrasound-assisted extraction (UAE)	- Low equipment cost - Low operating temperatures (protection against thermal degradation)	- Possibility of partial degradation of the analyte biocompounds - Additional purification of the extract required - Large amount of solvents

Figure 5. Comparison of the SFE, PHWE and UAE extraction techniques.

3.1.1. Green Corrosion Inhibitors Active Compounds

Metals and their alloys have been widely used in the building and construction industries as base materials for various equipment (e.g., pipes and water tanks). However, factors such as moisture, salts, acidic and alkaline solutions, gases, etc. can lead to numerous damages to the material, known as the corrosion process [92]. The corrosion products (i.e., rust) significantly affect the construction elements and, hence, generate serious impacts on human safety and the overall economy of the construction process. Various methods are used to protect metal surfaces from corrosion. One of them is the use of substances that inhibit the corrosion process, i.e., corrosion inhibitors [93–95]. Recently, most of the commercially used corrosion inhibitors have been synthetic inorganic molecules containing, e.g., copper, zinc, arsenic, nickel or arsenic salts [96]. However, the use of most of them (e.g., toxic phosphate or chromates) raises concerns regarding the safety of living organisms and the natural environment (e.g., surface water) [94,97,98].

Nowadays, great emphasis in the construction industry is given to the use of natural, non-toxic, readily available and biodegradable products; therefore, new sources of substances that will be effective and inexpensive are considered [94]. Active molecules from AP extracts (phenolic compounds with antioxidant properties) are tested as potential corrosion inhibitors, due to their electron-donating properties and active sites [94]. Their anti-corrosion mechanism of action consists in the creation/adsorption of the protective film layer on the metal surface by blocking active sites on the metal surface (in response to the Langmuir adsorption isotherm). The inhibitory effectiveness is associated with the chemical composition of AP extracts and chemical structure of active phenolic compounds—the presence of heteroatoms, such as sulfur (S), oxygen (O), phosphorus (P) and nitrogen (N) in their polar functional groups (e.g., -OH, -COOH, $-OCH_3$ -CN and $-NO_2$). These heteroatoms favour the adsorption processes via an interaction between the metal surface and the π-electrons clouds in the conjugated system, or by the formation of the bonds with the non-bonding electron pairs of the heteroatoms [97,99–101]. Moreover, the corrosion inhibition efficiency (IE) of plant extracts is related to the electron density sites of the inhibitor molecules [96,97].

There are several reports in the literature on the use of AP extracts and individual active substances, that can be isolated from AP, as green anti-corrosion agents (Table 3) [97,100,102–106]. In the work of Vera et al. [100], the phenolic antioxidants occurring in the Fuji apple peel extract turned out to be highly effective (IE = 89.88% at an inhibitor/extract concentration of 1000 ppm) anticorrosive agents of carbon steel. The major components of the AP extract were 3,5,2′-trihydroxy-7,8,4′-trimethoxyflavone 5-glucosyl-(1->2)-galactoside (44.33%), 5-methoxy-6″,6″-dimethyl-3′,4′-methylenedioxypyrano (2″,3″,7,8) flavone (38.49%), quercetin-5-glucoside (3.27%) and quercetin-3-α-L-arabinopyranoside (3.15%). Other phenolic antioxidants, such as caffeic acid, chlorogenic acid, rutin, kaempherol and isoquercetin, were detected in lower concentrations [100]. In the study by Nazari et al. [97], an AP-based green inhibitor was found to exhibit high efficiency in reducing the carbon steel corrosion in 3.5% NaCl brine. 1-Linoleoyl-sn-glycero-3-phosphocholine ($C_{26}H_{50}NO_7P$), containing N, P and O heteroatoms, was found to be a major constituent of AP extract (19.3 wt.%). The inhibition action mechanism of AP extract molecules was based on blocking the anode active sites on the steel surface and transforming Fe_3O_4 into a more corrosion-resistant Fe_2O_3. The highest IE (98%) was obtained on the seventh day of the measurement at the highest concentration of AP extract used (3%). In addition, the above-mentioned AP-derived inhibitor was synthesised without generating any waste [97]. In another study, pectin, which is abundant in AP, was used as an anti-corrosion coating for carbon steel. The protective effect (PE) increased with the increasing pectin concentration. For the lowest applied pectin concentration (100 ppm), PE = 83.62%, while for the highest (500 ppm), PE was equal 89.31% [102]. The influence of pectin on corrosion of metals in hydrochloric acid solution was also studied in the work of Fiori-Bimbi et al. [103]. In their work, the maximum value of pectin's mild steel corrosion inhibition efficiency was equal to 94.2% (T = 318 K, inhibitor concertation = 2 g L^{-1}) [102]. Pectin may also be a promising anti-corrosion agent for carbon steel in a neutral aqueous solution. Prabakaran et al. [104] developed an inhibitor composed of pectin (250 ppm), propyl phosphonic acid (50 ppm) and Zn(II) ions (20 ppm). The corrosion IE value for this mixture was 94%, indicating an excellent synergistic effect of components [104]. Procyanidin B2 and quercetin are major AP components. Procyanidin B2 was reported to be an effective corrosion inhibitor of carbon steel in 1 M HCl. The corrosion IE reached 94.21% at 30 °C (500 mg/L) after 24 h [105]. However, 800 ppm of quercetin was found to reduce 92% of mild steel corrosion in 1 M HCl after 1 h [106].

Table 3. Selected green corrosion inhibitors from AP.

Source	The Most Frequently Occurring Active Compounds	Metal and Electrolyte	Ref.
Fuji apple peel	3,5,2′-Trihydroxy-7,8,4′-trimethoxyflavone 5-glucosyl-(1->2)-galactoside, 5-Methoxy-6″,6″-dimethyl-3′,4′-methylenedioxypyrano(2″,3″,7,8)flavone	Carbon steel, 0.1 M NaCl	[100]
Apple pomace	1-Linoleoyl-sn-glycero-3-phosphocholine	Carbon steel, 3.5% NaCl	[97]
Pectin	-	Carbon steel, 1.0 M HCl	[102]
Pectin	-	Mild steel, 1.0 M HCl	[103]
Pectin	-	Carbon steel, H_2O	[104]
Procyanidin B2	-	Carbon steel, 1.0 M HCl	[105]
Quercetin	-	Mild steel, 1.0 M HCl	[106]

3.1.2. Green Wood Protectors' Active Compounds

Wood is a frequently used natural, renewable, relatively inexpensive and readily available building material used in the construction of structural beams, facilities, structures and wood objects (e.g., furniture and home decors). The use of wood in construction brings several benefits, e.g., wood is resistant to high temperatures, stretching (tensile strength) and electrical currents, it can absorb unwanted sounds (especially desirable in the construction of concert halls) and is highly machinable. Generally, wood can be divided into two types: hardwood (e.g., maple, oak, mahogany, beech and teak) and softwood (e.g., birch, pine and ash). Depending on the type, they differ in physical properties, such as density, strength, moisture content, etc. [107,108]. However, all of the types are exposed to factors such as weather conditions (moisture), fungi and insects, which contribute to its degradation [109].

Biological corrosion of wood causes significant changes in its structure, as well as in its chemical and psychical properties and can lead to complete material deconstruction (wood decay). To prevent wood and wood-based materials from these damages, various chemical wood preservatives are used. However, most of the traditional biocides used for wood protection are often highly toxic (Table 4) and can leach out from the preservative-treated wood, posing a serious risk to the environment, human and animal health [108,110]. For example, conventional synthetic wood preservative—CCA (Copper Chromium Arsenate)—contains arsenic and chromium (VI), which are easily leached from the wood surface and contaminate the surrounding soil. Arsenic is also known to be carcinogenic and, therefore, the use of CCA for wood conservation has been restricted since 2003 by the U.S. Environmental Protection Agency (EPA) [107,111]. To protect the environment and society, new alternative wood preservatives based on non-toxic and biodegradable natural substances should be developed. AP contains huge quantities of unused active substances, especially phenolic compounds, which are known to be potential antifungal and antibacterial agents [66,112].

Some types of wood (e.g., Alaska cedar, redwood) show natural resistance to insects, microorganisms and decay, due to the presence of extractives in hardwood [113–115]. Benzoic and cinnamic acids as well as their phenolic derivatives were found to be one of the extractive components responsible for the natural resistance of wood [116]. These compounds are plant secondary metabolites responsible for plant protection against biotic (insects, bacteria and fungi) and abiotic (drought, cold, heat and UV light) environmental stress [117,118]. Numerous studies have investigated the use of benzoic, cinnamic acids and their phenolic derivatives recovered from plant sources as potential natural and non-toxic wood protection agents (Table 5) [119–124]. The influence of benzoic, salicyli, syringic and vanillic acids on oil palm diseases caused by *Ganoderma boninense* was investigated in the study of Surendran et al. [119]. *G. boninense* is the major pathogen for basal stem rot (BSR) disease. Among all studied compounds, benzoic acid turned out to be the best *G. boninense* inhibitor. During all days of the measurement, benzoic acid at a concentration of 5 mM inhibited the growth of the tested pathogen. On the 120th day, the following weight loss was observed in the woodblocks treated successively with salicylic (≈34%), syringic (≈40%) and vanillic acids (≈75%) (C = 5 mM). For comparison, the mass loss of the untreated control woodblocks was 71.8% [119]. Sekine et al. [120], investigated the bioactivity of latifolin and its derivatives (Table 4) against termites and white- and brown-rot fungi. The results showed that latifolin exhibited significantly higher antifungal and anti-termite activity than the other tested compounds. For example, the value of inhibition rate of *T. versicolor* for latifolin was 79.1%, while for its derivatives, this was in the range of 13.2% to 21.8% [120]. In the studies of Little et al. [121], three flavonoids (quercetin, morin and catechin) and tannic acid were investigated as potential termite repellers. The results showed that wood treated with 3% tannic acid and 4% catechin caused high termite mortality—75% and 50%, respectively [121]. The anti-termite activity of flavonoids (apigenin, quercetin, biochanin A, genistein and taxifolin) was also reported in other works [122,123]. Efhamisisi et al. [124] impregnated 3-ply beech plywood with a mixture of 20% tannin solution and 1% boric acid (to enhance the crosslinking properties and prevent tannin loss). The results showed that such treatment significantly increased the resistance of panels against termites (*R. flavipes*) and fungal (*T. versicolor*) attack [124].

Table 4. Examples of toxic substances used in wood preservatives [110].

Active Ingredient	Toxicity Class	Lethal Dose (LD_{50}) * (mg/kg)	Main Use
Azaconazole	II. Moderately hazardous	308	Fungicide
Copper hydroxide	II. Moderately hazardous	1000	Fungicide
Copper oxychloride	II. Moderately hazardous	1440	Fungicide
Copper sulphate	II. Moderately hazardous	300	Fungicide
Chlorpyrifos	II. Moderately hazardous	135	Insecticide
Fipronil	II. Moderately hazardous	92	Insecticide
Thiamethoxam	II. Moderately hazardous	871	Insecticide
Disodium tetraborate (Borax)	III. Slightly hazardous	4500	Fungicide
Fenpropimorph	III. Slightly hazardous	3515	Fungicide
Tebuconazole	III. Slightly hazardous	1700	Fungicide
Dichlofluanid	U. Unlikely to present acute hazard in normal use	>5000	Fungicide
Fenoxycarb	U. Unlikely to present acute hazard in normal use	>10,000	Fungicide

* LD50—the amount of toxic substance (mg) per kg of body weight, which causes the death of 50% of a group of the tested animals.

Table 5. Phenolic compounds as natural wood preservatives.

Active Compounds	Wood Protected	Protection against	Results	Ref.
Benzoic acid Salicylic acid Syringic acid Vanillic acid	Oil palm (*Elaeis guineensis* Jacq.)	*Ganoderma boninense*	Controlled BSR disease.	[119]
Latifolin 2′-O-methyllatifolin Latifolin dimethyl ether Latifolin diacetate	Studies on paper discs	*Trametes versicolor* *Fomitopsis palustris* *Reticulitermes speratus* (Kolbe)	A significant activity of Latifolin against tested termites and fungi.	[120]
Quercetin Morin Catechin Tannic acid	*Pinus* sp.	*Reticulitermes flavipes*	A significant activity of catechin and tannic acid against tested termites.	[121]
Condensed tannin	European beach (*Fagus sylvatica* L.)	*Trametes versicolor* *Reticulitermes flavipes*	Increased resistance against tested termites and fungi.	[124]

4. Conclusions

To improve food and environmental safety, it is important to properly manage agri-food waste so that it can be reintegrated into the existing bioeconomy [15]. The production of bio-waste, including apple pomace (AP), is expected to increase every year. Therefore, it is necessary to develop safe and effective methods for the processing and disposal of AP in accordance with the idea of sustainable development. The research results will allow for the acquisition of new knowledge as well as new bio-materials and technological solutions that could have a big economic impact in the future. Some of them are currently in use, e.g., green extraction techniques for the recovery of active substances from AP in an environmentally friendly manner, i.e., by lowering the energy consumption and reducing the amounts of harmful chemicals. Active compounds extracted from AP (including benzoic and cinnamic acid derivatives) can further replace the widespread synthetic chemicals and reduce the amounts of generated waste, e.g., they can be used as non-toxic, readily available and biodegradable anticorrosion agents or wood protectors in different industrial sectors. Through various processes (e.g., fermentation, anaerobic digestion and pyrolysis), AP can be transformed into fuels and/or fuel intermediates in solid (e.g., biochar-based hard carbon that can be used in Na-ion battery production), liquid (e.g., bioethanol, biodiesel, pyrolysis oil, etc.) and gaseous (e.g., biogas/biomethane, etc.) forms. Such transformations of apple waste into environmentally friendly energy and materials can not only reduce the consumption of conventional fossil feedstocks but also reduce the amount of GHGs emitted into the atmosphere. Finally, the AP can be considered as an environmentally safe biopolymer that can be applied as an innovative additive, e.g., in the production of structural or building elements or packaging materials in many other industries.

Author Contributions: Conceptualization M.K.; Visualization, E.G.; methodology, E.G., M.K. and G.Y.; Formal analysis E.G., M.K. and G.Y.; Writing—original draft preparation, E.G.; Writing—review and editing, M.K. and G.Y.; Funding acquisition, M.K. All authors have read and agreed to the published version of the manuscript.

Funding: The source of funding for the study was the Polish Ministry of Higher Education and Science nr WZ/WB-IIŚ/5/2020.

Institutional Review Board Statement: Not applicable.

Informed Consent Statement: Not applicable.

Data Availability Statement: The data presented in this study are available on request from the corresponding author.

Conflicts of Interest: The authors declare no conflict of interest.

References

1. Guerrero, L.A.; Maas, G.; Hogland, W. Solid waste management challenges for cities in developing countries. *Waste Manag.* **2013**, *33*, 220–232. [CrossRef] [PubMed]
2. Mukhtar, E.; Williams, I.; Shaw, P.; Ongondo, F. A Tale of Two Cities: The Emergence of Urban Waste Systems in a Developed and a Developing City. *Recycling* **2016**, *1*, 254–270. [CrossRef]
3. Kaza, S.; Yao, L.C.; Bhada-Tata, P.; Van Woerden, F. *What a Waste 2.0: A Global Snapshot of Solid Waste Management to 2050*; World Bank: Washington, DC, USA, 2018.
4. Shalini, R.; Gupta, D.K. Utilization of pomace from apple processing industries: A review. *J. Food Sci. Technol.* **2010**, *47*, 365–371. [CrossRef] [PubMed]
5. Lyu, F.; Luiz, S.F.; Azeredo, D.R.P.; Cruz, A.G.; Ajlouni, S.; Ranadheera, C.S. Apple Pomace as a Functional and Healthy Ingredient in Food Products: A Review. *Processes* **2020**, *8*, 319. [CrossRef]
6. Płotka-Wasylka, J.; Rutkowska, M.; Owczarek, K.; Tobiszewski, M.; Namieśnik, J. Extraction with environmentally friendly solvents. *TrAC Trends Anal. Chem.* **2017**, *91*, 12–25. [CrossRef]
7. Molinuevo-Salces, B.; Riaño, B.; Hijosa-Valsero, M.; González-García, I.; Paniagua-García, A.I.; Hernández, D.; Garita-Cambronero, J.; Díez-Antolínez, R.; García-González, M.C. Valorization of apple pomaces for biofuel production: A biorefinery approach. *Biomass Bioenergy* **2020**, *142*, 105785. [CrossRef]
8. Olech, E.; Sikora, J.; Kuboń, M. Biogas production from corn silage and apple pomace. *J. Res. Appl. Agric. Eng.* **2017**, *62*, 155–157.
9. Xu, R.; Ferrante, L.; Hall, K.; Briens, C.; Berruti, F. Thermal self-sustainability of biochar production by pyrolysis. *J. Anal. Appl. Pyrolysis* **2011**, *91*, 55–66. [CrossRef]
10. Wu, L.; Buchholz, D.; Vaalma, C.; Giffin, G.A.; Passerini, S. Apple-Biowaste-Derived Hard Carbon as a Powerful Anode Material for Na-Ion Batteries. *ChemElectroChem* **2016**, *3*, 292–298. [CrossRef]
11. Koryś, K.A.; Latawiec, A.E.; Grotkiewicz, K.; Kuboń, M. The Review of Biomass Potential for Agricultural Biogas Production in Poland. *Sustainability* **2019**, *11*, 6515. [CrossRef]

12. Liu, H.; Kumar, V.; Jia, L.; Sarsaiya, S.; Kumar, D.; Juneja, A.; Zhang, Z.; Sindhu, R.; Binod, P.; Bhatia, S.K.; et al. Biopolymer poly-hydroxyalkanoates (PHA) production from apple industrial waste residues: A review. *Chemosphere* **2021**, *284*, 131427. [CrossRef] [PubMed]
13. Gustafsson, J.; Landberg, M.; Bátori, V.; Åkesson, D.; Taherzadeh, M.J.; Zamani, A. Development of bio-based films and 3D objects from apple pomace. *Polymers* **2019**, *11*, 289. [CrossRef] [PubMed]
14. Śliwa, K.; Malinowska, M.A.; Sikora, E.; Ogonowski, J. Comparison of extraction methods used for separation of active compounds from waste generated during the production of apple juice. *Przem. Chem.* **2015**, *94*, 357–360.
15. Campos, D.A.; Gómez-García, R.; Vilas-Boas, A.A.; Madureira, A.R.; Pintado, M.M. Management of Fruit Industrial By-Products—A Case Study on Circular Economy Approach. *Molecules* **2020**, *25*, 320. [CrossRef] [PubMed]
16. Fruit: World Production by Type 2019 Statista. Available online: https://www.statista.com/statistics/264001/worldwide-production-of-fruit-by-variety/ (accessed on 22 March 2021).
17. Rozkrut, D. (Ed.) *Statistical Yearbook of Agriculture 2019*; Zakład Wydawnictw Statystycznych: Warsaw, Poland, 2020.
18. PKO Bank Polski. Available online: https://www.pkobp.pl/centrum-analiz/analizy-sektorowe/#subcategory=165947&page=2 (accessed on 27 March 2021).
19. Bhushan, S.; Kalia, K.; Sharma, M.; Singh, B.; Ahuja, P.S. Processing of apple pomace for bioactive molecules. *Crit. Rev. Biotechnol.* **2008**, *28*, 285–296. [CrossRef] [PubMed]
20. Vukušić, J.L.; Millenautzki, T.; Cieplik, R.; Obst, V.; Saaid, A.M.; Clavijo, L.; Zlatanovic, S.; Hof, J.; Mösche, M.; Barbe, S. Reshaping Apple Juice Production into a Zero Discharge Biorefinery Process. *Waste Biomass Valorization* **2021**, *12*, 3617–3627. [CrossRef]
21. Wang, H.; Wang, J.; Fang, Z.; Wang, X.; Bu, H. Enhanced bio-hydrogen production by anaerobic fermentation of apple pomace with enzyme hydrolysis. *Int. J. Hydrogen Energy* **2010**, *35*, 8303–8309. [CrossRef]
22. Martău, G.A.; Teleky, B.E.; Ranga, F.; Pop, I.D.; Vodnar, D.C. Apple Pomace as a Sustainable Substrate in Sourdough Fermentation. *Front. Microbiol.* **2021**, *12*, 3850. [CrossRef]
23. FAO. World Fertilizer Trends and Outlook to 2018. Available online: http://www.fao.org/3/a-i4324e.pd (accessed on 12 January 2022).
24. Ishangulyyev, R.; Kim, S.; Lee, S.H. Understanding Food Loss and Waste—Why Are We Losing and Wasting Food? *Foods* **2019**, *8*, 297. [CrossRef] [PubMed]
25. Jaeger, S.R.; Machín, L.; Aschemann-Witzel, J.; Antúnez, L.; Harker, F.R.; Ares, G. Buy, eat or discard? A case study with apples to explore fruit quality perception and food waste. *Food Qual. Prefer.* **2018**, *69*, 10–20. [CrossRef]
26. Matei, E.; Râpă, M.; Predescu, A.M.; Țurcanu, A.A.; Vidu, R.; Predescu, C.; Bobirica, C.; Bobirica, L.; Orbeci, C. Valorization of Agri-Food Wastes as Sustainable Eco-Materials for Wastewater Treatment: Current State and New Perspectives. *Materials* **2021**, *14*, 4581. [CrossRef] [PubMed]
27. Abiad, M.G.; Meho, L.I. Food loss and food waste research in the Arab world: A systematic review. *Food Secur.* **2018**, *10*, 311–322. [CrossRef]
28. Santeramo, F.G. Exploring the link among food loss, waste and food security: What the research should focus on? *Agric. Food Secur.* **2021**, *10*, 26. [CrossRef]
29. Marks-Bielska, R.; Bielski, S.; Pik, K.; Kurowska, K. The Importance of Renewable Energy Sources in Poland's Energy Mix. *Energies* **2020**, *13*, 4624. [CrossRef]
30. Richards, B.K.; Herndon, F.G.; Jewell, W.J.; Cummings, R.J.; White, T.E. In situ methane enrichment in methanogenic energy crop digesters. *Biomass Bioenergy* **1994**, *6*, 275–282. [CrossRef]
31. Bozym, M.; Florczak, I.; Zdanowska, P.; Wojdalski, J.; Klimkiewicz, M. An analysis of metal concentrations in food wastes for biogas production. *Renew. Energy* **2015**, *77*, 467–472. [CrossRef]
32. Magyar, M.; da Costa Sousa, L.; Jin, M.; Sarks, C.; Balan, V. Conversion of apple pomace waste to ethanol at industrial relevant conditions. *Appl. Microbiol. Biotechnol.* **2016**, *100*, 7349–7358. [CrossRef] [PubMed]
33. Riggio, V.; Comino, E.; Rosso, M. Energy production from anaerobic co-digestion processing of cow slurry, olive pomace and apple pulp. *Renew. Energy* **2015**, *83*, 1043–1049. [CrossRef]
34. Karimi, K.; Chisti, Y. Bioethanol Production and Technologies. In *Encyclopedia of Sustainable Technologies*; Elsevier: London, UK, 2017; Volume 3, pp. 273–284.
35. Rodrussamee, N.; Lertwattanasakul, N.; Hirata, K.; Suprayogi; Limtong, S.; Kosaka, T.; Yamada, M. Growth and ethanol fermentation ability on hexose and pentose sugars and glucose effect under various conditions in thermotolerant yeast Kluyveromyces marxianus. *Appl. Microbiol. Biotechnol.* **2011**, *90*, 1573–1586. [CrossRef] [PubMed]
36. Demiray, E.; Kut, A.; Ertuğrul Karatay, S.; Dönmez, G. Usage of soluble soy protein on enzymatically hydrolysis of apple pomace for cost-efficient bioethanol production. *Fuel* **2021**, *289*, 119785. [CrossRef]
37. Kut, A.; Demiray, E.; Ertuğrul Karatay, S.; Dönmez, G. Second generation bioethanol production from hemicellulolytic hydrolyzate of apple pomace by Pichia stipitis. *Energy Sources Part A Recover. Util. Environ. Eff.* **2020**, *42*, 1–12. [CrossRef]
38. Bharathiraja, B.; Sudharsana, T.; Jayamuthunagai, J.; Praveenkumar, R.; Chozhavendhan, S.; Iyyappan, J. Biogas production–A review on composition, fuel properties, feed stock and principles of anaerobic digestion. *Renew. Sustain. Energy Rev.* **2018**, *90*, 570–582. [CrossRef]
39. Kim, S.; Sung, T.; Kim, K.C. Performance and Greenhouse Gas Reduction Analysis of Biogas-Fueled Solid-Oxide Fuel Cells for a Sewage Sludge and Food Waste Treatment Facility. *Energies* **2018**, *11*, 600. [CrossRef]

40. Claes, A.; Melchi, L.; Uludag-Demirer, S.; Demirer, G.N. Supplementation of Carbon-Based Conductive Materials and Trace Metals to Improve Biogas Production from Apple Pomace. *Sustainability* **2021**, *13*, 9488. [CrossRef]
41. Saletnik, B.; Bajcar, M.; Saletnik, A.; Zaguła, G.; Puchalski, C. Effect of the Pyrolysis Process Applied to Waste Branches Biomass from Fruit Trees on the Calorific Value of the Biochar and Dust Explosivity. *Energies* **2021**, *14*, 4898. [CrossRef]
42. Kumar, G.; Eswari, A.P.; Kavitha, S.; Kumar, M.D.; Kannah, R.Y.; How, L.C.; Muthukaruppan, G.; Banu, J.R. Thermochemical conversion routes of hydrogen production from organic biomass: Processes, challenges and limitations. *Biomass Convers. Biorefinery* **2020**, *2*, 1–26. [CrossRef]
43. Kosakowski, W.; Bryszewska, M.A.; Dziugan, P. Biochars from Post-Production Biomass and Waste from Wood Management: Analysis of Carbonization Products. *Materials* **2020**, *13*, 4971. [CrossRef] [PubMed]
44. Guerrero, M.R.B.; Salinas Gutiérrez, J.M.; Meléndez Zaragoza, M.J.; López Ortiz, A.; Collins-Martínez, V. Optimal slow pyrolysis of apple pomace reaction conditions for the generation of a feedstock gas for hydrogen production. *Int. J. Hydrogen Energy* **2016**, *41*, 23232–23237. [CrossRef]
45. Zhang, S.; Ji, Y.; Dang, J.; Zhao, J.; Chen, S. Magnetic apple pomace biochar: Simple preparation, characterization, and application for enriching Ag(I) in effluents. *Sci. Total Environ.* **2019**, *668*, 115–123. [CrossRef] [PubMed]
46. Dou, X.; Geng, C.; Buchholz, D.; Passerini, S. Research Update: Hard carbon with closed pores from pectin-free apple pomace waste for Na-ion batteries. *APL Mater.* **2018**, *6*, 047501. [CrossRef]
47. Talekar, S.; Patti, A.F.; Vijayraghavan, R.; Arora, A. Complete Utilization of Waste Pomegranate Peels to Produce a Hydrocolloid, Punicalagin Rich Phenolics, and a Hard Carbon Electrode. *ACS Sustain. Chem. Eng.* **2018**, *6*, 16363–16374. [CrossRef]
48. Nita, C.; Zhang, B.; Dentzer, J.; Matei Ghimbeu, C. Hard carbon derived from coconut shells, walnut shells, and corn silk biomass waste exhibiting high capacity for Na-ion batteries. *J. Energy Chem.* **2021**, *58*, 207–218. [CrossRef]
49. Arie, A.A.; Kristianto, H.; Demir, E.; Cakan, R.D. Activated porous carbons derived from the Indonesian snake fruit peel as anode materials for sodium ion batteries. *Mater. Chem. Phys.* **2018**, *217*, 254–261. [CrossRef]
50. Muruganantham, R.; Wang, F.-M.; Yuwono, R.A.; Sabugaa, M.; Liu, W.-R. Biomass Feedstock of Waste Mango-Peel-Derived Porous Hard Carbon for Sustainable High-Performance Lithium-Ion Energy Storage Devices. *Energy Fuels* **2021**, *35*, 10878–10889. [CrossRef]
51. Meenatchi, T.; Priyanka, V.; Subadevi, R.; Liu, W.-R.; Huang, C.-H.; Sivakumar, M. Probe on hard carbon electrode derived from orange peel for energy storage application. *Carbon Lett.* **2021**, *31*, 1033–1039. [CrossRef]
52. Górka, J.; Vix-Guterl, C.; Matei Ghimbeu, C. Recent Progress in Design of Biomass-Derived Hard Carbons for Sodium Ion Batteries. *C* **2016**, *2*, 24. [CrossRef]
53. Saavedra Rios, C.D.M.; Simonin, L.; De Geyer, A.; Ghimbeu, C.M.; Dupont, C. Unraveling the properties of biomass-derived hard carbons upon thermal treatment for a practical application in na-ion batteries. *Energies* **2020**, *13*, 3513. [CrossRef]
54. Moriwake, H.; Kuwabara, A.; Fisher, C.A.J.; Ikuhara, Y. Why is sodium-intercalated graphite unstable? *RSC Adv.* **2017**, *7*, 36550–36554. [CrossRef]
55. Stevens, D.A.; Dahn, J.R. The Mechanisms of Lithium and Sodium Insertion in Carbon Materials. *J. Electrochem. Soc.* **2001**, *148*, A803–A811. [CrossRef]
56. Saurel, D.; Orayech, B.; Xiao, B.; Carriazo, D.; Li, X.; Rojo, T. From Charge Storage Mechanism to Performance: A Roadmap toward High Specific Energy Sodium-Ion Batteries through Carbon Anode Optimization. *Adv. Energy Mater.* **2018**, *8*, 1703268. [CrossRef]
57. Dou, X.; Hasa, I.; Hekmatfar, M.; Diemant, T.; Behm, R.J.; Buchholz, D.; Passerini, S. Pectin, Hemicellulose, or Lignin? Impact of the Biowaste Source on the Performance of Hard Carbons for Sodium-Ion Batteries. *ChemSusChem* **2017**, *10*, 2668–2676. [CrossRef] [PubMed]
58. Mellinas, C.; Ramos, M.; Jiménez, A.; Garrigós, M.C. Recent Trends in the Use of Pectin from Agro-Waste Residues as a Natural-Based Biopolymer for Food Packaging Applications. *Materials* **2020**, *13*, 673. [CrossRef] [PubMed]
59. Production of Bio-Plastic Materials from Apple Pomace: A New Application for the Waste Material. Available online: https://www.diva-portal.org/smash/get/diva2:1327184/FULLTEXT01.pdf/ (accessed on 19 December 2021).
60. Rani, G.U.; Sharma, S. *Biopolymers, Bioplastics and Biodegradability: An Introduction*; Elsevier: Amsterdam, The Netherlands, 2021; pp. 2–10.
61. Crutchik, D.; Franchi, O.; Caminos, L.; Jeison, D.; Belmonte, M.; Pedrouso, A.; Val del Rio, A.; Mosquera-Corral, A.; Campos, J.L. Polyhydroxyalkanoates (PHAs) Production: A Feasible Economic Option for the Treatment of Sewage Sludge in Municipal Wastewater Treatment Plants? *Water* **2020**, *12*, 1118. [CrossRef]
62. Ten, E.; Jiang, L.; Zhang, J.; Wolcott, M.P. Mechanical performance of polyhydroxyalkanoate (PHA)-based biocomposites. In *Biocomposites*; Woodhead Publishing Series in Composites Science and Engineering; Woodhead Publishing: Sawston, UK, 2015; pp. 39–52.
63. Pereira, J.R.; Araújo, D.; Freitas, P.; Marques, A.C.; Alves, V.D.; Sevrin, C.; Grandfils, C.; Fortunato, E.; Reis, M.A.M.; Freitas, F. Production of medium-chain-length polyhydroxyalkanoates by Pseudomonas chlororaphis subsp. aurantiaca: Cultivation on fruit pulp waste and polymer characterization. *Int. J. Biol. Macromol.* **2021**, *167*, 85–92. [CrossRef]
64. Rebocho, A.T.; Pereira, J.R.; Freitas, F.; Neves, L.A.; Alves, V.D.; Sevrin, C.; Grandfils, C.; Reis, M.A.M. Production of medium-chain length polyhydroxyalkanoates by Pseudomonas citronellolis grown in apple pulp waste. *Appl. Food Biotechnol.* **2019**, *6*, 71–82.

65. Łata, B.; Tomala, K. Apple peel as a contributor to whole fruit quantity of potentially healthful bioactive compounds. Cultivar and year implication. *J. Agric. Food Chem.* **2007**, *55*, 10795–10802. [CrossRef] [PubMed]
66. Kalinowska, M.; Gryko, K.; Wróblewska, A.M.; Jabłońska-Trypuć, A.; Karpowicz, D. Phenolic content, chemical composition and anti-/pro-oxidant activity of Gold Milenium and Papierowka apple peel extracts. *Sci. Rep.* **2020**, *10*, 14951. [CrossRef]
67. Awasthi, M.K.; Ferreira, J.A.; Sirohi, R.; Sarsaiya, S.; Khoshnevisan, B.; Baladi, S.; Sindhu, R.; Binod, P.; Pandey, A.; Juneja, A.; et al. A critical review on the development stage of biorefinery systems towards the management of apple processing-derived waste. *Renew. Sustain. Energy Rev.* **2021**, *143*, 110972. [CrossRef]
68. Chemat, F.; Vian, M.A.; Cravotto, G. Green extraction of natural products: Concept and principles. *Int. J. Mol. Sci.* **2012**, *13*, 8615–8627. [CrossRef] [PubMed]
69. Oklu, N.K.; Matsinha, L.C.; Makhubela, B.C. Bio-Solvents: Synthesis, Industrial Production and Applications. In *Solvents, Ionic Liquids and Solvent Effects*; BoD–Books on Demand: Norderstedt, Germany, 2020; pp. 1–24.
70. König, G.; Reetz, M.T.; Thiel, W. 1-Butanol as a Solvent for Efficient Extraction of Polar Compounds from Aqueous Medium: Theoretical and Practical Aspects. *J. Phys. Chem. B* **2018**, *122*, 6975–6988. [CrossRef] [PubMed]
71. Saini, A.; Panesar, P.S.; Bera, M.B. Valorization of fruits and vegetables waste through green extraction of bioactive compounds and their nanoemulsions-based delivery system. *Bioresour. Bioprocess.* **2019**, *6*, 26. [CrossRef]
72. Wong, V.; Wyllie, S.G.; Cornwell, C.P.; Tronson, D. Supercritical fluid extraction (SFE) of monoterpenes from the leaves of Melaleuca alternifolia (Tea Tree). *Molecules* **2001**, *6*, 92–103. [CrossRef]
73. Khaw, K.-Y.; Parat, M.-O.; Shaw, P.N.; Falconer, J.R. Solvent Supercritical Fluid Technologies to Extract Bioactive Compounds from Natural Sources: A Review. *Molecules* **2017**, *22*, 1186. [CrossRef]
74. Girotra, P.; Singh, S.K.; Nagpal, K. Supercritical fluid technology: A promising approach in pharmaceutical research. *Pharm. Dev. Technol.* **2013**, *18*, 22–38. [CrossRef] [PubMed]
75. Woźniak, Ł.; Szakiel, A.; Pączkowski, C.; Marszałek, K.; Skąpska, S.; Kowalska, H.; Jędrzejczak, R. Extraction of Triterpenic Acids and Phytosterols from Apple Pomace with Supercritical Carbon Dioxide: Impact of Process Parameters, Modelling of Kinetics, and Scaling-Up Study. *Molecules* **2018**, *23*, 2790. [CrossRef] [PubMed]
76. Ferrentino, G.; Morozova, K.; Mosibo, O.K.; Ramezani, M.; Scampicchio, M. Biorecovery of antioxidants from apple pomace by supercritical fluid extraction. *J. Clean. Prod.* **2018**, *186*, 253–261. [CrossRef]
77. De la Peña Armada, R.; Bronze, M.R.; Matias, A.; Mateos-Aparicio, I. Triterpene-Rich Supercritical CO2 Extracts from Apple By-product Protect Human Keratinocytes Against ROS. *Food Bioprocess Technol.* **2021**, *14*, 909–919. [CrossRef]
78. Plaza, M.; Turner, C. Pressurized hot water extraction of bioactives. *TrAC Trends Anal. Chem.* **2015**, *71*, 39–54. [CrossRef]
79. Teo, C.C.; Tan, S.N.; Yong, J.W.H.; Hew, C.S.; Ong, E.S. Pressurized hot water extraction (PHWE). *J. Chromatogr. A* **2010**, *1217*, 2484–2494. [CrossRef] [PubMed]
80. Plaza, M.; Abrahamsson, V.; Turner, C. Extraction and Neoformation of Antioxidant Compounds by Pressurized Hot Water Extraction from Apple Byproducts. *J. Agric. Food Chem.* **2013**, *61*, 5500–5510. [CrossRef] [PubMed]
81. Chemat, F.; Rombaut, N.; Sicaire, A.G.; Meullemiestre, A.; Fabiano-Tixier, A.S.; Abert-Vian, M. Ultrasound assisted extraction of food and natural products. Mechanisms, techniques, combinations, protocols and applications. A review. *Ultrason. Sonochem.* **2017**, *34*, 540–560. [CrossRef] [PubMed]
82. Al Jitan, S.; Alkhoori, S.A.; Yousef, L.F. Phenolic Acids from Plants: Extraction and Application to Human Health. In *Studies in Natural Products Chemistry*; Elsevier: Amsterdam, The Netherlands, 2018; Volume 58, pp. 389–417.
83. Wang, L.; Boussetta, N.; Lebovka, N.; Vorobiev, E. Effects of ultrasound treatment and concentration of ethanol on selectivity of phenolic extraction from apple pomace. *Int. J. Food Sci. Technol.* **2018**, *53*, 2104–2109. [CrossRef]
84. Malinowska, M.; Śliwa, K.; Sikora, E.; Ogonowski, J.; Oszmiański, J.; Kolniak-Ostek, J. Ultrasound-assisted and micelle-mediated extraction as a method to isolate valuable active compounds from apple pomace. *J. Food Process. Preserv.* **2018**, *42*, e13720. [CrossRef]
85. Zhang, Z.; Poojary, M.M.; Choudhary, A.; Rai, D.K.; Tiwari, B.K. Comparison of selected clean and green extraction technologies for biomolecules from apple pomace. *Electrophoresis* **2018**, *39*, 1934–1945. [CrossRef] [PubMed]
86. Pollini, L.; Cossignani, L.; Juan, C.; Mañes, J. Extraction of Phenolic Compounds from Fresh Apple Pomace by Different Non-Conventional Techniques. *Molecules* **2021**, *26*, 4272. [CrossRef] [PubMed]
87. Liu, B.; Ma, Y.; Liu, Y.; Yang, Z.; Zhang, L. Ultrasonic-assisted extraction and antioxidant activity of flavonoids from Adinandra nitida leaves. *Trop. J. Pharm. Res.* **2013**, *12*, 1045–1051. [CrossRef]
88. Altemimi, A.; Watson, D.G.; Choudhary, R.; Dasari, M.R.; Lightfoot, D.A. Ultrasound Assisted Extraction of Phenolic Compounds from Peaches and Pumpkins. *PLoS ONE* **2016**, *11*, e0148758. [CrossRef] [PubMed]
89. Pingret, D.; Fabiano-Tixier, A.S.; Bourvellec, C.L.; Renard, C.M.G.C.; Chemat, F. Lab and pilot-scale ultrasound-assisted water extraction of polyphenols from apple pomace. *J. Food Eng.* **2012**, *111*, 73–81. [CrossRef]
90. Khan, Z.; Troquet, J.; Vachelard, C. Sample preparation and analytical techniques for determination of polyaromatic hydrocarbons in soils. *Int. J. Environ. Sci. Technol.* **2005**, *2*, 275–286. [CrossRef]
91. Panzella, L.; Moccia, F.; Nasti, R.; Marzorati, S.; Verotta, L.; Napolitano, A. Bioactive Phenolic Compounds from Agri-Food Wastes: An Update on Green and Sustainable Extraction Methodologies. *Front. Nutr.* **2020**, *7*, 60. [CrossRef] [PubMed]
92. Raja, P.B.; Ismail, M.; Ghoreishiamiri, S.; Mirza, J.; Ismail, M.C.; Kakooei, S.; Rahim, A.A. Reviews on Corrosion Inhibitors: A Short View. *Chem. Eng. Commun.* **2016**, *203*, 1145–1156. [CrossRef]

93. Vorobyova, V.; Skiba, M. Peach Pomace Extract as Efficient Sustainable Inhibitor for Carbon Steel Against Chloride-Induced Corrosion. *J. Bio Tribo-Corros.* **2021**, *7*, 11. [CrossRef]
94. Dewangan, A.K.; Dewangan, Y.; Verma, D.K.; Verma, C. *Synthetic Environment-Friendly Corrosion Inhibitors*; Elsevier: Amsterdam, The Netherlands, 2022.
95. Ahmed, S.K.; Ali, W.B.; Khadom, A.A. Synthesis and investigations of heterocyclic compounds as corrosion inhibitors for mild steel in hydrochloric acid. *Int. J. Ind. Chem.* **2019**, *10*, 159–173. [CrossRef]
96. Thakur, A.; Kumar, A. Sustainable Inhibitors for Corrosion Mitigation in Aggressive Corrosive Media: A Comprehensive Study. *J. Bio Tribo-Corros.* **2021**, *7*, 67. [CrossRef]
97. Honarvar Nazari, M.; Shihab, M.S.; Havens, E.A.; Shi, X. Mechanism of corrosion protection in chloride solution by an apple-based green inhibitor: Experimental and theoretical studies. *J. Infrastruct. Preserv. Resil.* **2020**, *1*, 7. [CrossRef]
98. Ikhmal, W.M.K.W.M.; Yasmin, M.Y.N.; Maria, M.F.F.; Syaizwadi, S.M.; Rafizah, W.A.W.; Sabri, M.G.M.; Zahid, B.M. Evaluating the performance of andrographis paniculata leaves extract as additive for corrosion protection of stainless steel 316l in seawater. *Int. J. Corros. Scale Inhib.* **2020**, *9*, 118–133.
99. Brycki, B.E.; Kowalczyk, I.H.; Szulc, A.; Kaczerewska, O.; Pakiet, M. *Organic Corrosion Inhibitors*; Itors, Principles and Recent Applications, Mahmood Aliofkhazraei; IntechOpen: London, UK, 2017; pp. 1–32.
100. Vera, R.; Figueredo, F.; Díaz-Gómez, A.; Molinari, A. Evaluation of Fuji Apple Peel Extract as a Corrosion Inhibitor for Carbon Steel in a Saline Medium. *Int. J. Electrochem. Sci.* **2018**, *13*, 4139–4159. [CrossRef]
101. Vorobyova, V.I.; Skiba, M.I.; Shakun, A.S.; Nahirniak, S.V. Relationship between the inhibition and antioxidant properties of the plant and biomass wastes extracts-A Review. *Int. J. Corros. Scale Inhib* **2019**, *8*, 150–178.
102. Abdallah, M.; Fawzy, A.; Hawsawi, H.; Hameed, R.S.A.; Al-Juaid, S.S. Estimation of Water-Soluble Polymers (Poloxamer and Pectin) as Corrosion Inhibitors for Carbon Steel in Acidic Medium. *Int. J. Electrochem. Sci.* **2020**, *15*, 8129–8144. [CrossRef]
103. Fiori-Bimbi, M.V.; Alvarez, P.E.; Vaca, H.; Gervasi, C.A. Corrosion inhibition of mild steel in HCL solution by pectin. *Corros. Sci.* **2015**, *92*, 192–199. [CrossRef]
104. Prabakaran, M.; Ramesh, S.; Periasamy, V.; Sreedhar, B. The corrosion inhibition performance of pectin with propyl phosphonic acid and Zn2+ for corrosion control of carbon steel in aqueous solution. *Res. Chem. Intermed.* **2015**, *41*, 4649–4671. [CrossRef]
105. Huang, L.; Yang, K.P.; Zhao, Q.; Li, H.J.; Wang, J.Y.; Wu, Y.C. Corrosion resistance and antibacterial activity of procyanidin B2 as a novel environment-friendly inhibitor for Q235 steel in 1 M HCl solution. *Bioelectrochemistry* **2022**, *143*, 107969. [CrossRef]
106. Dehghani, A.; Mostafatabar, A.H.; Bahlakeh, G.; Ramezanzadeh, B.; Ramezanzadeh, M. Detailed-level computer modeling explorations complemented with comprehensive experimental studies of Quercetin as a highly effective inhibitor for acid-induced steel corrosion. *J. Mol. Liq.* **2020**, *309*, 113035. [CrossRef]
107. Abdelhafez, A.; Awad, Y.; El-Azeem, S.A.; Kim, M.; Ham, K.-J.; Lim, K.; Yang, J.-E.; Ok, Y. Leaching of Chromium, Copper and Arsenic in Soils and Rapid Identification of CCA-Treated Woods Using Modified PAN Stain. *Korean J. Soil Sci. Fert* **2010**, *43*, 60–67.
108. Schiopu, N.; Tiruta-Barna, L. *Wood Preservatives*; Woodhead Publishing: Cambridge, UK, 2012; pp. 138–165.
109. Woźniak, M.; Mania, P.; Roszyk, E.; Ratajczak, I. Bending Strength of Wood Treated with Propolis Extract and Silicon Compounds. *Materials* **2021**, *14*, 819. [CrossRef]
110. WHO. The WHO Recommended Classification of Pesticides by Hazard and Guidelines to Classification. 2019. Available online: https://apps.who.int/iris/bitstream/handle/10665/332193/9789240005662-eng.pdf?ua=1 (accessed on 21 January 2022).
111. Morais, S.; Fonseca, H.M.A.C.; Oliveira, S.M.R.; Oliveira, H.; Gupta, V.K.; Sharma, B.; de Lourdes Pereira, M. Environmental and Health Hazards of Chromated Copper Arsenate-Treated Wood: A Review. *Int. J. Environ. Res. Public Health* **2021**, *18*, 5518. [CrossRef] [PubMed]
112. Kalinowska, M.; Gołębiewska, E.; Świderski, G.; Męczyńska-Wielgosz, S.; Lewandowska, H.; Pietryczuk, A.; Cudowski, A.; Astel, A.; Świsłocka, R.; Samsonowicz, M.; et al. Plant-Derived and Dietary Hydroxybenzoic Acids—A Comprehensive Study of Structural, Anti-/Pro-Oxidant, Lipophilic, Antimicrobial, and Cytotoxic Activity in MDA-MB-231 and MCF-7 Cell Lines. *Nutrients* **2021**, *13*, 3107. [CrossRef] [PubMed]
113. Grace, J.; Yamamoto, R. Natural resistance of Alaska-cedar, redwood, and teak to Formosan subterranean termites. *For. Prod. J.* **1994**, *44*, 41–45.
114. Taylor, A.M.; Gartner, B.L.; Morrell, J.J.; Tsunoda, K. Effects of heartwood extractive fractions of Thuja plicata and Chamaecyparis nootkatensis on wood degradation by termites or fungi. *J. Wood Sci.* **2006**, *52*, 147–153. [CrossRef]
115. Borges, C.C.; Tonoli, G.H.D.; Cruz, T.M.; Duarte, P.J.; Junqueira, T.A. Nanoparticles-based wood preservatives: The next generation of wood protection? *CERNE* **2018**, *24*, 397–407. [CrossRef]
116. Nascimento, M.S.; Santana, A.L.B.D.; Maranhão, C.A.; Oliveira, L.S.; Bieber, L. Phenolic Extractives and Natural Resistance of Wood. In *Biodegradation-Life of Science*; IntechOpen: London, UK, 2013; pp. 349–370.
117. Kumar, S.; Abedin, M.M.; Singh, A.K.; Das, S. Role of Phenolic Compounds in Plant-Defensive Mechanisms. In *Plant Phenolics in Sustainable Agriculture*; Springer: Singapore, 2020; pp. 517–532.
118. Böttger, A.; Vothknecht, U.; Bolle, C.; Wolf, A. *Plant Secondary Metabolites and Their General Function in Plants*; Springer: Berlin/Heidelberg, Germany, 2018; pp. 3–17.
119. Surendran, A.; Siddiqui, Y.; Ahmad, K.; Fernanda, R. Deciphering the physicochemical and microscopical changes in ganoderma boninense-infected oil palm woodblocks under the influence of phenolic compounds. *Plants* **2021**, *10*, 1797. [CrossRef] [PubMed]

120. Serine, N.; Ashitani, T.; Murayama, T.; Shibutani, S.; Hattori, S.; Takahashi, K. Bioactivity of latifolin and its derivatives against termites and fungi. *J. Agric. Food Chem.* **2009**, *57*, 5707–5712.
121. Little, N.S.; Schultz, T.P.; Nicholas, D.D. Termite-resistant heartwood. Effect of antioxidants on termite feeding deterrence and mortality. *Holzforschung* **2010**, *64*, 395–398. [CrossRef]
122. Boué, S.M.; Raina, A.K. Effects of plant flavonoids on fecundity, survival, and feeding of the Formosan subterranean termite. *J. Chem. Ecol.* **2003**, *29*, 2575–2584. [CrossRef] [PubMed]
123. Chen, K.; Ohmura, W.; Doi, S.; Aoyama, M. Termite feeding deterrent from Japanese larch wood. *Bioresour. Technol.* **2004**, *95*, 129–134. [CrossRef] [PubMed]
124. Efhamisisi, D.; Thevenon, M.F.; Hamzeh, Y.; Pizzi, A.; Karimi, A.; Pourtahmasi, K. Tannin-boron complex as a preservative for 3-ply beech plywoods designed for humid conditions. *Holzforschung* **2020**, *71*, 249–258. [CrossRef]

materials

MDPI

Article

Gravity and Electrostatic Separation for Recovering Metals from Obsolete Printed Circuit Board

Camila Mori de Oliveira [1], Rossana Bellopede [1,*], Alice Tori [2], Giovanna Zanetti [1] and Paola Marini [1]

1 Department of Environment, Land and Infrastructure Engineering (DIATI), Politecnico di Torino, 10129 Turin, Italy; camila.morideoliveira@studenti.polito.it (C.M.d.O.); giovanna.zanetti@polito.it (G.Z.); paola.marini@polito.it (P.M.)
2 OSAI Automation System S.p.A., 10010 Parella, Italy; a.tori@osai-as.it
* Correspondence: rossana.bellopede@polito.it

Abstract: This study proposed an evaluation of enrichment processes of obsolete Printed Circuit Boards (PCBs), by means of gravity and electrostatic separation, aiming at the recovery of metals. PCBs are the most important component in electronic devices, having high concentrations of metals and offering a secondary source of raw materials. Its recycling promotes the reduction in the environmental impacts associated with its production, use, and disposal. The recovery method studied started with the dismantling of the PCB, followed by a comminution and granulometric classification. Subsequent magnetic, gravity, and electrostatic separations were performed. After the separations, a macroscopic visual evaluation and chemical analysis were carried out, determining the metal content in the concentrate products. The results obtained from gravity separation showed a product with metallic concentrations of 89% and 76% for particle sizes of 0.3–0.6 mm and 0.6–1.18 mm, respectively. In electrostatic separation, the product obtained was 88% for the lower particle size (<0.3 mm) and 62% for particles sizes >1.18 mm.

Keywords: PCB; recycling; metal recovery; mechanical pre-treatment

Citation: de Oliveira, C.M.; Bellopede, R.; Tori, A.; Zanetti, G.; Marini, P. Gravity and Electrostatic Separation for Recovering Metals from Obsolete Printed Circuit Board. *Materials* **2022**, *15*, 1874. https://doi.org/10.3390/ma15051874

Academic Editors: Dimitra Vernardou and Damon Kent

Received: 2 December 2021
Accepted: 24 February 2022
Published: 2 March 2022

1. Introduction

It is difficult to imagine life today without technology, especially in the pandemic scenario, in which online meetings, classes, and appointments have become routine. Therefore, technology has taken up space, incorporating indispensably into everyday life very quickly and intensely.

The current digital context generates an extensive number of electronic products and, due to the advancement of technology, Electrical and Electronic Equipment (EEE) becomes obsolete faster and, consequently, their disposal also increases. Viewing them from the perspective of an exploitation potential for use can promote urban mining [1].

Urban mining, in contrast to traditional mining, consists of the process of obtaining raw materials derived from waste, being recycled and reused by the industry. The materials obtained from the recycling of the devices are called secondary raw materials [2].

The electronics industry is estimated to generate 57.4 million tons of waste electrical and electronic equipment (WEEE) in 2021, which represents 7.6 kg of WEEE per inhabitant, and in 2019, only 17.4% of the generated amount was officially documented as properly collected and recycled [3]. According to that same source, by the end of 2030, the mark of 74.7 million tons of WEEE is estimated to have been reached worldwide.

An effective recycling of these materials is essential to keep them available for the manufacturing of new products, conserving natural resources, and being a great contribution to the circular economy, removing waste from its disposal and reinserting it in the production cycle. Thus, the proper management of electronic waste is essential to guarantee access for future generations of electronic products, to preserve natural resources and human health, to protect working conditions, to reduce the environmental impacts associated with production, and to use and dispose of electronic equipment [4].

Plenty of research has been carried out in recent years to characterize the electronic waste generated, consisting mainly of household appliances, computers, televisions, and other goods that are damaged or broken [5]. WEEE encompasses up to 69 elements from the periodic table, becoming a diverse and complex type of waste having both hazardous and nonhazardous compounds [3]. The presence of metals, such as copper, gold, silver, and critical raw materials, such tantalum, makes the WEEE extremely attractive to recovery. Some elements have concentrations significantly higher than those usually found in corresponding mineral ores [6].

Printed Circuit Boards, also known as PCBs, are generally not visible, but they are part of everything that involves technology. They represent about 3% by weight of WEEE and, taking into account that their composition can reach approximately 40% of metals, the recycling of obsolete PCBs has a high economic importance [2]. A common structure of a PCB is made of layers of glass fibers and copper clads, usually held together by halogenated epoxy resins (HERS) or brominated epoxy resin (BES) in which electrical components (e.g., resistors, capacitors, and inductors) are soldered onto the top layer [7,8]. They can be structurally classified according to the number of layers: single-sided, having a conducting layer (copper) on one side; double-sided, having a conducting layer on both sides; and multi-layers, having metallized holes to connect different layers. The heterogeneous PCB composition hinders the process of recovering the materials, making it slow and expensive [8]. Thus, much research has been developed to optimize the efficiency of a sustainable recycling process of these components.

In recent years, the mainstream of the recycling approaches of PCBs has focused on chemical process approaches [9], including co-pyrolysis [10], hydrometallurgy with nitric acid [11], and bio-metallurgical processes by biosorption and bioleaching [12]. These processes are very time-consuming, high energy-demanding, and may release significant pollutants into the atmosphere [13]. Despite the progress of these chemical techniques, at present, there is a lack of studies focusing on sophisticated solutions to obtain physicomechanical improvements. Therefore, there is still a demand to identify low-cost and eco-friendly methods to improve the reach of a high concentration of metals and decrease the rate of metal loss during these operations.

In this context, the objective of the article is to evaluate green mechanical pre-treatments available for the sustainable recycling of PCBs, intending to obtain highly concentrated material before the chemical recovery processes. Treatment through eco-friendly processes will contribute not only to the solution from an environmental point of view, but also from an economical one, to increase the metal recovery rate during the operations and establishing an advanced industrial recycling sector.

To achieve that, the characterization of an obsolete PCB and the evaluation of the efficiency of gravimetric and electrostatic separation was performed. A pre-treatment composed of comminution, granulometric, and magnetic separations was performed before. Then, gravity separation by means of a shaking table and electrostatic separation by means of corona electrostatic separation were accomplished. To ascertain the metal content existing in the concentrated fractions after applying the treatments, samples were collected for visual analysis with a macroscope, and chemical analysis by inductively coupled plasma–optical emission spectroscopy was performed.

2. Materials and Methods

2.1. Materials

Printed circuit boards are composed of ceramic, polymers, and metals. The composition can vary significantly depending on the model of equipment and the age of the boards. For example, before the year 2003, equipment was fabricated containing solders with tin and lead alloys; however, after the Directive 2002/95/CE, where there is a restriction on the use of these substances, there is currently is a trend toward replacing this with SnAgCu solder alloys, which are considered "lead-free" [1].

PCBs typically consist of more than 60 different types of elements, having a content of base metals (Cu, Zn), precious metals (Au, Ag, Pd), and heavy metals even higher than those in natural minerals [8]. The chemical composition is also used to distinguish the PCBs among them, with scrap being a low-grade PCB that has a low concentration of precious metal (i.e., gold) [14].

Due to this variation in the composition, the material characterization step is essential to define the treatments aiming to recover and recycle materials. Several studies have been performed on printed circuit boards, and part of these studies also involved the characterization of the boards. As many recyclers receive a mix of low- and high-grade PCBs, not having a specific composition selection [9], a data compilation from several authors [15–30] and different types of PCBs (computers, smartphones, etc.) have been elaborated (Table 1) in order to obtain a mean value (and a standard deviation) for each metal of the PCBs, thus reaching an average reference value of PCB composition. The results differ, as the boards come from different products and periods. On average, 34.7% of a PCB's mass is metals. Cu is the metal with the highest concentration present in printed circuit boards, an average of 21.44%. Other metals with a significant mean percentage weight are Al (3.02%), Fe (3.28%), and Sn (3.14%). The quantities of valuable metals are significant considering, for example, that the average rate of gold in PCBs (0.04%) is higher than that in raw gold ore (0.0005%) [19]. It is possible to notice that some elements have a mean and standard deviation about the same size, such as tantalum, and the paucity of data gives a very skewed distribution, having a grossly inflated standard deviation.

Table 1. Representative composition of a PCB (by wt.%) considering several types of PCBs from [15–30] and market price (USD/kg).

Metal	Weight Average (%)	Stdev (%)	Price (US$/kg)
Al	3.02	1.70	2.68
Cu	21.44	9.14	9.65
Fe	3.28	2.79	0.0994
Sn	3.14	1.65	36.79
Au	0.04	0.04	57,598.06
Pd	0.05	0.11	65,040.96
Ag	0.13	0.15	767.28
Ni	1.03	1.29	19.824
Pb	1.86	0.94	2.40
Zn	0.73	0.76	3.34
Ta	0.01	0.01	220 *
Total	34.7%		

Note: Metal price according to the London Metal Exchange (LME) (November 2021) [31]; * Tantalum price (USD 100/pound), (August 2021) [32].

It is also possible to observe the commodity metals market prices, from the London Metal Exchange [31], for 2 November 2021. Combining the average composition with the price, it is possible to quantify the economic value of the obsolete PCBs. These data reveal that a ton of waste PCBs can reach up to USD 60,000. Copper, representing 21.4% by weight, creates around of 3.5% of the product price. Another highlight is for the precious metals, gold or palladium, whose mass in a PCB represents less than 1%; however, from an economical point of view, they can represent more than 90% of the price market.

The printed circuit board used in this research was supplied by OSAI Automation System S.p.A, Parella, Italy, and represented a PCB of a server. It has a multi-layer structure, and its total weight, containing all the electronical components, was about 1.8 kg. Each electronical component was recognized and analyzed on the SEM (FEI Company, Hillsboro, OR, USA), coupled with an energy dispersive X-ray spectroscopy EDS detector (EDAX (Ametek Inc.), Mahwah, NJ, USA), to identify the chemical elements composition. The analysis showed that a significant number of metal elements of the periodic table were present: Sn, Pb, Au, Cu, Ni, Pd, Ag, Al, Fe, Ta, Ba, and Mn. The lack of data for some metals (e.g., barium and manganese) in Table 1 may be explained by the very low concentrations.

2.2. Methods

In order to recover metals from the obsolete PCB, the methodology presented in Figure 1 was applied. Prior to the mechanical separation, a preliminary work was carried out consisting of the identification of the PCB and its constituent composition. Then, the dismantling of the plates followed by a comminution and granulometric classification were carried out. Next, according to the particle size obtained by sieving, magnetic and gravity methods were used for the medium classes, while magnetic and electrostatic methods were used for the coarser and the finest classes. A visual evaluation of the quality of the products by means of a macroscope and chemical analyses were carried out, in order to assess the potential of the method in the recovery of metals present in printed circuit boards.

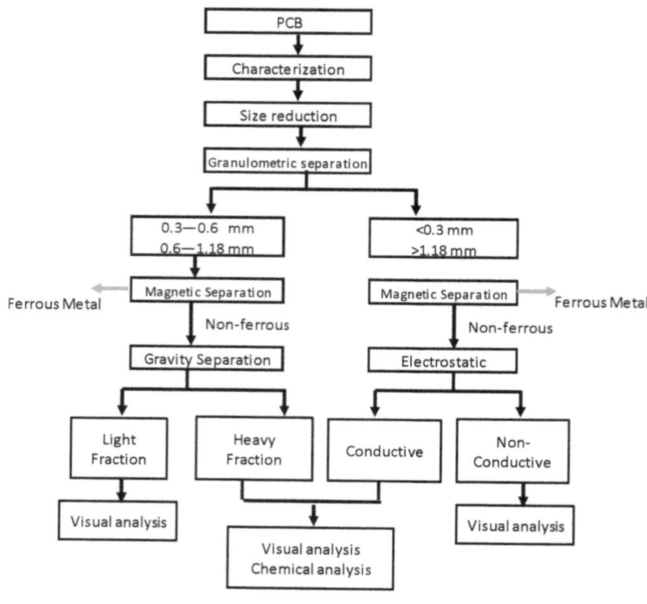

Figure 1. Flowsheet of the process applied in this study.

- Disassembly and size reduction

The disassembly of the board was carried out manually in the Raw Materials Laboratory of Politecnico di Torino, Turin, Italy using different types of tools, in which elements of greater volume and without interest such as liquid electrolytic capacitors and the central process unit were removed. About 80 g of material was removed.

Subsequently, the first board was cut into pieces with a maximum size of 2.5 × 4.0 cm, enabling an adequate feed to the shredding operation performed by means of a cutting mill RETSCH SM100 (Retsch GmbH, Haan, Germany) with a rotor made up of three nonaligned blades, whose action fragments the material introduced. The size reduction is an important phase of mechanical processing because it allows the release of materials from printed circuit boards, enabling the optimization of metal recovery.

- Granulometric classification

Considering the heterogeneous composition of the PCB, in order to minimize errors, the fragments generated should be classified granulometrically, facilitating the characterization of materials through chemical analysis, and enabling the identification of fractions of concentrated metals and nonmetals. The fragmented material was classified by means of a sieve shaker model FTL-0200 (OMM, Busnago, Italy), and a series of standard ASTM sieves (Controls, Milano, Italy) (1.18 mm; 0.6 mm, 0.3 mm and bottom) were used. The samples were weighed in order to calculate the retained mass in each granulometric range.

After these steps, which aimed to prepare the material for separation tests, the enrichment operations began.

- Magnetic separation

In magnetic separation, the materials can be classified according to their responses to the magnetic field. There are ferromagnetic materials, which are strongly attracted by the magnetic field; paramagnetic materials, which are weakly attracted; and diamagnetic materials, which are repulsed [33]. In this study, this separation was performed manually, using a AlNiCo magnet with an intensity of 1 T. The experimental parameters included material particle size and distance between magnet and fragments, established at 1 cm. The products obtained were a ferromagnetic and a nonferromagnetic portion.

- Gravity separation

Separation by gravity allows the classification of materials based on their different densities. In these processes, particles are separated thanks to different sedimentation velocities when falling into a fluid (air, water). This speed will simultaneously depend on its density and size. In the present work, the separation was performed by a wet shaking table model Krupp (Humboldt Wedag GmbH, Köln, Germany) on particle classes 0.6–1.18 mm and 0.3–0.6 mm. The adjustable parameters of the device included its inclination, the frequency of movement, the water flow, and the feed speed. The frequency was 300 cycles per minute, with 1° of inclination angle and a water flow of 10 L/min. In the concentrate collection and tailing discharge areas, gutters were placed, with a partition separating these same areas, to collect the products resulting from the separation. After the passage, the products collected in the different fractions were filtered and placed to dry in an oven, at 40 °C. The products obtained were classified as heavy fraction (metals concentrate) and light fraction (tailing).

- Electrostatic separation

The corona electrostatic separation was carried out on particle size classes >1.18 mm and <0.3 mm using the separator Dings Coronatron (Prodecologia, Rivne, Ukraine). This process is based on the electrical conductivity of some elements. In this way, a fraction rich in conductive metals, such as copper, and another one consisting of polymeric and ceramic materials were obtained [1]. After some tests, it was established to perform two passages. The first passage with a voltage of 20 kV and a rotation speed of 30 Hz was performed. After that, a second passage, with the same parameters, was performed only to the conductive product obtained from the first passage, increasing the quality of the products.

- Visual characterization

After the mechanical treatments, an inspection of the quality of all products was performed by means of a visual analysis using the optical macroscope Leica/Wild M420 (Leica Microsystems, Wetzlar, Germany).

- Chemical characterization

On the metal-concentrated products, a chemical analysis using the Inductively Coupled Plasma–Optical Emission Spectrometry instrument (Perkin Elmer, Optima 2000DV, Waltham, MA, USA) was carried out.

The chemical analysis was divided into microwave digestion and optical spectroscopy. For each product, two samples were performed.

The metals leaching was obtained by the microwave digestion system Milestone MLS-1200 Mega (Milestone, Sorisole, Italy) laboratory unit with aqua regia (nitric acid 65%/hydrochloric acid 37%) and HF. The analyzed product was added in the proportion of

0.25 g for 6 mL of aqua regia and 1 mL of HF. The mixture was then subjected to microwave heating to complete the digestion. After that, the content of each tube was filtered directly into a 50 mL volumetric flask, which was brought to volume with distilled water.

Each volumetric flask was analyzed by ICP–OES (Perkin Elmer, Optima 2000DV, Waltham, MA, USA). A calibration line was prepared at increasing concentrations, containing the following metals: lead, copper, tantalum, gold, tin, nickel, and aluminum, which was used to determine the concentration of metals in each sample.

3. Results and Discussion

During the treatment steps, losses in material are common. In the following sections, the yield of each product is reported in terms of percentage disregarding the losses (which are 15% of the PCB total weight).

3.1. Size Reduction and Classification

The size reduction performed by the cutting mill was performed in two stages. The screen used was 2 mm. To enhance the liberation of particles, the >1.18 mm size fraction size was re-shredded. A significant release of fine particles was observed. When compared to the initial feed, a loss of 11% was already accounted.

The grain size classification step provided the separation of the previously fragmented material in different classes. It can be observed from the Figure 2 that only about 16% of the material mass was obtained in particle sizes over 1.18 mm. An amount of 50% of material was particles between 0.6 and 1.18 mm, 13.2% of material was particles between 0.3 and 0.6 mm, and 21% of material was particles less than 0.3 mm.

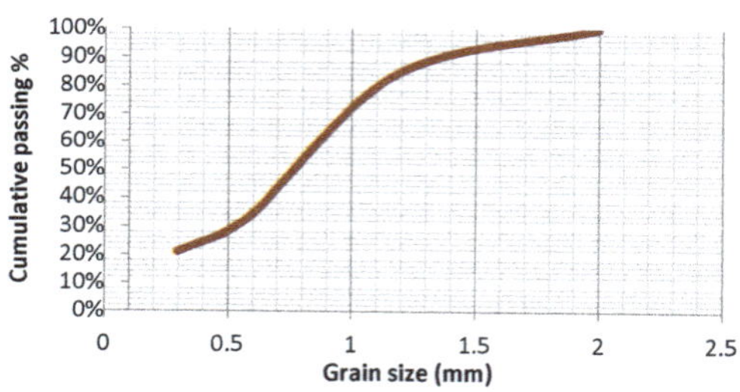

Figure 2. Particle size distributions of the PCB after size reduction.

3.2. Magnetic Separation

The magnetic separation was the first separation step to be performed as it is a common and easy classification that isolates the magnetic particles. Table 2 shows the results for each class size fraction.

Table 2. Weight percentage of the products obtained by magnetic separation and their yield.

Class Size	Products	% Class	Yield
>1.18 mm	Magnetic fraction Nonmagnetic fraction	5% 95%	0.9% 16%
0.6–1.18 mm	Magnetic fraction Nonmagnetic fraction	2% 98%	1.0% 49%
0.3–0.6 mm	Magnetic fraction Nonmagnetic fraction	2.4% 97.6%	0.3% 14%
<0.3 mm	Magnetic fraction Nonmagnetic fraction	0.3% 99.7%	0.1% 19%

The greater amount of ferromagnetic material was observed in the >1.8 mm fraction, which may be due to the element being used in larger components, such as supports, and remains in the larger fractions due to its mechanical properties, making it more difficult to grind than polymeric materials.

Table 2 shows that around 2.3% of the total product weight obtained from the separations was separated in this step. The presence of copper in the magnetic products was also observed, as despite not being magnetic, it is the predominant element in PCBs and its presence in these fractions can be justified due to them being dragged by iron and nickel particles when attracted by the magnet.

Even though a low concentration was found, the main objective of this treatment was to separate iron, having a higher efficiency in the obtaining of copper in conductive fractions from electrostatic separation.

3.3. Gravity Separation

The fraction 0.6–1.18 and 0.3–0.6 mm of nonferrous material went to the wet shaking table separation.

Regarding the grain size particles >1.18 mm, despite a possible efficient separation, it was not carried out, because the volume of material was not enough to perform this treatment. Nonetheless, the particles <0.3 mm were not performed, due to the difficult collection of products after separation. Both were taken to electrostatic separation.

The products obtained were weighed and the data are shown in Table 3. The yield of each product in terms of percentage was also calculated. Assuming that the heavy-fraction products obtained were mostly made up of metallic elements, the efficiency of the treatment was obtained through the weight in percentage of the products. The 0.6–1.18 mm class was the one with the highest efficiency of 59%, followed by the granulometric class 0.3–0.6 mm with 33%.

Table 3. Weight percentage of the products and losses obtained by gravity separation and their yield.

Class Size	Products	%Class	Yield
0.6–1.18 mm	Heavy fraction	59%	29%
	Light fraction	41%	20%
0.3–0.6 mm	Heavy fraction	34%	5%
	Light fraction	66%	9%

3.4. Electrostatic Separation

From Table 4, as expected, it is possible to observe a small amount of conductive product for the finest grain size (<0.3 mm). This low efficiency can be explained due the fact that metals have a high mechanical strength and the fine material resulting from comminution mainly consisted of fiberglass.

Table 4. Weight percentage of the products and losses obtained by electrostatic separation and their yield.

Class Size	Products	Class	Yield
>1.18 mm	Conductive	64%	10%
	Nonconductive	36%	6%
<0.3 mm	Conductive	16%	3%
	Nonconductive	84%	16%

For the class size >1.18 mm, the conductive product obtained was copper grains attached to the fiberglass and the epoxy resin. The content of conductive elements (copper) reached more than 64% (in mass). The nonconductive product had an appearance with a predominance of polymer and fiberglass.

3.5. Visual Characterization

For the gravity separation, the macroscope visual observation of the products allowed us to conclude that the heavy fraction (Figure 3b) was rich in metallic elements (many filaments from electronic connectors) and had a low quantity of nonmetallic elements. However, metallic elements were also visible in the light fraction (Figure 3a), but in reduced amounts. In both cases, the separations were satisfactory.

(a) (b)

Figure 3. Light fraction (**a**) and the heavy fraction (**b**) resulting from gravity separation for the class size of 0.3–0.6 mm.

Nonetheless, for the electrostatic separation, it can be seen in Figure 4b that the product obtained was composed mostly of metals in copper color and some in gray/silver color. On the other hand, the nonconductive fraction, as can be observed in Figure 4a, was mainly made up of a mix of resin and fiberglass, having a green dark color.

(a) (b)

Figure 4. Nonconductive (**a**) and conductive (**b**) products resulting from electrostatic separation for the class size <0.3 mm.

3.6. Chemical Characterization

Table 5 shows the mean concentrations of metals in the metal-enriched products resulting from gravity (heavy fraction) and electrostatic separation (conductive), and it is possible to also observe the yield of each product in terms of percentage considering all PCB weight. The chemical analysis was performed to determine the concentrations of Cu, Pb, Ta, Au, Sn, and Al. Ni was found in connectors having a double layer metallic coating with gold; therefore, to evaluate the liberation between Ni and Au, this element was also included in the chemical analysis whose results showed a presence of nickel in all the grain size fractions.

Table 5. Metal content (%) of the enriched products. (n.i. = not identified).

Class Size (mm)	Operation (Product)	Yield *	Concentration (%)							
			Cu	Pb	Ta	Au	Sn	Al	Ni	Others
>1.18	Electrostatic (Conductive)	10%	56.1 ± 4.3	0.001 ± 0.001	n.i.	0.01 ± 0.005	3.1 ± 2.6	3.1 ± 0.6	0.3 ± 0.1	37.3 ± 7.6
0.6–1.18	Gravity (Heavy Fraction)	29%	64.5 ± 0.4	n.i.	0.15 ± 0.15	0.05 ± 0.04	8.8 ± 0.7	1.6 ± 0.1	0.42 ± 0.02	24.4 ± 1.3
0.3–0.6	Gravity (Heavy Fraction)	5%	75.3 ± 1.25	0.03 ± 0.02	0.6 ± 0.22	0.17 ± 0.06	11.10 ± 0.92	1.58 ± 0.05	0.68 ± 0.2	10.5± 2.7
<0.3	Electrostatic (Conductive)	3%	80.5 ± 0.42	0.06 ± 0.005	0.6 ± 0.07	0.17 ± 0.02	4.55 ± 0.7	2.12 ± 0.08	0.65 ± 0.05	11.4 ± 1.32

* Yield on the total weight of products.

From Table 4, it is evident that the finer the particle size resulting from the comminution, the higher the probability of obtaining a greater amount of "free grain," and therefore a higher efficiency in the separation processes.

Copper is the metal with the most abundant concentrations in the PCB, having been found with a content in the conductive product of 80% for the class size of <0.3 mm and 56% for >1.18 mm. The concentration by gravity separation was between 64% and 75%, for 0.6–1.18 and 0.3–0.6 mm, respectively.

The relationship between particle size and metal concentration after separation processes for lead and tantalum showed that the greater the particle size, the lower the concentration amount. Gold was found in similar quantities in the two smaller particle sizes from the different separation processes, as well for nickel. Regarding the influence of particle size on tin concentration, it was higher in intermediate particle sizes. However, aluminum had the greater concentration grade in the coarser particle size resulting from electrostatic separation.

Considering the analyzed metallic element of Table 4, and their weight for the 75% on the enriched products mass, they were constituted by 65% of copper, 7.5% of tin, 2% of aluminum, 0.4% of nickel, 0.3% of tantalum, 0.08% of gold, and 0.01% of lead. The remaining 25% may contain fiber glass, epoxy resin, plastic, ceramic, and other metals not analyzed. In terms of price, gold represents 82% of the enriched product total price, followed by Cu (11%), Sn (5%), and Ta (1.5%). The presence in high concentrations of metals and their market value are the impetus for the development of an enrichment and recovery system of metals in a PCB.

The metallic fraction, also considering the magnetic product, represents 35.3% of the products yield, and according to Table 1: the composition of a PCB, it probably means a high-grade PCB type, having higher concentrations of metals.

4. Conclusions

Printed circuit boards are present in almost all WEEE, being a material with high metals concentrations. They represent a significant fraction of the economic value of the total electronic waste, making PCB scrap economically attractive for recycling. It conserves natural resources as it prevents new minerals from being extracted, and it is a great contribution to the circular economy.

The studied process was composed of pre-processing, formed by dismantling, comminution, and classification of the material in different granulometric sizes. Then, magnetic, gravity, and electrostatic separation were performed and chemically evaluated.

The found weight loss during all the processes was about 15% of the initial weight. During physical separation processes, losses can reach amounts of 10–35% [34]. To increase the efficiency of the mechanical pre-treatment plants, a material loss evaluation during the shredding and separation steps would be necessary to analyze in further works.

From gravity separation, the heavy fraction showed a metallic concentration of 76% for the class size of 0.6–1.18 mm and 89% for the 0.3–0.6 mm class size.

The conductive material, from electrostatic separation, had metal concentrations of about 88% for the class size <0.3 mm and 62% for the >1.18 mm one. This lower content for the coarsest grain size may be explained by the nonmetallic particles being attached to the metallic elements, in other words, the grains were not free.

This work presents interesting results in terms of the application in industrial materials recovery processes, specifically for metals present in PCBs. From a research point of view, the integration between this first pre-treatment validation and refining chemical processes will be explored in the future with a combined approach.

Author Contributions: Conceptualization, C.M.d.O., R.B. and P.M.; methodology, C.M.d.O., R.B., P.M.; validation, A.T., G.Z., R.B. and P.M.; formal analysis, C.M.d.O. and G.Z.; investigation, R.B., P.M.; resources, A.T.; data curation, G.Z.; writing—original draft preparation, C.M.d.O.; writing—review and editing, R.B. and P.M.; supervision, R.B. and P.M. All authors have read and agreed to the published version of the manuscript.

Funding: This research received no external funding.

Institutional Review Board Statement: Not applicable.

Informed Consent Statement: Not applicable.

Data Availability Statement: Not applicable.

Acknowledgments: The authors would thank Claudio De Regibus for technical support during the experiments.

Conflicts of Interest: The authors declare no conflict of Interest.

References

1. Vermeșan, H.; Tiuc, A.-E.; Purcar, M. Advanced Recovery Techniques for Waste Materials from IT and Telecommunication Equipment Printed Circuit Boards. *Sustainability* **2019**, *12*, 74. [CrossRef]
2. Arya, S.; Patel, A.; Kumar, S.; Pau-Loke, S. Urban mining of obsolete computers by manual dismantling and waste printed circuit boards by chemical leaching and toxicity assessment of its waste residues. *Environ. Pollut.* **2021**, *283*, 117033. [CrossRef] [PubMed]
3. Forti, V.; Baldé, C.; Kuehr, R.; Bel, G. *The Global E-Waste Monitor 2020. Quantities, Flows, and the Circular Economy Potential*; United Nations University (UNU)/United Nations Institute for Training and Research (UNITAR)—Co-Hosted SCYCLE Programme; International Telecommunication Union (ITU): Geneva, Switzerland; International Solid Waste Association (ISWA): Vienna, Austria, 2020.
4. Bilesan, M.R.; Makarova, I.; Wickman, B.; Repo, E. Efficient separation of precious metals from computer waste printed circuit boards by hydrocyclone and dilution-gravity methods. *J. Clean. Prod.* **2020**, *286*, 125505. [CrossRef]
5. He, Y.; Xu, Z. The status and development of treatment techniques of typical waste electrical and electronic equipment in China: A review. *Waste Manag. Res. J. Sustain. Circ. Econ.* **2014**, *32*, 254–269. [CrossRef]
6. Vats, M.; Singh, S. E-Waste characteristic and its disposal. *Int. J. Ecol. Sci. Environ. Eng.* **2014**, *1*, 49–61.
7. Kang, K.D.; Ilankoon, I.M.S.K.; Dushyantha, N.; Chong, M.N. Assessment of Pre-Treatment Techniques for Coarse Printed Circuit Boards (PCBs). *Recycl. Miner.* **2021**, *11*, 1134. [CrossRef]
8. Hao, J.; Wang, Y.; Wu, Y.; Guo, F. Metal recovery from waste printed circuit boards: A review for current status and perspectives. *Resour. Conserv. Recycl.* **2020**, *157*, 104787. [CrossRef]
9. Copani, G.; Colledani, M.; Brusaferri, A.; Pievatolo, A.; Amendola, E.; Avella, M.; Fabrizio, M. Integrated Technological Solutions for Zero Waste Recycling of Printed Circuit Boards (PCBs). In *Factories of the Future*; Springer: Cham, Switzerland, 2019; pp. 149–169. [CrossRef]
10. Chen, W.; Shu, Y.; Li, Y.; Chen, Y.; Wei, J. Co-pyrolysis of waste printed circuit boards with iron compounds for Br-fixing and material recovery. *Environ. Sci. Pollut. Res.* **2021**, *28*, 64642–64651. [CrossRef]
11. Gande, V.V.; Vats, S.; Bhatt, N.; Pushpavanam, S. Sequential recovery of metals from waste printed circuit boards using a zero-discharge hydrometallurgical process. *Clean. Eng. Technol.* **2021**, *4*, 100143. [CrossRef]
12. Kaur, P.; Sharma, S.; Albarakaty, F.M.; Kalia, A.; Hassan, M.M.; Abd-Elsalam, K.A. Biosorption and Bioleaching of Heavy Metals from Electronic Waste Varied with Microbial Genera. *Sustainability* **2022**, *14*, 935. [CrossRef]
13. Kaya, M. Recovery of metals and nonmetals from electronic waste by physical and chemical recycling processes. *Waste Manag.* **2016**, *57*, 64–90. [CrossRef]
14. Wang, X.; Gaustad, G. Prioritizing material recovery for end-of-life printed circuit boards. *Waste Manag.* **2012**, *32*, 1903–1913. [CrossRef]
15. Shuey, S.A.; Vildal, E.E.; Taylor, P.R. Pyrometallurgical Processing of Electronic Waste. In Proceedings of the SME Annual Meeting, St. Louis, MO, USA, 27–29 March 2006.

16. Oguchi, M.; Murakami, S.; Sakanakura, H.; Kida, A.; Kameya, T. A preliminary categorization of end-of-life electrical and electronic equipment as secondary metal resources. *Waste Manag.* **2011**, *31*, 2150–2160. [CrossRef]
17. Goosey, M.; Kellner, R. Recycling technologies for the treatment of end of life printed circuit boards (PCBs). *Circuit World* **2003**, *29*, 33–37. [CrossRef]
18. Veit, H.M.; De Pereira, C.C.; Bernardes, A.M. Using mechanical processing in recycling printed wiring boards. *JOM* **2002**, *54*, 45–47. [CrossRef]
19. Bizzo, W.A.; Figueiredo, R.A.; De Andrade, V.F. Characterization of Printed Circuit Boards for Metal and Energy Recovery after Milling and Mechanical Separation. *Materials* **2014**, *7*, 4555–4566. [CrossRef]
20. Zhao, Y.; Wen, X.; Li, B.; Tao, D. Recovery of copper from waste printed circuit boards. *Mining Metall. Explor.* **2004**, *21*, 99–102. [CrossRef]
21. Das, A.; Vidyadhar, A.; Mehrotra, S. A novel flowsheet for the recovery of metal values from waste printed circuit boards. *Resour. Conserv. Recycl.* **2009**, *53*, 464–469. [CrossRef]
22. Guo, C.; Wang, H.; Liang, W.; Fu, J.; Yi, X. Liberation characteristic and physical separation of printed circuit board (PCB). *Waste Manag.* **2011**, *31*, 2161–2166. [CrossRef]
23. Veit, H.; Diehl, T.; Salami, A.; Rodrigues, J.; Bernardes, A.M.; Ten'orio, J. Utilization of magnetic and electrostatic separation in the recycling of printed circuit boards scrap. *Waste Manag.* **2005**, *25*, 67–74. [CrossRef]
24. Wen, X.; Zhao, Y.; Duan, C.; Zhou, X.; Jiao, H.; Song, S. Study on Metals Recovery from Discarded Printed Circuit Boards by Physical Methods. In Proceedings of the 2005 IEEE International Symposium on Electronics and the Environment, New Orleans, LA, USA, 16–19 May 2005. [CrossRef]
25. Kim, B.-S.; Lee, J.-C.; Seo, S.-P.; Park, Y.-K.; Sohn, H.Y. A process for extracting precious metals from spent printed circuit boards and automobile catalysts. *JOM* **2004**, *56*, 55–58. [CrossRef]
26. Yokoyama, S.; Iji, M. Recycling of Printed Wiring Boards with Mounted Electronic Parts. In Proceedings of the 1997 IEEE International Symposium on Electronics and the Environment, San Francisco, CA, USA, 5–7 May 1997; pp. 109–114. [CrossRef]
27. Yoo, J.-M.; Jeong, J.; Yoo, K.; Lee, J.-C.; Kim, W. Enrichment of the metallic components from waste printed circuit boards by a mechanical separation process using a stamp mill. *Waste Manag.* **2009**, *29*, 1132–1137. [CrossRef]
28. Vasile, C.; Brebu, M.A.; Totolin, M.; Yanik, J.; Karayildirim, T.; Darie, H. Feedstock Recycling from the Printed Circuit Boards of Used Computers. *Energy Fuels* **2008**, *22*, 1658–1665. [CrossRef]
29. Castro, J.; Edenir, R.P.-F. Chemical exploratory analysis of printed circuit board (PCB) using inductively coupled plasma optical emission spectrometry (ICP OES): Data treatment and elements correlation. *Detritus* **2020**, *13*, 131–139. [CrossRef]
30. Santaolalla, A.; Lens, P.N.L.; Barona, A.; Rojo, N.; Ocio, A.; Gallastegui, G. Metal Extraction and Recovery from Mobile Phone PCBs by a Combination of Bioleaching and Precipitation Processes. *Minerals* **2021**, *11*, 1004. [CrossRef]
31. LME (London Metal Exchange). LME. 2021. Available online: https://www.lme.com/en/Metals/ (accessed on 2 November 2021).
32. Passive Electronic Components: Global Market Update, August 2021. Available online: https://passive-components.eu/passive-electronic-components-global-market-update-august-2021/ (accessed on 2 November 2021).
33. Suponik, T.; Franke, D.; Nuckowski, P. Electrostatic and magnetic separations for the recovery of metals from electronic waste. *IOP Conf. Ser. Mater. Sci. Eng.* **2019**, *641*, 012017. [CrossRef]
34. Tuncuk, A.; Atazi, V.; Akcil, A.; Yazici, E.Y.; Deveci, H. Aqueous metal recoverytechniques from e-scrap: Hydrometallurgy in recycling. *Miner. Eng.* **2012**, *25*, 28–37. [CrossRef]

materials

MDPI

Article

The Comparison of the Environmental Impact of Waste Mineral Wool and Mineral in Wool-Based Geopolymer

Beata Łaźniewska-Piekarczyk [1,*], Monika Czop [2] and Dominik Smyczek [1,3]

1 Department of Building Processes and Building Physics, Faculty of Civil Engineering, The Silesian University of Technology, Akademicka 5, 44-100 Gliwice, Poland; dominik.smyczek@saint-gobain.com

2 Department of Technologies and Installations for Waste Management, Faculty of Energy and Environmental Engineering, The Silesian University of Technology, Konarskiego 18, 44-100 Gliwice, Poland; monika.czop@polsl.pl

3 Saint Gobain Construction Products Polska Sp z o.o., ul. Okrężna 16, 44-100 Gliwice, Poland

* Correspondence: beata.lazniewska@polsl.pl; Tel.: +48-32-237-15-67

Abstract: Waste generated in fine wool production is homogeneous and without contamination, which increases its chances of reuse. Waste mineral wool from demolition sites belongs to the specific group of waste. However, the storage and collection require implementing restrictive conditions, such as improper storage of mineral wool, which is highly hazardous for the environment. The study focuses on the leachability of selected pollutants (pH, Cl^-, $SO_4{}^{2-}$) and heavy metals (Ba, Co, Cr, Cu, Ni, Pb, Zn) from the waste mineral wool. As a solution to the problem of storing mineral wool waste, it was proposed to process it into wool-based geopolymer. The geopolymer, based on mineral wool, was also assessed regarding the leaching of selected impurities. Rock mineral wool is very good for geopolymerisation, but the glass wool needs to be completed with additional components rich in Al_2O_3. The research involved geopolymer prepared from mineral glass wool with bauxite and Al_2O_3. So far, glass wool with the mentioned additives has not been tested. An essential aspect of the article is checking the influence of wool-based geopolymer on the environment. To investigate the environmental effects of the wool-based monolith and crushed wool geopolymers were compared. Such research has not been conducted so far. For this purpose, water extracts from fragmented geopolymers were made, and tests were carried out following EN 12457-4. There is no information in the literature on the influence of geopolymer on the environment, which is an essential aspect of its possible use. The research results proved that the geopolymer made on the base of mineral wool meets the environmental requirements, except for the pH value. As mentioned in the article, the geopolymerisation process requires the dissolution of the starting material in a high pH (alkaline) solution. On the other hand, the pH minimum 11.2 value of fresh geopolymer binder is required to start geopolymerisation. Moreover, research results analysed in the literature showed that the optimum NaOH concentration is 8 M. for the highest compressive strength of geopolymer. Therefore, the geopolymer strength decreases with NaO concentration in the NaOH solution. Geopolymers glass wool-based mortars with Al_2O_3 obtained an average compressive strength of 59, the geopolymer with bauxite achieved about 51 MPa. Thus, Al_2O_3 is a better additional glass wool-based geopolymer than bauxite. The average compressive strength of rock wool-based geopolymer mortar was about 62 MPa. The average compressive strength of wool-based geopolymer binder was about 20–25 MPa. It was observed that samples of geopolymers grout without aggregate participation are characterised by cracking and deformation.

Keywords: production waste; rock wool; glass wool; loss on ignition; pollutants washout; geopolymer

Citation: Łaźniewska-Piekarczyk, B.; Czop, M.; Smyczek, D. The Comparison of the Environmental Impact of Waste Mineral Wool and Mineral in Wool-Based Geopolymer. *Materials* **2022**, *15*, 2050. https://doi.org/10.3390/ma15062050

Academic Editors: Rossana Bellopede and Lorena Zichella

Received: 28 January 2022
Accepted: 8 March 2022
Published: 10 March 2022

1. Introduction

Proper waste management should be directed on priority circular economy. The main principle of circular economy is to maintain the value of raw materials, ready-made goods, and finished products if possible while minimising waste. The generated waste, however, should be reused, i.e., recycled. In 2020, 123 million tons of waste was generated, including 109.47 thousand tons from various economic activity branches; as for the construction industry, 7.3 million tons was generated in 2020. It is an increase of 60% compared to 2019 [1].

Mineral wool is a popular construction material, so it often functions in waste management obtained at the construction site. The producer is responsible for the entire waste management process (usually the contractor on the construction site) and becomes the owner of the waste at the time of its production. In addition, he is responsible for the costs associated with managing and storing the waste [2].

According to the applicable regulations, construction waste management should be based on the waste management hierarchy. At the very beginning, the institution causing waste production should prevent its generation at every production stage (in the case of mineral wool, it is a factory producing it for the construction of buildings). In the next step, the waste owner should be prepared to reuse it, applying the principle of proximity (process them at the point of origin). In practice, many waste materials are reused at the construction site to save construction costs. However, it should be noted that the recycled material to be reused for construction purposes needs to undergo control tests determining its mechanical and physical properties. Ultimately, it all comes down to recycling, which is currently crucial to waste management. Unfortunately, wool as waste is generally put into storage. For the environment, it is the least favourable solution. It occupies space in landfills, thus leaving less and less room for people and all other creatures inhabiting the Earth.

As part of the search for a replacement, fewer emission binders for concrete production developed geopolymers in construction. However, the practical application of these materials is still minimal. The production of every ton of cement puts one tonne of CO_2 into the atmosphere [1,3,4]. According to various estimates, the synthesis of geopolymers absorbs 2–3 times less energy than Portland cement and causes liberation of 4–8 times less CO_2 [1,5–7].

The first geopolymer applications can be dated to the beginning of the 1970s. At that time, fire-resistant chipboards were developed with a wooden core covered with two geopolymer coatings. In 1978–1980, a metakaolin-based liquid geopolymer binder containing a soluble alkali silicate was developed at the Cordis-Laboratory. It was the first mineral resin produced [8,9]. Rocla has developed a technology for producing sewage pipes from steel-reinforced geopolymers [9]. They have successfully produced pipes with diameters ranging from 37.5 cm to 1.8 mm and adapted the conventional concrete prestressing technology to make railway sleepers from GPC prestressed geopolymer concrete. Geopolymer sleepers have been alternated with conventional sleepers on the main track line since 2002, and no problems were identified [10]. Currently, low-profile sleepers made of GPC are being developed as an alternative to wooden sleepers. The Czech Republic has developed a new inorganic binder system for self-hardening masses based on a geopolymer binder, a viscous liquid with a low degree of polymerisation used to prepare self-hardening masses. As a result of the hardener, polymerisation increases and a polymer with a high binding capacity is formed. These materials enable foundries to produce cores from self-hardening materials ecologically and economically [11]. Attempts have also been made to use geopolymer materials in drilling, mining, and hydro-technical construction for strengthening and sealing [12,13]. Alkaline activated geopolymer concrete has been commercialised in Australia under E-Crete (TM) name and is very popular among customers. E-Crete (TM) concrete reduced greenhouse gas emissions in concrete production by nearly 80% compared to concretes made of Portland cement [1]. Until now, many different types of surfaces have been made of geopolymers, such as pavements, playgrounds slabs, building foundations, and many others. Rocla was the first company globally to launch geopolymer concrete products on a commercial scale [3]. In Australia, at the University of Global Change Queensland (GCI), the world's first public utility building was designed and constructed in 2013; the structural elements made of geopolymer concrete were used in building constructions.

Geopolymer mixtures have good fluidity, which allows for an accurate representation of the form and the possibility of obtaining surfaces with various degrees of smoothness [3,4,12–20]. The concrete produced on the base of the geopolymer binder does not undergo the phenomenon of shrinkage cracks [12,18,19]. Geopolymer concrete is characterised by high compressive strength, minimal shrinkage and negligible creep, better adhesion and high resistance to acid and sulphate corrosion [1,3–6,11–13,16–19,21–29]. Some researchers also found that this concrete is also resistant to carbonate corrosion and possesses a very high fire resistance [15] and a high resistance to UV radiation [30]. Therefore, it is possible to the printing of geopolymers [22]. Geopolymers have excellent pouring properties and perfectly match even complex forms. In addition to the applications mentioned above, they are also used to produce various types of small-sized products, such as decorative elements. They have also found their use as injections strengthening the soil. In addition, geopolymer concrete is used to immobilise hazardous waste, including radioactive waste. The availability of several geopolymer-based products will increase rapidly, mainly due to the properties of this material and the fact that an increasing number of enterprises and part of the society are beginning to become convinced of this type of product.

Geopolymers are inorganic polymers formed by combining aluminosilicate with an activator and water. A geopolymer recipe could be based on a variety of aluminosilicate materials. The geopolymer microstructure considers an aluminosilicate polymer, synthetically produced using silicon (Si) and aluminium (Al), geologically acquired from minerals. The chemical composition of the geopolymer is similar to the chemical composition of zeolite, but the structure of the geopolymer is different [27]. The process of geopolymerisation is a chemical reaction between an alkali solution and a source material containing aluminosilicate (FA). It gives a three-dimensional polymeric chain and ring structure consisting of Si–O–Al–O bonds, as reported by Scheme 1 [30,31]:

- PSDS Si-O-Al-O-Si-O-Si-O—poly(sialate-disiloxo),
- PSS Si-O-Al-O-Si-O—poly(sialate-siloxo),
- PS Si-O-Al-O—polysialate.

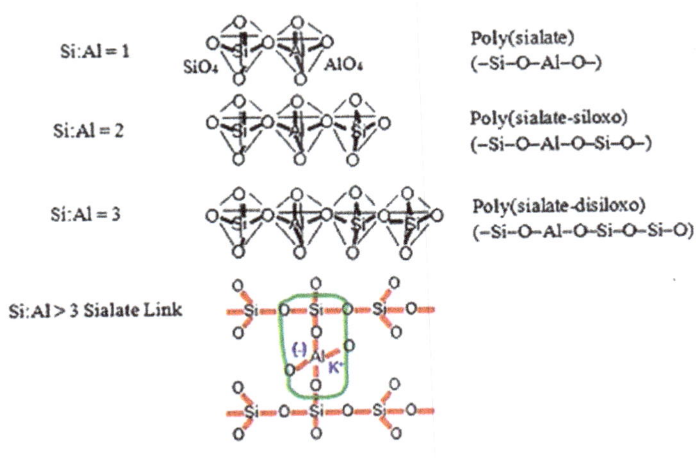

Scheme 1. Structure of polysialates [29].

The reactions can occur at room temperature. Therefore, it can be considered energy and source efficient and much cleaner. Two main categories of the division of geopolymers are accepted. The first one considers the elementary units of polymeric chains [5,28,29,32].

The second classification of geopolymers concerns their origin and, more specifically, the type of pozzolanic aluminosilicate material. The oxides of silicon and aluminium constitute the bases of the geopolymer composition, and the additives of metal cations such as sodium or potassium constitute the stabilising material here.

Table 1 presents the chemical composition of the three most popular and most readily available materials from which geopolymeric cement can be produced. As it is evident, the contents of individual oxides are similar. All three are built on the base of silicon and aluminium oxides.

Table 1. Chemical composition of the geopolymeric cement [33,34].

Mineral Composition	SiO_2	Al_2O_3	Fe_2O_3	TiO_2	CaO	K_2O	MgO	SO_3	Na_2O	P_2O_5
Fly ash	53.70	32.90	5.50	2.10	1.84	1.76	0.92	0.46	0.37	0.15
Metakaolin	52.00	45.00	0.50	1.50	0.40	0.05	0.20	0	0.15	0
Meta-clay	64.80	13.70	4.90	0	1.00	4.00	2.60	0	0.20	0

Due to this criterion, we distinguish geopolymers formed on the base of the following: fly ash, metakaolin, various types of rocks, volcanic agglomerates, silicas, fossil materials. Geopolymers are also from rock materials such as phosphorites [34] or pozzolanic volcanic ashes [35,36]. An important step forward was the use of geopolymer technology in waste management. Establishing the geopolymer recipe on industrial waste brings many environmental benefits and reduces costs related to waste management [3,4,12–14,17–20]. In producing geopolymers, such wastes as fly ash, diabase scrubbers [37] or bio-carbon resulting from biomass [38] are used.

From a chemical and mineralogical point of view, mineral rock wool is an ideal material for geopolymerisation (Table 2).

Table 2. Chemical composition of the mineral wool.

Mineral Composition	SiO_2	Al_2O_3	Fe_2O_3	TiO_2	CaO	K_2O	MgO	SO_3	Na_2O	P_2O_5
Glass wool	65.3	2.1	0.1	0.0	8.1	0.7	2.4	0.0	16.4	0.0
Rock wool	43.8	16.4	5.4	0.5	21.83	0.21	9.7	0.0	1.9	0.1

However, in the case of glass wool, Al_2O_3 is too low, and it is necessary to add another waste or natural component rich in Al_2O_3. Moreover, the essential factors which affect the process of geopolymerisation are given below [34,35]:

- Type of raw materials containing aluminosilicate
- The surface area of solid raw materials
- Glassy phase content in the raw material
- Amount of aluminium and reactive silicon
- Presence of iron, calcium, and inert particles in FA
- Curing temperature and pressure
- Duration of curing
- Type of curing (conventional heating or microwave heating)
- Type and concentration of alkalies
- Alkaline liquid-to-raw material ratio
- H_2O to Na_2O molar ratio
- Water to Geopolymer solids ratio
- Na_2O to SiO_2 ratio
- SiO_2 to Al_2O_3 ratio

The properties of geopolymers depend on the type of base material, the type and amount of activator used and their production technology, i.e., mixing time, material fragmentation, hardening temperature, humidity, hardening time and the amount of added water [3,4,9,12–14,17–20]. In addition, types of used alkaline activators, such as NaOH solution, KOH solution, a mixture of NaOH solution-water glass, and a mixture of

KOH solution-water glass, are also essential to achieve a high compressive strength the geopolymers [15].

Moreover, there is the possibility of adjusting the porosity (it is possible to attain low porosity below 2.5% or high porosity above 20%, depending on needs, up to foaming geopolymers (as insulating materials) [33,37]. If the geopolymer is foamed in the production process, it might become lightweight and partly insulating [6,36]. On the other hand, in an almost pore-free form, the geopolymer may be used to protect walls against moisture and fungi formation [37–39].

Geopolymers are becoming essential in waste neutralisation technologies, especially in the neutralisation of hazardous waste [3,5,29,40]. Similarly, as material approved for use in construction, geopolymers should meet environmental regulations regarding the environment and human health safety.

The research involved geopolymer prepared from mineral rock and glass wool. The authors have already attempted to create a geopolymer from the mineral wool of the publication [6,7]. Depending on the production method, wool geopolymers may have different properties. For example, due to the high degree of grinding, the wools might have high strength as of 50 MPa [6] or, due to the addition of other components, they may have a porous structure which causes the geopolymer to develop different properties [6]. The rock wool has an adequate proportion of Si and Al to geopolymerisation, but glass wool has got to a low amount of Al. Therefore, the glass wool needs an additional component rich in Al_2O_3. As a source of Al, the bauxite was used. So far, glass wool with the mentioned additives has not been tested.

The leachability of wool and wool-based geopolymers was checked. Such research has not been conducted so far. Moreover, that is also new, and the article compares the vulnerability of mineral wool and geopolymer made of mineral wool to the leaching of soluble components. The geopolymer was washed out in a comminated form (crushed geopolymer pieces <10 mm). The study allowed us to determine the size and type of harmful substances that may negatively affect soil, groundwater, and surface waters.

The second part of the research investigated the mechanical properties of glass or rock wool-based binder and mortar.

2. Materials and Methods

2.1. Materials

2.1.1. Mineral Rock and Glass Wool

The materials used for the tests were waste generated in the mineral wool production cycle. Rock wool (Figure 1a) and glass wool (Figure 1b) were tested.

According to the current Regulation [41], the examined wastes were assigned the 10 12 99 for rock wool and 10 11 03 for glass wool. These are non-hazardous waste. The advantage of the tested waste is its homogeneity and high purity. The manufacturer has introduced waste segregation at the "source" in the production cycle. However, the production cycle cannot use the tested waste due to its form. So far, it has been mainly deposited in landfills for non-hazardous and inert waste and stored in separate quarters due to their character and properties.

The wool density was 10 kg/m^3. Mineral wool has a low coefficient of thermal conductivity (so-called lambda, λ). Products made of glass or rock mineral wool, most found on the market, have a thermal conductivity coefficient in the range of 0.030–0.045 $W/(m \cdot K)$. In the next stage of research, the chemical composition was checked. The investigated compositions of tested wools are given in Table 2. The chemical compositions of the rock and glass wool were also determined using the ICP-OES method [7].

(a)

(b)

Figure 1. Shredded rock wool (**a**) glass wool (**b**) used for tests.

2.1.2. The Methodology of Making Wool-Based Geopolymers

The materials used for the tests were glass and rock mineral wool grounded in a ball mill as the target geopolymer powder (Figure 2). The wool was ground for one hour in a Los Angeles drum equipped with additional grinding balls. The density of green wool in the bulk state was 0.935 kg/dm^3.

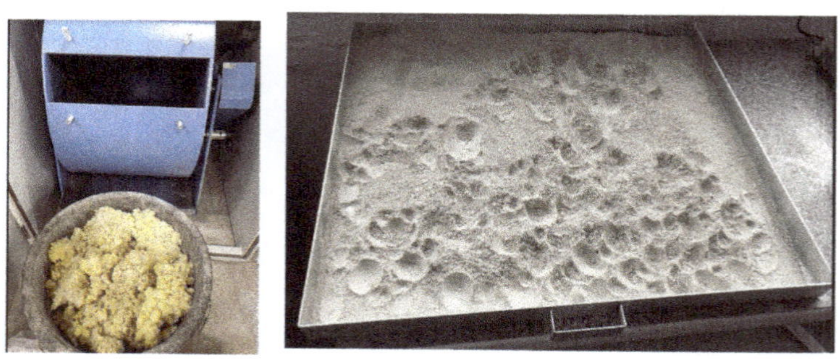

Figure 2. A view of a Los-Angeles drum and ground wool.

In case of an increase of Al in ground glass wool, a bauxite or Al_2O_3 power was used. In both cases, Geopolymers were made with ground bauxite or Al_2O_3 in 15% of grounded mineral glass wool. The geopolymers from ground rock wool were prepared without adding another component.

Bauxite contains overwhelmingly: hydrous aluminium oxides, aluminium hydroxides, clay minerals, and insoluble materials such as quartz, hematite, magnetite, siderite, and goethite. The aluminium minerals in bauxite can include gibbsite $Al(OH)_3$, boehmite AlO(OH), and diaspore, AlO(OH). Table 3 presents the chemical composition of tested bauxite.

Table 3. The chemical composition of bauxite, %.

Al_2O_3	SiO_2	Fe_2O_3	TiO_2	Silica Module; Al_2O_3/SiO_2
36.0	15.6	20.7	1.5	2.32

The geopolymerisation process requires the dissolution of the starting material in a high pH (alkaline) solution, and thus pH values of fresh geopolymer pastes are usually 11.2–13.2 [37,38]. Sodium oxide reacts exothermically with cold water to produce a sodium hydroxide solution. A concentrated solution of sodium oxide in water will have pH 14. $Na_2O + H_2O \rightarrow 2NaOH$. When the OH- concentration of the aqueous solution is reduced by ten times, pH is decreased by only one. According to publications [1,3,7,19,24–31,37], NaOH and soda-silicon water glass are adequate alkane sources to achieve the high compressive strength of geopolymers. The study [37] analysed the effect of NaOH concentration (6–14 M) on the mechanical properties of kaolin geopolymers. Compressive strength results showed that the optimum NaOH concentration is 8 M. Moreover, according to research results [3,4,11–14,16–19,37–40,42,43], the soluble glass to 8 M of NaOH solution ratio should be around 2.5 to give geopolymers high compressive strength [37,40]. Therefore, such recommendations were adopted to design the ratio of the alkaline solution in the analysed studies.

For the preparation of paste and mortar was used 450 g of ground wool were and 225 g of alkaline solution were mixed previously in a magnetic stirrer (Figure 3b). The grounded wool to alkali weight ratio was 0.5. The alkaline solution was cooled down to 20 degrees Celsius before ground wool. The paste and mortar were mixed according to the methodology described in EN-196-1 [44]. The value of pH of fresh geopolymers was also tested (Figure 3a).

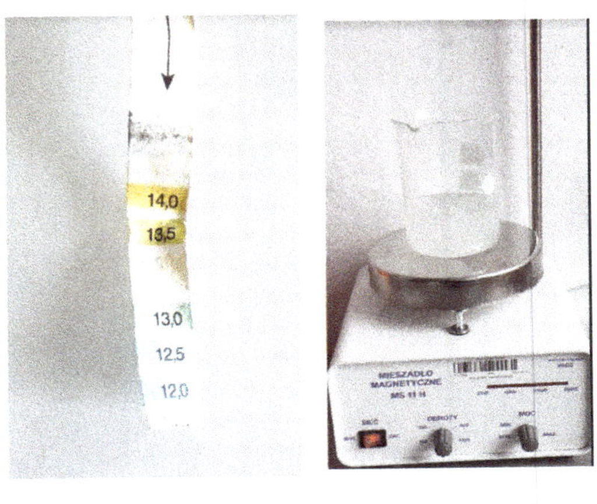

(**a**) (**b**)

Figure 3. (**a**) The view of pH-value of fresh geopolymers; (**b**) alkaline solution in magnetic stirrer.

In the case of geopolymer mortar, normalised sand according to EN-196-1 was used in an amount of 1350 g.

The slump flow of mortars was measured before and after the table was jolted (according to EN 1015-3 [45]).

The geopolymer paste and mortar was formed as 20 mm × 20 mm × 160 mm (Figure 4a) and 40 mm × 40 mm × 160 mm (Figure 4b) specimens, respectively, and were heat-treated for 48 h at 70 °C.

(**a**) (**b**)

Figure 4. Geopolymer binder (**a**) mortar (**b**) with rock wool with bauxite (top sample) and with Al_2O_3 (downsample).

The tested geopolymers had been matured in laboratory conditions for 26 days. The air temperature of 20 °C and humidity was about 50%. After that time, they were crushed (Figure 5a,b), then water extracts were made.

(**a**)

(**b**)

Figure 5. Geopolymer rock wool binder with the addition of bauxite (SG_B), fragmented form (**a**), Geopolymer Rock wool binder with the addition of Al_2O_3 (SG_Al_2O_3), fine form (**b**).

The leachability of pollutants from geopolymers produced on the base of waste mineral wool might relate to the form in which they occur, translating into a harmful environmental footprint. In the case of monolithic structures, the leachability level may be determined by the surface release process and diffusion. However, in the case of fragmented forms, the leachability of pollutants determines the percolation process. Therefore, the article also presents.

2.2. Research Methodology

2.2.1. Leachability of Mineral Wool

The research procedure included the preparation of water extracts from waste mineral wool (rock and glass), executing chemical tests, and assessing criteria to allow the waste to be stored. The tested waste was mechanically milled for chemical analysis; also, water extracts were carried out on the pre-prepared samples. The samples were tested for the content of organic carbon (TOC)—EN 1484:1999 [46]. Loss on ignition (LOI) was determined according to EN 15935:2013-02 [47]. However, the determination of calorific value by calorimetric bomb combustion to PN-ISO 1928:2020-05 [48]. The procedure for preparing water extracts from solid waste was carried out according to EN 12457-2:2006 standard [49]. Water extracts for waste were prepared at a liquid to solid ratio of L/S = 10 dm^3/kg (basic test). The prepared water extracts were shaken on a laboratory shaker for 24 h, after which time the obtained extracts were filtered. The leaching liquid was distilled water with a pH of 7.1 and a specific conductivity of 61.18 μS/cm. The analysis of water extracts from waste included several determinations. First, the pH of water extracts was determined using the potentiometric method using the Elmetron CPC-501 apparatus available in Silesian University of Technology in Gliwice, Poland [50]. Chloride (Cl^-) contents were determined by the Mohr method with the use of silver nitrate (V) as a titration reagent and potassium chromate (VI) as an indicator (PN-ISO 9297:1994) [51]. Determination of sulphates (VI) (SO_4^{2-}) was carried out using the gravimetric method with barium chloride according to PN-ISO 9280:2002 [52]. Determination of phosphorus was according to EN ISO 6878:2006 [53]. The sodium, calcium, potassium, lithium, and barium content in water extracts were determined by flame emission spectrometry—following PN-ISO 9964-3:1994 [54]. Using the GBC AVANTA PM spectrometer available in Silesian University of Technology in Poland, trace element concentrations were determined by flame atomic absorption spectrometry.

2.2.2. The Leachability of the Wool-Based Geopolymer

The execution of water extracts from monolithic and fragmented geopolymers was executed following the EN 12457-4:2006 standard [49]. Monolithic geopolymers were mechanically grounded to grain sizes <10 mm. Water extracts were prepared from the prepared samples with a liquid to solid ratio of L/S = 10 L/kg (basic test). The leaching liquid was distilled water with a pH of 7.1 and a specific conductivity of 62.18 μS/cm. The extracts were then shaken on a laboratory shaker for 24 h, and the suspension was filtered. The analysis of water extracts from fragmented geopolymers covered several determiners. First, the pH of the solutions was carried out using an Elmetron CPC-501 apparatus [50]. Chloride contents were determined by the Mohr method with the use of silver nitrate (V) as a titration reagent and potassium chromate (VI) as an indicator (PN-ISO 9297:1994) [51]. In addition, the determination of sulphates (VI) (SO_4^{2-}) was carried out with a gravimetric method with barium chloride, according to PN-ISO 9280:2002 standard [52]. The flame emission spectrometry method determined sodium, calcium, potassium, and barium content in water extracts from ashes and concrete according to the PN-ISO 9964-3:1994 [54] standard. Using the GBC AVANTA PM spectrometer, trace element concentrations were determined by flame atomic absorption spectrometry.

2.2.3. The Mechanical Properties of the Wool-Based Geopolymer

The geopolymers were matured in laboratory conditions for 26 days. Ambient temperature 20 °C and humidity around 50%. After 28 days, the flexural and compressive strength of the wool-based geopolymers binders and mortars were tested (Figure 6) acc. EN 196-1:2016 [44].

Figure 6. The view of the sample of geopolymer mortar under testing of the compressive strength.

3. The Results and Their Discussion

3.1. Mineral Wool Leachability Test Results

Table 4 shows the leachability levels of harmful substances that may become an environmental hazard. The analysed mineral wool was slightly alkaline and had a relatively high leachability of chlorides and sulphates.

Table 4. Leachability of harmful substances and heavy metals from the tested mineral wool, expressed in mg/kg dry weight (except for pH).

	Rock Wool	Glass Wool	Criteria for Waste Landfilling Acceptance [55]		
			Neutral	Other Than Neutral and Hazardous	Hazardous
pH	9.3	10.2	-	-	-
Chlorides, Cl^-	1382.40	2073.60	800	15,000	25,000
Sulphates, SO_4^{2-}	522.48	2678.21	1000	20,000	50,000
Barium, Ba	12.10	248	20	100	300
Zink, Zn	<0.05	<0.05	4	50	200
Copper, Cu	0.20	0.40	2	50	100
Lead, Pb	<0.30	<0.30	0.5	10	50
Cadmium, Cd	0.14	0.05	0.04	1	5
Chromium, Cr	<0.50	<0.50	0.5	10	70
Cobalt, Co	<0.50	<0.50	-	-	-
Iron, Fe	<0.20	0.55	-	-	-
Manganese, Mn	<0.20	<0.02	-	-	-
Nickel, Ni	<0.40	<0.40	0.4	10	40

Since the following values are exceeded: chloride (in both types of wool), sulphates (in glass wool), barium (in glass wool) and cadmium (in rock wool), it can be said that mineral wool does not qualify for storage at a landfill for inert waste. However, the situation is different in the case of non-hazardous and inert waste, which is not municipal waste. Based on the results, it can be concluded that the values obtained using the primary test for Rock wool (the ratio of liquid to solids equal: 10 L/kg) did not exceed the permissible leaching limits. Thus, rock wool can be stored in a landfill for non-hazardous and inert waste. On the other hand, too much barium in a tested sample of glass wool discriminates against it for storage in this landfill. Therefore, in this case, it must be deposited in a hazardous waste landfill to ensure that all results do not exceed the leaching limit values. As a curiosity in the use of waste mineral wool, the effect of using the addition of mineral wool can be cited from crops under cover and municipal sewage sludge on water retention and heavy metals. The average soil's leaching efficiency (Pb, Zn and Cd) was assessed. Sewage and mineral sludge Wool, widely used in many soil remediation technologies, has been found to have a beneficial and diversified influence on soil water properties and the mobility of heavy metals [56].

Table 5 presents the parameters considered as additional criteria allowing waste to be stored in non-hazardous and inert landfills. Again, a much more significant loss on ignition of glass wool relative to rock wool can be seen. On the other hand, the opposite situation occurs in the case of organic carbon, where rock wool has a much more significant amount (Rock wool contains more organic compounds than glass wool). Therefore, both glass wool and rock wool meet the requirements for waste destined for storage in landfills other than neutral and hazardous.

Table 5. Chemical properties of mineral wool that allow for depositing in landfills for non-hazardous and inert waste.

Component	Unit	Rock Wool	Glass Wool	Limit Values [55]
Loss on ignition (LOI)	% dry mass	5	7	8
Total Organic Carbon (TOC)	% dry mass	4	3	5
Gross calorific value (GCV)	MJ/kg dry mass	0	2	6

Determination of the ignition loss at 950 °C is an essential parameter for waste in construction. In addition, it is a fire safety parameter. Loss on ignition at 950 °C temperature for rock wool was at a loss of 4.59%, while it was at 7.88% for glass wool. The rock wool mineral residue had a loose consistency, and it can be stated that it was like sand (Figure 4a). On the other hand, glass wool at the temperature of 950 °C had a liquid consistency; it underwent vitrification (Figure 7b).

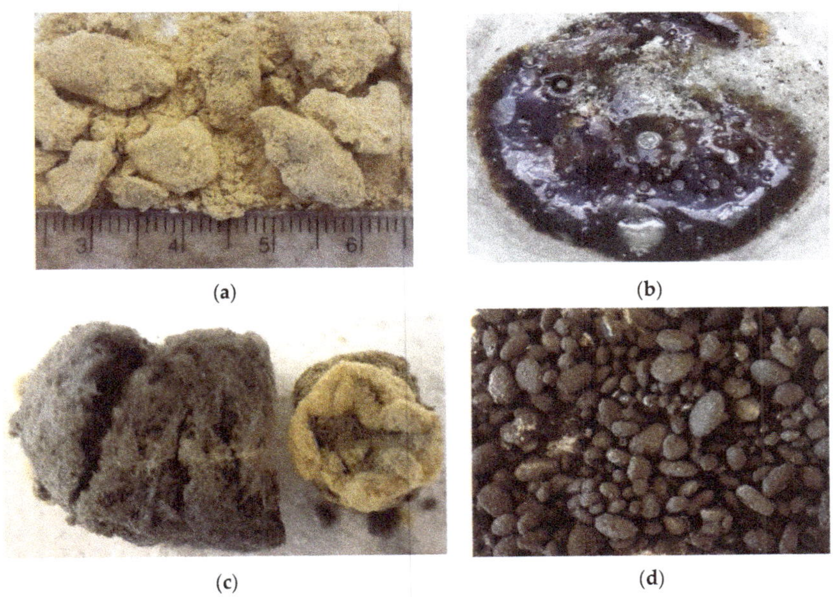

Figure 7. Residue from Loss on Ignition (LOI) at 950 °C, (**a**) rock wool, (**b**) glass wool; wool residue after thermal analysis at 815 °C without oxygen, (**c**) rock, (**d**) glass.

As part of the research, the content of volatile elements was determined. The volatile element content for Rock wool was at the level of 3.74%, while about 5.82% for glass wool. When analysing the samples after denoting, carbonisation of the samples was observed (Figure 4c,d). The conducted analyst in thermal transformation confirmed that the material is non-flammable rock wool is characterised by a higher leachability of selected elements than glass wool.

Only the content of phosphorus was higher for glass wool. The difference can be seen in the lithium content, almost 17 times higher in rock wool (Table 6). All the results met the required conditions regarding the acceptable limit values for waste intended to be stored in landfills.

Table 6. Leachability of selected elements (mg/L) from the tested waste.

	Rock Wool	Glass Wool	Permissible Limit Values [57]
Sodium, Na	87.60	85.18	800
Potassium, K	11.60	3.55	80
Lit, Li	2.20	0.13	-
Calcium, Ca	315.20	37.78	-
Phosphorus, P	0.07	0.71	2

3.2. *The Results of the Study of Leaching Mineral Wool-Based Geopolymer*

Similar studies of the leachability of geopolymers made from mineral wool have not been carried out. This is the first study to assess the environmental effects of mineral wool-based geopolymer. The leachability of metals depends on the pH of the material in which they are immobilised. The paper [58] presents the results obtained in the pH leaching test. It assesses the influence of pH changes and occurring processes on releasing heavy metals (Cd, Ni, Cr total, Pb, Cu and Zn) from metallurgical slag in a zinc smelter. Based on test results obtained in the pH test, a strong dependence of heavy metals leaching on the pH was found. The highest concentrations of the analysed elements were observed in an acidic environment. For most metals, except for lead, an increase in the pH of the solution caused a decrease in their concentration. Lead showed an upward trend of release under alkaline conditions. A sharp increase of copper leaching at pH 10.5 was also observed. Based on the study results, cadmium can be considered the most mobile element from metallurgical slag. Chromium indicated the lowest degree of release.

The results of the leachability test (Table 6) were compared to the highest permissible rates of pollution in sewage discharged into water or soil. The water extract of geopolymer binders (SG_B and SG_Al_2O_3) was characterised by a strongly alkaline pH value, exceeding the permissible values. The remaining ions are eluting themselves in amounts lower than the limit values. The analysis of the test results presented in Tables 7 and 8 proves that the geopolymer made of mineral wool meets the requirements in terms of chloride, sulphate, phosphorus, potassium, calcium, lithium, and barium concerning pollution indicators in wastewater discharged into water or soil. Unfortunately, the pH value and sodium content exceed the acceptable environmental standards.

Table 8 summarises the results of studies on the leachability of heavy metals from the geopolymer. The research results on the content of heavy metals in water extracts showed that the rock wool geopolymers in the comminute form meet all the requirements in this regard.

Table 9 shows the leachability of harmful substances from the monolithic geopolymer mortar. The obtained results were related to the leachability of monolithic cement mortars based on Portland cement. The geopolymer mortar was characterised by a high reaction value (pH > 11). The exceedance was also noted for barium, and its content exceeds four times the permissible value. However, the remaining substances do not exceed the allowable values of water and soil.

Table 7. Leachability of pollutants from ground rock wool-based geopolymer binders expressed as mg/L dry weight (except for pH).

Parameter	SG_B	SG_Al_2O_3	Highest Permissible Value [57]
pH	12.2	12.3	6.0–9.0
Chlorides, Cl^-	BLQs *	BLQs *	1000
Sulphates, SO_4^{2-}	298.40	341.19	500
General phosphorus, P	0.07	0.07	2
Potassium, K	180.00	79.17	80
Calcium, Ca	965.83	982.50	-
Lit, Li	12.50	12.50	-
Sodium, Na	2715.83	2745.17	800
Barium, Ba	BLQs *	BLQs *	7
The sum of chlorides and sulphates, (Cl + SO_4)	298.40	341.19	1500

BLQs *—below the limit of quantification.

Table 8. Leachability of heavy metals (mg/L) from geopolymer rock wool binders in a comminated form.

Parameter	SG_B	SG_Al_2O_3	Highest Permissible Value [57]
Zinc, Zn	0.04	0.08	2
Copper, Cu	0.03	0.04	0.5
Lead, Pb	<0.30	<0.30	0.5
Cadmium, Cd	<0.005	<0.005	0.2
Chromium, Cr	0.07	0.07	0.1
Cobalt, Co	<0.05	<0.05	1
Iron, Fe	0.03	0.05	10
Molybdenum, Mn	<0.02	<0.020	1
Nickel, Ni	<0.04	<0.04	0.1

Table 9. Leachability of harmful substances for monolithic rock wool geopolymer mortar, expressed in mg/L dry weight (except for pH).

Parameter	CEM I	ZG_B	Highest Permissible Value [57]
pH value	10.7	11.5	6.0–9.0
Chlorides, Cl^-	529.92	161.28	1000
Sulphates, SO_4^{2-}	109.84	63.36	500
General phosphorus, P.	0.73	0.33	2
Potassium, K.	23.20	10.50	80
Calcium, Ca	546.20	158.50	-
Lit, Li	1.80	3.00	-
Sodium, Na	1803.40	371.50	800
Bar, Ba	24.40	30.50	7
Sum of chlorides and sulphates, (Cl + SO_4)	639.76	224.64	1500

The leachability of heavy metals from the monolithic form of geopolymers is presented in Table 10. The content of heavy metals in water extracts from geopolymer mortar did not exceed the permissible values. Therefore, it can be said that they do not pose a threat to the natural environment.

The rock wool and glass wool show different leachability performances for each ion because their base composition is very different [7]. Table 11 The chemical composition of rock wool and glass wool, determined by ICP-OES.

Table 10. Heavy metal content (mg/L) in water extracts for monolithic rock wool geopolymers.

Parameter	CEM I	ZG_B	Highest Permissible Value [57]
Zinc, Zn	0.17	0.13	2
Copper, Cu	0.09	<0.02	0.5
Lead, Pb	<0.30	<0.30	0.5
Cadmium, Cd	<0.005	<0.005	0.2
Chromium, Cr	2.33	<0.05	0.1
Cobalt, Co	<0.05	<0.05	1
Iron, Fe	0.53	<0.02	10
Molybdenum, Mn	<0.02	<0.02	1
Nickel, Ni	0.13	<0.04	0.1

Table 11. The chemical composition of rock wool (RW) and glass wool (GW), determined by ICP-OES [7].

Chemical Component	Rock Wool (RW)	Glass Wool (GW)
Ca, ICP (g/kg)	125	20.1
Si, ICP partial solution (g/kg)	0.4	0.4
Al, ICP (g/kg)	85.6	0.3
Fe, ICP (g/kg)	61.3	0.9
Na, ICP (g/kg)	10.4	68.2
K, ICP (g/kg)	3.7	3.7
Mg, ICP (g/kg)	75.8	3.9
P, ICP (g/kg)	0.3	<0.020
Ti, ICP (g/kg)	3.5	<0.050
S, ICP (g/kg)	0.1	3.8
Ba, ICP (g/kg)	0.2	1.3
Mn, ICP (g/kg)	1	0.5
As, ICP (mg/kg)	<3	<3
Cd, ICP (mg/kg)	<0.3	<0.3
Cr, ICP (mg/kg)	280	2.3
Cu, ICP (mg/kg)	34	8.8
Hg, CVAAS (mg/kg)	<0.04	<0.04
Ni, ICP (mg/kg)	49	1.8
Pb, ICP (mg/kg)	<3	3.7
Zn, ICP (mg/kg)	47	430
B, ICP (mg/kg)	8.9	6260
Be, ICP (mg/kg)	<1	<1
Co, ICP (mg/kg)	21	2
Mo, ICP (mg/kg)	<1	<1
Sb, ICP (mg/kg)	<3	<3
Se, ICP (mg/kg)	<3	<3
Sn, ICP (mg/kg)	<3	<3
V, ICP (mg/kg)	170	<2

Note: CVAAS represents "cold-vapor atomic absorption spectrometry".

The presented test results proved that it is possible to store rock wool in a landfill for non-hazardous and inert waste. However, too much barium in glass wool discriminates against it for storage in such landfills; therefore, it should be stored in a hazardous waste landfill. Moreover, the geopolymer is characterised by too high alkalinity due to the environmental requirements.

The proposed reuse of waste mineral wool in the form of a geopolymer is a solution that is beneficial for the environment, climate, human health, and economic reasons.

The results of the research on the quality of water extracts made of comminute geopolymer binders based on rock wool show that:

- the pH is at the level of 12.2,
- the content of sulphates (SO_4^{2-}) is present in amounts from 298 to 342 mg/L, depending on the added recycling material,

- sodium content was at the level of 2700 mg/L.

On the other hand, geopolymer products are used in the industry, such as PCI-Geofug geopolymer grout by Basf [59], GeoLite geopolymer mortar by Kerakoll [60], ASTRA GKB geopolymer concrete [61], also the geopolymer injection as a quick and non-invasive method of strengthening the ground. It is used in linear (infrastructure) and cubature construction (industrial, commercial, and residential). It has been successfully used in Western Europe and Scandinavia for over 40 years. Geopolymer is gaining popularity due to its exceptional performance parameters and convenience of use: speedy repair time and minimal nuisance during geoengineering works. Unfortunately, the results of the leachability test of hazardous substances and heavy metals do not pose a potential threat to the environment. However, a strongly alkaline reaction and the excess sodium content are open for further consideration. The authors will lower the alkaline reaction of geopolymer binders in further research steps. The geopolymerisation process requires the dissolution of the starting material in a high pH (alkaline) solution, and thus, pH values of fresh geopolymer pastes are usually 11.2–13.2 [37,38]. Like products based on Portland cement, the pH must not be too low due to reinforcement corrosions [55]. The concrete is acceptable for use, although the pH of the water extract is from 9 to 11 (Table 9). Due to environmental requirements, the value of the concrete solution is also too high, but not as high as in the case of the geopolymer solution. The authors do not know the pH of the mentioned above geopolymer products. The authors want to check out it is possible to lower the pH geopolymer add an acid component to it, but this will drastically reduce its strength. The authors will analyse this problem in the future. The problem is very significant in terms of the environment. In the future, the authors will decrease the alkalinity of wool-based geopolymers to 11.2 according to the suggestion of publication [25,42].

3.3. *The Fresh and Mechanical Results of Wool-Based Geopolymer*

The slump-flow of glass wool-based geopolymer fresh mortar with Al_2O_3 or bauxite was comparable and was about 13 cm (Figure 8). Whereas the slump flow of geopolymer mortar with rock wool was more significant by up to 3 cm. thus, the type of wool is essential for the rheological properties of fresh geopolymer. It has been noticed that geopolymers, due to their high viscosity value, are very resistant to segregation, which is undoubtedly their great advantage.

Figure 8. The view of slump-flow of glass wool-based geopolymer fresh mortar with Al_2O_3.

The geopolymer was characterised by high adhesion strength to the forms in which they stayed. The phenomenon of geopolymer adhesion was mentioned in the publication [18]. It was noticed that geopolymers show more adhesion than materials at the entrance to cement. This phenomenon may be due to work with the fibres or reinforcement of the geopolymer. Moreover, the geopolymers were highly viscous, especially with sodium-potassium water glass, which requires careful mechanical concentration. In Figure 9, pores in the structure of wool-based geopolymers are presented, probably from under dumping the sample. Despite this, geopolymers glass wool-based mortars with

Al_2O_3 (Figure 9) obtained an average compressive strength of 59 MPa and tensile strength of 4.5 MPa. The geopolymer with bauxite achieved about 51 MPa, and flexural strength 4.1 MPa. Thus, Al_2O_3 is a better additional glass wool-based geopolymer than bauxite.

Figure 9. The view of pores of glass wool-based geopolymer mortar with Al_2O_3.

The average compressive strength of their binders (tested on 20 mm × 20 mm × 160 mm samples, Figure 10) was about 20 MPa (in the case of glass wool-based geopolymer with bauxite). It was observed that samples of geopolymeric binders without aggregate participation are characterised by cracking and deformation of samples due to shrinkage. The reason was that the samples were too slender. The shrinkage value is greatly influenced by the geometry of the samples, especially without aggregate. The compressive strength of binders with AL_2O_3 was about 25 MPa.

Figure 10. The view of pores of glass wool-based geopolymer mortar with bauxite.

The average compressive strength of rock wool-based geopolymer mortar was about 62 MPa, and their compressive strength of binder was about 25 MPa and flexural strength 4.8 MPa. It was also noticed that the rock wool thickens better in the moulds, which translated into a higher value of their strength.

The studies [6,7] show that maximum compressive strengths of 48.7 and 30.0 MPa were measured for wool-based binders. The binder matrix consisted of aluminosilicate gel with partly dissolved mineral wool fibres. The maximum flexural strength was 13.2 MPa for GW and 20.1 MPa for RW. This study shows that high strength can be obtained without additional co-binders by activating alkali with sodium aluminate solution. Furthermore, the research carried out by the authors of this publication proved that it is possible to obtain the strength of the geopolymer, resistant to shrinkage deformation, but after modification of its composition. The latest achievement of the authors is a wool-based geopolymer with a strength of 100 MPa, which will be the subject of the following publication.

4. Conclusions

Due to the increasing necessity to reduce CO_2 emissions and the energy consumption of buildings, the consumption of mineral wool in construction is increasing. Mineral wool is a waste material that meets the requirements for storage; however, it is not the direction to be followed when considering the future. In line with the closed-loop policy and the increasing need to reuse materials, the goal is to recycle them, also due to the decreasing availability of the waste storage area.

The carried-out research results proved that:

- It is possible to store rock wool in a landfill for non-hazardous and inert waste. However, in this case, too much barium in a glass wool sample discriminates against it for storage in a non-hazardous landfill; storing it in a hazardous waste landfill seems necessary. Leachability tests were performed to investigate the environmental effects of the wool-based monolith and crushed forms of wool-based geopolymers. The geopolymer was washed out in a comminated form (crushed geopolymer pieces <10 mm). The study allowed us to determine the size and type of harmful substances that may negatively affect soil, groundwater, and surface waters.
- The proposed use of wool for geopolymer binders is a correct solution for environmental reasons. The geopolymer meets the guidelines of the Regulation of the Minister of Maritime Economy and Inland Navigation of 12 July 2019, on substances particularly harmful to the aquatic environment and the conditions to be met when discharging sewage into waters or soil. As well as during the discharge of rainwater or meltwater into waters or water facilities (Journal of Laws 2019, item 1311), except for the pH, the maximum permissible value of which should be 9.0, and the sodium content, the ultimate value of which should be 800.
- Due to their high viscosity value, the wool-based geopolymers are resistant to segregation, which is undoubtedly their great advantage. Moreover, the geopolymer is characterised by high strength of adhesion. Therefore, the type of wool is essential for the rheological properties of fresh geopolymer. Moreover, the geopolymers were characterised by high viscosity, requiring careful mechanical concentration.
- Geopolymers glass wool-based mortars with Al_2O_3 obtained an average compressive strength of 59, the geopolymer with bauxite achieved about 51 MPa, and flexural strength 4.1 MPa. Thus, Al_2O_3 is a better additional glass wool-based geopolymer than bauxite. Moreover, Al_2O_3 may also be used as a waste material, which contributes to the eco-friendliness of the geopolymer.
- The average compressive strength of rock wool-based geopolymer binder was about 20 MPa. It was observed that samples of geopolymers grout without aggregate participation are characterised by cracking and deformation. The average compressive strength of rock wool-based geopolymer mortar was about 62 MPa, their binder's compressive strength was about 25 MPa, and flexural strength was 4.8 MPa.

The compressive strength of geopolymer results [37] showed that the optimum NaOH concentration is 8 M. The geopolymer strength decreases with NaO concentration in the NaOH solution. As mentioned in the article, the geopolymerisation process requires the dissolution of the starting material in a high pH (alkaline) solution. Thus, the pH values of fresh geopolymer pastes are usually 11.2–13.2 [37,38]. Therefore, an attempt will be made to lower the wool-based geopolymer alkalinity to 11.2. Although it still does not meet the environmental requirements [1,57]. However, the problem needs future research.

In the following research steps, Al_2O_3 will be replaced with a waste material that predominantly contains the same compound to create a glass wool geopolymer from waste materials only.

Further modifications aim to replace the sand of geopolymer mortar with material derived from waste, in line with the goals of sustainable development and the protection of natural resources.

Author Contributions: Conceptualisation, B.Ł.-P., M.C. and D.S.; Methodology, B.Ł.-P., M.C. and D.S.; validation B.Ł.-P., M.C. and D.S.; formal analysis, B.Ł.-P., M.C. and D.S.; writing—original draft preparation, B.Ł.-P., M.C. and D.S.; writing—review and editing, B.Ł.-P., M.C. and D.S. All authors have read and agreed to the published version of the manuscript.

Funding: Publishing supported by the pro-quality grant: 03/030/RGJ21/0122; The Silesian University of Technology, Department of Building Processes and Building Physics, Faculty of Civil Engineering: BK-223/RB3/2022; The research of publication was funded by a subsidy allocated for 2021. The research reported in this paper was co-financed by the European Union from the European

Social Fund in the framework of the project "Silesian University of Technology as a Centre of Modern Education based on research and innovation" POWR.03.05.00-00-Z098/17.

Institutional Review Board Statement: Not applicable.

Informed Consent Statement: Not applicable.

Data Availability Statement: Not applicable.

Conflicts of Interest: The authors declare no conflict of interest. The funders did not play any role in the design of the study plan; in collecting the resulting data, analysing its results, or interpreting data, and formatting the content of the manuscript or in the decision to publish the results of the study.

References

1. Environment 2021. Statistics Poland. Available online: https://stat.gov.pl/en/topics/environment-energy/environment/environment-2021,1,13.html (accessed on 27 January 2022).
2. *Announcement of The Seym of the Republic of Poland of 15 April 2021 on the Publication of the Uniform Text of the Act on Waste (Journal of Laws of 2021, Item 779, Art. 22)*; The Seym of the Republic of Poland: Warszawa, Poland, 2021.
3. Tomasz, Z.; Błaszczyński, M.; Król, R. Geopolymers in construction; Civil and environmental engineering reports. *CEER* **2015**, *16*, 25–40.
4. Pacheco-Torgal, F.; Castro-Gomes, J.; Jalal, S. Alkali-activated binders: A review: Part 1. Historical background, terminology, reaction mechanisms and hydration products. *Constr. Build. Mater.* **2022**, *7*, 1305–1314.
5. Manso, M.; Castro-Gomes, J. Green wall systems: A review of their characteristics. *Renew. Sustain. Energy Rev.* **2015**, *41*, 863–871. [CrossRef]
6. Yliniemi, J. Mineral wool waste-based geopolymers. *IOP Conf. Ser. Earth Environ. Sci.* **2019**, *297*, 012006. [CrossRef]
7. Yliniemi, J.; Kinnunen, P.; Karinkanta, P.; Illikainen, M. Utilization of Mineral Wools as Alkali-Activated Material Precursor. *Materials* **2016**, *9*, 312. [CrossRef] [PubMed]
8. Erofeev, V.; Rodin, A.; Bochkin, V.; Yakunin, V.; Ermakov, A.A. Lightweight geopolymers made of mineral wool production waste. *Mag. Civ. Eng.* **2020**, *93*, 3–12. [CrossRef]
9. Błaszczyński, T.; Król, M. Durability of Green-Concretes. In Proceedings of the 8th International Conference AMCM 2014, Wrocław, Poland, 27–30 October 2014; pp. 530–540.
10. Available online: http://www.rocla.com.au/ViewNews.php?id=40 (accessed on 27 January 2022).
11. World's First Public Building with Structural Geopolymer Concrete. Available online: https://www.geopolymer.org/news/worlds-first-public-building-with-structural-geopolymer-concrete/ (accessed on 27 January 2022).
12. Kejkar, R.; Wanjari, S.; Sharma, D.; Rajankar, R. Experiment Investigation and Physical Performance of Geopolymer Fly Ash. *Bricks IOP Conf. Ser. Mater. Sci. Eng.* **2018**, *431*, 092005. [CrossRef]
13. Srividya, T.; Kannan Rajkumar, P.R.; Sivasakthi, M.; Sujitha, A.; Jeyalakshmi, R. A state-of-the-art on development of geopolymer concrete and its field applications. *Case Stud. Constr. Mater.* **2022**, *16*, e00812. [CrossRef]
14. Purbasari, A.; Samadhi, T.W.; Bindar, Y. The Effect of Alkaline Activator Types on Strength and Microstructural Properties of Geopolymer from Co-Combustion Residuals of Bamboo and Kaolin. *Indones J. Chem.* **2018**, *18*, 397–402. [CrossRef]
15. Pacheco-Torgal, F.; Castro-Gomes, J.; Jalali, S. Alkali-activated binders: A review. Part 2. About materials and binders manufacture. *Constr. Build. Mater.* **2007**, *22*, 1315–1332. [CrossRef]
16. Pacheco-Torgal, F.; Castro-Gomes, J.P.; Jalali, S. Adhesion characterisation of tungsten mine waste geopolymeric binder. In-fluence of OPC concrete substrate surface treatment. *Constr. Build. Mater.* **2008**, *22*, 154–161. [CrossRef]
17. Pacheco-Torgal, F.; Castro-Gomes, J.; Jalali, S. Investigations about the effect of aggregates on strength and microstructure of geopolymeric mine waste mud binders. *Cem. Concr. Res.* **2007**, *37*, 933–941. [CrossRef]
18. Pacheco-Torgal, F.; Gomes, J.P.; Jalali, S. Properties of tungsten mine waste geopolymeric binder. *Constr. Build. Mater.* **2008**, *22*, 1201–1211. [CrossRef]
19. Zribi, M.; Baklouti, S. Phosphate-based geopolymers: A critical review. *J. Polym. Bull.* **2021**. [CrossRef]
20. Elmahdoubi, F.; Mabroum, S.; Hakkou, R.; Ibnoussina, M. Geopolymer Materials Based on Natural Pozzolans from the Moroccan Middle Atlas. *Minerals* **2021**, *11*, 1344. [CrossRef]
21. Davidovits, J. Chemistry of geopolymer systems, terminology. In Proceedings of the Geopolymer 99 International Conferences, Saint-Quentin, France, 30 June–2 July 1999.
22. Panda, B.; Mohamed, N.A.; Tan, M.J. Effect of 3D Printing on Mechanical Properties of Fly Ash-Based Inorganic Geopolymer. In Proceedings of the International Congress on Polymers in Concrete, Washington, DC, USA, 29 April–1 May 2018; pp. 509–515.
23. Kriven, W.M.; Bell, J.L.; Gordon, M.; Mallicoat, S. Microstructure and Microchemistry of Fully Reacted Geopolymers and Geopolymer Matrix Composites. *Ceram. Trans.* **2003**, *153*, 227–252.
24. Davidovits, J. The state of the Geopolimer R&D. In Proceedings of the Geopolymer Camp 2011 Conferences, Gazipur City, Bangladesh, 5–6 July 2011; Université de Picardie: Saint-Quentin, France, 2011.
25. Davidovits, J. The state of the Geopolimer. In Proceedings of the Geopolymer Camp 2010 Conferences, Gazipur City, Bangladesh, 6–7 July 2010; Université de Picardie: Saint-Quentin, France, 2010.

26. Błaszczyński, T.; Łowińska-Kluge, A. Experimental investigations and assessment of damages in the case of swimming-pool repairs. *Arch. Civ. Mech. Eng.* **2007**, *1*, 5–20. [CrossRef]
27. Gupta, S. *Strength of Flyash Based Geopolymer Concrete*; National University of Singapore: Singapore, 2009.
28. Metha, P.K. Reducing the environmental impact of concrete. *ACI Concr. Int.* **2001**, *23*, 61–66.
29. Nakshatra, B.S. Fly Ash-Based Geopolymer Binder: A future construction material. *Minerals* **2018**, *8*, 299.
30. Abdullah, M.M.; Hussin, K.; Bnhussain, M.; Ismail, K.N.; Ibrahim, W.M. Mechanism and chemical reaction of fly ash geopolymer cement—A Review. *Int. J. Pure Appl. Sci. Technol.* **2011**, *6*, 35–44.
31. Škvára, F.; Doležal, J.; Svoboda, P.; Kopecký, L.; Pawlasová, S.; Lucuk, M.; Dvořáček, K.; Beksa, M.; Myšková, L.; Šulc, R. *Concrete Based on Fly Ash Geopolymers*; IBAUSIL: Weimar, Germany, 2006.
32. Kaps, C. Geopolymer Formation via Metaclays and using Ferrihydrite. In Proceedings of the Geopolymer Camp 2010 Conferences, Gazipur City, Bangladesh, 6–7 July 2010; Université de Picardie: Saint-Quentin, France, 2010.
33. Radwan, A.B.; Sliem, M.H.; Yusuf, N.S.; Alnuaimi, N.A.; Abdullah, A.M. Enhancing the corrosion resistance of reinforcing steel under aggressive operational conditions using behentrimonium chloride. *Sci. Rep.* **2019**, *9*, 18115. [CrossRef] [PubMed]
34. Malhotra, V.M. Making concrete 'greener' with fly ash. *ACI Concr. Int.* **1999**, *12*, 61–66.
35. Szechyńska-Hebda, M.; Marczyk, J.; Ziejewska, C.; Hordyńska, N.; Mikuła, J.; Hebda, M. Optimal Design of pH-neutral Geopolymer Foams for Their Use in Ecological Plant Cultivation Systems. *Materials* **2019**, *12*, 2999. [CrossRef] [PubMed]
36. Rashad, A.M. The effect of polypropylene, polyvinyl-alcohol, carbon and glass fibres on geopolymers properties. *Mater. Sci. Technol.* **2019**, *35*, 127–146. [CrossRef]
37. Yong, H.C.; Hussin, K.; Abdullah, M.M.A.B.; Bnhussain, M.; Musa, L.; Ismail, K.N.; Ming, L.Y. Effect of alkali concentration on mechanical properties of kaolin geopolymers. *Rom. J. Mater.* **2012**, *42*, 179–186.
38. Spanou, M.; Luhar, S.; Savva, P.; Ioannou, S.; Petrou, M.F.; Luhar, I.; Nicolaides, D. Diabase Mud-Based Geopolymer Paste: Formulation and Properties. *Mater. Proc.* **2021**, *5*, 77. [CrossRef]
39. Piccolo, F.; Andreola, F.; Barbieri, L.; Lancellotti, I. Synthesis and Characterisation of Biochar-Based Geopolymer. *Mater. Appl. Sci.* **2021**, *11*, 10945. [CrossRef]
40. Pacheco-Torgal, F.; Castro-Gomes, J.P.; Jalali, S. Investigations on mix design of tungsten mine waste geopolymeric binder. *Constr. Build. Mater.* **2008**, *22*, 1939–1949. [CrossRef]
41. Regulation of the Minister of Climate of 2 January 2020 on the Waste Catalogue (Journal of Laws of 2020, Item 10). Available online: https://isap.sejm.gov.pl/isap.nsf/DocDetails.xsp?id=WDU20200000010 (accessed on 27 January 2022).
42. Skvara, R.; Šulc, R.; Tišler, Z.; Skricik, P.; Šmilauer, V.; Zlámalová Cílová, Z. Preparation and properties of fly ash-based geopolymer foams. *Ceram. Silik.* **2014**, *58*, 188–197.
43. Zhang, Z.; Provis, J.L.; Reid, A.; Wang, H. Geopolymer foam concrete: An emerging material for sustainable construction. *Constr. Build. Mater.* **2014**, *56*, 113–127. [CrossRef]
44. *EN 196-1:2016*; Methods of Testing Cement Determination of Strength. European Committee for Standardization (CEN): Brussels, Belgium, 2016.
45. *EN 1015-3:1999*; Methods of Test for Mortar for Masonry. Determination of Consistence of Fresh Mortar (by Flow Table). European Committee for Standardization (CEN): Brussels, Belgium, 1999.
46. *EN 1484:1999*; Guidelines for the Determination of Total Organic Carbon (TOC). European Committee for Standardization (CEN): Brussels, Belgium, 1999.
47. *EN 15935:2013-02*; Determination of Loss on Ignition (LOI). European Committee for Standardization (CEN): Brussels, Belgium, 2013.
48. *PN-ISO 1928:2020-05*; Determination of Calorific Value by Calorimetric Bomb Combustion and Calorific Value Calculation. Polish Committee for Standardization: Warszawa, Poland, 2020.
49. *EN 12457-2:2006*; Characterisation of Waste—Leaching—Compliance Test for Leaching of Granular Waste Materials and Sludges. Part 2: One Stage Batch Test at a Liquid to Solid Ratio of 10 l/kg for Materials with Particle Size below 4 mm (without or with Size Reduction). European Committee for Standardization (CEN): Brussels, Belgium, 2006.
50. *EN ISO 10523:2012*; Determination of pH. European Committee for Standardization (CEN): Brussels, Belgium, 2012.
51. *ISO 9297:1994*; Determination of Chloride Ion Concentration by Titration (Mohr's Method). International Organization for Standardization: Geneva, Switzerland, 1994.
52. *PN-ISO 9280:2002*; Determination of Sulphates (VI). GRAVIMETRIC Method with Barium Chloride. Polish Committee for Standardization: Warszawa, Poland, 2002.
53. *EN ISO 6878:2006*; Determination of Phosphorus. Ammonium Molybdate Spectrometric Method. European Committee for Standardization (CEN): Brussels, Belgium, 2006.
54. *PN-ISO 9964-3:1994*; Determination of Sodium, Potassium, Calcium, Lithium and Bar by Flame Photometry. Polish Committee for Standardization: Warszawa, Poland, 1994.
55. Council Decision of 19 December 2002 Establishing Criteria and Procedures for the Acceptance of Waste at Landfills Pursuant to Article 16 of and Annex II to Directive 1999/31/EC. Available online: https://eur-lex.europa.eu/legal-content/PL/TXT/PDF/?uri=CELEX:32003D0033&from=EN (accessed on 27 January 2022).
56. Żukowska, G.; Bik-Małodzińska, M.; Myszura, M.; Pawłowki, A.; Pawłowska, M. Effect of sewage sludge and mineral wool on water retention and heavy metal content in medium agronomic category soil. *Environ. Prot. Eng.* **2021**. [CrossRef]

57. Regulation of the Minister of Maritime Economy and Inland Navigation of 12 July 2019 on Substances Particularly Harmful to the Aquatic Environment and the Conditions to Be Met When Discharging Sewage into Waters or Soil, as well as Discharging Rainwater or Meltwater into Waters or for Water Devices (Journal of Laws 2019, Item 1311). Available online: https://isap.sejm.gov.pl/isap.nsf/DocDetails.xsp?id=WDU20190001311 (accessed on 27 January 2022).
58. Król, A.; Mizerna, K.; Bożym, M. An assessment of pH-dependent release and mobility of heavy metals from metallurgical slag. *J. Hazard. Mater.* **2020**, *384*, 121502. [CrossRef] [PubMed]
59. Available online: www.chemiabudowlana.info/wiadomosci,art,3397,,pcfug_fuga?PHPSESSID=1e0333a4fabcd3819a1edaed10d7cd39 (accessed on 27 January 2022).
60. Available online: www.products.kerakoll.com/pl-PL/p/geolite-40 (accessed on 27 January 2022).
61. Available online: www.astra-polska.com (accessed on 27 January 2022).

materials

MDPI

Article

Research on Dewatering Characteristics of Waste Slurry from Pipe Jacking Construction

Libing Jiang [1], Liang Zhen [2,*], Jianfeng Wang [2], Tao Zhang [2] and Xianwen Huang [3,*]

1 Water Supply and Drainage Management Office, Zhenjiang 212008, China; lbjiang1985@yahoo.com
2 Shanghai Road and Bridge (Group) Co., Ltd., Shanghai 200433, China; wjf7907@126.com (J.W.); zhangtao554@126.com (T.Z.)
3 School of Civil Engineering and Architecture, Anhui University of Science and Technology, Huainan 232001, China
* Correspondence: drzhenl@126.com (L.Z.); huangxianwen194@163.com (X.H.)

Abstract: A large amount of waste slurry is produced during the construction of pipe jacking projects. To avoid the waste slurry occupying too much urban land, it needs to be rapidly reduced. Due to the complex composition of waste slurry, the existing dewatering methods face the problem of low efficiency, and the soil after dewatering is difficult to recycle as soil materials due to high water content and low strength. There is currently a lack of research on dewatering and resource utilization of waste slurry from pipe jacking projects. In response to this problem, this paper studies the flocculation-settling characteristics of waste slurry and the mechanical properties of solidified sediment. It was found that the anionic polyacrylamide (APAM) 7126 obtained the best separation effect if the waste slurry contains bentonite, which increases the zeta potential, resulting in poor separation. Thus, $FeCl_3 \cdot 6H_2O$ and APAM 7126 can be used as compound conditioners. The sediment after settling was further added with 20–30% sulphate aluminum cement (SAC), and the unconfined compressive strength of the solidified sediment for 3 days could exceed 30 kPa. After flocculation-settling and solidification treatment, the waste pipe jacking slurry can be quickly dewatered into a soil material with a certain strength, which provides a reference for engineering applications.

Keywords: waste slurry; flocculation; solidification; pipe jacking

Citation: Jiang, L.; Zhen, L.; Wang, J.; Zhang, T.; Huang, X. Research on Dewatering Characteristics of Waste Slurry from Pipe Jacking Construction. *Materials* **2022**, *15*, 2242. https://doi.org/10.3390/ma15062242

Academic Editors: Rossana Bellopede and Lorena Zichella

Received: 6 February 2022
Accepted: 15 March 2022
Published: 18 March 2022

1. Introduction

As a trenchless technology, pipe jacking is increasingly used in the construction of urban sewage and rainwater pipe networks [1,2]. Slurry is an indispensable material in pipe jacking projects, which plays the role of balancing the pressure of the excavation surface, discharging the slag, and lubricating [1,3]. During the construction process, a large amount of waste slurry is continuously produced, which is likely to contain bentonite, carboxymethyl cellulose (CMC) and other components [1,4,5]. These components may affect the separation results, thus requiring more site on which to dispose of the slurry. With the development of urbanization, there are fewer sites that can be used as slurry storage. It is necessary to progress the rapid reduction and resource utilization of waste slurry.

Flocculation was initially used in water treatment [6–8], and was gradually applied in the dewatering of slurry, such as waste dredging slurry [9,10] and waste tailings slurry [11,12]. Suitable flocculation pretreatment can increase the floc size in slurry and improve the separation efficiency [9]. Wang et al. [13] used various types of polyacrylamide (PAM) to treat construction waste slurry. After 7 days of settlement, the water content of the slurry can be reduced to about 80%. He et al. [14] used cationic PAM to treat waste slurry and, after 10 min of setting, the volume of slurry can be reduced by 60%. However, the properties of pipe jacking slurry are special, and it is likely to contain bentonite, CMC, and so forth. These special components may have a potential impact on the flocculation and separation of pipe jacking slurry, and corresponding research needs to be carried out.

In addition, there are many types of commercial flocculants on the market—it is unclear which flocculant has the best flocculation effect and needs to be studied.

During the flocculation process of slurry, soil particles aggregate to form flocs [15], which increases the soil–water separation rate. However, due to the limitation of storage site space, the time of settlement cannot last too long. The water content of sediment remains high after a short period of settling [13,16], resulting in poor mechanical properties and cannot be used as soil materials. In some projects, plate and frame filter presses have been used to dewater the slurry, which can greatly reduce the water content of slurry [17,18]. However, the dewatering efficiency of the filter press is low in some cases, and the problem of untimely treatment often occurs [19]. Untimely reduction leads to no place to store excess waste slurry, resulting in the stagnation of construction. Solidification technology is often used to increase the strength of soft soil [20–23] and has been applied in some projects [24]. The curing agent converts the excess free water into mineral water through hydration, which improves the mechanical properties of soft soil. It may be possible to further treat the flocculated sediment by means of solidification, which could increase the strength of sediment to make it a reusable soil material. The solidification characteristics of sediment obtained by slurry flocculation and settling are still lacking, and need to be studied further.

Based on the problem of unclear flocculation, the sedimentation characteristics of waste pipe jacking slurry and the unclear strength growth law of the solidification sediment after sedimentation, this paper carried out relevant laboratory experiments. The waste slurry produced in two different stages of pipe jacking construction was selected as the experimental material. Five different conditioners were used to pretreat pipe jacking slurry, and the sedimentation characteristics were studied. For the sediment obtained by settling, two solidification agents were used to conduct solidification experiments to study mechanical properties of the solidified sediment. Based on the above tests, the flocculation-settling characteristics of pipe jacking waste slurry were studied, and the feasibility of using a flocculation–sedimentation and solidification combined method to treat waste pipe jacking slurry was discussed, to provide a reference for related projects.

2. Materials and Methods

2.1. Slurry

In order to study the flocculation–sedimentation and solidification characteristics of waste pipe jacking slurry, waste slurry from a pipe jacking project in Zhenjiang, Jiangsu Province, China, was used as the test material. The sampling of the waste slurry at the site was carried out twice. The first sampling time was on 4 January 2021 (Type I slurry), and the second was on 26 March 2021 (Type II slurry). According to the construction report of the pipe jacking project, the construction was excavated to a special stratum in March 2021. To maintain the stability of the excavation surface, more bentonite was added to the slurry. Therefore, Type II waste slurry was mixed with some bentonite. Table 1 shows the basic properties of the pipe jacking waste slurry. Particle size distribution of the slurry was measured by a laser particle size analyzer, the Malvern Mastersizer 2000. The slurry was diluted to a water content of 300%. The sample was dropped into the Hydro MU dispersion unit through an eyedropper, and was dispersed by ultrasound and stirring. The obscuration parameter was kept between 10% and 20% to ensure good quality of the signal. The pump speed was set as 2000 r/min. The particle size distribution of waste pipe jacking slurry can be seen in Figure 1. Type II slurry has a smaller particle size because it contains bentonite, with an average particle size of 7.84 μm, while the average particle size of Type I slurry is 15.78 μm.

Table 1. Basic properties of the pipe jacking slurry.

Pipe Jacking Slurry	Specific Gravity	Liquid Limit (%)	Plastic Limit (%)	Organic Matter Content (%)
Type I	2.58	42.16	22.18	4.68
Type II	2.61	71.22	32.76	4.12

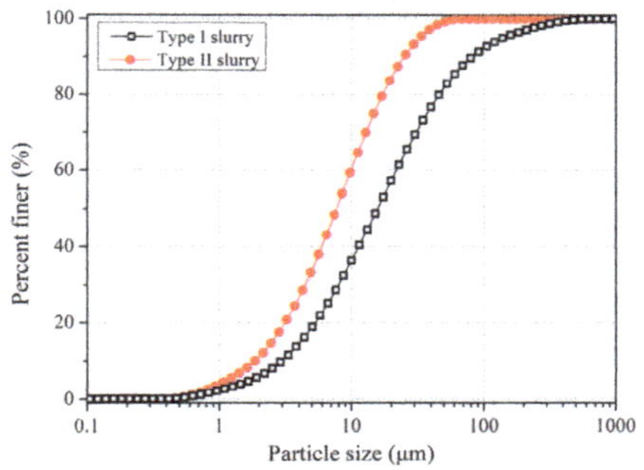

Figure 1. Particle size distribution of the pipe jacking waste slurry.

2.2. Conditioners

Four types of commercial polyacrylamide (PAM) were used for flocculation, manufactured by Shanghai Wshine Chemical Co., Ltd., Shanghai, China. Four types of PAM were recommended by the manufacturer because of their good effect in the reduction of tailings slurry. Wshinefloc 412VS and 611HN are cationic polyacrylamide (CPAM). Wshinefloc 7126 and 720VJ are anionic polyacrylamide (APAM). The specific parameters of PAM are shown in Table 2. The flocculant was reconstituted every day, at a concentration of 0.1% (w/w).

Table 2. Chemical properties of PAM used in this study.

Flocculants	PAM Characteristics	Molecular Weight	Solid Content (%)	Charge Density (meq/g)
412VS	cationic	≥10 million	≥89	1.8–2.2
611HN	cationic	≥10 million	≥91	2.5
7126	anionic	16 million	≥89	/
720VJ	anionic	12 million	≥89	/

Ferric chloride ($FeCl_3{\cdot}6H_2O$) was used as a coagulant in composite conditioning, which is produced by Shanghai Yuanye Biological Co., Ltd., Shanghai, China, and its molecular weight is 270.3 g/mol.

2.3. Solidification Agent

The 425# ordinary Portland cement (OPC) and sulphate aluminum cement (SAC) were selected as the solidification agent. SAC has a faster hydration rate than OPC, which can make the solidified soil achieve a higher strength at a shorter curing age. SAC was used to test whether it can make the sediment reach the required strength earlier in solidification experiments.

2.4. Flocculation-Settling Experiment

The water content of slurry was initially adjusted to 300% by adding tap water, which was used to simulate the initial state of slurry produced by pipe jacking construction. Afterwards, flocculation tests were carried out. The flocculation scheme is shown in Table 3.

The flocculation test was carried out in a 500 mL beaker. After adding flocculant, the slurry was stirred at 450 rpm for 2 min. The slurry after flocculation was then subjected to a sedimentation test. The clear sediment–water interface was used to evaluate the result of separation during settling, which was recorded by the scale value on the side wall of the beaker. The sediment below the sediment–water interface is considered to be the product after slurry separation. The sedimentation test lasted for 600 s, representing a short period of concentration. As a comparative test, the sedimentation test of the slurry without adding flocculant was also carried out.

Table 3. Conditioning scheme of waste slurry.

Pipe Jacking Slurry	Water Content (%)	Conditioners	PAM Characteristics	Dosage [1] (%)
Type I	300	412VS	cationic	0.10, 0.15, 0.20
Type I	300	611HN	cationic	0.10, 0.15, 0.20
Type I	300	7126	anionic	0.03, 0.06, 0.09
Type I	300	720VJ	anionic	0.04, 0.07, 0.10
Type II	300	412VS	cationic	0.60, 0.65, 0.70
Type II	300	611HN	cationic	0.55, 0.60, 0.65
Type II	300	7126	anionic	0.20, 0.25, 0.30
Type II	300	720VJ	anionic	0.20, 0.25, 0.30
Type II	300	$FeCl_3$ + 0.25% 412VS	cationic	1, 3, 5 [2]
Type II	300	$FeCl_3$ + 0.10% 7126	anionic	1, 3, 5

[1] The dosage is calculated as the ratio of the dry mass of conditioner to the dry mass of slurry. The dosage is determined by pre-experiment. [2] 1%, 3%, and 5% are the dosage of $FeCl_3{\cdot}6H_2O$, which are determined by pre-experiment.

During the pre-experiment, it was found that the flocculation-settling effect of Type II slurry was very poor. Thus, two types of composite conditioning experiments were designed, as shown in Table 3. Type II slurry was first added with $FeCl_3{\cdot}6H_2O$ and was stirred at 450 rpm for 2 min, and then added with PAM and stirred at 450 rpm for 2 min to test sedimentation results.

2.5. Solidification Experiment

After the flocculation-settling experiment, the sediment with the lowest water content after settling was subjected to the solidification experiment. The supernatant of the experimental group was poured out, and a solidification agent was added to the sediment for stirring, wherein the stirring speed was 450 rpm, and the stirring was performed for 2 min. Subsequently, the sediment was layered into the mold. The inner diameter and height of the mold are 39.1 mm and 80.0 mm, respectively.

According to results with PAM pretreatment in Section 3, the sediment pretreated with APAM 7126 (0.25%) had the lowest water content. For the composite pretreatment experiments, the combined addition of $FeCl_3{\cdot}6H_2O$ (3%) and APAM 7126 (0.10%) obtained the sediment with lower water content. Therefore, the sediment obtained from these two sets of experiments was used for solidification experiments. The solidification scheme is shown in Table 4.

Table 4. Solidification scheme of sediment.

Conditioners	Solidification Agent	Dosage [1] (%)	Curing Time (day)
0.25% 7126	OPC	20, 30, 40	3, 7, 28
0.25% 7126	SAC	20, 30, 40	3, 7, 28
3% $FeCl_3$ + 0.10% 7126	OPC	20, 30, 40	3, 7, 28
3% $FeCl_3$ + 0.10% 7126	SAC	20, 30, 40	3, 7, 28

[1] The dosage is calculated as the ratio of the dry mass of solidification agent to the dry mass of slurry.

The moist room YH-60B type (produced by Beijing central North Road Instrument Equipment Co., Ltd., Beijing, China) was used in the solidification experiment. The curing conditions were a temperature of 20 °C and humidity of >95%. After a set curing time, an unconfined compressive strength experiment was performed.

2.6. Unconfined Compressive Strength Experiment

After the designated curing time was reached, specimens were tested for the unconfined compressive strength (q_u). A strain-controlled apparatus (YYW-2 type, Nanjing Soil Instrument Factory CO., LTD., Jiangsu, China) was used to test unconfined compressive strength. Each test was performed thrice in parallel.

2.7. Zeta Potential Experiment

After the settling experiment, the supernatant was taken out to test the zeta potential of the slurry after flocculation pretreatment. Zeta potential is often used to characterize the electrical characteristics of particles and can be used to reveal the degree to which particles are affected by conditioners. The zeta potential was measured through electrophoretic light scattering by using a Zetasizer Nano ZSP (Malvern Instruments, Malvern, UK). DTS1070 cell was used to perform the zeta potential test. The supernatant was injected into the cell, then the cell was pressed into the measurement chamber to test the zeta potential. Each test was performed thrice in parallel.

3. Results

3.1. Sedimentation Results without Flocculation

The pipe jacking waste slurry with an initial water content of 300% was subjected to a self-weight sedimentation test. Figure 2 demonstrates that the sedimentation and separation of two types of slurry without pretreatment are very slow, and the water content of slurry drops from 300% to 292% and 296% respectively within the settling time of 600 s. It can be inferred that if the pipe jacking waste slurry is not treated, it could occupy a large area of the site for storage. Most pipe jacking projects are in the underground areas of a city. It is obviously difficult to find a large storage space in the city to store the slurry. Therefore, it is meaningful to quickly reduce the amount of slurry.

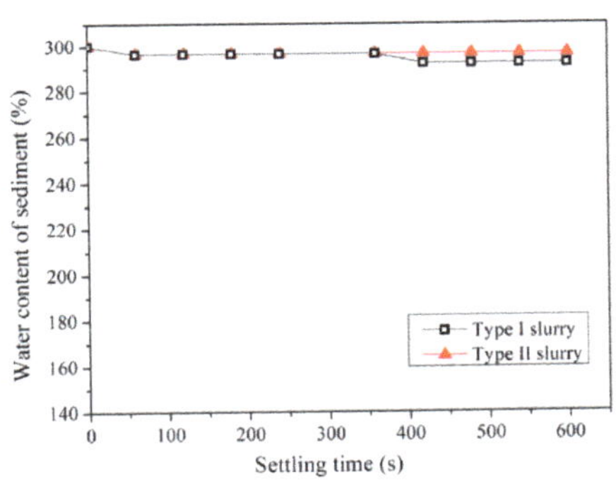

Figure 2. Sedimentation results of pipe jacking waste slurry without flocculation.

3.2. Sedimentation Results with PAM Pretreatment

Figures 3 and 4 show the sedimentation results of the waste pipe jacking slurry after adding CPAM. Compared with the slurry without pretreatment, the addition of two types of CPAM can make the waste slurry settle and separate rapidly.

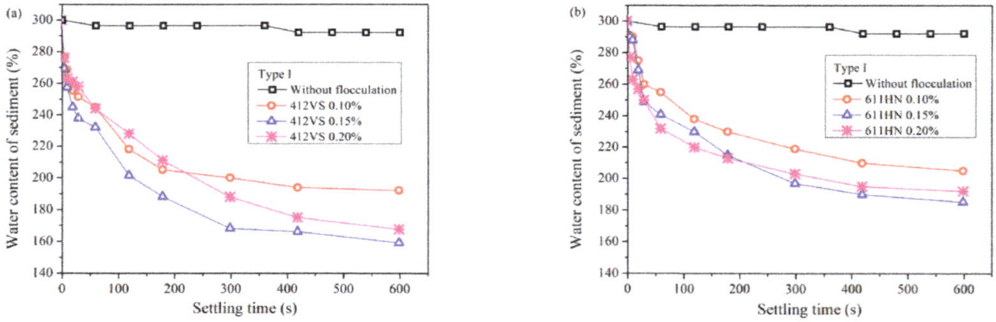

Figure 3. Sedimentation results of slurry (Type I) after pretreatment with CPAM. (**a**) 412VS. (**b**) 611HN.

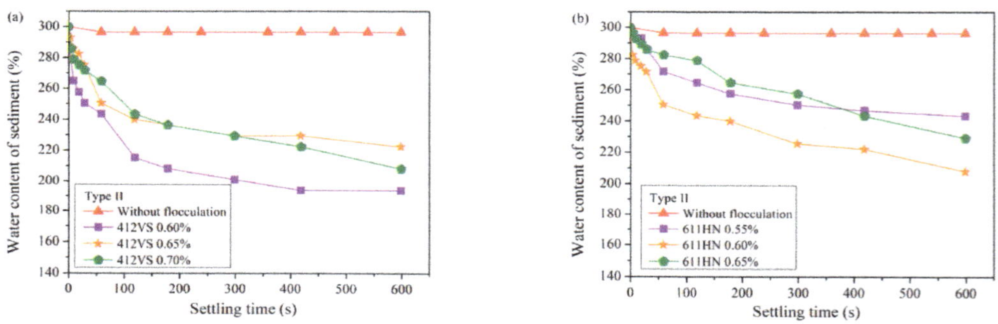

Figure 4. Sedimentation results of slurry (Type II) after pretreatment with CPAM. (**a**) 412VS. (**b**) 611HN.

Figure 3 shows the settling results of Type I slurry. The optimal dosage of CPAM 412VS was 0.15%, and the water content of the sediment dropped rapidly from 300% to 159% within the settling time of 600 s. The CPAM 611HN also obtained good treatment results. The optimum dosage was 0.15%, and the water content of the sediment finally dropped to 185%.

Figure 4 reveals the settling results of the Type II pipe jacking waste slurry. Compared with Type I slurry, the sedimentation effect of Type II slurry containing bentonite was worse, and a higher dosage of CPAM was required for pretreatment. Using the same type of CPAM, the optimal addition amount was increased by three times, and the optimal dosage of 412VS and 611HN were both increased to 0.60%. In addition, in the same settling time, the water content of the sediment was higher, and the water content of the sediment after pretreatment with the two types of flocculants was 194% and 208%, respectively.

Comparing two types of CPAM, 412VS has a better pretreatment effect than 611HN, and the water content of the sediment can be reduced to a lower level under the same addition amount.

Figures 5 and 6 show the sedimentation results after pretreatment of waste pipe jacking slurry using APAM. The addition of APAM also increased the speed of soil–water separation. Unlike CPAM, the optimal dosage of APAM was much lower than that of CPAM, no matter for Type I slurry or Type II slurry.

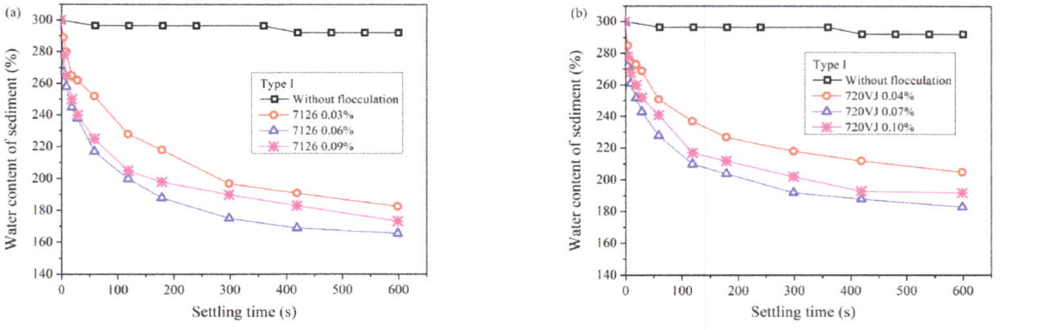

Figure 5. Sedimentation results of slurry (Type I) after pretreatment with APAM. (**a**) 7126. (**b**) 720VJ.

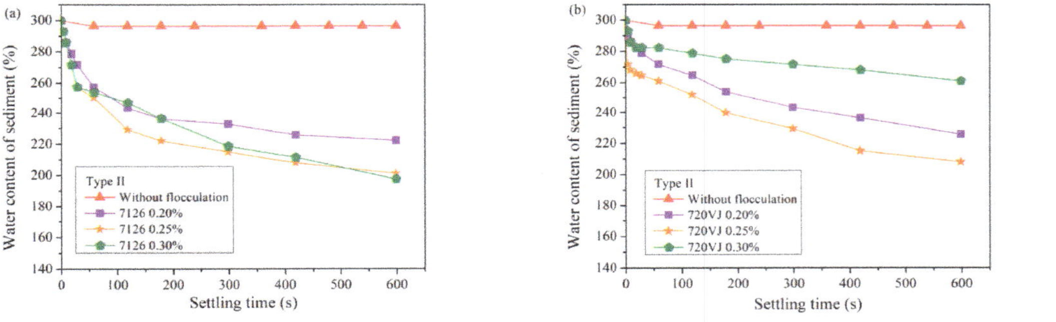

Figure 6. Sedimentation results of slurry (Type II) after pretreatment with APAM. (**a**) 7126. (**b**) 720VJ.

Figure 5 reveals the sedimentation results of Type I slurry. The optimal addition amount of APAM 7126 was 0.06%. After 600 s of sedimentation, the water content of sediment finally dropped to 166%. The optimum dosage of APAM 720VJ was 0.07%, and the final water content of the sediment was 183%.

Figure 6 shows the settling results of bentonite-rich slurry (Type II slurry). Similar to the results of CPAM pretreatment, the treatment of Type II slurry with APAM requires a larger amount of flocculant than Type I slurry. The optimum addition amount of APAM 7126 reached 0.25%, and the final water content of the sediment dropped to 201%. The optimum dosage of APAM 720VJ reached 0.25%, and the water content of the sediment was 208%.

3.3. Sedimentation Results with Compound Pretreatment

Figure 7 reveals the sedimentation results of Type II slurry after compound pretreatment. Comparing Figure 7a with Figure 4a, after the compound treatment of $FeCl_3 \cdot 6H_2O$ and CPAM 412VS, the settling and separation effect became better. After adding 3% $FeCl_3 \cdot 6H_2O$, the optimal addition of CPAM 412VS decreased from 0.60% to 0.25%. The water content of the sediment can be reduced to 169% in 600 s, which is lower than the result of the single addition of CPAM 412VS.

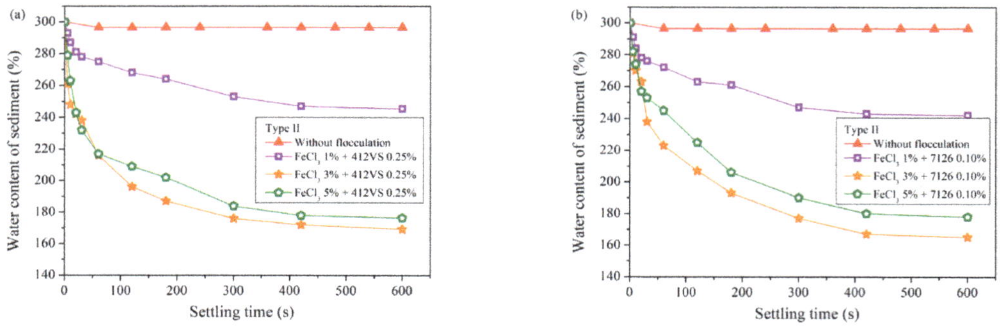

Figure 7. Sedimentation results of slurry (Type II) after compound pretreatment. (**a**) $FeCl_3$ + 412VS. (**b**) $FeCl_3$ + 7126.

Similarly, Type II slurry pretreated with $FeCl_3{\cdot}6H_2O$ and APAM 7126 also obtained better separation results, and the optimal combination was 3% $FeCl_3{\cdot}6H_2O$ and 0.10% APAM 7126. Comparing Figure 7b with Figure 6b, after the compound pretreatment, the water content of the sediment can be reduced to 165% in 600 s, and the added amount of the APAM 7126 used was also reduced from 0.25% to 0.10%.

3.4. Solidification Results

In experiments where the slurry was only pretreated with PAM, the slurry treated with APAM 7126 achieved the best settling effect. The sediment obtained by settling had a lower water content, so the sediment was used for the solidification experiment. The cured samples were tested for unconfined compressive strength and the results are shown in Figure 8. Figure 8a shows the unconfined compressive strength of cured sediments with different dosages of OPC at different curing ages. With the increase of curing age, the unconfined compressive strength increased continuously. When the OPC dosage was 30%, the strength was significantly improved, the 7-day strength was 49.23 kPa, and the 28-day strength was 83.09 kPa. In the initial stage, the strength of the 3-day was only 8.95 kPa.

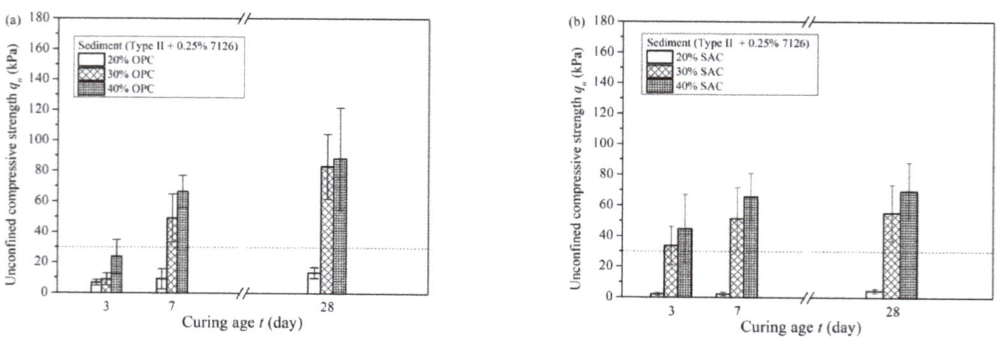

Figure 8. Unconfined compressive strength of the solidified sediment after APAM pretreatment. (**a**) OPC. (**b**) SAC.

Figure 8b shows the unconfined compressive strength of sediments added with SAC at various curing ages. Compared with OPC, SAC could improve the strength of the cured sediment in less curing time. For the solidified sediment with 30% SAC, the 3-day strength was 33.70 kPa, and the strength nearly reached the peak value of 51.46 kPa after curing for 7 days, and the strength was 55.30 kPa at the age of 28 days.

The slurry had better sedimentation results after the compound pretreatment of $FeCl_3{\cdot}6H_2O$ and APAM 7126, and the sediments were also subjected to solidification experiments. Figure 9 shows the unconfined compressive strength of the sediments. Compared with the samples obtained after the pretreatment with APAM 7126, the samples with the compound pretreatment have higher strength under the same amount of curing agent. For example, when OPC was used as the curing agent, the unconfined compressive strength at 30% was 15.2 kPa at 3 days, 55.8 kPa at 7 days, and 100.44 kPa at 28 days. In contrast, SAC can make the sample have a certain strength earlier under the same addition amount, as shown in Figure 9b. When the dosage of SAC was 30%, the strength reached 45.80 kPa, 64.2 kPa and 64.1 kPa at 3 days, 7 days, and 28 days, respectively.

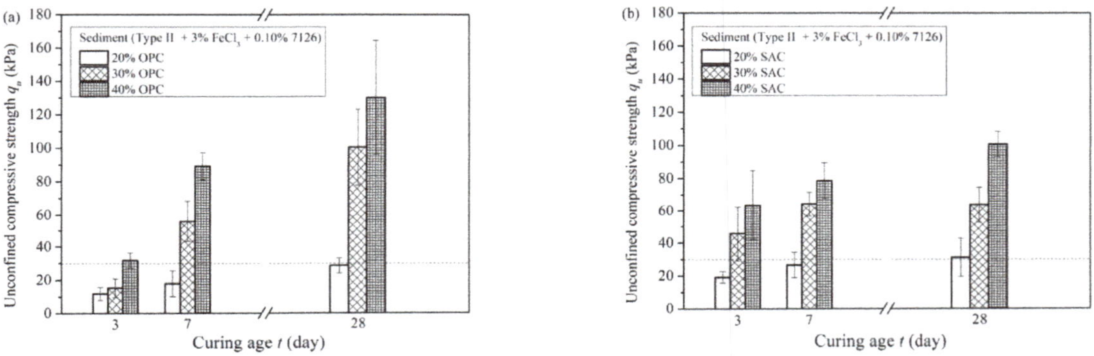

Figure 9. Unconfined compressive strength of the solidified sediment after compound pretreatment. (**a**) OPC. (**b**) SAC.

In summary, the sediment separated by sedimentation has a certain strength after solidification. Both types of curing agents can improve the strength of sediment and modify the sediment into a soil material with a certain strength. The use of SAC as a curing agent can rapidly improve the mechanical properties of solidified sediment in a short period of time. The unconfined compressive strength of cured sediment is greater than 30 kPa in 3 days.

3.5. Zeta Potential after Pretreatment

The supernatants of all sedimentation experiments were used to test the zeta potential, and the test results are shown in Figure 10. The initial zeta potentials of Type I slurry and Type II slurry were quite different, which were −15.4 mV and −31.6 mV, respectively. Figure 10a,b show the zeta potentials of the slurry pretreated with CPAM and APAM, respectively. After adding CPAM or APAM, the zeta potential of the slurry did not change significantly, which indicated that the four types of polymer flocculants used in the experiment did not basically change the electrical properties of the particles in the slurry. Especially when APAM was added, the zeta potential of the slurry was basically unchanged. Combining with Table 2, this could be because APAM has a relatively low charge density, and the electrical properties of the particles in the slurry basically do not change.

Figure 10c shows the results of the zeta potential of the waste pipe jacking slurry treated with $FeCl_3{\cdot}6H_2O$ and PAM. With the addition of $FeCl_3{\cdot}6H_2O$, the zeta potential of the slurry changed significantly. The zeta potential gradually approached 0 from −31.6 mV, and when the addition of $FeCl_3{\cdot}6H_2O$ reached 5%, the zeta potential changed to about 10 mV.

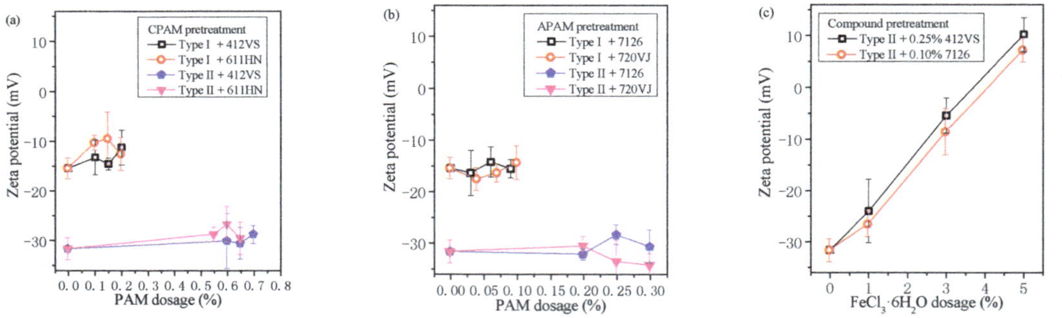

Figure 10. Zeta potential after pretreatment. (**a**) CPAM. (**b**) APAM. (**c**) $FeCl_3$ + PAM.

4. Discussion

4.1. Flocculation-Settling Characteristics of Pipe Jacking Waste Slurry

If the waste slurry from pipe jacking is not pretreated, it can be seen from Figure 2 that the sedimentation effect is very poor, and it is difficult to separate soil and water. When PAM is used as a pretreatment agent, all four flocculants can significantly improve the efficiency of slurry self-weight separation from Figures 3–6. The water content of slurry can be rapidly reduced from 300% to 159–208%. For Type I slurry, the optimal flocculant is APAM 7126, and the optimal addition is 0.06%. For Type II slurry, APAM 7126 is also the optimal flocculant, but the optimal addition is 0.25%. The properties of waste slurry produced in different periods of the same pipe jacking project are different, resulting in great differences in the optimal dosage of flocculant and sedimentation results.

From the test results of zeta potential in Section 3.5, the zeta potential of Type II slurry after PAM pretreatment is about −30 mV, and the corresponding optimal dosage of PAM is high. The zeta potential of Type I slurry after PAM pretreatment is about −15 mV, and the optimal additional amount of PAM is low.

The zeta potential of the slurry changes dramatically when the slurry is pretreated with compound conditioners (Figure 10c), and the optimal additional amount of PAM decreases. It can be inferred that the zeta potential of the waste slurry from pipe jacking is closely related to the sedimentation and flocculation results of the slurry. Therefore, the above experiments pretreated with PAM are analyzed, and the relationship between the water content of sediment and zeta potential after sedimentation is discussed, as shown in Figure 11.

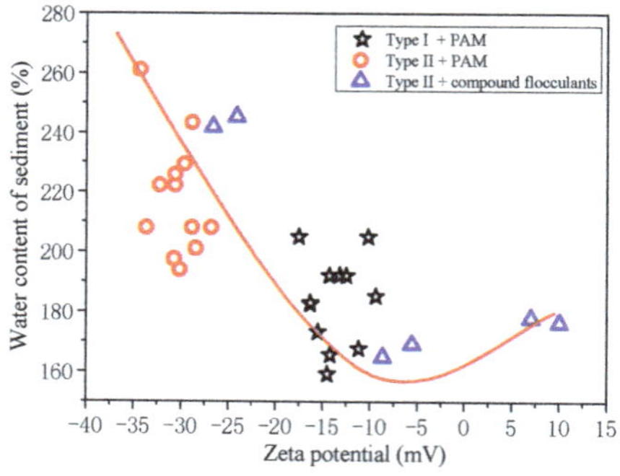

Figure 11. Relationship between water content of sediment and zeta potential.

Figure 11 clearly reveals that the water content of sediment is significantly correlated with zeta potential. After flocculant pretreatment, the closer the zeta potential is to 0, the lower the water content of the sediment is. The initial zeta potential of Type II slurry is high. After PAM pretreatment, the zeta potential is still about −30 mV, so the water content of the sediment is high. The initial zeta potential of Type I slurry is relatively low. After PAM pretreatment, the water content of sediment obtained by sedimentation is relatively low. Therefore, it is speculated that Type II slurry contains more bentonite, which increases the zeta potential of the slurry, making it difficult to flocculate and separate. In other studies [9,25], slurries rich in montmorillonite were also found to be less effective at flocculation, explained by lower zeta potential, which confirmed the findings of this paper. According to these test results, it can be inferred that the clay minerals contained in the waste slurry produced by the construction of different formations is different, and it is necessary to further study the influence of clay minerals on flocculation in detail in the future.

It is worth noting that, for Type II slurry, after pretreatment with compound conditioners, the zeta potential of the slurry changes rapidly and the water content of the corresponding sediment decreases rapidly.

Based on the above discussion, the flocculation mechanism of different PAM can be further discussed. All four PAM shown in Figures 3–6 can accelerate the settlement of slurry. However, the addition of PAM has a very small change in the zeta potential of the slurry in Figure 10. Therefore, it can be speculated that electrical neutralization is not the main mechanism of the four PAM. The initial zeta potential of the slurry is negative, but both APAM can make slurry flocculate and settle rapidly, indicating that bridging plays a major role in flocculation. APAM 7126 achieves a better sedimentation effect than APAM 720VJ; the higher molecular weight of 7126 could be the main reason, which confirms that bridging is the main flocculation mechanism. Two CPAM have a certain charge density, and compared with the two APAMs, they have a better performance in adjusting the zeta potential as shown in Figure 10. However, the optimal amount of flocculant added is much higher than that of APAM, and its lower molecular weight may be the main reason. These results confirm that bridging is the main mechanism for PAM to flocculate particles.

The zeta potential of the slurry decreases rapidly after the addition of $FeCl_3 \cdot 6H_2O$, which indicates that the $FeCl_3 \cdot 6H_2O$ significantly increases the strength of the ions in the slurry, thereby reducing the electrostatic repulsion between soil particles. After the subsequent addition of PAM, the PAM can be better interacted with soil particles in the slurry, thereby improving the effect of flocculation and separation. Therefore, under the action of compound conditioning, the electric neutralization and bridging mechanism work together, which greatly reduces the amount of PAM added. It can be seen from Figure 10c that when excess $FeCl_3 \cdot 6H_2O$ is added, the zeta potential begins to increase again, and the electrostatic repulsion between particles increases, and the effect of flocculation begins to decrease again.

In summary, in the case of slurry with difficulty in separation, compound conditioning can be used for pretreatment to reduce zeta potential and improve separation efficiency.

4.2. New Method for Rapid Dewatering

Aiming at the problem that a large amount of waste pipe jacking slurry with high water content needs to be treated and utilized, a new method based on the combination of flocculation–sedimentation and solidification is proposed; the schematic diagram is shown in Figure 12.

According to the amount of bentonite added to the slurry during the construction of pipe jacking projects, $FeCl_3 \cdot 6H_2O$ is added appropriately to adjust the zeta potential of slurry. Then, the waste slurry is pretreated with APAM (7126), and after a short period of sedimentation (600 s or more), the water content of the sediment would be less than 165%. It is estimated that, after 600 s of sedimentation and separation, the volume of slurry is reduced to 60% of the original, which greatly reduces the occupation of the storage site by the waste slurry. The separated sediment is cured by adding a solidification agent. According to Section 3.4, adding 20–30% SAC can achieve an unconfined compressive strength of more than 30 kPa within 3 days, which meets the requirements of walking and vehicle transportation. The cured sediment can be transported to other areas for disposal and utilization.

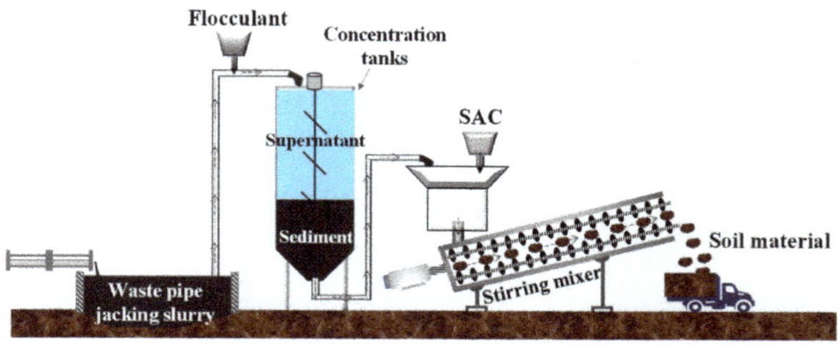

Figure 12. Schematic diagram of flocculation-sedimentation and solidification combined method.

The cost of flocculant and solidification agent by this method can be estimated. The cost depends on the nature of the waste slurry. If more bentonite remains in the waste slurry, the dewatering and reduction of the slurry will become difficult, and the addition of flocculant and curing agent will increase, resulting in the increase of the cost. Considering the most unfavorable conditions, such as Type II slurry, $FeCl_3 \cdot 6H_2O$ and APAM 7126 are selected as conditioners and SAC as a solidification agent. The price of conditioners and solidification agent per cubic slurry is 1.19 € and 3.02 € respectively, and the total is 4.21 €. It can be seen that the cost of adding SAC is slightly higher, which is mainly because the water content of the sediment after sedimentation is still high, which leads to the need to add more SAC to ensure the strength of solidified sediment. Therefore, if the slurry storage site in the project allows, the sedimentation time could be extended, thereby further reducing the water content of the sediment, which can reduce the amount of SAC added and reduce the construction cost.

It is worth noting that the addition of $FeCl_3 \cdot 6H_2O$ results in the presence of some chloride ions in the solidified sediment, and chloride ion erosion can lead to the deterioration of cement-based materials and reduce the strength of the soil [26]. The effect of chloride ions on solidified sediments, as well as finding other agents to replace ferric chloride, will be the subject of future research.

5. Conclusions

(1) A method for the rapid reduction and reuse of waste slurry from pipe jacking projects has been studied and proposed. Through the combined treatment of flocculation–sedimentation and solidification, the water content of the waste slurry can be reduced to about 159% in a very short time (10 min), and it can be quickly transformed into a soil material with a strength greater than 30 kPa.

(2) The properties of waste slurry from pipe jacking change with the change of excavated stratum, which affects the results of flocculation and sedimentation. The zeta potential of waste slurry containing bentonite is high, which leads to the poor result of flocculation. Through the compound pretreatment of $FeCl_3 \cdot 6H_2O$ and APAM, the zeta potential of slurry can be reduced, and better dewatering results can be obtained.

(3) The conditioners and solidification agents suitable for the treatment of waste pipe jacking slurry were proposed. It was found that the optimal flocculation effect was achieved when APAM 7126 (0.06–0.10%) was added. If the waste slurry contains bentonite, 3% $FeCl_3 \cdot 6H_2O$ can be added in advance. For sediment after flocculation and sedimentation, adding 20–30% SAC can greatly improve the strength of solidified sediment in a short period of time, which is convenient for subsequent treatment and utilization.

Author Contributions: Conceptualization, L.J. and L.Z.; methodology, J.W. and X.H.; software, J.W.; validation, L.J., L.Z. and T.Z.; formal analysis, T.Z. and X.H.; investigation, T.Z.; resources, T.Z.; data curation, L.J.; writing—original draft preparation, L.J.; writing—review and editing, L.J.; visualization, L.J.; supervision, L.Z.; project administration, L.Z.; funding acquisition, J.W. and X.H. All authors have read and agreed to the published version of the manuscript.

Funding: This research was funded by Shanghai Rising-Star Program(No. 19QB1401800) and supported by China Scholarship Council.

Institutional Review Board Statement: Not applicable.

Informed Consent Statement: Not applicable.

Data Availability Statement: The data presented in this study are available on request from the corresponding author. The data are not publicly available due to data which also forms part of an ongoing study.

Conflicts of Interest: The authors declare no conflict of interest.

References

1. Zhang, P.; Ma, B.; Zeng, C.; Xie, H.M.; Li, X.; Wang, D.W. Key techniques for the largest curved pipe jacking roof to date: A case study of Gongbei tunnel. *Tunn. Undergr. Space Technol.* **2016**, *59*, 134–145. [CrossRef]
2. Cheng, W.C.; Ni, J.C.; Shen, J.S.; Huang, H.W. Investigation into factors affecting jacking force: A case study. *ICE Proc. Geotech. Eng.* **2017**, *170*, 322–334. [CrossRef]
3. Ji, X.B.; Zhao, W.; Ni, P.P.; Barla, M.; Han, J.Y.; Jia, P.J.; Chen, Y.; Zhang, C.Z. A method to estimate the jacking force for pipe jacking in sandy soils. *Tunn. Undergr. Space Technol.* **2019**, *90*, 119–130. [CrossRef]
4. Reilly, C.C.; Orr, T.L. Physical modelling of the effect of lubricants in pipe jacking. *Tunn. Undergr. Space Technol.* **2017**, *63*, 44–53. [CrossRef]
5. Yao, Z.S.; Wang, C.; Xue, W.P.; Zhang, P.; Fang, Y. Experimental study on the dynamic mechanical properties of high-performance hybrid fiber-reinforced concrete of mine shaft lining. *J. Mater. Res. Technol.* **2021**, *14*, 888–900. [CrossRef]
6. Lee, K.E.; Morad, N.; Teng, T.T.; Poh, B.T. Development, characterization and the application of hybrid materials in coagulation/flocculation of wastewater: A review. *Chem. Eng. J.* **2012**, *203*, 370–386. [CrossRef]
7. Igwegbe, C.A.; Ighalo, J.O.; Onukwuli, O.D.; Obiora-Okafo, I.A.; Anastopoulos, I. Coagulation-flocculation of aquaculture wastewater using green coagulant from Garcinia kola seeds: Parametric studies, kinetic modelling and cost analysis. *Sustainability* **2021**, *13*, 9177. [CrossRef]
8. Maćczak, P.; Kaczmarek, H.; Ziegler-Borowska, M.; Węgrzynowska-Drzymalska, K.; Burkowska-But, A. The use of chitosan and starch-based flocculants for filter backwash water treatment. *Materials* **2022**, *15*, 1056. [CrossRef]
9. Wu, S.L.; Zhu, Y.Y.; Zhu, W.; Shu, S.; Cui, Y. A study on the differences in the flocculation of dredged slurries and the influencing mechanisms. *Water Sci. Technol.* **2019**, *80*, 1751–1762. [CrossRef]
10. Song, Z.Z.; Zhang, W.J.; Gao, H.Y.; Wang, D.S. Comprehensive assessment of flocculation conditioning of dredged sediment using organic polymers: Dredged sediment dewaterability and release of pollutants. *Sci. Total Environ.* **2020**, *739*, 139884. [CrossRef]
11. Botha, L.; Soares, J.B. The influence of tailings composition on flocculation. *Can. J. Chem. Eng.* **2015**, *93*, 1514–1523. [CrossRef]
12. Wang, D.L.; Zhang, Q.L.; Chen, Q.S.; Qi, C.C.; Feng, Y.; Xiao, C.C. Temperature variation characteristics in flocculation settlement of tailings and its mechanism. *Int. J. Miner. Metall. Mater.* **2020**, *27*, 1438–1448. [CrossRef]
13. Wang, D.X.; Di, S.J.; Wu, L.F.; Tan, Y.Z.; Tang, Y.K. Sedimentation behavior of organic, inorganic, and composite flocculant-treated waste slurry from construction works. *J. Mater. Civil Eng.* **2021**, *33*, 04021134. [CrossRef]
14. He, J.; Chu, J.; Tan, S.K.; Vu, T.T. Sedimentation behavior of flocculant-treated soil slurry. *Mar. Georesources Geotechnol.* **2017**, *35*, 593–602. [CrossRef]
15. Wei, H.; Gao, B.Q.; Ren, J.; Li, A.M.; Yang, H. Coagulation/flocculation in dewatering of sludge: A review. *Water Res.* **2018**, *143*, 608–631. [CrossRef]
16. Wang, S.; Song, X.P.; Wang, X.J.; Chen, Q.S.; Qin, J.C.; Ke, Y.X. Influence of coarse tailings on flocculation settlement. *Int. J. Miner. Metall. Mater.* **2020**, *27*, 1065–1074. [CrossRef]
17. Cui, Y.; Zhu, W.; Wu, S.L.; Liu, J.M.; Hou, H.; Lin, N.X. The role of lime in dredged mud dewatered by a plate and frame filter press and potential substitutes. *Environ. Sci. Pollut. Res. Int.* **2021**, *28*, 17331–17342. [CrossRef]
18. Ding, A.; Qu, F.S.; Liang, H.; Guo, S.D.; Ren, Y.H.; Xu, G.R.; Li, G.B. Effect of adding wood chips on sewage sludge dewatering in a pilot-scale plate-and-frame filter press process. *RSC Adv.* **2014**, *4*, 24762–24768. [CrossRef]
19. Oh, J.Y.; Park, H.; Chang, S.; Lee, S. Enhancement of slurry treatment efficiency in slurry shield TBM tunneling by optimal application of coagulant agent. In Proceedings of the 2017 World Congress on Advances in Structural Engineering and Mechanics (ASEM17), Seoul, Korea, 28 August–1 September 2017.
20. Yoobanpot, N.; Jamsawang, P.; Poorahong, H.; Jongpradist, P.; Likitlersuang, S. Multiscale laboratory investigation of the mechanical and microstructural properties of dredged sediments stabilized with cement and fly ash. *Eng. Geol.* **2020**, *267*, 105491. [CrossRef]

21. Huang, X.W.; Yao, Z.S.; Cai, H.B.; Li, X.W.; Chen, H.Q. Performance evaluation of coaxial borehole heat exchangers considering ground non-uniformity based on analytical solutions. *Int. J. Therm. Sci.* **2021**, *170*, 107162. [CrossRef]
22. Reddy, V.A.; Solanki, C.H.; Kumar, S.; Reddy, K.R.; Du, Y.J. Stabilization/solidification of Zinc-and lead-contaminated soil using limestone calcined clay cement (LC^3): An environmentally friendly alternative. *Sustainability* **2020**, *12*, 3725. [CrossRef]
23. Li, Y.; Ni, W.; Gao, W.; Zhang, S.; Fu, P.; Li, Y. Study on solidification and stabilization of antimony-containing tailings with metallurgical slag-based binders. *Materials* **2022**, *15*, 1780. [CrossRef] [PubMed]
24. Sato, T.; Kato, H. Application of the pneumatic flow mixing method to land development for Central Japan International Airport. In Proceedings of the 2002 International Symposium on Underwater Technology (Cat. No.02EX556), Tokyo, Japan, 16–19 April 2002.
25. McLaughlin, R.A.; Bartholomew, N. Soil factors influencing suspended sediment flocculation by polyacrylamide. *Soil Sci. Soc. Am. J.* **2007**, *71*, 537–544. [CrossRef]
26. Liu, J.J.; Zha, F.S.; Xu, L.; Yang, C.B.; Tan, X.H. Effect of chloride attack on strength and leaching properties of solidified/stabilized heavy metal contaminated soils. *Eng. Geol.* **2018**, *246*, 28–35. [CrossRef]

materials

MDPI

Article

Aerobic Biostabilization of the Organic Fraction of Municipal Solid Waste—Monitoring Hot and Cold Spots in the Reactor as a Novel Tool for Process Optimization

Sylwia Stegenta-Dąbrowska [1], Peter F. Randerson [2] and Andrzej Białowiec [1,]*

[1] Department of Applied Bioeconomy, Wrocław University of Environmental and Life Sciences, 37a, Chełmońskiego Str., 51-630 Wrocław, Poland; sylwia.stegenta@upwr.edu.pl

[2] School of Biosciences, Cardiff University, Sir Martin Evans Building, Museum Avenue, Cardiff CF10 3AX, UK; randerson@cardiff.ac.uk

* Correspondence: andrzej.bialowiec@upwr.edu.pl

Abstract: The process of aerobic biostabilization (AB) has been adopted for treatment of the organic fraction of municipal solid waste (OFMSW). However, thermal gradients and some side effects in the bioreactors present difficulties in optimization of AB. Forced aeration is more effective than natural ventilation of waste piles, but "hot and cold spots" exist due to inhomogeneous distribution of air and heat. This study identified the occurrence of hot and cold spots during the OFMSW biostabilization process at full technical scale. It was shown that the number of hot and cold spots depended on the size of the pile and aeration rate. When the mass of stabilized waste was significantly lower and the aeration rate was two-fold higher the number of anaerobic hot spots decreased, while cold spots increased. In addition, the results indicated that pile construction with sidewalls decreased the number of hot spots. However, channelizing the airflow under similar conditions increased the number of cold spots. Knowledge of the spatial and temporal distribution of process gases can enable optimization and adoption of the OFMSW flow aeration regime. Temperature monitoring within the waste pile enables the operator to eliminate undesirable "hot spots" by modifying the aeration regime and hence improve the overall treatment efficiency.

Keywords: air flow rate; composting; municipal solid waste; monitoring; optimization; spatial distributions

Citation: Stegenta-Dąbrowska, S.; Randerson, P.F.; Białowiec, A. Aerobic Biostabilization of the Organic Fraction of Municipal Solid Waste—Monitoring Hot and Cold Spots in the Reactor as a Novel Tool for Process Optimization. *Materials* **2022**, *15*, 3300. https://doi.org/10.3390/ma15093300

Academic Editors: Rossana Bellopede and Lorena Zichella

Received: 6 April 2022
Accepted: 2 May 2022
Published: 4 May 2022

Publisher's Note: MDPI stays neutral with regard to jurisdictional claims in published maps and institutional affiliations.

1. Introduction

One of the crucial strategies for waste management both in Poland and around the world is the treatment of the organic fraction of municipal solid waste (OFMSW). Basic assumptions and objectives in the area of municipal solid waste (MSW) management, adopted both in the EU [1–3] and in national legal acts, aim to limit the use of landfilling as a means of reducing the organic fraction [4]. One of the solutions is the mechanical and biological treatment (MBT) of mixed municipal waste [5]. In Poland, at the end of 2016, there were 192 MBT plants with a total mechanical capacity of around 11 million tons of waste per year [6]. Mostly, the biological process of aerobic biostabilization (AB) has been adopted for OFMSW treatment [7]. However, optimization AB of OFMSW is difficult, due to its heterogeneity, thermal gradients, and some side effects in the bioreactors. Additionally, bioreactors treating many tons of OFMSW are poorly equipped with sensors for temperature, oxygen, or moisture, so that the plant operator has little control over most of the waste mass.

Gas flow patterns within the waste have a large influence on heat and mass transfers. Consequently, O_2 supply, moisture and temperature distribution have a large impact on the end-product quality (kinetics of biodegradation; stage of stabilization; hygienization of the compost), as well as on the environmental impact of the treatment (gaseous emissions and odors) [8]. Within the AB process, gases in the pores are heated due to microbial activity [9], which reduces air density inside the pile pores and increases partial pressure, creating a flux from the bottom to the top layers in naturally aerated piles [10]. However, the gas production rate differs with spatial gradients because of ineffective mixing (inhomogeneity) and compaction effects, resulting in concentration gradients which drive gas diffusion and transfer inside the pores [11]. The effect can be visible even in well-mixed waste, especially in municipal solids which are characterized by huge inhomogeneity. Anaerobic areas (so-called "hot spots") need to be diagnosed and eliminated because gaseous biostabilization products (CO, CO_2, CH_4) are substantial threats to both low treatment efficiency and human and environmental protection.

The technique of forced aeration has been adopted in MSW treatment, as it is more effective than natural (passive) ventilation for the aerobic metabolism of microorganisms, removal of water, and control of the temperature of the system [12]. It has been reported that the O_2 content in the air space has no significant effect on the biological degradation efficiency until it falls below 5% in the composting matrix [13]. During the aeration process O_2 content may rise above 15%, gradually decreasing again after the air blower stops, although O_2 content may support aerobic bioactivity for an extended period. Furthermore, it has been demonstrated that intermittent aeration could reduce NH_3 loss from the composting system compared to continuous aeration [14]. Our previous research showed that low O_2 concentration could also favor CO production [15,16]. Gaseous emissions during biostabilization not only reduce the compost quality, but also cause atmospheric pollution [17]. Therefore, an improved AB process is urgently needed.

Even if forced aeration is more effective than natural ventilation, "hot spots" in OFMSW exist due to its heterogeneity [15], as well as aeration rate and reactor design. The proper design and operation of a biostabilization project requires an understanding of the dynamics of biostabilization [18]. In particular, the process depends on the abundance and activity of microorganisms, which are mainly affected by temperature, moisture, readily degradable organic content, O_2 level and its diffusion in the matrix, and presence of inhibiting compounds. Without frequent turning in a static composting system, or in the absence of dynamic aeration, significant spatial differences in these parameters resulting from one-directional air flow will impact the spatio-temporal dynamics and hence the uniformity of the compost product [16,19]. An appropriate reactor, adapted to the characteristics of the waste, should not only maintain appropriate levels of O_2 and temperature, but also allow for uniform distribution within the pile [20]. Hence, monitoring the spatial and temporal distribution of pore gas concentrations is an important method for evaluating and optimizing the aeration strategy and reactor design and operation [21].

Adoption of a flow aeration regime, together with knowledge of the spatial and temporal distribution of process gases, and temperature can enable optimization and control of parameters such as temperature, O_2 or even CO. Monitoring the anaerobic "hot spots" during AB of OFMSW may be a useful tool in mitigating emissions of gaseous pollutants and optimizing the biostabilization processes.

The aim of this study was to investigate the spatial and temporal distribution of temperature and pore gas (O_2, CO_2, and CO) concentrations in relation to anaerobic "hot spots". Spatial and temporal variability of gas concentrations and temperatures were determined at full technical scale in a municipal biostabilization plant.

2. Materials and Methods

2.1. Characteristics of Organic Fraction of Municipal Solid Waste

The material used in the research was the Organic Fraction (undersize fraction < 80 mm) of Municipal Solid Waste (OFMSW) originated from Warsaw, Poland. The waste was characterized by: moisture content in accordance with Polish standard PN-EN 14346:2011, volatile solids (VS) in accordance with PN-EN 15169:2011, the Total Organic Carbon (TOC) in accordance with Polish standard PN-EN 15936:2013, pH in accordance with Polish standard PN-EN-15011-3:2001, and morphological composition in accordance with Polish standard PN-93/Z-15006. The tests were performed each time before (6 samples) and after the biostabilization process was completed (6 samples). This gave 12 samples in total.

Total mass of OFMSW loaded into the bioreactor was also recorded before and after the biostabilization process. Each OFMSW sample was collected according to the following procedure: from 4 places along the length of the reactor, in each place, ~10 samples were taken. All collected samples, from one reactor, at the same collection moment, were mixed and using the quartering method were assigned to a representative sample. Properties of the waste before and after treatment are summarized in Figure 1. Detailed analytical results are described in Section 3.1 and in Supplementary Tables S3 and S4 in our previous paper [22].

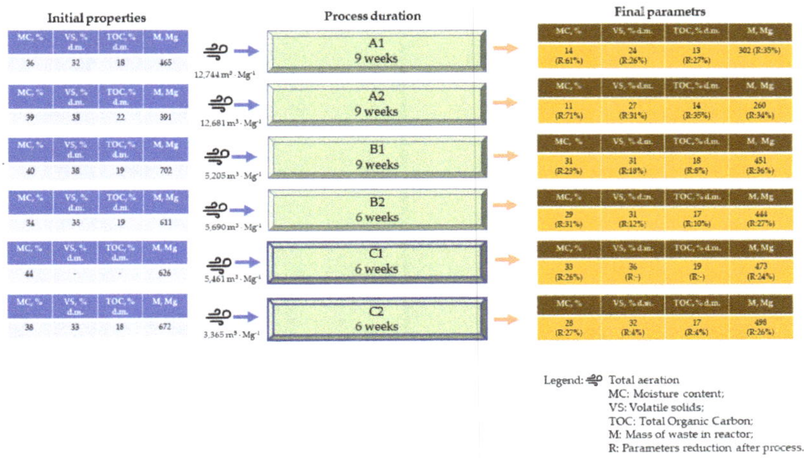

Figure 1. Process configuration and initial and final waste properties. Reactors A1, A2, B1, and B2 were covered by semipermeable membrane. Reactors C1 and C2 were constructed with concrete side walls and covered by semipermeable membrane.

2.2. Process Configuration

Experimental monitoring of the AB process of OFMSW was performed between 24 April and 9 September at the Municipal Cleaning Company in Warsaw, Poland at full industrial scale. For each of the six static pile bioreactors, dynamic aeration was provided through three aeration channels (Table 1). The procedure for pile building was to collect waste with a loader (two–three days); cover the waste with a membrane; turn on blower to start aeration process. The membrane provides full protection against the weather, including precipitation. Bioreactor pile configuration differed in terms of initial mass (350–400 Mg in piles A1 and A2, and 600–700 Mg in piles B1, B2, C1, and C2) duration of the process (six weeks in piles A1, A2, and B1 or nine weeks in piles B2, C1, and C2) as well as reactor construction (the use of concrete sidewalls in piles C1 and C2) and the total amount of air forced into the pile (Figure 1). Weekly measurements of O_2, CO_2, and CO concentrations and temperature measurements inside the piles were performed. Outside temperature measurements were also taken.

Table 1. Technical specifications of aeration fan (blower).

Parameter	Value
Rotation, $1{\cdot}min^{-1}$	3480
Back pressure, Pa	3800
Rate of delivery, $m^3{\cdot}min^{-1}$	51
Power, kW	2.64

The total experimental configurations are shown in Figure 1. Experimental details and the raw data obtained from the measurements can be found in [22].

2.3. Gas Concentrations and Temperature Measurements

Measurements were taken along the length of each reactor at four locations: 2.5 m, 17.5 m, 32.5 m, and 47.5 m from the fan (Figure 2A). At each location (cross-section of the pile), measurements were made at three heights on both sides of the pile (Figure 2B). Additionally, on one side of pile one "deep measurement" in each cross-section was made. The height and depth of sampling were determined individually for each bioreactor according to their dimensions, as shown in Table 2. Measurements of gas concentration and temperature were made with a 3.5 m long steel probe (lance), perforated at the end. As shown in Figure 2, in piles A1, A2, B1, and B2 (Figure 2B), the gas/temperature sampling probe was inserted parallel to the ground, whereas in piles C1 and C2 (Figure 2C) the lowest measurement (H1) was made at an angle of 45° to avoid the sidewall. Each measurement was done outside the insulating layer of biostabilization waste, at least 1.25 m deep. In addition, a so-called deep measurement was taken on one side of the reactor at each distance point at the middle height (Figure 2B,C). The probe was connected to an electrochemical analyzer Kigaz 300 by Kimo (Kimo Instruments, Chevry-Cossigny, France) with a plastic hose as well as a thermocouple. The procedures for gas and temperature were as follows: analyzer was started; 2-min autocalibration; placing the probe in the measuring location; waiting for concentration values stabilization (typically within ~5 min). After each measurement, the probe was removed from the biostabilization waste for ~1 min, allowing the measured gas values to return to ambient air (gas concentrations equal to 0%/ppm). O_2 and CO_2 volumetric contents in piles were measured in % ($\pm$0.1%), but CO contents were measured in volumetric ppm $\pm$1 ppm. The temperature inside the piles was measured with $\pm$ 1 °C precision. The total amount of sampling is shown in Table 3. Detailed measurements of gas concentrations and temperature in the reactors have been previously shown [22].

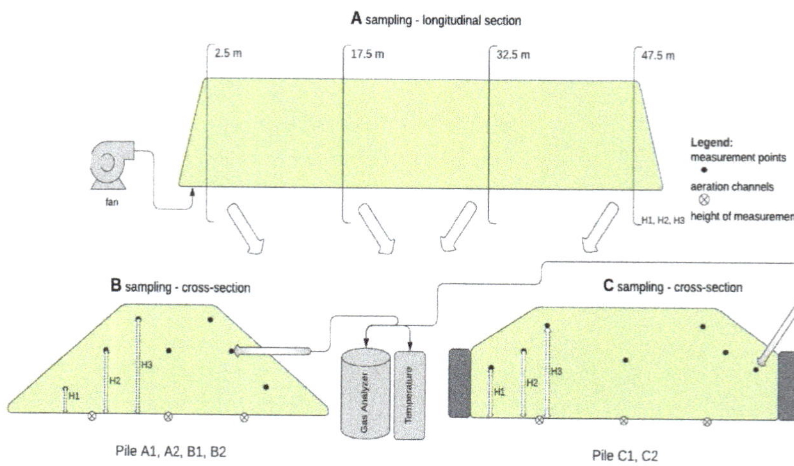

Figure 2. Process configuration and location of sampling points, (**A**) longitudinal repeats for each cross-section sampling point: (**B**) in piles A1, A2, B1, and B2; (**C**) in piles C1 and C2.

Table 2. The dimensions of tested reactors [22].

Pile	Probing Height H1, m	Probing Height H2, m	Probing Height H3, m	Reactor Height, m	Reactor Width, m	Reactor Length, m
A1	0.625	1.250	1.875	2.5	8.1	50.0
A2	0.625	1.250	1.800	2.1	8.0	50.0
B1	0.750	1.500	2.250	2.8	8.4	50.0
B2	0.750	1.500	2.250	2.9	9.0	50.0
C1	0.750	1.500	2.250	2.6	8.0	50.0
C2	0.750	1.500	2.250	2.5	8.0	50.0

Table 3. Configurations of biostabilization piles, data acquisition cycles, and numbers of collected gaseous concentration samples and temperature measurements.

Pile	Process Start Date	Process Time, Days	Reactor Design	Number of Temperature and Gas Sampling Cycles	Number of Sampling Cross-Sections	Number of Sampling Pointsin Each Cross-Section	Number of Temperature and Gas Samples Collected
A1	24 April 2015	69	Membrane-covered pile	10	4	7	280
A2	27 April 2015	66	Membrane-covered pile	9	4	7	252
B1	13 May 2015	62	Membrane-covered pile	9	4	7	252
B2	15 July 2015	44	Membrane-covered pile	6	4	7	168
C1	22 July 2015	44	Membrane-covered pile with sidewalls	6	4	7	168
C2	30 July 2015	42	Membrane-covered pile with sidewalls	6	4	7	168
Total	-	327	-	46	24	42	1288

2.4. Spatial Distribution Modeling

The Surfer 10 program (Golden Software, version 8.0, Cracow, Poland) was used to visualize the raw data recorded in Supplementary Material Table S3 [22], for spatial and temporal distributions of CO, O_2, and CO_2 gases and temperature in the piles in each cross-section and each measurement time, using a color scale.

Gas compositions and temperature values at each sample location within the piles (Figure 2 and Table 2) enabled spatial distribution modeling of each parameter, using the Natural Neighbor mathematical function method for cross-sections and Radial Basic Function for longitudinal cross-sections. Boundary conditions external to the piles were defined by the composition of atmospheric air for gases and external temperature, as well as corresponding measurements in the aeration channels below the piles (Figure 2B,C). A total of 172 Figures were generated, to represent the spatial distribution models Figures 5–8 (Supplementary Material Figures S1–S169: 1012 drawings, including 668 cross-sections and 344 longitudinal sections). Each figure presented 4 cross-sections: (a) 2.5 m, (b) 17.5 m, (c) 32.5 m, (d) 47.5 m; and 2 longitudinal sections: (e) left, (f) right. As proposed in the previous article [22], incomplete and uncertain data were rejected in the modeling process, hence the total number of Figures generated is fewer than the collected data (Table 3).

3. Results and Discussion

3.1. Waste Properties

As shown by other authors, the composition of the waste used for composting and its mixing has a great impact on the chemical and microbiological changes during the process [23,24]. In all OFMSW reactor piles, moisture contents (36–44%; Figure 1), were below the optimal range of 50–60%, according to Liang et al. (2003), but removal of moisture was more than two-times higher in piles A1 and A2, with a lower initial mass and a high rate of airflow > 12,000 $m^3 \cdot Mg^{-1}$ (Figure 1) [25]. In reactors B1, B2, C1, and C2, the reduced airflow (Figure 1), and the relatively comparable ambient temperature (Figure 3) greatly reduced the loss of moisture in the stabilized material (23–31% of initial value compared to 61–71% in piles A1/A2). Greater removal of moisture in a smaller composting pile was also obtained by Ermolaev et al. (2012) [26]. However, this was much higher than the small decrease (~10%) obtained by Mulbry and Ahn, (2014), using much smaller piles (volume ~1.9 m^3) [27]. The effectiveness of the process of moisture removal apparently depends on the amount of air blown in relation to the total waste mass of the pile.

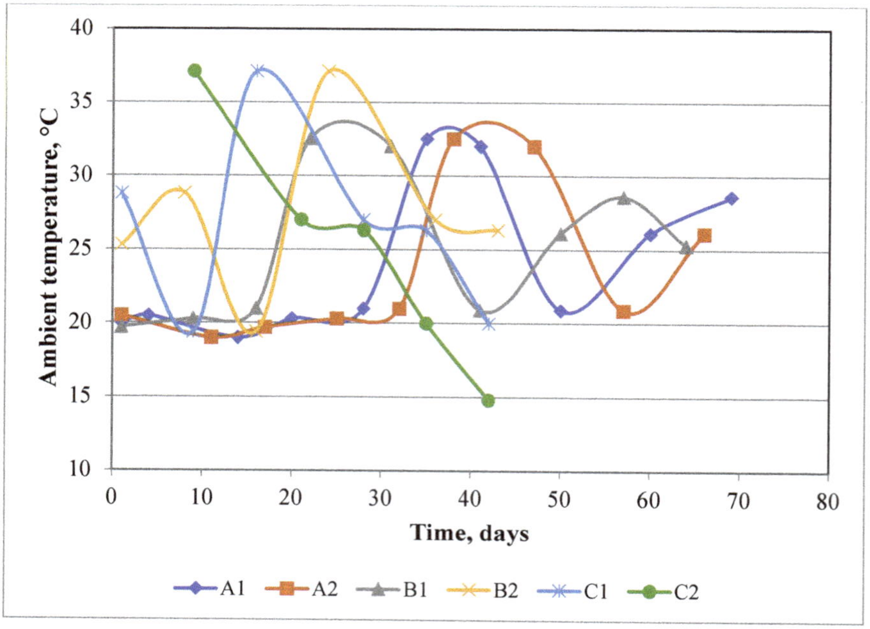

Figure 3. Ambient temperature around piles during the biostabilization process of Organic Fraction of Municipal Solid Waste (OFMSW).

Reduction in total mass of the OFMSW was greater in piles A1, A2, and B1 (34–36%), where the process time was longer (9 weeks), compared to the other piles (6 weeks). The final content of organic material (VS: 32–38% d.m.; TOC: 18–22% d.m. (Figure 1) was about half that recorded by [28] during MBT composting of municipal solid waste, and similar values were observed in Komilis et al. (2012) in a larger-scale installation (capacity 250,000 Mg·year^{-1}) [29]. The content of organic substances (both VS and TOC) is typical for MBT composting plants [30]. As with moisture removal, organic matter was most effectively reduced in piles A1/A2 with relatively small size and high airflow. Similarly, Hu et al. (2003) concluded that, for the decomposition of TOC, the moisture and fraction size have a greater influence than the process temperature [31].

The pH value of initial samples of OFMSW was very similar in all analyzed piles (5.4–5.6) (Supplementary Material Table S3 [22]), whereas at the end of the process the pH increased towards neutrality. These values and their change during the biostabilization process are typical for MSW [32].

The fine fraction < 20 mm comprises a large proportion (66–77%) of the OFMSW (Figure 4), with variable amounts, typical for municipal waste, of kitchen waste, paper and plastic and small proportions of textiles (0.3–2.2%), glass (3.8–5.3%), metals (0.8–2.9%), other organic (0.9–2.4%), other minerals (1.7–3.8%), and other materials (0.3–2.6%). After the process, a decrease in the content of kitchen waste and paper was observed in A1, A2, and B1, whereas in B2, C1, and C2, the content remained unchanged or increased, which suggests that the degradation process is more effective with a longer operation time.

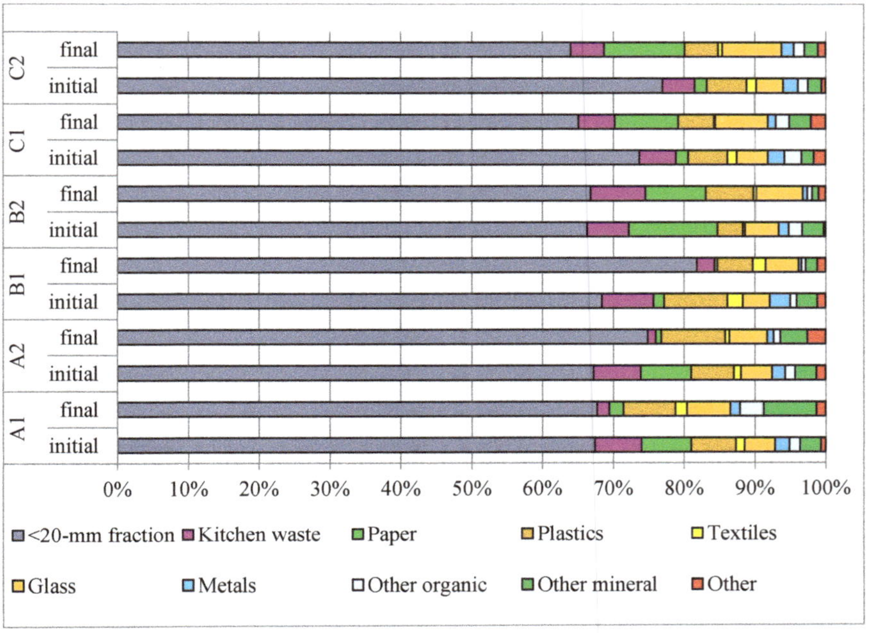

Figure 4. Changes in the morphological composition of OFMSW—samples were collected during the first and last day of the AB process and represent the entire reactor.

The < 20 mm fraction in the biostabilized material was at least 65% (Figure 4), higher than values of 50%, typical of Polish conditions [33]. In other studies, the content of organic waste was 60–67%, paper 15–17%, glass 7–8%, plastic 6–7%, and metal 3–5% [34]. The efficiency of waste mass removal, ranging from about 20 to 30% (Figure 1), was close to the average values obtained in other similar installations, operating biostabilization technology for municipal solid waste in Poland [35].

3.2. Spatial and Temporal Distribution Changes

Figures S1–S172 show the spatial and temporal changes in piles as follows: temperature (Figures S1–S41) and the concentration of O_2 (Figures S41–S85), CO_2 (Figures S86–S129) and CO (Figures S130–S169) in individual sections, at weekly intervals during the biostabilization process in individual piles. The right side of the pile was facing to the south and after the semi-permeable membrane was removed, it was exposed to direct sunlight on the day of the measurement, but measurements were taken inside the pile. No heating effect of exposure to the sun was observed inside the pile.

As the terms "hot" and "cold" spots in waste reactors were not found in the literature, they are defined here as localized areas within the pile where all three measured parameters (Table 4) were met in the same place and at the same time. All hot and cold spots recorded in this study are shown in Table 5 (cold spots) and Table 6 (hot spots). Locating these points is of great practical importance.

Hot spots—areas of elevated temperature, liable to loss of stability of the biocenosis (high temperature > 60 °C with low oxygen < 15% can cause the loss of valuable microorganisms). Cold spots—areas where conditions for sanitizing the waste by inactivating potential pathogens and parasites cannot be maintained (low temperature < 30 °C reduce the efficiency of microorganisms or induce the switch to spore formation).

Table 4. Characterization of hot and cold spots identified in this study.

Type of Spot	Parameter		
	Temperature, °C	O_2, %	CO_2, %
hot spot	>60	<15	>5
cold spot	<30	>15	<5

In either case, the quality of stabilized waste may be reduced, or the process time increased.

3.2.1. Spatial and Temporal Distribution of Temperature

The pattern of temperature change during the waste composting process was expected to follow the "classic" pattern described by Cooperband, (2002) [36] and Kowal et al. (2017) [37], with an initial rapid increase in temperature (1st phase of composting) followed by a gradual drop in temperature during maturation/cooling (2nd phase of composting). In contrast, temperatures in piles A1 and A2 (Supplementary Material Figures S1–S3, S10 and S11) remained close to ambient (Figure 3) throughout the entire prism profile (~20 °C) during the first two weeks of the process. This lag phase resulted from the high aeration rate (>12,000 $m^3 \cdot Mg^{-1}$) and the small mass weight in the reactors (300–400 Mg: ~40% less than normal for this treatment plant). From day 20, in both piles, internal temperatures increased to ~60–70 °C (Supplementary Material Figures S4, S5 and S12), three hot spots where found (high temperature coupled with low O_2: Table 5), and further small increases were observed from weeks 5 to 7. Similar temperature patterns were observed in a reactor without forced aeration by Jiang et al. (2015) [38], during composting of green waste [39], kitchen waste [40], and vermicomposting of duck manure [41]. However, Mulbry and Ahn (2014) showed that, in much larger scale piles, passive aeration allows them to heat up as quickly as during forced aeration [27].

In the other piles, an increase to 60 °C was observed at the beginning of the 2nd week. This high temperature was maintained in the piles until the end of the process ~day 50 for piles A1, A2, and B1 (Supplementary Material Figures S8, S16 and S23), ~day 35–40 for B2 and C1 (Supplementary Material Figures S29 and S35), and ~day 30 for C2 (Supplementary Material Figure S39). Temperatures in piles B1, B2, C1, and C2 followed the expected 2-phase pattern.

An interesting trend was observed in pile A1—the highest temperatures, even >70 °C, were found at the end furthest from the fan (Supplementary material section 47.5 m Figures S4d and S5c,d) from where the heat spread towards the front of the pile. This could be due to a poorer air supply to the far end of the pile, resulting in self-heating and poor heat removal. In pile A2 this did not occur (Supplementary Material Figures S13–S16), despite a similar air load and a mass of waste greater by about 15% greater. In addition, lower aeration values were noted by about 1.5 $m^3 \cdot Mg^{-1} \cdot h^{-1}$ in the first 3 weeks of the research (Supplementary Material Table S1 Stegenta-Dąbrowska et al. (2020)) [22]. The opposite effect was noticed in pile B2 (Supplementary Material Figure S30a) which showed greater cooling close to the fan on the final day. This may be the result of faster decomposition of waste located near the fan, which at the end of the process significantly reduced endothermic processes due to the lower activity of microorganisms.

At the same time a few cold spots were observed (Table 6), located mainly at the bottom of pile, due to the forced aeration. The cooling effect of aeration by the three channels below each pile is clear in the visualization (Supplementary Material Figures S13a, S47b and S134b). Typically, this occurs in both the first week of biostabilization when internal heat takes longer to accumulate, and in the final week when the need for O_2 is lower.

The observation that, in piles A1 and A2 high temperatures > 50 °C remained during the final days of the process despite the continuous high air stream (Supplementary Material Figures S9 and S17) indicates high bio-activity of the stabilized waste and consequent high

O_2 demand. However, it could result from the unexpected initial two week lag phase in heating the pile.

The highest observed temperatures occurred in pile B1 (from day 16: >70 °C inside the pile, >60 °C at the edge), which contained the highest TOC and organic substances (VS) of all analyzed samples, as well as having twice the mass of piles A1 and A2, but half the rate of aeration.

In piles C1 and C2 the sidewalls apparently contributed to an overall reduction of temperature (Supplementary material Figures S35–S38, S41 and 42). The cooling of the waste at the border with the walls is visible, especially at the beginning and at the end of the process—a few cold spots were identified near to the sidewalls and at the bottom (Table 6). Despite favorable conditions (large mass of waste; less aeration), no elevation of temperature was noticed.

Experiments carried out under similar conditions by Ermolaev et al. (2012) showed that, despite continuous operation, fans may be unable to maintain an appropriate temperature [26]. Our observations of cold spots (low temperatures) at the base of the pile (e.g., Figure 5; Supplementary Material Figure S30 and Table 6) indicates the impact of pumping large amounts of air, resulting in removal of warm air from the center and overall cooling of the pile. Similar effects of lowering temperature by increased aeration were observed by Shen et al. (2011) and Sołowiej et al. (2010) [42,43]. In bigger piles, B1–2 and C1–2, the optimal temperature for the biostabilization process, around 60 °C, was reached much faster, an effect also noted by [44]. Although heat contributes to elimination of pathogenic organisms (Stentiford, (1996) it does not ensure maximum mass removal [45]. According to many authors, greater mass reduction can be achieved at temperatures between 40 and 60 °C [45–47]. On the other hand, Richard, (1993) claims that maintaining the temperature in the range of 56–70 °C for too long (over a week) reduces biodiversity and increases the intensity of odor compounds [47].

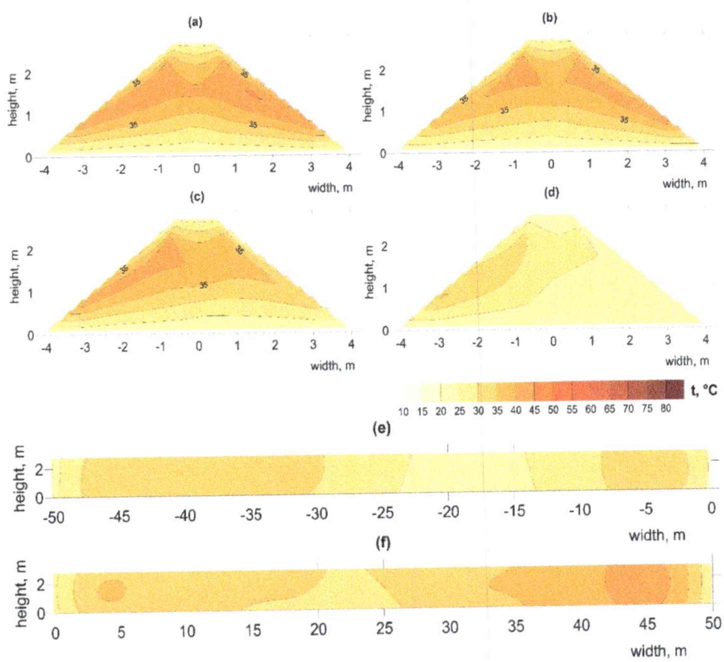

Figure 5. Spatial distribution of temperature changes on day 1 in pile B1, at distances from the aeration fan (**a**) 2.5 m, (**b**) 17.5 m, (**c**) 31.5 m, (**d**) 47.5 m, longitudinal sections (**e**) left (**f**) right. Illustration of the problem with irregular temperature in pile.

3.2.2. Spatial Distribution of O_2 and CO_2 Concentration

Despite the high aeration in piles A1 and A2 during the first two weeks, low O_2 concentrations were recorded, together with low temperatures (Supplementary Material Figures S43, S44, S52 and S53). The high rate of airflow would have created optimal conditions for the activity of psychro- and mesophilic microorganisms at that time, leading to depletion of oxygen by respiration. However, the inhomogeneity of the OFMSW and differences in bulk density could also affect airflow through the pores creating local "hot spots". O_2 levels were particularly low in most cases at the center of the pile early in the process (e.g., Supplementary Material Figures S46, S47, S54 and S62), but hot spots were discovered in pile A2 at 2.5 m length on day 25 (Table 5). In pile A1, reduced O_2 concentrations (~15% lower) occurred on the left side of the pile (Supplementary Material Figures S46–S48), corresponding to the higher temperatures in the same areas (Supplementary Material Figures S4–S6). The above examples were all identified as a hot spot (Table 5). Low O_2 concentrations result from the high intensity of biodegradation early in the process (Jiang et al. (2015) especially if aeration is insufficient [38]. Mohajer et al. (2010) note that the O_2 consumption strongly increases in the first 4 days of composting to as much as 40 mmol$\cdot h^{-1} \cdot kg^{-1}$ d.m., and then decreases with the duration of the process [48].

O_2 concentrations increased during the process, related to decreasing O_2 consumption, while supply of forced air remained stable (Supplementary Material Table S3, [22]). From day 30, O_2 concentrations > 17% were observed in most piles, indicating excellent oxygenation of the waste and sufficient air forced into the pile to promote aerobic digestion.

Data obtained in this experiment shows a generally better aeration system (few locations with less than 15% O_2) compared to composting in active aerated and static piles reported by Szanto et al. (2007) [49]. Although the total aeration intensity per waste mass $m^3 \cdot Mg^{-1}$ (Supplementary Material Table S1, [22]) was below the recommended level, >10 $m^3 \cdot Mg^{-1} \cdot h^{-1}$ [4], this did not appear to affect the O_2 concentration observed in the piles.

Extremely low O_2 concentrations < 5% were observed only at the beginning of the process in pile B1 (Supplementary Material Figure S60). Low concentrations (<10% O_2) also occurred in the center of this pile up to day 40 (Figures 6 and 7; Supplementary Material Figures S60–S64) coinciding with high temperatures (Figure 5; Supplementary Material Figures S19–S23). Compared to other piles, B1 contained the greatest number of such hot spots (Table 5). This may indicate that, with an increased amount of waste and high TOC content, the air flow was insufficient during the most intense phase, up to the 4th week of the decomposition process [50].

In pile B2, lower concentrations of O_2 were observed at its sides (Supplementary Material Figures S69–S72) with hot spots mainly in the center of the pile (Table 5), which may also confirm insufficient air supply where the total waste mass exceeds 600 Mg. The influence of reactor design during biological waste treatment process has been noted by Mason and Milke, (2005) [20]. Another explanation could be the structure of the waste, consisting mainly of waste fractions < 20 mm (Supplementary Material Figure S3), which could reduce the free air spaces and obstruct the air from the aeration channels. Whether aeration is passive or active, airspace within the substrate plays an important role in the composting process [51]. Air porosity influences not only air permeability, but also determines O_2 transport and the removal of water and heat from the pile.

Very high concentrations of O_2 > 18% were observed in C1 and C2, where sidewalls were constructed. Low concentrations of O_2 < c8% (Supplementary Material Figures S84b and S85b) and high concentrations of CO_2 (Figures S128b and S129b) occurred at only a few points (e.g., the cross-section of 17.5 m) indicating hot spots (Table 5). CO_2 concentrations are consistent with results obtained by Clemens and Cuhls, (2003) from various types of piles composting municipal solid waste [52]. The spatial distribution of CO_2 showed an inverse relationship with O_2 (e.g., in pile A1, higher CO_2 concentrations occurred on the left side of the pile together with high temperature and lower O_2), which is typical for aerobic waste treatment [53,54]. The highest content of CO_2 > 10% was observed in B1, mainly up to week 2, but levels were mostly very low (2 to 3%) and occurred in the center of the pile as single hot spots. Similarly, the highest CO_2 concentrations were observed in the first phase of composting a mixture of manure and sawdust, and then its gradual reduction as the compost matures [55], and at a small scale during home composting [56].

The influence of the sidewalls in piles C1 and C2 was also noticed as an increase in the occurrence of hypoxic zones near to the border of the walls. This could have resulted from poorly located aeration channels, which were originally designed for non-compacted piles, or from the small amount of air supplied to the pile. Despite such issues of reactor design, an active aeration system is essential, since CO_2 may increase to over 25% with inadequate aeration [49].

Table 5. Localization of hot spots during biostabilization process.

Pile	Place (Length), m	Time, Day	Hot Spots Localization	Figures (Supplementary Material)
A1	2.5; 17.5, 32.5	20	In the center	Figures S4, S45 and S89
A1	2.5, 32.5	28	In the center	Figures S5, S47 and S90
A1	32.5	41	In the top left corner	Figures S7, S49 and S92
A2	2.5	25	In the center of pile	Figures S13, S55, S98
A2	2.5	32	In the top left corner	Figures S14, S56 and S99
A2	17.5	47	In the top left corner	Figures S16, S58 and S101
B1	17.5	9	In the center of pile	Figures S18, S61 and S104
B1	2.5, 17.5, 32.5, 47.5	16	In the center and left side	Figures S19, S62 and S105
B1	2.5, 17.5, 32.5, 47.5	22	In the center and left side	Figures S20 and S106
B1	2.5, 17.5, 32.5, 47.5	31	In the center and left side	Figures S21, S63 and S107
B1	2.5, 17.5, 32.5, 47.5	41	In the center and left side	Figures S22, S64 and S107
B1	17.5	57	In the left bottom corner	Figures S23, S66 and S109
B2	2.5, 32.5, 47.5	8	In the center and left side	Figures S26, S69 and S112
B2	32.5	16	In the left bottom corner	Figures S27, S70 and S113
B2	17.5, 32.5, 47.5	24	In the center and right side	Figures S28, S71 and S114
B2	32.5	43	Right down corner	Figures S30, S73 and S116
C1	32.5	28	In the center	Figures S34, S77 and S120
C2	17.5	9	On the right side	Figures S38, S81 and S124
C2	32.5	21	On the left side	Figures S39, S82 and S125
C2	17.5	35	On the left and right corner	Figures S41, S84 and S127

Table 6. Localization of cold spots during biostabilization process (excluding data from first three weeks of biostabilization in piles A1 and A2, due to low temperature in all piles).

Pile	Place (Length), m	Time, day	Hot Spots Localization	Figures (Supplementary Material)
A1	2.5; 17.5, 32.5	20	In the left side in the bottom	Figures S4, S46 and S89
A1	2.5, 32.5	28	In the bottom	Figures S5, S47 and S90
A2	2.5, 17.5	17	In the bottom	Figures S12, S54 and S97
A2	2.5, 17.5, 32.5	25	In the bottom	Figures S13, S55 and S98
A2	2.5, 17.5	38	In the bottom	Figures S15, S57 and S100
A2	2.5, 17.5, 32.5, 47.5	57	In the bottom	Figures S17, S59 and S102
B1	32.5, 47.5	1	In the bottom and in the center	Figures S60 and S103
B2	2.5, 17.5, 32.5, 47.5	1	In the bottom	Figures S25, S69 and S111
B2	2.5	43	In the bottom and in the center	Figures S30, S74 and S116
C1	2.5, 17.5	28	left side, near to sidewalls and in the bottom	Figures S34, S77 and S120
C1	2.5, 32.5, 47.5	35	left side, near to sidewalls	Figures S35, S78 and S121
C1	2.5, 32.5, 47.5	42	Left and right side, near to sidewalls	Figures S36, S79 and S122
C2	2.5, 17.5, 32.5, 47.5	1	In the bottom	Figures S37, S80 and S123
C2	32.5, 47.5	28	In the bottom	Figures S40, S83 and S126
C2	2.5, 32.5, 47.5	35	In the bottom, left side, near to sidewalls	Figures S41, S84 and S127
C2	2.5, 32.5, 47.5	42	In the bottom, left side, near to sidewalls	Figures S42, S85 and S128

It has been shown that the number of hot and cold spots depended on the size of the pile and aeration rate. Comparison between piles A1–2 and B1–2 shows that when the mass of stabilized waste is significantly lower and the aeration rate is two-fold higher the number of hot spots decreases, while cold spots increase. In the case of piles A1–2, the number of hot spots was 6 and 3, respectively, while in piles B1–2, 16 and 8. The opposite situation was in the case of cold spots 5–11 (A1–2) and 2–5 (B1–2) (Table 5). It shows that the application of hot and cold spot monitoring may be a useful tool for the optimization of the AB process. The elimination of hot and cold spots should be the aim, to achieve proper conditions for an efficient process, however, it requires further investigation. Additionally, our results indicated that construction with sidewalls (piles C1–2) decreased the number of hot spots to just 1 and 3, respectively (Table 4). This decrease may be due to improved airflow through the waste by eliminating air escape near the base of the pile. However, in piles B1–2, with similar waste mass and airflow rate, the channelized airflow increased the number of cold spots to 8 and 12, respectively (Table 5), indicating that differing reactor constructions also requires an optimal airflow rate to avoid inadequate aeration.

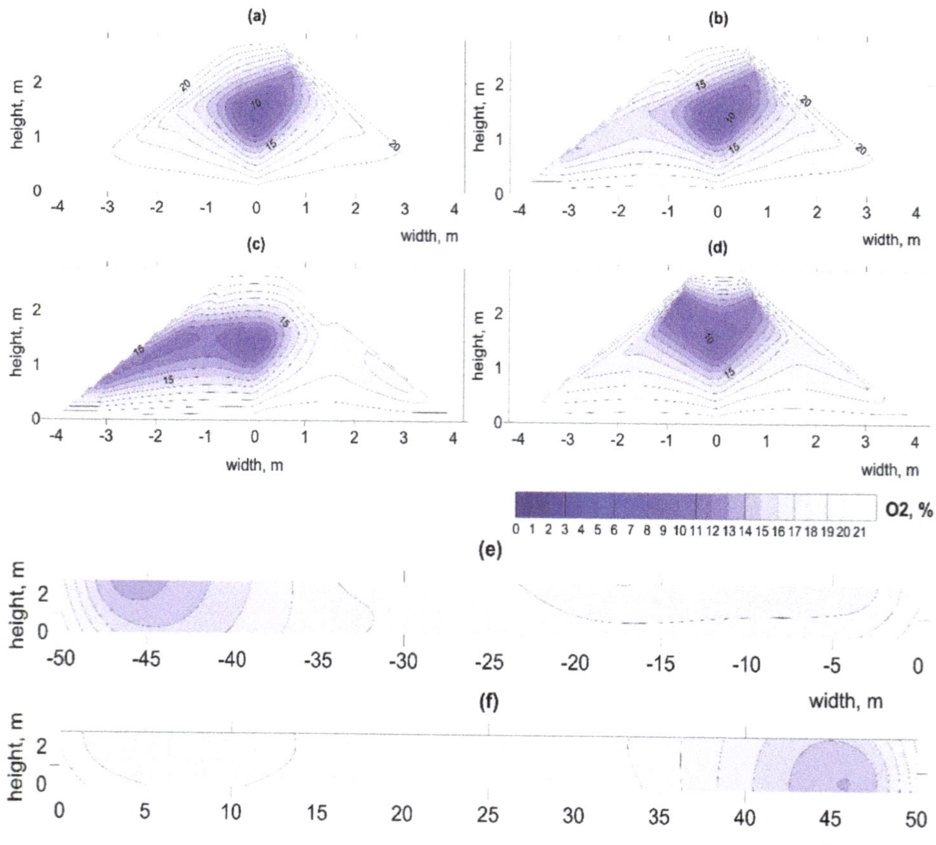

Figure 6. Spatial distribution of O_2 changes on day 22 in pile B1, at distances from aeration fan (**a**) 2.5 m, (**b**) 17.5 m, (**c**) 31.5 m, (**d**) 47.5 m, longitudinal sections (**e**) left (**f**) right. Illustration of low O_2 concentration in center of pile.

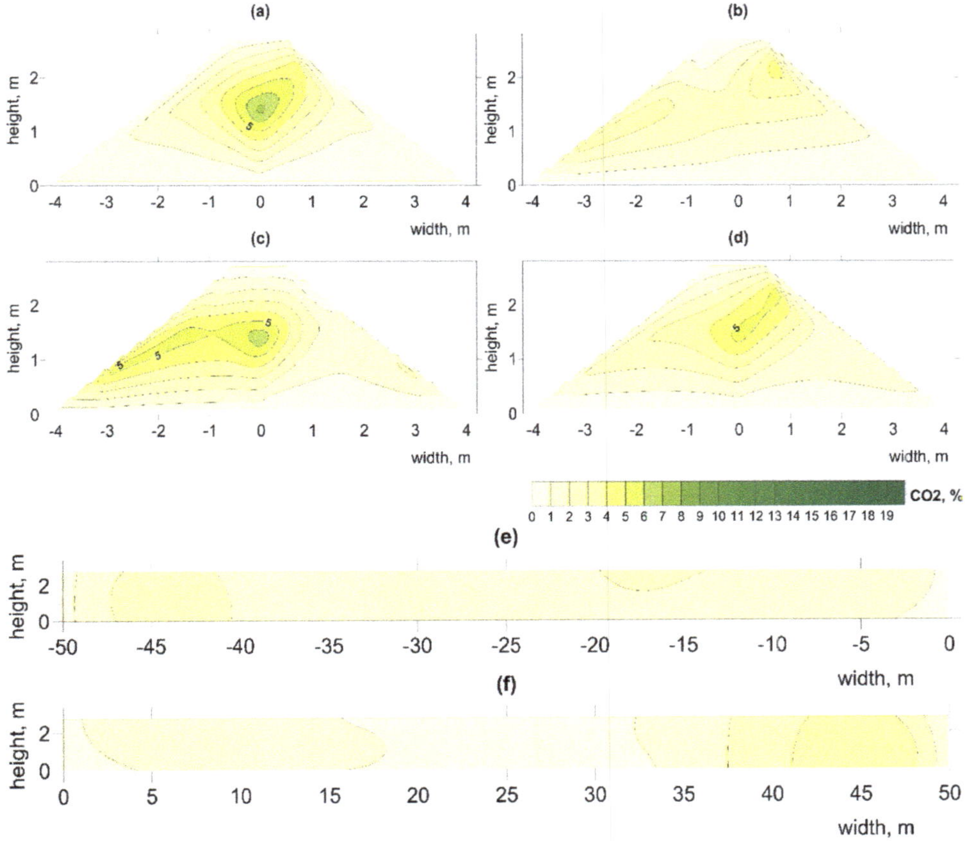

Figure 7. Spatial distribution of CO_2 changes on day 22 in pile B1, at distances from aeration fan (**a**) 2.5 m, (**b**) 17.5 m, (**c**) 31.5 m, (**d**) 47.5 m, longitudinal sections (**e**) left (**f**) right. Illustration of high CO_2 concentration in center of pile.

3.2.3. Spatial Distribution of CO Concentration

CO was widely distributed throughout the entire process in all piles. In most cases the highest concentrations were observed in the center of the piles and during the first two weeks, in accordance with research carried out on other materials such as organic waste [57], green waste with sewage sludge [16,58], and municipal waste [15]. Similar relationships were noted by Boldrin et al. (2009) [59] during the composting of green waste, and Andersen et al. (2010) [39] who observed increased CO concentration in the early stages, which continued in composted material for a year.

Low concentrations, up to about 200 ppm, were observed during day 1 of the process in piles A1, A2, B1, and C1 (Supplementary Material Figures S130, S139, S146 and S159), while in B2 and C2 there were unusually high values from the start, even over 1000 ppm, in the whole volume of the pile (Supplementary Material Figures S153 and S165). High concentrations of CO throughout the pile during the whole process indicate the dynamic nature of CO production, resulting from the relatively large inhomogeneity of the waste material used.

High CO concentrations were clearly associated with locations of highest temperature. In the case of pile A1, following the initial lag phase, significant increases in both CO and temperature occurred from day 20 of the process (Supplementary Material Figures S4 and S133).

The highest concentration of CO was observed first at the beginning of the process (1st week, near the fan), which could be associated with the low O_2 content, and at the far end of the pile over the whole duration. Around week 3, increased CO concentration was also noticed. In piles B1 and C1, from the first days of the process, temperatures > 60 °C were associated with very high concentrations of CO (>1000 ppm) (Figure 8; Supplementary Material Figures S149, S158 and S160). Similarly, Phillip et al. (2011) observed high CO concentrations during the first 6 h [60]. Also, in pile B1, very low O_2 concentrations (<5%) were observed in the early stages (Supplementary Material Figure S60), while in pile C1, only single hot spots were observed near the sidewalls, together with a lower O_2 concentration (Supplementary Material Figure S73a–d). Hellebrand, (1999) showed that intensive aeration stimulates microorganisms to produce CO, greatly increasing its concentration [61]. In the present and previous studies [16,57], CO production coincides with the highest temperatures in the piles (up to 1800 ppm at 75 °C), which implies a thermochemical basis for the process.

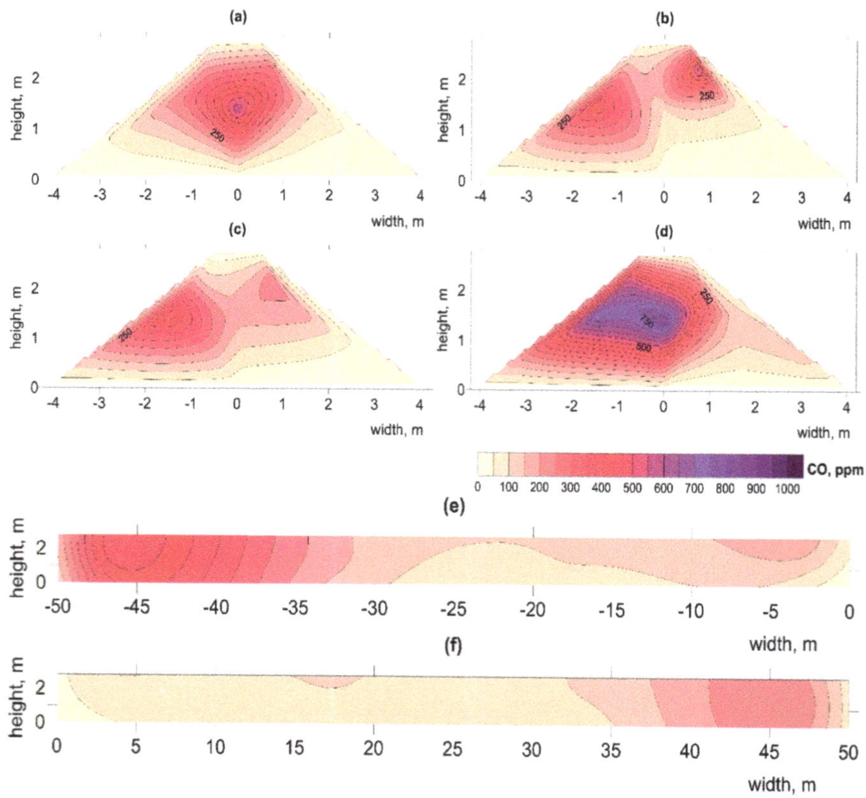

Figure 8. Spatial distribution of CO changes on day 16 in pile B1, at distances from aeration fan (**a**) 2.5 m, (**b**) 17.5 m, (**c**) 31.5 m, (**d**) 47.5 m, longitudinal sections (**e**) left (**f**) right. Illustration of high CO concentration in center of pile.

The zones of lower oxygenation and higher temperatures near the sidewalls of the reactor in piles C1 and C2 coincided with higher concentrations of CO (Supplementary Material Figures S158, S161, S166, S167 and S169). Research by Hellebrand and Kalk, (2001) linked the release of CO directly to the availability of O_2 in the pile, indicating that CO production is favored under both hypoxic and high temperature conditions [62]. As shown in previous studies, CO is also formed under aerobic conditions, but biotic production is more favorable at lower temperatures <40 °C, whereas at >60 °C, CO production is more thermochemical [57]. This clearly indicates the need for a waste treatment technology that will not form hot spots. The use of technologies that homogenize municipal solid waste, before placing it in the reactor may prove effective [63].

In piles where sidewalls were not used (A1, A2, B1, and B2), the biggest concentrations of CO increased in the center of the piles (Figure 8; e.g., Supplementary Material Figures S133, S134, S141, S142, S145 and S148) or at the top of the pile, but smaller concentrations were observed, mainly at the end of the process (Supplementary Material Figures S132, S138 and S154). This may result from the aeration channels, which promote faster decomposition inside the pile, followed by decomposition in the upper regions of the material.

Several studies conclude that, in the decomposition of organic materials, microbial activity has a significant influence on CO production [60,62,64,65]. High temperatures > 60 °C prevailing in most of the piles, combined with a pH of about 7 at the end of the process, were optimal conditions for colonization by CO-metabolizing microorganisms [66]. Such microorganisms could then reduce CO production after the 4th week of the process, despite the high temperatures still prevailing.

Research by Moxley and Smith, (1998) showed the importance of moisture content for CO concentrations from various types of soils, with an optimum of 15 to 25% moisture [67]. Above and below these levels, CO concentration was reduced. Piles B1, C1, and C2 recorded the highest concentrations of CO but the lowest moisture removal, suggesting that the optimal value of moisture (in relation to CO production) of the OFMSW is rather higher, between 30 and 35%.

4. Summary

Spatial analysis of concentrations of key gases within the mass of OFMSW enabled the efficient localization of all hot and cold spots in time and space, regardless of the experimental variant tested or the reactor construction. It was observed that the localization of hot and cold spots depends on biostabilization process parameters including, aeration rate, and mass of OFMSW, or the type of reactor modification applied. It was shown that to reduce the appearance of cold and hot spots, it is necessary to:

- increase the mass of the stabilized waste as it provides greater stability of the process to external conditions;
- increase aeration of waste to remove anaerobic zones.

In addition, the use of sidewalls in pile construction reduced the occurrence of hot spots, and may have the effect of increasing the frequency of cold spots near to walls.

It has been shown that simple research on spatial and temporal distribution of temperature and gas concentration during the OFMSW biostabilization process is advisable, especially in the case of introducing new systems for processing municipal waste. Performing the tests allows quick and easy localization of all hot and cold spots, discovery of possible design mistakes, and adjustment of the parameters of the biostabilization process to shorten it and optimize its final products.

The identification of "hot spots" requires action to eliminate them by modifying waste aeration or its mechanical turning. This is important, especially to eliminate harmful gases such as CO, which have been seen clearly in hot areas. It may indicate the domination of thermochemical processes over biological ones, as already observed in green waste. Less importance in the production of CO is ascribed to the concentration of O_2 and CO_2, which implies that temperature measurement, together with spatial simulation, may be more effective in finding process irregularities. Locating areas of increased temperature within the pile will enable their elimination and the reduction of harmful gases. The identification of hot and cold spots during AB of OFMSW may be a useful tool for process optimization and indication of problems related to reactor construction, which also opens a new approach for research.

Supplementary Materials: The following supporting information can be downloaded at: https://www.mdpi.com/article/10.3390/ma15093300/s1, Figures.zip contains files (Figures S1–S168.pdf) with a graphical visualization of temperature and gas distributions during composting.

Author Contributions: Conceptualization, S.S.-D. and A.B.; methodology, S.S.-D. and A.B.; validation, A.B. and P.F.R.; formal analysis, S.S.-D.; resources, S.S.-D. and A.B.; data curation, S.S.-D.; writing—original draft preparation, S.S.-D.; writing—review and editing, A.B., S.S.-D. and P.F.R.; visualization, S.S.-D.; supervision, A.B. and P.F.R.; project administration, A.B.; funding acquisition, A.B. All authors have read and agreed to the published version of the manuscript.

Funding: This research was funded by the Municipal Cleaning Company in Warsaw, Poland, "The investigation on the intensity of the aerobic biological processes occurring in the prisms for biostabilization of the municipal solid waste undersize fraction". The APC is financed/co-financed by Wroclaw University of Environmental and Life Sciences.

Institutional Review Board Statement: Not applicable.

Informed Consent Statement: Not applicable.

Data Availability Statement: Not applicable.

Acknowledgments: The presented article was prepared as part of the activity of the leading research team Waste and Biomass Valorization Group (WBVG), Department of Applied Bioeconomy, Wrocław University.

Conflicts of Interest: The authors declare no conflict of interest.

References

1. European Commission Closing the Loop-An EU Action Plan for the Circular Economy. Communication from the Commission to the European Parliament, the Council, the European Economic and Social Committee and the Committee of the Regions. Available online: https://eur-lex.europa.eu/legal-content/EN/ALL/?uri=CELEX%3A52015DC0614 (accessed on 28 February 2019).
2. EU Council Directive 1999/31/EC of 26 April 1999 on the Landfill of Waste. Available online: https://eur-lex.europa.eu/legal-content/EN/TXT/?uri=CELEX:31999L0031 (accessed on 28 February 2019).
3. European Union. The European Parliament and the Council of the European Union Directive 2008/98/EC of the European Parliament and of the Council of 19 November 2008. *Off. J. Eur. Union* **2008**, *51*, 3–30.
4. Połomka, J.; Jędrczak, A. Efficiency of Waste Processing in the MBT System. *Waste Manag.* **2019**, *96*, 9–14. [CrossRef]
5. Montejo, C.; Tonini, D.; Márquez, M.; del, C.; Fruergaard Astrup, T. Mechanical-Biological Treatment: Performance and Potentials. An LCA of 8 MBT Plants Including Waste Characterization. *J. Environ. Manag.* **2013**, *128*, 661–673. [CrossRef] [PubMed]
6. Jędrczak, A. Properties of the Organic Fraction Directed to Biostabilization in MBT Installations During the Heating Season. *Civil. Environ. Eng. Rep.* **2018**, *28*, 65–78. [CrossRef]
7. Den Boer, E. Nowoczesne Systemy Mechaniczno-Biologicznego Przetwarzania Odpadów. In *Innowacje w Gospodarce Odpadami Zagadnienia Wybrane*; Wydawnictwo Uniwersytetu Przyrodniczego We Wrocławiu: Wrocław, Poland, 2018; ISBN 9788377172780.
8. Henon, F.; Tremier, A.; Debenest, G.; Martel, J.L.; Quintard, M. A Method to Characterize the Influence of Air Distribution on the Composting Treatment: Monitoring of the Thermal Fields. *Glob. Nest J.* **2009**, *11*, 172–180. [CrossRef]
9. Zeng, J.; Shen, X.; Han, L.; Huang, G. Dynamics of Oxygen Supply and Consumption during Mainstream Large-Scale Composting in China. *Bioresour. Technol.* **2016**, *220*, 104–109. [CrossRef] [PubMed]
10. Oudart, D.; Robin, P.; Paillat, J.M.; Paul, E. Modelling Nitrogen and Carbon Interactions in Composting of Animal Manure in Naturally Aerated Piles. *Waste Manag.* **2015**, *46*, 588–598. [CrossRef] [PubMed]
11. Ni, J. Mechanistic Models of Ammonia Release from Liquid Manure: A Review. *J. Agric. Eng. Res.* **1999**, *72*, 1–17. [CrossRef]
12. Gao, M.; Li, B.; Yu, A.; Liang, F.; Yang, L.; Sun, Y. The Effect of Aeration Rate on Forced-Aeration Composting of Chicken Manure and Sawdust. *Bioresour. Technol.* **2010**, *101*, 1899–1903. [CrossRef]
13. Haug, R.T. *The Practical Handbook of Compost Engineering*; Lewis Publishers: Boca Raton, FL, USA, 1993; p. 717.
14. Korner, I.; Braukmeier, J.; Herrenklage, J.; Leikam, K.; Ritzkowski, M.; Schlegelmilch, M.; Stegmann, R. Investigation and Optimization of Composting Processes-Test Systems and Practical Examples. *Waste Manag.* **2003**, *23*, 17–26. [CrossRef]
15. Stegenta, S.; Dębowski, M.; Bukowski, P.; Randerson, P.F.; Białowiec, A. The Influence of Perforation of Foil Reactors on Greenhouse Gas Emission Rates during Aerobic Biostabilization of the Undersize Fraction of Municipal Wastes. *J. Environ. Manag.* **2018**, *207*, 355–365. [CrossRef]
16. Stegenta, S.; Sobieraj, K.; Pilarski, G.; Koziel, J.A.; Białowiec, A. Analysis of the Spatial and Temporal Distribution of Process Gases within Municipal Biowaste Compost. *Sustainability* **2019**, *11*, 2340. [CrossRef]
17. Yang, F.; Li, Y.; Han, Y.; Qian, W.; Li, G.; Luo, W. Performance of Mature Compost to Control Gaseous Emissions in Kitchen Waste Composting. *Sci. Total Environ.* **2019**, *657*, 262–269. [CrossRef]

18. De Guardia, A.; Petiot, C.; Rogeau, D.; Druilhe, C. Influence of Aeration Rate on Nitrogen Dynamics during Composting. *Waste Management* **2008**, *28*, 575–587. [CrossRef]
19. Wang, K.; Li, W.; Guo, J.; Zou, J.; Li, Y.; Zhang, L. Spatial Distribution of Dynamics Characteristic in the Intermittent Aeration Static Composting of Sewage Sludge. *Bioresour. Technol.* **2011**, *102*, 5528–5532. [CrossRef]
20. Mason, I.G.; Milke, M.W. Physical Modelling of the Composting Environment: A Review. Part 1: Reactor Systems. *Waste Manag.* **2005**, *25*, 481–500. [CrossRef]
21. Poulsen, T.G. Oxygen and Carbon Dioxide Distribution And Movement in Passively Aerated Compost Piles. *Compost. Sci. Util.* **2011**, *19*, 25–32. [CrossRef]
22. Stegenta-Dąbrowska, S.; Rogosz, J.; Bukowski, P.; Dębowski, M.; Randerson, P.F.; Bieniek, J.; Białowiec, A. The Fluctuation of Process Gasses Especially of Carbon Monoxide during Aerobic Biostabilization of an Organic Fraction of Municipal Solid Waste under Different Technological Regimes. *Data* **2020**, *5*, 40. [CrossRef]
23. Cayuela, M.L.; Mondini, C.; Insam, H.; Sinicco, T.; Franke-Whittle, I. Plant and Animal Wastes Composting: Effects of the N Source on Process Performance. *Bioresour. Technol.* **2009**, *100*, 3097–3106. [CrossRef] [PubMed]
24. Eiland, F.; Klamer, M.; Lind, A.-M.; Leth, M.; Bååth, E. Influence of Initial C/N Ratio on Chemical and Microbial Composition during Long Term Composting of Straw. *Microb. Ecol.* **2001**, *41*, 272–280. [CrossRef]
25. Liang, C.; Das, K.C.; McClendon, R.W. The Influence of Temperature and Moisture Contents Regimes on the Aerobic Microbial Activity of a Biosolids Composting Blend. *Bioresour. Technol.* **2003**, *86*, 131–137. [CrossRef]
26. Ermolaev, E.; Pell, M.; Smårs, S.; Sundberg, C.; Jönsson, H.; Smars, S.; Sundberg, C.; Jonsson, H. Greenhouse Gas Emission from Covered Windrow Composting with Controlled Ventilation. *Waste Manag. Res.* **2012**, *30*, 155–160. [CrossRef] [PubMed]
27. Mulbry, W.; Ahn, H. Greenhouse Gas Emissions during Composting of Dairy Manure: Influence of the Timing of Pile Mixing on Total Emissions. *Biosyst. Eng.* **2014**, *126*, 117–122. [CrossRef]
28. Evangelou, A.; Gerassimidou, S.; Mavrakis, N.; Komilis, D. Monitoring the Performances of a Real Scale Municipal Solid Waste Composting and a Biodrying Facility Using Respiration Activity Indices. *Environ. Monit. Assess.* **2016**, *188*, 302. [CrossRef] [PubMed]
29. Komilis, D.; Evangelou, A.; Giannakis, G.; Lymperis, C. Revisiting the Elemental Composition and the Calorific Value of the Organic Fraction of Municipal Solid Wastes. *Waste Manag.* **2012**, *32*, 372–381. [CrossRef]
30. Montejo, C.; Costa, C.; Márquez, M.C. Influence of Input Material and Operational Performance on the Physical and Chemical Properties of MSW Compost. *J. Environ. Manag.* **2015**, *162*, 240–249. [CrossRef]
31. Hu, T.J.; Zeng, G.M.; Huang, G.H.; Xie, G.X.; Chen, Y.N.; Huang, D.L.; Liu, H.L. Main Influencing Factors of TOC Biodegradation in Aerobic Composting. In Proceedings of the Energy & environment-A World of Challenges and Opportunities, Proceedings; Science Press: Monmouth Junction, NJ, USA; Changsha, China, 2003; pp. 815–822. Available online: https://sdr.gdos.gov.pl/Documents/GO/Ekspertyzy/Ekspertyza%20MBP_III%20etap_%2022-06-2015%20+%20za%C5%82aczniki.pdf (accessed on 28 April 2019).
32. Beck-Friis, B.; Smårs, S.; Jönsson, H.; Kirchmann, H. SE—Structures and Environment: Gaseous Emissions of Carbon Dioxide, Ammonia and Nitrous Oxide from Organic Household Waste in a Compost Reactor under Different Temperature Regimes. *J. Agric. Eng. Res.* **2001**, *78*, 423–430. [CrossRef]
33. Den Boer, E.; Jędrczak, A.; Kowalski, Z.; Kulczycka, J.; Szpadt, R. A Review of Municipal Solid Waste Composition and Quantities in Poland. *Waste Manag.* **2010**, *30*, 369–377. [CrossRef]
34. Yousefi, J.; Younesi, H.; Ghasempoury, S.M. Co-Composting of Municipal Solid Waste with Sawdust: Improving Compost Quality. *Clean-Soil Air Water* **2013**, *41*, 185–194. [CrossRef]
35. Jędrczak, A.; DenBoer, E. *Raport Końcowy III Etapu Ekspertyzy Mającej Na Celu Przeprowadzenie Badań Odpadów w 20 Inst*; Uniwesytet Zielonogórski: Zielona Góra, Poland, 2015.
36. Cooperband, L. *The Art and Science of Composting A Resource for Farmers and Compost Producers*; University of Wisconsin-Madison: Madison, WI, USA, 2002; pp. 1–14.
37. Kowal, K.; Libudzisz, Z.; Żakowska, Z. *Mikrobiologia Techniczna Tom 2-Praca Zbiorowa-Książka-Księgarnia Internetowa PWN*; Wydawnictwo Naukowe PWN: Warszawa, Poland, 2017.
38. Jiang, T.; Li, G.; Tang, Q.; Ma, X.; Wang, G.; Schuchardt, F. Effects of Aeration Method and Aeration Rate on Greenhouse Gas Emissions during Composting of Pig Feces in Pilot Scale. *J. Environ. Sci.* **2015**, *31*, 124–132. [CrossRef]
39. Andersen, J.K.; Boldrin, A.; Christensen, T.H.; Scheutz, C. Mass Balances and Life-Cycle Inventory for a Garden Waste Windrow Composting Plant (Aarhus, Denmark). *Waste Manag. Res.* **2010**, *28*, 1010–1020. [CrossRef] [PubMed]
40. Beck-Friis, B.; Pell, M.; Sonesson, U.; Jonsson, H.; Kirchmann, H. Formation and Emission of N2O and CH4 from Compost Heaps of Organic Household Waster. *Environ. Monit. Assess.* **2000**, *62*, 317–331. [CrossRef]
41. Wang, J.Z.; Hu, Z.Y.; Xu, X.K.; Jiang, X.; Zheng, B.H.; Liu, X.N.; Pan, X.B.; Kardol, P. Emissions of Ammonia and Greenhouse Gases during Combined Pre-Composting and Vermicomposting of Duck Manure. *Waste Manag.* **2014**, *34*, 1546–1552. [CrossRef] [PubMed]
42. Shen, Y.J.; Ren, L.M.; Li, G.X.; Chen, T.B.; Guo, R. Influence of Aeration on CH_4, N_2O and NH_3 Emissions during Aerobic Composting of a Chicken Manure and High C/N Waste Mixture. *Waste Manag.* **2011**, *31*, 33–38. [CrossRef]
43. Sołowiej, P.; Piechocki, J.; Neugebauer, M. Wpływ Napowietrzania Złoża Na Przebieg Pierwszej Fazy Procesu Kompostowania. *Inżynieria Rol.* **2010**, *14*, 193–198.

44. Bongochgetsakul, N.; Ishida, T. A New Analytical Approach to Optimizing the Design of Large-Scale Composting Systems. *Bioresour. Technol.* **2008**, *99*, 1630–1641. [CrossRef]
45. Stentiford, E.I. Composting Control: Principles and Practice. In *The Science of Composting*; Springer: Dordrecht, The Netherlands, 1996; pp. 49–59.
46. Hamoda, M.F.F.; Abu Qdais, H.A.A.; Newham, J. Evaluation of Municipal Solid Waste Composting Kinetics. *Resour. Conserv. Recycl.* **1998**, *23*, 209–223. [CrossRef]
47. Richard, T.L. *Municipal Solid Waste Composting: Biological Processing*; Cornell University: New York, NY, USA, 1996.
48. Mohajer, A.; Tremier, A.; Barrington, S.; Teglia, C. Compost Mixture Influence of Interactive Physical Parameters on Microbial Kinetics and Substrate Fractionation. *Waste Manag.* **2010**, *30*, 1464–1471. [CrossRef] [PubMed]
49. Szanto, G.L.; Hamelers, H.M.; Rulkens, W.H.; Veeken, A.H.M. NH_3, N_2O and CH_4 Emissions during Passively Aerated Composting of Straw-Rich Pig Manure. *Bioresour. Technol.* **2007**, *98*, 2659–2670. [CrossRef]
50. Martínez-Valdez, F.J.; Martínez-Ramírez, C.; Martínez-Montiel, L.; Favela-Torres, E.; Soto-Cruz, N.O.; Ramírez-Vives, F.; Saucedo-Castañeda, G. Rapid Mineralisation of the Organic Fraction of Municipal Solid Waste. *Bioresour. Technol.* **2015**, *180*, 112–118. [CrossRef]
51. Su, D.; McCartney, D.; Wang, Q. Comparison of Free Air Space Test Methods. *Compost. Sci. Util.* **2006**, *14*, 103–113. [CrossRef]
52. Clemens, J.; Cuhls, C. Greenhouse Gas Emissions from Mechanical and Biological Waste Treatment of Municipal Waste. *Environ. Technol.* **2003**, *24*, 745–754. [CrossRef] [PubMed]
53. Petric, I.; Mustafic, N. Dynamic Modeling the Composting Process of the Mixture of Poultry Manure and Wheat Straw. *J. Environ. Manag.* **2015**, *161*, 392–401. [CrossRef]
54. Zeng, J.; Shen, X.; Sun, X.; Liu, N.; Han, L.; Huang, G. Spatial and Temporal Distribution of Pore Gas Concentrations during Mainstream Large-Scale Trough Composting in China. *Waste Manag.* **2018**, *75*, 297–304. [CrossRef] [PubMed]
55. Chang, C.-H.; Chen, I.C.; Yang, S.-S. Methane and Carbon Dioxide Emissions from Different Composting Periods. *Terr. Atmos. Ocean. Sci.* **2009**, *20*, 511–520. [CrossRef]
56. Adhikari, B.K.; Tremier, A.; Barrington, S.; Martinez, J.; Daumoin, M. Gas Emissions as Influenced by Home Composting System Configuration. *J. Environ. Manag.* **2013**, *116*, 163–171. [CrossRef] [PubMed]
57. Stegenta-Dąbrowska, S.; Drabczyński, G.; Sobieraj, K.; Koziel, J.A.; Białowiec, A. The Biotic and Abiotic Carbon Monoxide Formation During Aerobic Co-Digestion of Dairy Cattle Manure With Green Waste and Sawdust. *Front. Bioeng. Biotechnol.* **2019**, *7*, 283. [CrossRef]
58. Stegenta, S.; Sobieraj, K.; Pilarski, G.; Koziel, J.; Białowiec, A. The Spatial and Temporal Distribution of Process Gases within the Biowaste Compost. *Data* **2019**, *4*, 37. [CrossRef]
59. Boldrin, A.; Andersen, J.K.; Moller, J.; Christensen, T.H.; Favoino, E. Composting and Compost Utilization: Accounting of Greenhouse Gases and Global Warming Contributions. *Waste Manag. Res.* **2009**, *27*, 800–812. [CrossRef] [PubMed]
60. Phillip, E.A.; Clark, O.G.; Londry, K.; Yu, S.; Leonard, J. Emission of Carbon Monoxide during Composting of Municipal Solid Waste. *Compost. Sci. Util.* **2011**, *19*, 170–177. [CrossRef]
61. Hellebrand, H.J. Carbon Monoxide Emissions during Composting. *Ber. Uber Landwirtsch.* **1999**, *77*, 109–112.
62. Hellebrand, H.J.; Kalk, W.D. Emission of Carbon Monoxide during Composting of Dung and Green Waste. *Nutr. Cycl. Agroecosystems* **2001**, *60*, 79–82. [CrossRef]
63. Vaverková, M.D.; Elbl, J.; Voběrková, S.; Koda, E.; Adamcová, D.; Mariusz Gusiatin, Z.; Al Rahman, A.; Radziemska, M.; Mazur, Z. Composting versus Mechanical–Biological Treatment: Does It Really Make a Difference in the Final Product Parameters and Maturity. *Waste Manag.* **2020**, *106*, 173–183. [CrossRef] [PubMed]
64. Rich, J.J.; King, G.M. Carbon Monoxide Consumption and Production by Wetland Peats. *FEMS Microbiol. Ecol.* **1999**, *28*, 215–224. [CrossRef]
65. Voordouw, G. Carbon Monoxide Cycling by Desulfovibrio Vulgaris Hildenborough. *J. Bacteriol.* **2002**, *184*, 5903–5911. [CrossRef]
66. Techtmann, S.M.; Colman, A.S.; Robb, F.T. 'That Which Does Not Kill Us Only Makes Us Stronger': The Role of Carbon Monoxide in Thermophilic Microbial Consortia. *Environ. Microbiol.* **2009**, *11*, 1027–1037. [CrossRef]
67. Moxley, J.; Smith, K. Carbon Monoxide Production and Emission by Some Scottish Soils. *Tellus Ser. B Chem. Phys. Meteorol.* **1998**, *50*, 151–162. [CrossRef]

materials

MDPI

Article

Synthesise and Characterization of Cordierite and Wollastonite Glass—Ceramics Derived from Industrial Wastes and Natural Raw Materials

Gamal A. Khater *, Amany A. El-Kheshen and Mohammad M. Farag

Glass Research Department, National Research Centre, Cairo 12622, Egypt; aelkheshen1@yahoo.com (A.A.E.-K.); mmfaragnrc@gmail.com (M.M.F.)
* Correspondence: j.khater@yahoo.com

Abstract: Industrial waste is one of the primary sources that harm the environment, and this topic has occupied many scientists on how to take advantage of these wastes or dispose of them and create a clean environment. By-pass cement dust is considered one of the most dangerous industrial wastes due to its fine granular size and its volatilization in the air, which causes severe environmental damage to human and animal health, and this is the reason for choosing the current research point. In this article, eight samples of glass–ceramics were prepared using by-pass cement dust and natural raw materials known as silica sand, magnesite, and kaolin. Then melted by using an electric furnace which was adjusted at a range of temperatures from 1550 to 1600 °C for 2 to 3 h; the samples were cast and were subjected to heat treatment at 1000 °C for 2 h based on the DTA results in order to produce crystalline materials. Various techniques were used to study the synthesized glass–ceramic samples, including differential thermal analysis (DTA), X-ray diffraction (XRD), scanning electron microscope (SEM), and thermal expansion coefficient (CTE). X-ray analysis showed that the phases formed through investigated glass–ceramic samples consisted mainly of β- wollastonite, parawollastonite, diopside, anorthite, and cordierite. It was noticed that β- the wollastonite phase was formed first and then turned into parawollastonite, and also, the anorthite mineral was formed at low temperatures before the formation of the diopside mineral. SEM showed that the formed microstructure turned from a coarse grain texture to a fine-grained texture, by increasing the percentage of cordierite. It also showed that the increase in time at the endothermic temperature significantly affected the crystalline texture by giving a fine-grained crystalline texture. The linear thermal expansion measurements technique used for the studied glass–ceramic samples gives thermal expansion coefficients ranging from 6.2161×10^{-6} to 2.6181×10^{-6} C^{-1} (in the range of 20–700 °C), and it decreased by increasing cordierite percent.

Keywords: crystallization; glass–ceramics; diopside; β-wollastonite; parawollastonite; anorthite; by-pass cement

Citation: Khater, G.A.; El-Kheshen, A.A.; Farag, M.M. Synthesise and Characterization of Cordierite and Wollastonite Glass—Ceramics Derived from Industrial Wastes and Natural Raw Materials. *Materials* **2022**, *15*, 3534. https://doi.org/10.3390/ma15103534

Received: 23 March 2022
Accepted: 12 May 2022
Published: 14 May 2022

1. Introduction

Since the dawn of time, the industry has been considered one of the most important pillars that depend on the consumption of materials, the production of all that humanity needs for the well-being of life, and also is assumed as one of the most crucial reasons for the existence of waste materials.

From the financial and ecological points of view, the recycling of manufacturing wastes is well thought-out to be a significant issue. The increase of industrial waste materials, due to the rapid increase in populations and growth in the prosperity of humanity, has become a major social and environmental problem. Management of all industrial waste has become the most significant environmental issue in numerous expanding countries. Primary sources of solid waste are human liveliness, such as building and technology innovation. These wastes cause many complex problems concerning storage space, transportation, and environmental or atmosphere contamination. Utilizing manufacturing waste to produce beneficial resources via green chemical methods will be an essential step to having a hygienic and safe environment [1,2].

Cement is one of the most remarkable binding materials. It is manufactured in massive amounts all over the world. The cement industry is considered one industry that generates an enormous amount of solid waste. These wastes should be managed to guarantee a clean and harmless environment [3].

In Egypt, several cement factories create significant amounts of cement kiln dust (CKD), the by-product formed in cement kilns and associated procedures. It has a destructive outcome on all living organisms [4]. The chemical composition of CKD mainly depends on the composition of raw materials, speed of the gas in the kiln, and type of procedure. Usually, CKD contains calcined materials, un-reacted raw feed, clinker dust, ash, alkali sulfates, halides, and other volatile materials [5–8].

Efforts have been made by several researchers to recycle that harmful CKD waste through different valuable applications such as removal of heavy metals [9], treating contaminated soil [10], ceramic production [11], glass–ceramic fabrication [12–14], and so on.

The chemical composition of by-pass cement dust qualifies it for the glass–ceramic and glass industries [15]. As the glass–ceramics industry has various applications, such as in the microelectronics industry and construction. Utilizing cement dust to produce glass–ceramic materials is highly considered a technological, scientific, and cost-effective concept.

Glass–ceramic is one of the majority multipurpose and valuable materials needed in diverse industrial applications. Glass-ceramic has varieties of properties according to its composition, such as hardness, insulation, thermal shock resistance, low thermal expansion coefficient, toughness, and optical properties [16]. It is usually prepared by scheming glass crystallization. It also has fine-grained microstructure, low or no porosity, and lots of different properties that may be synchronized by changes in sample composition and the regime of heat treatment.

Wollastonite ($CaSiO_3$) has been conventionally applied as a raw material for tiles, paint, paper industries [17], insulators [18], ceramics [19], and so on. It has good mechanical, thermal, and optical properties such as slight shrinkage, high strength, whiteness in color, and bending properties. Nowadays, the increasing demands for wollastonite are followed by the proportional increases in its fabrication worldwide [20]. Natural wollastonite is limited by its deposits condition so that it can be synthesized artificially [17]. Synthesized wollastonite glass–ceramics [21] are used for various purposes and must be studied.

Cordierite ($2MgO–2Al_2O_3–5SiO_2$) is one of the most motivating phases of glass–ceramics; and has a broad collection of applications in many industrial areas [22]. It possesses features such as low thermal expansion, high chemical durability, high thermal resistance, low dielectric constant, and good mechanical properties, which give scientists the facility to use this compound to produce traditional or advanced products [23,24]. Cordierites have lower production costs and superior electrical properties, they can be used as substrate more than alumina in electronic industries and manufacturing multilayer circuit boards and thermal insulators.

This research aims to take advantage of cement dust as one of the industrial wastes produced in huge quantities from cement factories with some raw materials available in the earth's crust to produce high-performance glass–ceramic materials that can be used for construction applications. The purpose of the research is to take advantage of by-pass cement dust, one of the solid industrial wastes which harm the environment, in obtaining high-performance glass–ceramic materials based on the wollastonite–cordierite system that can be used for various industrial and construction purposes by using many techniques such as differential thermal analysis (DTA), X-ray diffraction (XRD), scan electron microscope (SEM: JEOL, XL30, Philips, Amsterdam, The Netherlands)), and thermal expansion coefficient (CTE).

2. Experimental Techniques

2.1. Batch Calculation and Glass Preparation

Eight glass compositions were designed based on cordierite ($Mg_2Al_4Si_5O_{18}$) and wollastonite ($CaSiO_3$) systems selected for the current project. The cordierite content in these compositions varies from 10 to 80 at ten wt% intervals, and consequently, the wollastonite content ranges from 90% to 20%. These studied samples were G10, G20, G30, G40, G50, G60, G70, and G80. These numbers denote the wt% of the cordierite and the rest being the wollastonite components. For example, G40 means that this glass composition contains 40wt% cordierite component and the rest, 60wt% wollastonite component. The batches related to these compositions were designed by calculating the appropriate proportions of solid wastes represented in by-pass cement dust, magnesite, kaolin, and silica sand. Table 1 shows the chemical compositions of the raw materials used. Table 2 shows glass compositions in the oxide percentages and raw materials used. These batches were ground in a ball mill for about 45 min, then they were melted in platinum crucibles in an electrically heated globar furnace at temperatures ranging from 1550 °C to 1600 for 2–3 h, according to the batch composition. It was noted that glasses rich in cordierite were highly viscous and needed higher melting temperatures. The melting temperature increased with an increase in the percentage of cordierite contents (that is, the degree of viscosity increases from G10 to G80).

Table 1. Chemical analyses of raw materials used in the batch preparation.

Oxide wt %	Magnesite (Al-Nasr Mining Co.)	Silica Sand Abu-Zenima (Sinai)	Kaolin El-Tih (Sinai)	By-Pass Cement
SiO_2	0.54	99.20	44.20	6.12
Al_2O_3	1.02	0.28	37.75	2.58
Fe_2O_3	0.48	0.03	0.93	3.37
TiO_2	trace	trace	1.85	0.21
CaO	6.31	0.10	0.82	55.96
MgO	40.35	trace	0.52	0.84
Na_2O	trace	trace	1.15	0.29
K_2O	trace	trace	0.72	0.73
LOI at 1000 °C	50.98	0.40	13.01	25.11

Table 2. Chemical compositions and the corresponding batches (wt%) of the investigated glasses.

Glas No.	Nominal Phase * Composition (wt%)	Oxide wt%				Batch Ingredients (wt%)			
		SiO_2	Al_2O_3	CaO	MgO	By-Pass	Mag.	Kaolin	Sand
G10	10%cord. + 90% wol.	51.68	3.49	43.45	1.38	60.2	1.4	3.21	35.19
G20	20%cord. + 80% wol.	51.65	6.97	38.62	2.76	53.19	4.15	11.11	31.54
G30	30%cord. + 70% wol.	51.61	10.46	33.8	4.13	46.29	6.87	18.95	27.9
G40	40%cord. + 60% wol.	51.57	13.95	28.97	5.51	39.43	9.57	26.68	24.31
G50	50%cord. + 50% wol.	51.54	17.43	24.14	6.89	32.63	12.22	34.28	20.87
G60	60%cord. + 40% wol.	51.54	20.92	19.31	8.27	25.98	14.87	41.89	17.26
G70	70%cord. + 30% wol.	51.47	24.4	14.48	9.65	19.38	17.47	49.34	13.8
G80	80%cord. + 20% wol.	51.43	27.89	9.66	11.02	12.84	20.29	56.35	10.53

Where, * cord = cordierite, wol. = wollastonite, Mag. = magnesite.

After melting and refining, the resulting bubble-free melts were cast onto a hot steel marver into steel molds in the form of buttons (about 20 mm and 9 mm thickness) as rods of 50 × 10 ×10 mm. The hot glass samples were transferred to preheated electric muffle furnace to avoid thermal shock.

2.2. Differential Thermal Analysis (DTA)

The differential thermal analyses of the studied glasses were achieved via a Labsys DSC-TG 1600 °C Setaram (France). About 60 mg of the pulverized glass of grain size less than 0.60 mm and more significant than 0.2 mm was used. Al_2O_3 powder was the reference material. A heating rate of 10 °C/min and a sensitivity setting of 8 µv/cm were maintained for all the DTA runs.

2.3. Heat Treatment

The glass samples were heat-treated from room temperature to the required temperature for a specific soaking time in a muffle furnace, then the furnace was switched off, and the samples were permitted to cool inside it to room temperature. Detection of the crystallization temperatures was guided by the DTA results. Soaking times were measured after the furnace reached the desired temperature. Figure 1 shows Schematic presentation for the production of glass and glass–ceramic materials from industrial wastes.

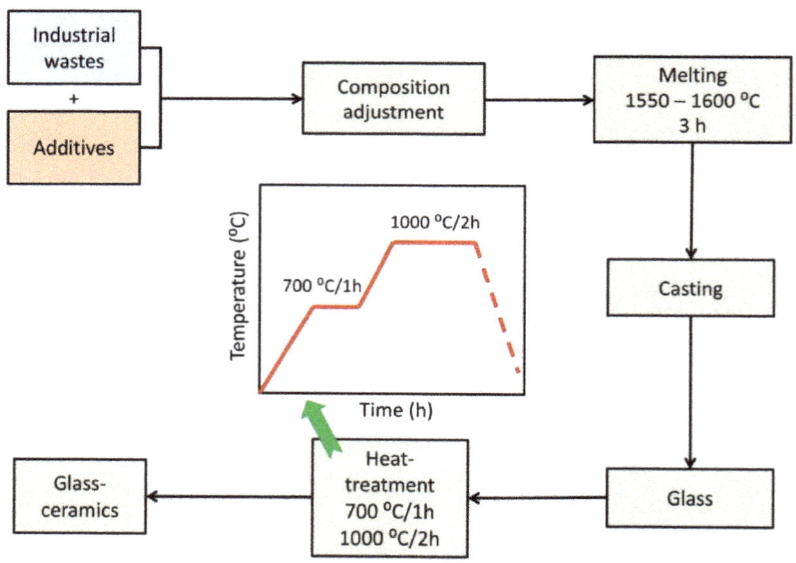

Figure 1. Schematic presentation for the production of glass and glass–ceramic materials from industrial wastes.

2.4. X-ray Diffraction (XRD)

The formed phases were recognized via the X-ray diffraction technique. The instrument used was Bruker AXS diffractometer (CD8—ADVANCE) with Cu-Ka radiation operating at 40 Kv and ten mA. The diffraction information was listed as 2θ values between 4° and 70°, and the scanning rate was 10°/min.

2.5. Scanning Electron Microscope (SEM)

The microstructure of the synthesized glass–ceramic samples was investigated by using a scanning electron microscope (SEM: JEOL, XL30, Philips, Amsterdam, The Netherlands), that operated at an acceleration voltage of 20 kV. The freshly fractured sample was coated with a layer of gold on the broken surface (to minimize any charging effect) to observe the internal microstructure.

2.6. Thermal Expansion Measurement

In this study, the mechanical dilatometry technique for thermal expansion measurement is applied. In this technique, a sample is heated in an oven, and the specimen ends; displacements are transmitted to a sensor utilizing pushrods. This test is usually functionalized to materials with CTE above 5×10^{-6}/K (2.8×10^{-6}/°F) over the temperature range of −180 to 900 °C (−290 to 1650 °F). Pushrods are of high-purity alumina type. Alumina systems can extend the temperature range up to 1600 °C (2900 °F). The instrument used was NETZSCH DIL 402 PC (made in Japan).

3. Results and Discussion

3.1. Differential Thermal Analysis (DTA)

DTA of the studied glasses is explained in Figure 2. The curves resulting from the investigated glasses showed endothermic peaks in the temperatures ranging from 749 to 797 °C. The differential thermal analysis curves show that the endothermic peaks of the investigated glasses are wholly affected by the cordierite contents in the glass. As the cordierite content increases, the endothermic peak temperatures shift towards higher values. The endothermic and exothermic peaks of the investigated glasses for each glass composition are shown in Table 3. The DTA of the glasses G10 to G50 (Figure 2) are, to some extent, similar. They are characterized by a broad exothermic peak with a temperature ranging from 954 to 987 °C. While the other glasses, G60 to G80, have characterized sharp exothermic peaks recorded at temperatures ranging from 954 to 1018 °C, respectively.

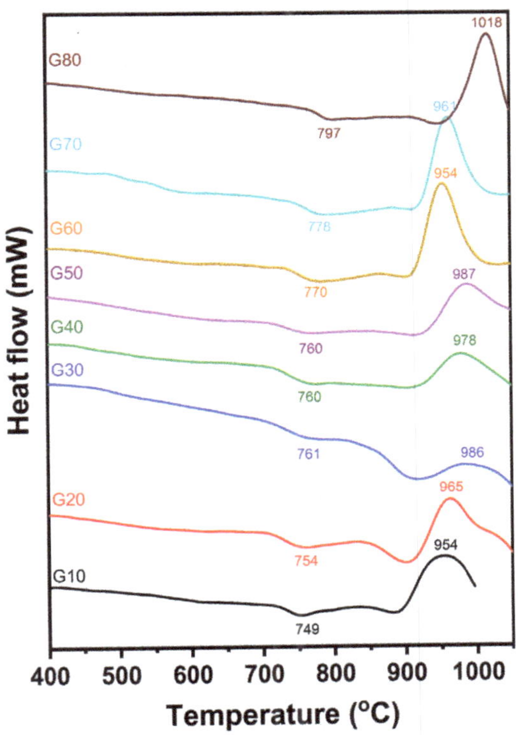

Figure 2. Differential thermal analysis of the investigated glasses.

Table 3. Thermal expansion coefficient and phases developed of the investigated samples.

Glass No.	Linear Expansion Coefficient (ὰ) × 10^{-6}/°C			Phases Developed
	50–300 °C	50–500 °C	50–700 °C	
G10	4.0838	5.9616	6.2161	B-woll.
G20	2.7820	3.4909	5.2155	B-woll.
G30	2.0008	3.0399	4.5930	Diop. and parawoll.
G40	1.9857	3.0385	4.5721	Parawo. and Diop
G50	1.3134	3.0764	4.5671	An. and Diop
G60	1.2896	2.7932	3.6538	An. and Diop
G70	1.2126	2.4522	3.3364	An. and Diop
G80	0.1478	2.2153	2.6181	Cord. and An.

Where. B-woll. = B-wollastonite, Diop. = Diopside, parawoll. = Parawollastonite, An. = Anorthite, Cord. = Cordierite.

It should be said that the data recorded from DTA are considered to be the responsible guide for the heat treatment schedule of the investigated glasses.

The DTA curves show a slight dip in the range 749 to 797 °C most probably, as confirmed above, owing to the transition temperature of the glass (T_g) or corresponding approximately, following Devekey and Majumdar [25], to the temperature range named as T_g and $T_{s.}$

These endothermic agree to initial crystallization (precrystallization), where the glass-creating oxides begin to arrange themselves at this period in beginning structure groups appropriate for the subsequent crystallization El-Shennawi [26]. This thermal absorption span will be, however, considered as the nucleation range. Exothermic peaks result from glass devitrification with a corresponding release of thermal energy.

3.2. X-ray Diffraction (XRD)

3.2.1. XRD of the Investigated Samples after Treatment at Temperature 1000 °C for 2 h

X-ray diffraction analysis of the studied glass–ceramic samples G10–G80 after heat treatment at a temperature of 1000 °C for 2 h is shown in Figure 3. The main crystalline minerals developed are β-wollastonite ($CaSiO_3$) (JSPDS No.27-1064), parawollastonite ($CaSiO_3$) (JCPDS No.27-88), diopside ($CaMgSi_2O_6$) (JCPDS No.22-534), anorthite ($CaAl_2Si_2O_8$) (JCPDS No.12-301) and cordierite ($Mg_2Al_4Si_5O_{18}$) (JCPDS No.13-294). Concerning samples G10 and G20, it is obvious that β- wollastonite ($CaSiO_3$) is the only monomineralic phase formed, characterized by the following lines 3.296, 3.807, 3.489, and 3.076 Å. While for samples G30 and G40, the β-wollastonite transferred into parawollastonite, characterized by lines 2.97, 3.318, and 2.699 Å with the formation of the diopside phases, which is characterized by the following lines 2.972, 3.182, and 2.503 Å. Whereas in glass–ceramic samples Nos (G50–G70), the parawollastonite transforms into anorthite, characterized by the following lines 3.202, 4.036, 3.739, and 2.501 Å. In sample G80, the low-cordierite phase formed, which is characterized by the following lines, 8.412, 3.128, 3.366, and 3.015 Å with an anorthite phase. Samples G10 and G20 show the formation of β-wollastonite only; this is consistent with Table 2, where the percentage of wollastonite ranges from 90 to 80 % in samples G10 and G20, respectively. It also agrees with many publications, such as Shaker et al. [27] and Khater et al. [28,29], who explained that β-wollastonite is formed at low temperatures and is unstable and then transformed into parawollastonite. Samples G30 and G40 show the formation of parawollastonite and diopside. This shows that β-wollastonite ($CaSiO_3$) is converted to parawollastonite, which agrees with Shaker et al. [27]. In samples G50, G60, and G70, the formation of diopside and anorthite is attributed to the aluminum elements present in cordierite with wollastonite, which help form the anorthite mineral while the magnesium and silica elements present in cordierite help in the formation of diopside. Sample G80 contains both minerals cordierite and anorthite, indicating the reach of a stage close to equilibrium phases.

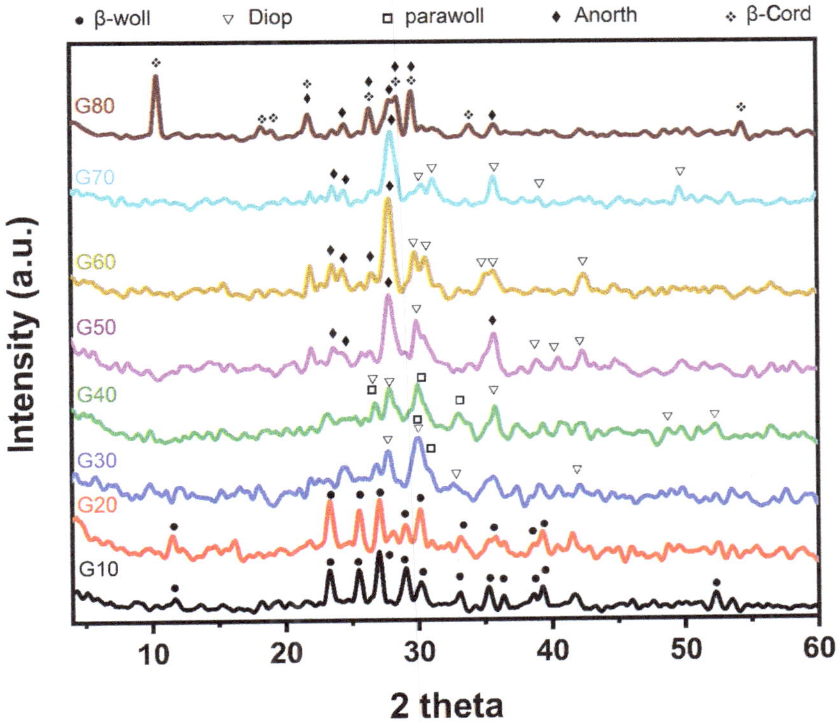

Figure 3. X-ray diffraction patterns of the investigated samples based on by-pass cement dust after heat-treatment at 1000 °C for 2 h.

3.2.2. XRD of the G60 after Treatment at a Temperature

Sample G60 is selected to study the effect of temperature on the crystalline phases formed by heat-treating. Sample G60 is heat-treated at a temperature of 950 °C for 2 h and also at 1000 °C for two hours (Figure 4). It is shown that after its heat treatment at 950 °C for 2 h, anorthite ($CaAl_2Si_2O_8$) (JCPDS No.18-1202) is formed with the following lines 3.173, 4.013, 2.875, and 2.138Å. When the temperature is raised to 1000 °C, the mineral diopside is formed with anorthite, and this shows that the anorthite phase is formed before the diopside phase and at low temperatures. This agrees with Juan Qin et al. [30].

3.3. *Scanning Electron Microscope (SEM)*

Figure 5 shows SEM photographs of the glass–ceramics samples (G10 to G80) after heat treatment at 1000 °C for 2 h. It can be noticed from the figure that all samples are generally marked with well-crystallized oriented tubular crystals except sample G60, which is characterized by randomly oriented crystal bundles. Moreover, the crystal lengths are decreased from G10 to G80. These tubular crystals are identical to wollastonite crystals. Rounded crystals are noticed in the G80 sample, which is attributed to the formation of cordierite crystals, as confirmed by XRD analysis. The cordierite ($Mg_2Al_4Si_5O_{18}$) is nominally increased from G10 to G80 at the expense of the wollastonite ($CaSiO_3$) phase in the cordierite–wollastonite system designed in this work. This increase in cordierite percentage increases the glass melt viscosity, as observed during the glass preparation process. This finding can be explained on the base of Al_2O_3 content, where it increased as the cordierite percentage increased in the glass composition. Al_2O_3 is capable of forming bonds with silica by forming bridging oxygens and generates Al–O–Si bonds causing an increase in glass viscosity [31]; hence, the glass crystallization affinity can be likely decreased, and the glassy phase is usually detected as a matrix for the formed crystals. In this study, the glassy phase can be observed in the crystal interstices, specifically in the G60 sample.

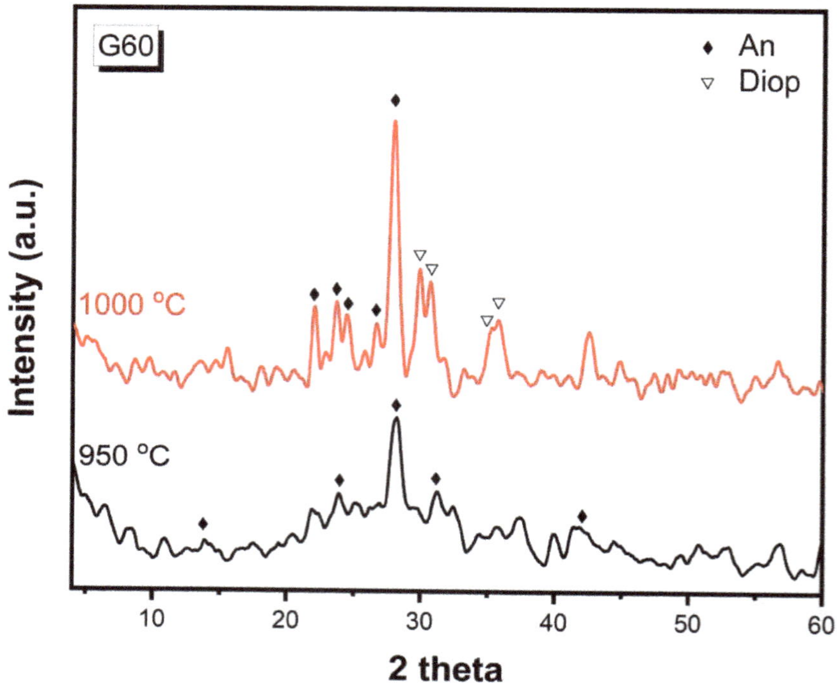

Figure 4. X-ray diffraction patterns of G60 after heat-treatment at 950 °C for 2 h and 1000 °C for 2 h.

On the other hand, a selected glass sample was heat-treated at a temperature close to the nucleation temperature to study the effect of temperature heat treatment on the formed microstructure of the glass–ceramic. Sample G60 is selected because it is characterized by optimum properties measured in this study. Figure 6 presents G60 glass–ceramic heat-treated at 700 °C for 3 h and 1000 °C for 1 h. It can be observed from the figure that fine-grained microstructure is noticed. While the sample, after being heat-treated at 700 °C for one h and then 1000 °C, has a coarse-grained microstructure.

3.4. *Theoretical Considerations of Thermal Expansion and Thermal Expansion Behavior of the Studied Samples*

The coefficient of thermal expansion (CTE) is a property of a substance indicating the degree of material to be expanded by heating; substances expand at different altitudes. In the temperature ranges, the thermal expansion of objects is proportional to temperature alteration. Thermal expansion is an essential property for detecting the targeted application of materials when a structural part is heated and kept at a constant length [32].

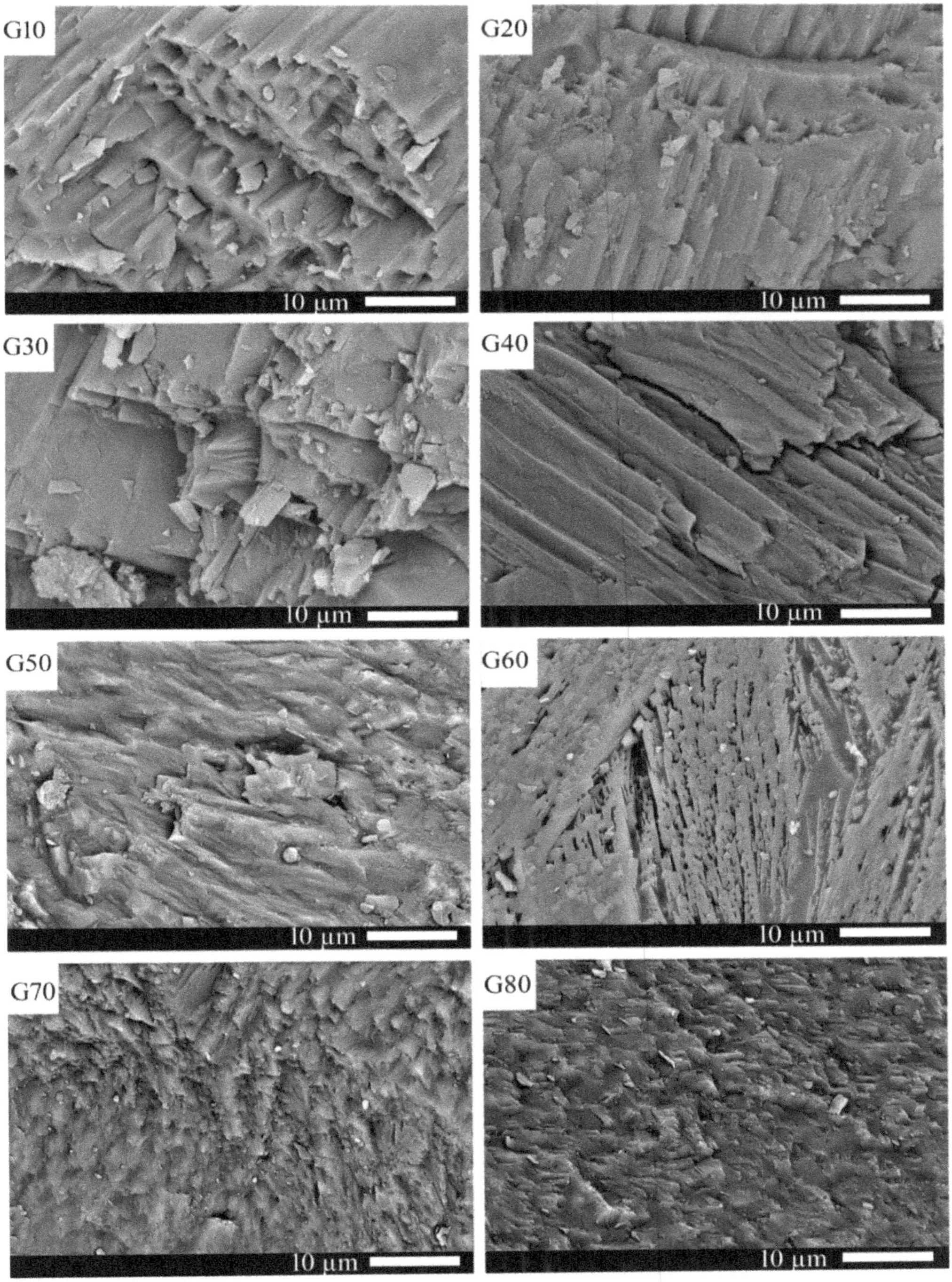

Figure 5. SEM micrographs of the investigated samples after heat-treatment at 1000 °C for 2 h.

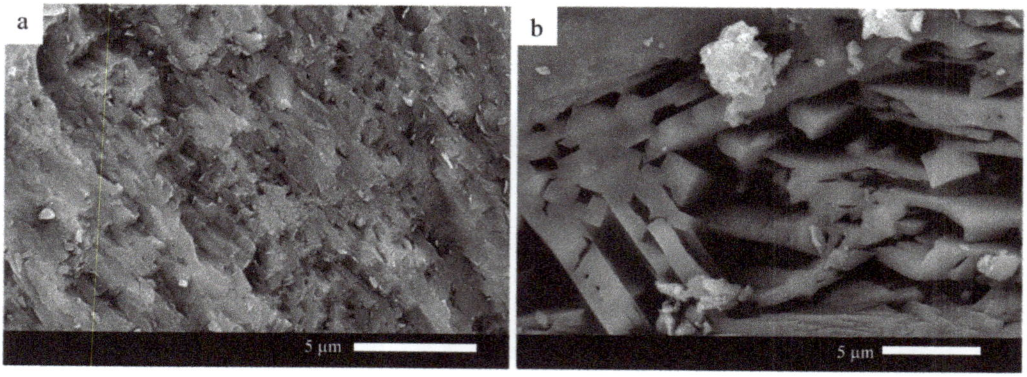

Figure 6. SEM micrographs of G60 heat-treated at (**a**) 700 °C 3 h and 1000 °C 1 h; (**b**) 700 °C 1 h and 1000 °C 1 h.

Most solid substances expand by heating and contract after cooling. The change in length with temperature for a solid material can be expressed as:

$$(L_f - L_0)/L_0 = \alpha_l\,(T_f - T_0) \quad (1)$$

$$\Delta L/L_0 = \alpha_l \Delta T \quad (2)$$

$$\alpha_l = 1/L(dL/dT) \quad (3)$$

where L_0 and L_f stand for the original and final lengths sample with the temperature change from T_0 to T_f, respectively. The parameter α_l (CTE) has units of reciprocal temperature (K^{-1}) such as μm/m K or 10^{-6}/K.

Heating or cooling has a significant effect on the dimensions of a material, with a considerable volume change. Volume changes can be calculated from:

$$\Delta V/V_0 = \alpha_v \Delta T \quad (4)$$

where ΔV and V_0 are the volume change and original volume of the sample, respectively, and α_v symbolizes the volume coefficient of thermal expansion. In many materials, the value of α_v is anisotropic; that is, it depends on the crystallographic orientation along which it is measured. For materials in which the thermal expansion is isotropic, α_v is approximately $3\alpha_l$ [33].

Thermal expansion leads to a change in the space between particles of a substance, which affects the volume of the substance while negligibly varying its mass (the negligible amount comes from energy–mass equivalence).

Depending on the above considerations, the results of thermal expansion coefficients of the studied samples can be discussed as follows:

From Figure 7 and Table 3, it is clear that all coefficients of thermal expansion (CTE) values decrease from G10 to G80 samples, i.e., by increasing the cordierite percent at the expense of wollastonite content.

Coefficient thermal expansion (CTE) values of the glass–ceramic depend on the nature and quantity of the developed crystalline phases and the residual glassy matrix [34,35]. A comprehensive range of thermal expansion coefficients is controlled by the crystal types and proportions of these phases, which are considered the main principle of producing glass–ceramics with restricted thermal expansion coefficients.

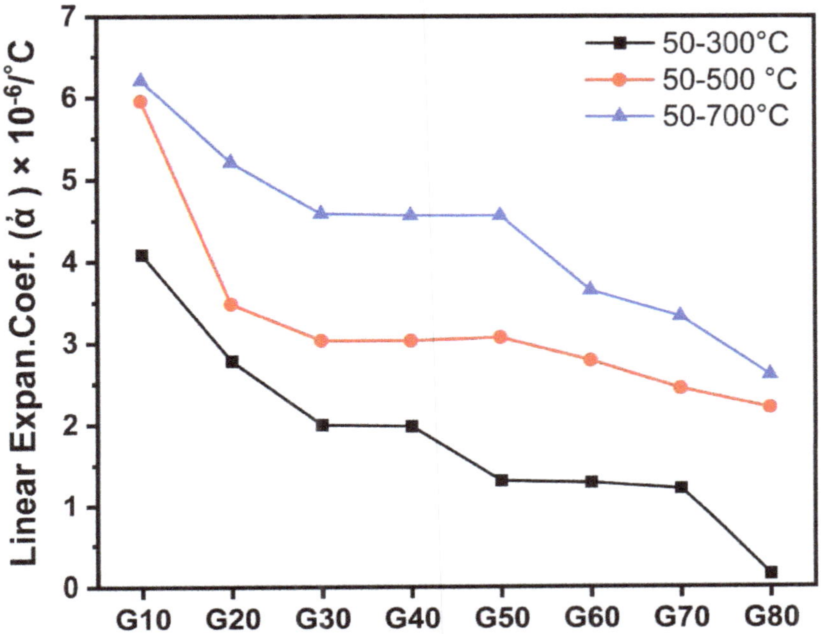

Figure 7. Thermal expansion coefficient of the studied glass–ceramics after heat-treatment at 1000 °C for 2 h.

Low-expansion glass–ceramics usually contain crystalline phases as well as a vitreous phase. Ion exchange always takes place in the vitreous phase. Schairer and Bown [36] found a solid solution between wollastonite and diopside in the CaO–MgO–SiO_2 system with the highest 22% diopside at eutectic. The produced Ca–Mg–silicate phase is a solid solution between diopside and wollastonite. Diopside constituent in this solid solution is about 49.6%. However, Omer et al. [37] mentioned that a wollastonite–diopside solid solution phase with about 66 % of diopside might be created. This Mg-rich wollastonite solid solution is eventually transformed by increasing time or raising the heat treatment temperature to wollastonite and diopside [38].

By increasing Al^{3+} ions (from G10 to G80), the formation of diopside, anorthite, and cordierite phases at the expense of wollastonite is noticed. Thus the values of the expansion coefficients (α) of glass–ceramic samples are decreased. Consequently, higher Tg temperatures could be expected.

The thermal expansion property of the crystalline solids is relatively different from that of the parent glasses. The thermal expansion coefficient (α) of the glass–ceramics is a function of the thermal expansion coefficients, and elastic properties of all precipitated crystalline phases, residual glass, and the resulting microstructure. An extensive range of thermal expansion coefficients is covered by the different crystal types [39].

Wollastonite has an α-value of 94×10^{-7} °C^{-1} [40], but anorthite has α-values of $51–64 \times 10^{-7}$ °C^{-1} [41].

Cordierite is well known for its very low thermal expansion coefficient of about 2.5×10^{-6}/°C [42]. However, the CTE of cordierite may depend on the nature and amount of phases that coexist with it in the glass–ceramic sample [43]. The above α-values give a good explanation of the noticeable decrease in thermal expansion of the studied samples (Table 3 and Figure 7).

The thermal expansion coefficient (CTE) of the diopside is registered to be 83.0×10^{-7} °C^{-1} [44,45].

The co-existence of diopside, which has a high thermal expansion coefficient, with anorthite, which has a comparatively lower value of α, leads to a lowering of the thermal expansion coefficient of the final glass–ceramic product (G50–G70).

In samples (G50–G80), it was observed that the presence of both CaO and Al_2O_3 containing crystalline phases leads to a decrease in the thermal expansion of the glass–ceramic samples. Al_2O_3 is mainly efficient in this concept. The α-values of samples G50–G80 (with CaO and Al_2O_3 containing phases) were lower than those of samples G10–G40, free from Al_2O_3 containing phases and composed of crystalline phases with high CTE values, which are β-wollastonite, para-wollastonite and diopside.

It is worth mentioning that β-wollastonite ($CaSiO_3$) contains triclinic wollastonite and monoclinic parawollastonite. These forms are not easily differentiated except for single-crystal X-ray investigation [46]. The stabilities of β- and para-wollastonite phases are likely to be very similar; because of the presence and intergrowth of both forms [47].

4. Conclusions

1. The high-performance glass–ceramic materials were successfully obtained through a wollastonite–cordierite system based on industrial waste and natural raw materials.
2. The present results showed that the by-pass cement dust could be used in quantities that may exceed 60% of the batch weight to produce glass–ceramic materials. This application may pave the way for the disposal of this waste in an environmentally friendly manner. By-pass cement and natural raw materials can be used successfully in preparing glass–ceramic materials without any chemical additives.
3. These glass–ceramic materials have been used for different purposes, such as floor and wall tiles, benchtops, sewer pipes, and many others.
4. Cleaning, environmental protection, and public health preservation by getting rid of by-pass cement dust through the preparation of glass–ceramic materials.
5. It was found that β-wollastonitte is formed at low temperatures, and this helps to save energy and then turns into parawollastonite at higher temperatures. Accordingly, glass–ceramic materials containing β-wollastonite have a better economic value due to their formation at low temperatures.
6. A fine-grained microstructure is obtained in the samples that are cordierite-rich.
7. Increasing the time of heat-treatment at endothermic temperature helped form fine-grained microstructures.
8. Obtaining glass–ceramic materials with low thermal expansion ranging from 6.216 to 2.618×10^{-6} (in the range of 20–700 ° C) decreased with an increasing percentage of cordierite.

Author Contributions: Conceptualization, G.A.K. and A.A.E.-K.; methodology, M.M.F.; validation, G.A.K.; software, M.M.F.; Formal analysis, A.A.E.-K. and M.M.F.; Project administration, G.A.K.; Supervision, G.A.K.; Writing–review & editing, G.A.K. All authors have read and agreed to the published version of the manuscript.

Funding: This article is based upon work supported by Science and Technology Development Fund Authority (STDF) under grant No. 41555.

Institutional Review Board Statement: Not applicable.

Data Availability Statement: The data of this work is available for first author after puplication.

Acknowledgments: This article is based upon work supported by Science and Technology Development Fund Authority (STDF) under grant No. 41555.

Conflicts of Interest: The authors declare that they have no known competing financial interests or personal relationships that could have influenced the work reported in this paper.

References

1. Awuchi, C.G.; Hannington, T.; Awuchi, C.G.; Igwe, V.S.; Amagwula, I.O. Industrial waste management, treatment, and health issues: Wastewater, solid, and electronic wastes. *Eur. Acad. Res.* **2020**, *8*, 1081–1119.

2. Fagariba, C.J.; Song, S. Assessment of impediments and factors affecting waste management: A case of Accra Metropolis. *Preprints* **2016**, 2016090012. [CrossRef]
3. El-Attar, M.M.; Sadek, D.M.; Salah, A.M. Recycling of high volumes of cement kiln dust in bricks industry. *J. Clean. Prod.* **2017**, *143*, 506–515. [CrossRef]
4. Ali, A.A.M.; Negm, A.M.; Bady, M.F.; Ibrahim, M.G.; Suzuki, M. Environmental impact assessment of the Egyptian cement industry based on a life-cycle assessment approach: A comparative study between Egyptian and Swiss plants. *Clean Technol. Environ. Policy* **2016**, *18*, 1053–1068. [CrossRef]
5. El-Sayed Seleman, M.M.; El-kheshen, A.A.; Kharbish, S.; Ebrahim, W.R. Utilization of Cement Kiln Dust for the Preparation of Borosilicate Glass. *Intercom Int. Ceram. Rev.* **2020**, *69*, 26–33. [CrossRef]
6. Abid, K.N.; Yasse, R.S. Use of Bypass Cement Dust from Al-Kufa Cement Factory for Production of Glass Ceramic Material. *Al Qadisiyah J. Eng. Sci.* **2019**, *12*, 118–126. [CrossRef]
7. Srivastava, L.; Paramkusam, B.; Prasad, A. Stabilisation of engine oil contaminated soil using cement kiln dust. In Proceedings of the Indian Geotechnical Conference, Delhi, India, 16–18 December 2010.
8. Hamann, B.; Schulz, M.; Schneider, C. Use of alkaline process dusts from the cement industry in glass production processes. *Veröffentlichung_DGG J.* **2008**.
9. Pigaga, A.; Juškenas, R.; Virbalytė, D.; Klimantavičiūtė, M.G.; Pakštas, V. The use of cement kiln dust for the removal of heavy metal ions from aqueous solutions. *Trans. IMF* **2005**, *83*, 210–214. [CrossRef]
10. Taha, R.; Al-Rawas, A.; Al-Harthy, A.; Al-Siyabi, H. *Use of Cement by-Pass Dust in Soil Stabilization*; Qatar University: Doha, Qatar, 2001.
11. Khater, G.; Nabawy, B.S.; El-Kheshen, A.A.; Abdel-Baki, M.; Farag, M.M.; Abd Elsatar, A.G. Preparation and characterization of low-cost wollastonite and gehlenite ceramics based on industrial wastes. *Constr. Build. Mater.* **2021**, *310*, 125214. [CrossRef]
12. Khater, G. Glass-ceramics in the CaO–MgO–Al_2O_3–SiO_2 system based on industrial waste materials. *J. Non Cryst. Solids* **2010**, *356*, 3066–3070. [CrossRef]
13. Khater, G. Use of by-pass cement dust for production of glass ceramic materials. *Adv. Appl. Ceram.* **2006**, *105*, 107–111. [CrossRef]
14. Rawlings, R.; Wu, J.; Boccaccini, A. Glass-ceramics: Their production from wastes—A review. *J. Mater. Sci.* **2006**, *41*, 733–761. [CrossRef]
15. William, W.; Njenga, H. Industrial Chemistry. 2007. Available online: https://54.225.189.63/handle/123456789/40 (accessed on 22 March 2022).
16. Karmakar, B. *Functional Glasses and Glass-Ceramics: Processing, Properties and Applications*; Butterworth-Heinemann: Oxford, UK, 2017.
17. Hayashi, S.; Sugai, M.; Nakagawa, Z.E.; Takei, T.; Kawasaki, K.; Katsuyama, T.; Yasumori, A.; Okada, K. Preparation of $CaSiO_3$ whiskers from alkali halide fluxes. *J. Eur. Ceram. Soc.* **2000**, *20*, 1099–1103. [CrossRef]
18. Zhang, W.; Liu, H. A low cost route for fabrication of wollastonite glass–ceramics directly using soda-lime waste glass by reactive crystallization–sintering. *Ceram. Int.* **2013**, *39*, 1943–1949. [CrossRef]
19. Hodges, C.A. Mineral resources, environmental issues, and land use. *Science* **1995**, *268*, 1305–1312. [CrossRef]
20. Boccaccini, A.R. Glass-ceramics from municipal incinerator fly ash. In *Fuel and Energy Abstracts*; Elsevier Science: Amsterdam, The Netherlands, 1998.
21. Chang, C.; Mao, D.; Wu, J. Characteristics of crystals precipitated in sintered apatite/wollastonite glass ceramics. *Ceram. Int.* **2000**, *26*, 779–785. [CrossRef]
22. Costa Oliveira, F.; Franco, J.A.; Cruz Fernandes, J.; Dias, D. Newly developed cordierite–zircon composites. *Br. Ceram. Trans.* **2002**, *101*, 14–21. [CrossRef]
23. Schneider, H.; Schreuer, J.; Hildmann, B. Structure and properties of mullite—A review. *J. Eur. Ceram. Soc.* **2008**, *28*, 329–344. [CrossRef]
24. Nadafan, M.; Malekfar, R.; Dehghani, Z. Structural and optical properties of cordierite glass-ceramic doped in polyurethane matrix. *AIP Adv.* **2015**, *5*, 067135. [CrossRef]
25. De Vekey, R.; Majumdar, A. Nucleation and crystallization studies of some glasses in the CaO-MgO-Al_2O_3-SiO_2 system. *Mineral. Mag.* **1970**, *37*, 771–779. [CrossRef]
26. El-Shennawi, A.W.A. The use of differential thermal analysis in the study of glass-ceramics. *Ain Shams Sci. Bull.* **1983**, *24*, 225–244.
27. Edrees, S.J.; Shukur, M.M.; Obeid, M.M. First-principle analysis of the structural, mechanical, optical and electronic properties of wollastonite monoclinic polymorph. *Comput. Condens. Matter* **2018**, *14*, 20–26. [CrossRef]
28. Khater, G.A.; Nabawy, B.S.; El-Kheshen, A.A.; Abdel-Baki, M.; Farag, M.M. Utilizing of solid waste materials for producing porous and lightweight ceramics. *Mater. Chem. Phys.* **2022**, *280*, 125784. [CrossRef]
29. Khater, G.A.; Nabawy, B.S.; El-Kheshen, A.A.; Abdel Latif, M.A.B.; Farag, M.M. Use of Arc Furnace Slag and Ceramic Sludge for the Production of Lightweight and Highly Porous Ceramic Materials. *Materials* **2022**, *15*, 1112. [CrossRef]
30. Qin, J.; Cui, C.; Cui, X.; Hussain, A.; Yang, C.; Yang, S. Recycling of lime mud and fly ash for fabrication of anorthite ceramic at low sintering temperature. *Ceram. Int.* **2015**, *41*, 5648–5655. [CrossRef]
31. Leko, V.; Mazurin, O. Analysis of regularities in composition dependence of the viscosity for glass-forming oxide melts: II. Viscosity of ternary alkali aluminosilicate melts. *Glass Phys. Chem.* **2003**, *29*, 16–27. [CrossRef]
32. Ho, C.Y.; Taylor, R.E. *Thermal Expansion of Solids*; ASM International: Almere, The Netherlands, 1998; Volume 4.

33. Nave, R. *Hyperphysics*; Georgia State University: Atlanta, Georgia, 2002; Available online: http://hyperphysics.phy-astr.gsu.edu (accessed on 22 March 2022).
34. Abo-Mosallam, H.A.; Darwish, H.; Salman, S.M. Crystallization characteristic and properties of some zinc containing soda lime silicate glasses. *J. Mater. Sci. Mater. Electron.* **2009**, *21*, 889–896. [CrossRef]
35. Al-Harbi, O.A. Crystallisation and stability of glass-ceramics within the $CaO–Li_2O–SiO_2$ system that contain wollastonite. *Glass Technol. Eur. J. Glass Sci. Technol. Part A* **2007**, *48*, 35–40.
36. Schairer, J.F.; Bowen, N.L. The binary system CaSiO 3-diopside, and the relations between CaSiO 3 and akermanite. *Am. J. Science* **1942**, *240*, 725–742. [CrossRef]
37. Omar, A.A.; Salman, S.M.; Mahmoud, M. Phase relations in the diopside-anorthite-akermanite system. *Ceram. Int.* **1986**, *12*, 53–59. [CrossRef]
38. Salman, S.M.; Darwish, H. Development of low cost glass-ceramic based on blast furnace slag and granite rock. *Ceram. Silik.* **2006**, *50*, 88.
39. Bach, H.; Krause, D. *Low Thermal Expansion Glass Ceramics*; Springer: Berlin/Heidelberg, Germany, 2005.
40. Strnad, Z. *Glass—Ceramic Materials: Liquid Phase Separation, Nucleation and Crystallization in Glasses*; Elsevier Science Publishers: Amsterdam, The Netherlands, 1986; Volume 268.
41. Peng, F.; Liang, K.-M.; Hu, A.-M. Nano-crystal glass—Ceramics obtained from high alumina coal fly ash. *Fuel* **2005**, *84*, 341–346. [CrossRef]
42. Valášková, M. Clays, clay minerals and cordierite ceramics-A review. *Ceram. Silikáty* **2015**, *59*, 331–340.
43. Li, Y.; Wang, J.; Sun, L.; Wang, J. Mechanisms of ultralow and anisotropic thermal expansion in cordierite $Mg_2Al_4Si_5O_{18}$: Insight from phonon behaviors. *J. Am. Ceram. Soc.* **2018**, *101*, 4708–4718. [CrossRef]
44. Rigby, G.; Green, A. The thermal expansion characteristics of some calcareous and magnesian minerals. *Trans. Br. Ceram. Soc.* **1942**, *41*, 123–143.
45. Khater, G.; Yue, Y.; Abu Safiah, M.O.; Mahmoud, M.A. Synthesis and characterization of anorthite and magnetite glass-ceramics from basaltic rocks. *Silicon* **2019**, *11*, 1763–1774. [CrossRef]
46. Tolliday, J. Crystal structure of β-wollastonite. *Nature* **1958**, *182*, 1012–1013. [CrossRef]
47. Ohsato, H.; Sugimura, T. Morphology of synthetic β-wollastonite and para-wollastonite. *J. Cryst. Growth* **1986**, *74*, 656–658. [CrossRef]

materials

Article

Effective Extraction of the Al Element from Secondary Aluminum Dross Using a Combined Dry Pressing and Alkaline Roasting Process

Han Lv [1,2], Mingzhuang Xie [1,2], Zegang Wu [1,2], Lili Li [1,2], Runjie Yang [1,2], Jinshan Han [1,2], Fengqin Liu [1,2,*] and Hongliang Zhao [1,2,*]

1 State Key Laboratory of Advanced Metallurgy, University of Science and Technology Beijing, Beijing 100083, China
2 School of Metallurgical and Ecological Engineering, University of Science and Technology Beijing, Beijing 100083, China
* Correspondence: liufq@ustb.edu.cn (F.L.); zhaohl@ustb.edu.cn (H.Z.)

Abstract: Secondary aluminum dross (SAD) is a hazardous solid waste discharged from aluminum electrolysis and processing and the secondary aluminum industries, which causes severe environmental pollution and public health disasters. The stable presence of the α-Al_2O_3 and $MgAl_2O_4$ phases in SAD makes it difficult for it to be efficiently utilized. A combined dry pressing and alkaline roasting process was proposed for extracting the valuable Al element from SAD. Two alkaline additives (NaOH and Na_2CO_3) were selected as a sodium source for extracting the aluminum source from SAD in order to perform the thermodynamic analysis and roasting experiments. The phase transition behavior and the leaching performance tests were conducted using X-ray diffraction, scanning electron microscopy, X-ray fluorescence, leaching kinetics and thermal analysis. The recovery of Al and Na reached the values of 90.79% and 92.03%, respectively, under the optimal conditions (roasting temperature of 1150 °C, Na_2CO_3/Al_2O_3 molar ratio of 1.3, roasting time of 1 h, leaching temperature of 90 °C, L/S ratio of 10 mL·g^{-1} and leaching time of 30 min). Meanwhile, the removal efficiency of N and Cl reached 98.93% and 97.14%, respectively. The leaching kinetics indicated that the dissolution of $NaAlO_2$ clinkers was a first-order reaction and controlled by layer diffusion process. The green detoxification and effective extraction of the Al element from SAD were simultaneously achieved without any pretreatments.

Keywords: aluminum dross; recycling; thermodynamic analysis; kinetics

Citation: Lv, H.; Xie, M.; Wu, Z.; Li, L.; Yang, R.; Han, J.; Liu, F.; Zhao, H. Effective Extraction of the Al Element from Secondary Aluminum Dross Using a Combined Dry Pressing and Alkaline Roasting Process. *Materials* **2022**, *15*, 5686. https://doi.org/10.3390/ma15165686

Academic Editors: Lorena Zichella and Rossana Bellopede

Received: 18 July 2022
Accepted: 10 August 2022
Published: 18 August 2022

1. Introduction

With the rapid development of the global aluminum industry, the accumulation of hazardous solid wastes has become a serious environmental problem. Aluminum dross (AD) is a hazardous solid waste discharged during the electrolysis, processing and regeneration of aluminum, which can usually be divided into primary aluminum dross (PAD) and secondary aluminum dross (SAD). Primary aluminum dross (PAD) is generated in primary smelters and contains 15–80% metallic aluminum [1]. On the other hand, SAD is a by-product of PAD produced after mechanical screening or remelting with flux to recover the metallic aluminum, and it contains less than 10% aluminum [2]. Currently, the recovery process for extracting metallic aluminum from PAD is comparatively mature and has been applied in most of the aluminum processing plants [3]. The composition of SAD depends on the sources of aluminum scrap, the remelting technology, the type of additives, and the employed process. Generally, SAD consists of alumina (Al_2O_3), aluminum nitride (AlN), metallic aluminum, magnesium aluminum spinel ($MgAl_2O_4$), other impurity oxides and salts [4]. Globally, the aluminum industry generates more than 3 million tons of SAD every year. Around 95% of the SAD is stockpiled in landfill sites due to its complex composition and the technical difficulties involved in processing it [5].

The toxic and hazardous substances in aluminum dross are mainly AlN and chloride salts. Aluminum nitride (AlN) is produced due to the reaction of molten aluminum with nitrogen during the formation of AD, which reacts very easily with water and moist air to release toxic and irritating ammonia [6]. One ton of SAD is capable of releasing approximately 109 m^3 of ammonia, which can burn the skin and the respiratory tract and even cause serious pulmonary diseases [7]. Ammonia also reacts with acidic gases in the atmosphere to produce aerosols of ammonium salts, posing severe air pollution risks. The arbitrary stockpiling of SAD can lead to the infiltration of soluble chlorides into the soil and groundwater, posing a serious environmental threat. Additionally, as the particles of SAD are extremely fine, they are easily dispersed in the air during processing, transportation and landfill, causing silicosis and bronchitis through excessive inhalation [8]. Therefore, SAD is considered as a hazardous solid waste in most of the countries [9].

The development of more sustainable processes to mitigate environmental pollution is absolutely essential. Considering hazardous wastes as alternative materials for preparing valuable products is a potential method for reducing the generation and management of wastes [10]. SAD is an excellent alternative resource for the development of the alumina industry, especially in view of the increasing alumina production and the growing scarcity of bauxite [11]. However, due to the complex composition of SAD and technical limitations, it is difficult to effectively extract Al elements from SAD.

In recent years, scholars have aimed to develop a variety of methods for the efficient recovery of Al element from SAD under different extraction conditions [12,13]. Some of them have achieved good results. However, there is no single method that can simultaneously achieve high Al extraction efficiency from SAD in a sustainable process [14]. Therefore, the large-scale utilization of SAD is severely restricted, and most studies are still conducted at the laboratory scale [15]. Current studies on the extraction of Al from SAD mainly focus on two processing routes: pyro-metallurgical and hydro-metallurgical processes [16]. The pyro-metallurgical processes mainly involve the alkaline roasting process with a leaching pretreatment and the plasma arc melting process. High Al extraction efficiency and high purity products can be achieved using the former process. However, this process requires a leaching pretreatment for the desalination and denitrification, which produces toxic ammonia and salt-containing waste liquid, thus greatly increasing the recovery costs [17]. The extremely high processing temperatures and energy consumption of the latter process make it unattractive for large-scale applications. The hydro-metallurgical routes are generally carried out by acidic or alkali leaching [18,19].

The acid leaching causes the impurity ions (Mg^{2+}, Fe^{3+}, Ca^{2+}, NH^{4+} and Cl^-) in SAD to enter the leaching solution, which makes it difficult to separate them from valuable aluminum ions and synthesize high-purity products [20,21]. Moreover, the alkaline leaching can yield high-purity $NaAlO_2$ solutions for the synthesis of high-value products; however, the recovery of Al obtained under either atmospheric or high pressure is too low [22]. The main problem with hydro-metallurgical routes is that a large amount of the stable α-Al_2O_3 and $MgAl_2O_4$ phases in SAD is difficult to dissolve in acid or alkali solution, resulting in the low recovery of Al [23,24]. Therefore, the key to developing a sustainable process for recycling SAD into a valuable product lies in the efficient recovery of the Al element [25,26].

In the present study, a novel promising method comprising of dry pressing and alkaline roasting is developed, as shown in Figure 1. The proposed process aims at economically achieving high Al recovery.

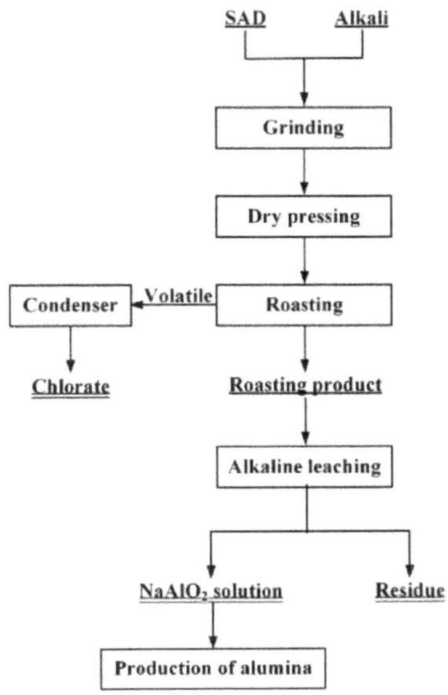

Figure 1. Flowchart of the preparation of alumina using SAD as the raw material.

2. Materials and Methods

The process presented in Figure 1 consists of three stages of dry molding, roasting and alkaline leaching. First, according to the content of Al element, SAD was mixed with a certain amount of alkali. The mixture was dry pressed under a constant pressure to produce cylindrical samples. Then, the clinkers were roasted at a high temperature and leached in alkaline solution to obtain $NaAlO_2$ solution, which can be used to produce alumina. During the roasting process, AlN was decomposed into harmless N_2, while the chlorate was separated and recovered after volatilization into the gaseous phase.

First, the thermodynamic analysis was carried out. Then, the roasting and leaching experiments were conducted to investigate the phase transition and to optimize the recovery efficiency. The leaching kinetics and the apparent activation energy of the $NaAlO_2$ leaching process were discussed. The developed extraction process described here has several innovations, some of which are as follows:

(1) The roasting process can achieve the sustainable treatment and extraction of the valuable Al element from SAD without any pretreatment. (2) Dry pressing was applied to avoid the generation of harmful ammonia and liquid waste and to decrease the energy consumption of the roasting process. (3) The reaction exhibited a high yield and was simple and sustainable. (4) The chlorate in the SAD was volatilized in the roasting process and could be recovered after condensation.

2.1. Materials

The SAD used in this study came from an aluminum remelting plant in Guangxi, China. The X-ray diffraction (XRD) analysis of the SAD showed that the aluminum-containing phases included the metallic aluminum, α-Al_2O_3, γ-Al_2O_3, AlN and $MgAl_2O_4$, as shown in Figure 2. These were the products of the reactions of the molten aluminum with the O_2, N_2 and Mg impurities during the formation and remelting of the SAD. Additionally, some impurity oxides (Fe_2O_3 and SiO_2) and salts, including chlorides (NaCl) and fluorides (CaF_2), were present in the SAD. The salts were added as a flux during the melting process to facilitate the heat transfer and reduce the oxidation of the molten aluminum. The chemical analysis of the SAD is presented in Table 1. Nearly 40 wt.% of the sample was made up of

Al element, indicating that the SAD was a solid waste with a high utilization value. All the nitrogen was transformed into the phase of AlN with a content of 19.62 wt.%.

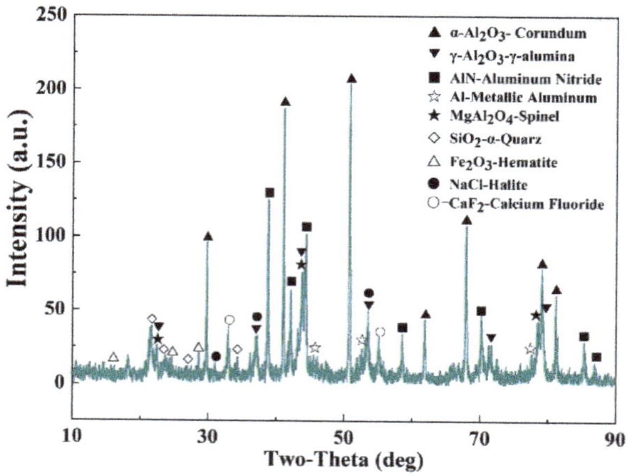

Figure 2. XRD pattern of the SAD sample.

Table 1. Chemical composition of SAD.

Element	Al	O	N	F	Na	Ca	Si	Mg	Fe	K	Cl
Content (wt.%)	38.23	38.14	6.7	0.25	2.16	0.32	2.04	6.18	1.04	0.57	4.2

2.2. Experimental Procedure

Figure 3 illustrates the schematic of the experimental setup. First, 100 g of SAD was accurately weighed and mixed thoroughly with a certain amount of alkali. In order to produce the specific compounds during the roasting process, the formulation of the SAD and alkali was controlled according to the molar ratio of Na_2O to Al_2O_3 (n(N/A)). The mixtures were dry pressed at 30 MPa to produce cylindrical samples with a diameter and length of 20 mm and 40 mm, respectively. The cylindrical samples were placed in a corundum crucible with dimensions of 120 mm × 80 mm × 30 mm and roasted in a muffle furnace under an atmospheric environment at a preset temperature for 1 h, with a constant heating rate of 10 °C min^{-1}. The crucible was taken out and placed in a desiccator to cool to room temperature. The roasting clinkers were crushed to 200 mesh in a mortar and leached in caustic liquor to obtain $NaAlO_2$ solution. The leaching residue was filtered and rinsed several times with boiling water. According to the industrial dissolution conditions of bauxite clinkers, the caustic concentration of the leaching solution was maintained at 60 g· L^{-1}.

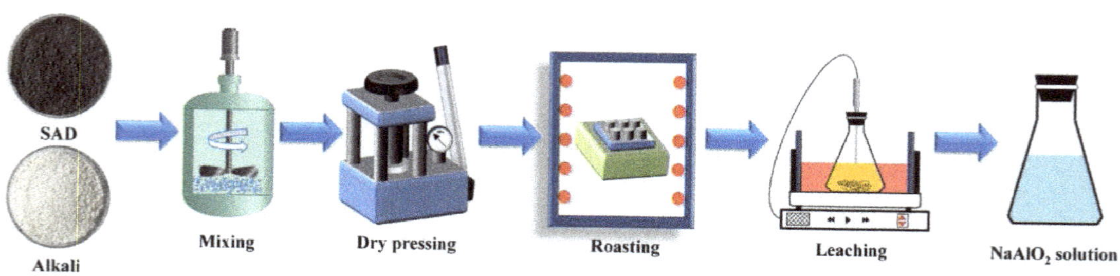

Figure 3. Schematic of the experimental setup and procedure.

2.3. Characterization Methods

The Gibbs free energy of the chemical reactions that can occur during roasting was analyzed using the reaction module of the software package Factsage (ver. 7.0, Thermfact, Montreal, Canada). The contents of Al and Na in the roasting clinkers and leaching residues were determined using X-ray fluorescence (XRF), and the recovery efficiencies of the Al and Na were calculated using Equations (1) and (2), respectively:

$$\eta_{Al} = [Al_1 - Al_2(Mg_1/Mg_2)/Al_1 \tag{1}$$

$$\eta_{Na} = [Na_1 - Na_2(Mg_1/Mg_2)/Na_1 \tag{2}$$

where Al_1, Mg_1 and Na_1 are the contents of Al, Mg and Na in the roasting clinkers, respectively, wt.%; and Al_2, Mg_2 and Na_2 are the contents of Al, Mg and Na in the leaching residues, respectively, wt.%. The content of soluble chlorate in the SAD samples was determined using a chloride ion activity meter (PCL-202, INESA Scientific Instruments, Shanghai, China). The content of AlN in the SAD samples was determined using the Kjeldahl method [27].

For the mineralogical study, an X-ray diffractometer (PW1710, Philips, Amsterdam, The Netherlands) was used with Cu Kα-radiation at 40.0 kV and 30.0 mA. The XRD tests were conducted within the 2θ range of 10–60° using a scanning speed of $0.1° \cdot min^{-1}$. Before testing, the samples were crushed to pass through a 200-mesh sieve and dried in an oven at 105 °C for 2 h. The diffractograms were identified with the help of the ICDD (International Centre for Diffraction Data) Powder Diffraction File (PDF-2) reference database and the Jade (vs. 6.0). The chemical composition of SAD was determined by XRF (mAX, AXIOS, Alamelo, The Netherlands). Micrographs of the roasting clinkers and leaching residues were observed using a scanning electron microscope (Regulus 8100, Hitachi, Japan) at an accelerating voltage of 15.0 kV. The microscopy samples were crushed to pass through a 200-mesh sieve and dried in an oven at 105 °C for 2 h, then sprayed evenly on the conductive adhesive and metal-lized with carbon in a vacuum evaporator (Q150T ES, Quorum, East Grinstead, Britain). An EDS (ul-tra-DLD, Shimadzu, Kyoto, Japan) connected with SEM was used to perform elemental anal-ysis on the sample particles. The elements with the content of 3%~20 wt.% have the accuracy of relative error <10%. And the elements with content greater than 20 wt.% have the higher accuracy of relative error <5%.

3. Results and Discussion

3.1. Thermodynamic Analysis

The amounts of the reactants and products under different roasting conditions were predicted using thermodynamic analysis. Two alkalis (NaOH and Na_2CO_3) were selected as additives to perform the thermodynamic analysis. The possible reactions and the changes in Gibbs free energy are shown in Equations (3)–(11) and Figure 4.

$$Al_2O_3 + 2NaOH = 2NaAlO_2 + H_2O \tag{3}$$

$$MgAl_2O_4 + 2NaOH = 2NaAlO_2 + MgO + H_2O \tag{4}$$

$$Fe_2O_3 + 2NaOH = 2NaFeO_2 + H_2O \tag{5}$$

$$2AlN + 2NaOH + 1.5O_2 = 2NaAlO_2 + H_2O + N_2 \tag{6}$$

$$Al_2O_3 + Na_2CO_3 = 2NaAlO_2 + CO_2 \tag{7}$$

$$MgAl_2O_4 + Na_2CO_3 = 2NaAlO_2 + MgO + CO_2 \tag{8}$$

$$Fe_2O_3 + Na_2CO_3 = 2NaFeO_2 + CO_2 \tag{9}$$

$$2AlN + Na_2CO_3 + 1.5O_2 = 2NaAlO_2 + CO_2 + N_2 \tag{10}$$

$$SiO_2 + NaAlO_2 = NaAlSiO_4 \tag{11}$$

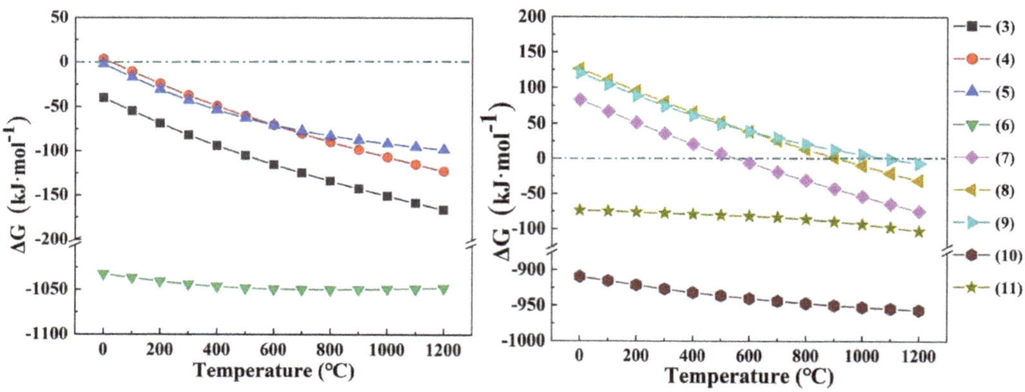

Figure 4. Relationship of the changes in Gibbs free energy with the temperature for the reactions during the roasting process.

From the reaction curves (3)–(6), it can be seen that the Gibbs free energy of the reactions of Al_2O_3, $MgAl_2O_4$, Fe_2O_3 and AlN with NaOH were negative at room temperature, which indicates that the thermodynamic reaction conditions are sufficient. The Al_2O_3 of $MgAl_2O_4$ phase was combined with NaOH to obtain $NaAlO_2$ and release MgO. Moreover, AlN was oxidized to generate non-toxic N_2. The reaction curves (7)–(9) show that the Gibbs free energy values of the reactions were negative only at higher temperatures, which indicates that Na_2CO_3 was less active compared to NaOH. The reaction curve (10) shows that the thermodynamic conditions for the reaction of AlN and Na_2CO_3 to produce $NaAlO_2$, CO_2 and N_2 were sufficient. From the reaction curve (11), it is clear that insoluble $NaAlSiO_4$ was produced by the combination of SiO_2 and $NaAlO_2$, resulting in the loss of alumina in the SAD.

These results demonstrate that the reactions of the SAD with the two alkali additives (NaOH and Na_2CO_3) were feasible. All the Al-containing compounds (Al, Al_2O_3, AlN and $MgAl_2O_4$) in SAD can react with alkali additives to form readily soluble $NaAlO_2$. In addition, the other impurities (Mg, Fe, and Si) generated insoluble species, thus achieving the separation and extraction of valuable aluminum elements from SAD.

3.2. Effects of Different Factors on the Roasting System

In order to ensure that the desired compounds are produced in the clinkers, the extent of the roasting of the mixed materials should be strictly controlled. Since both the alkali additives can thermodynamically react with SAD to produce the expected compounds, the effects of these alkali additives on the roasting system were examined.

The effects of different factors on the recoveries of Al and Na were studied by conducting a variety of experiments for each additive type, various roasting temperatures and ingredients depending on the n(N/A) (see Table 2 and Figure 5).

Table 2. List of roasting parameters studied in the experiments.

Parameters	Parameter Values	Fixed Roasting Parameters	Fixed Leaching Parameters
Roasting temperature (°C)	850, 900, 950, 1000, 1050	Additive of NaOH, t = 1 h, n (N/A) = 1.	$C(Na_2O)$ = 60 g/L, α_k = 3.0, T = 90 °C, t = 60 min
n(N/A)	1.0, 1.1, 1.2, 1.3, 1.4	Additive of NaOH, T = 950 °C. t = 1 h.	
Roasting temperature (°C)	1000, 1050, 1100, 1150, 1200	Additive of Na_2CO_3, t = 1 h, n (N/A) = 1.	
n (N/A)	1.0, 1.1, 1.2, 1.3, 1.4	Additive of Na_2CO_3, T = 1100 °C, t = 1 h.	

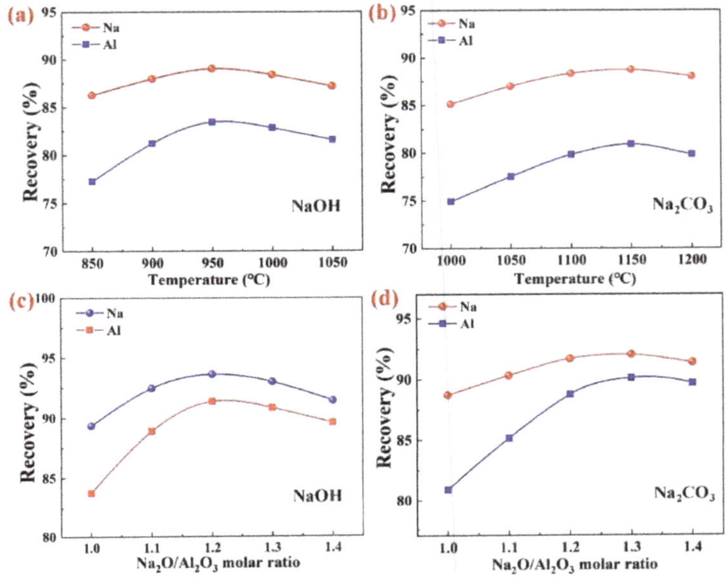

Figure 5. Recovery of Al and Na in roasting clinkers under different roasting conditions: roasting temperature with NaOH additive (**a**), n(N/A) with NaOH additive (**b**), roasting temperature with Na_2CO_3 additive (**c**), and n(N/A) with Na_2CO_3 additive (**d**).

3.2.1. Effects of Different Factors on the Recoveries of Al and Na

For the alkali additives, NaOH and Na_2CO_3, the recoveries of Al and Na in the clinkers under different temperatures are shown in Figures 5a and 5b, respectively. As is evident from Figure 5a,b, the roasting temperature had a significant effect on the recoveries of Al and Na. With the increase in the roasting temperature, the recoveries of both the Al and Na initially increased and then decreased. When NaOH was used as the additive, the recoveries of Al and Na reached the maximum values of 83.47% and 89.06%, respectively, at the roasting temperature of 950 °C. However, when Na_2CO_3 was used as the additive, the highest recoveries of Al and Na had values of 80.90% and 88.72%, respectively, at the roasting temperature of 1100 °C. Additionally, the recoveries of Al and Na were similar in the within ±50 °C interval of the optimum temperature, which indicates that the range of the optimum temperature was wide and simple to control.

On the one hand, the mixed materials primarily relied on the solid−solid reaction during the roasting process. The generation of less liquid phase in the clinkers at low roasting temperatures led to slow reaction rates, large porosity and poor dissolution performance. On the other hand, the generation of a large amount of liquid phase in the clinkers at high roasting temperatures resulted in the rapid volatilization of the alkali, low porosity and poor dissolution performance [28]. Therefore, the appropriate liquid phase and porosity of clinkers are available only at a certain roasting temperature, which offers a better reaction rate and dissolution performance.

The n(N/A) is a critical parameter which affects the extraction of the valuable Al element from SAD. The n(N/A) represents the amount of alkali additives which have been added to the system. For the additives NaOH and Na_2CO_3, the recoveries of Al and Na in the clinkers under different n(N/A) are shown in Figures 5c and 5d, respectively. As shown in Figure 5c, when the n(N/A) was increased from 1.0 to 1.2, the recovery of Al increased from 83.47% to 91.83%, whereas that of Na increased from 89.36% to 93.64%. When Na_2CO_3 was used as the additive, the n(N/A) of 1.3 was optimal for extracting $NaAlO_2$, while the Al and Na recoveries were 90.79% and 92.03%, respectively. This could be due to the fact that the low quantities of alkali were not sufficient enough to adequately support the reaction of the aluminum-containing compounds (Al, Al_2O_3, AlN, and $MgAl_2O_4$) in SAD with Na_2O to produce soluble $NaAlO_2$, thus resulting in a poor dissolution performance of the clinkers. With the increase in the n(N/A), the recoveries of Al and Na gradually declined. This can be attributed to the production of some insoluble substances due to excessive amounts of alkali, which prevented the further dissolution of $NaAlO_2$.

It is worth mentioning that the NaOH additive resulted in a lower roasting temperature and slightly higher Al and Na recoveries compared to the Na_2CO_3 additive under the optimal roasting conditions. However, due to the NaOH deliquescence, the mixed materials were able to easily absorb the moisture in the air during the experiment, thus causing the hydrolysis of the AlN in SAD and releasing a large amount of toxic ammonia. Additionally, the produced cylindrical samples quickly lost their strengths, forming into a slurry. Therefore, Na_2CO_3 was selected as the appropriate additive for the subsequent experiments.

3.2.2. Effects of Different Factors on the Mineralogical Phases

The mineralogical analysis of the roasting clinkers and leaching residues at various roasting temperatures is illustrated in Figure 6. The main phases in the roasting clinkers included $NaAlO_2$, MgO, $NaFeO_2$ and $NaAlSiO_4$, which was consistent with the results of the thermodynamic analysis. The diffraction peak intensity of $NaAlO_2$ in the roasting clinkers significantly increased from 1000 °C to 1050 °C, which was the critical solid phase of the Al recovery. Moreover, SiO_2 played a negative role in the recovery of Al because SiO_2 was able to react with soluble $NaAlO_2$ to produce insoluble $NaAlSiO_4$.

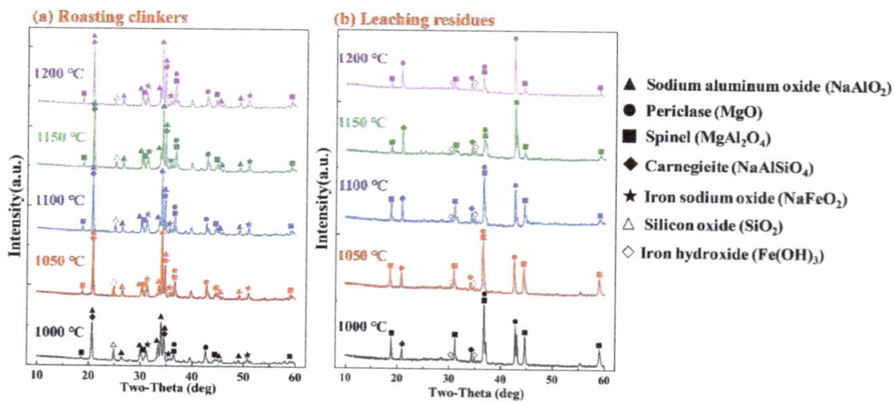

Figure 6. XRD patterns of the roasting clinkers (**a**) and leaching residues (**b**) under different roasting temperatures.

After the leaching process, the $NaAlO_2$ in the roasting clinkers was completely dissolved, and the $NaFeO_2$ was hydrolyzed to generate insoluble $Fe(OH)_3$ and release NaOH. Therefore, the main phases in the leaching residues were MgO, $MgAl_2O_4$, $Fe(OH)_3$ and $NaAlSiO_4$. The diffraction peak intensity of MgO in the leaching residues continued to increase from 1000 °C to 1150 °C, while the diffraction peak intensity of $MgAl_2O_4$ significantly decreased. This indicates that an increase in the temperature was beneficial to the reaction (8). A small amount of $MgAl_2O_4$ was still observed in the roasting clinkers and leaching residues due to the low amounts of Na_2CO_3. The phase composition of the roasting temperature above 1150 °C remained unchanged, indicating that the mineralogical phase of the roasting clinkers was more effective at 1150 °C, which is consistent with the results shown in Figure 5c.

The XRD patterns of the roasting clinkers and leaching residues at different n(N/A) are shown in Figure 7. As the n(N/A) increased, the diffraction peak intensities of $NaAlO_2$ and MgO in the roasting clinkers continued to increase, while the peak intensity of $MgAl_2O_4$

continued to decrease. This shows that the reaction (8) was facilitated by increasing the n(N/A). When the n(N/A) exceeded the value of 1.2, the SiO_2 disappeared and $NaAlSiO_4$ continued to decrease, while $Na_{1.95}Al_{1.95}Si_{0.05}O_4$ was generated as a result of the reaction (12), and this was easily dissolved in the alkali solution [29].

$$0.1SiO_2 + 1.95Na_2CO_3 + 1.95Al_2O_3 = 2Na_{1.95}Al_{1.95}Si_{0.05}O_4 + 1.95CO_2 \quad (12)$$

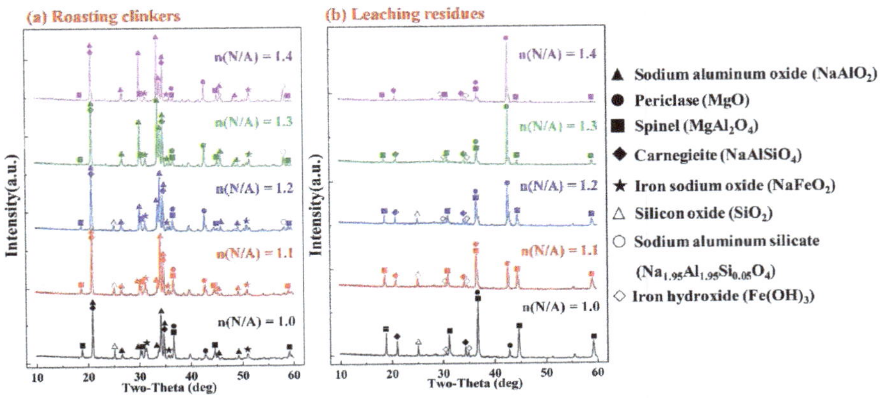

Figure 7. XRD patterns of the roasting clinkers (**a**) and leaching residues (**b**) under different Na_2CO_3/SAD mass ratios.

With the increase in the n(N/A) from 1.0 to 1.3, the MgO in the leaching residues significantly increased, while the $MgAl_2O_4$ continued to decrease, which is consistent with the XRD results of the roasting clinkers. The phase composition of the system with an n(N/A) of over 1.3 was essentially the same as that of the system with a lower molar ratio, indicating that the mineralogical phases of the roasting clinkers with the n(N/A) of 1.3 were more effective, which is consistent with the results shown in Figure 5d.

3.3. *Effects of Different Factors on the Leaching System*

Since the structure of solid $NaAlO_2$ is different from the structure of the aluminate ions in the solution, the dissolution of $NaAlO_2$ in the roasting clinkers is actually a chemical reaction, as given by the reaction (13) [30].

$$Na_2O{\cdot}Al_2O_3(s) + 4H_2O = Na^+ + 2Al(OH)_4^- \quad (13)$$

In this section, the effects of various leaching factors including the leaching temperature, leaching time and liquid-to-solid ratio on the recovery of Al were evaluated. These experiments were conducted at the roasting temperature of 1150 °C, n(N/A) of 1.3 and roasting time of 1 h.

3.3.1. Effects of the Leaching Temperature

Figure 8a shows the recovery efficiency of the Al element in the roasting clinkers at different leaching temperatures as a function of the leaching time. The reaction was severe within the first 15 min, and the $NaAlO_2$ in the roasting clinkers was continuously dissolved. However, no significant increase in the recovery of the Al was observed with the further extension of the time. When the leaching temperature was increased from 30 °C to 90 °C, the 15 min Al recovery increased from 68.79% to 88.11%. The leaching temperature had a significant impact on the recovery of the Al, because the increase in the temperature was beneficial in accelerating the mass transfer and diffusion.

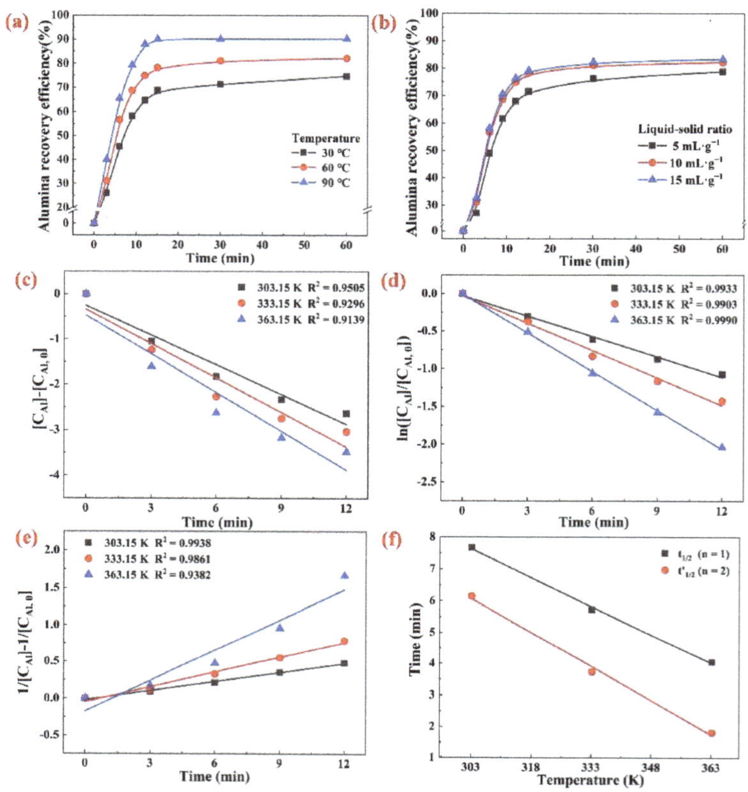

Figure 8. Recovery of Al in the roasting clinkers under different leaching temperatures (**a**) and L/S ratios (**b**), with fitting experimental data for different reaction orders: n = 0 (**c**), n = 1 (**d**) and n = 2 (**e**), and the half-life of leaching process at n = 1 and n = 2 (**f**).

3.3.2. Effect of the Liquid-to-Solid (L/S) Ratio

The effect of L/S ratio within the range of 5–15 mL·g^{-1} on the leaching process was investigated at a leaching temperature of 60 °C. As shown in Figure 8b, when the L/S ratio was increased from 5 mL·g^{-1} to 10 mL·g^{-1}, the recovery of the Al increased significantly. However, when the L/S ratio was increased to 15 mL·g^{-1}, the recovery of the Al became relatively stable. Therefore, the L/S ratio of the leaching process had a significant influence on the recovery of the Al within a certain range. Increasing the amount of leaching solution can help to reduce the slurry viscosity and increase the concentration difference of the aluminate ions on the solid–liquid interface, thus accelerating the dissolution of $NaAlO_2$ in the roasting clinkers. Therefore, an L/S ratio of 10 mL·g^{-1} was selected as the optimum leaching condition for further experimentation.

3.4. *Analysis of the Leaching Kinetics*

The leaching process of the $NaAlO_2$ in the roasting clinkers is a liquid-to-solid heterogeneous reaction system, which takes place at the phase interfaces. In order to obtain a more efficient process, the kinetics of the dissolution of $NaAlO_2$, including the reaction rate constant, reaction order, activation energy and the rate-determining step, should be thoroughly evaluated. For the $NaAlO_2$ dissolution in the roasting clinkers, the rate of reaction is given by Equation (14) [29].

$$-\frac{dC_{Al}}{dt} = k\, C_{Al}{}^{n} \tag{14}$$

where C_{Al} is the content of NaAlO2 in the roasting clinkers after leaching for time t, t is the leaching time, n is the reaction order and k is the reaction rate constant, s^{-1}.

The experimental results shown in Figure 8a were used to plot the different reaction orders (n = 0, 1, and 2) in Equation (14) and the corresponding results are shown in Figure 8c–e), respectively. The confidence level of the fitted straight line was high for the first order reaction (n = 1), while the linear relationship exhibited strong agreement in the second order reaction (n = 2), except for a slight deviation at 363.15 K. It has been shown that many reactions can be satisfactorily correlated using both the first and second order reactions. Therefore, it was not prudent to rely only on the linear correlation of the reaction equation to determine the reaction order. The half-life method was used to effectively determine the reaction order of the dissolution of $NaAlO_2$ in the roasting clinkers (Equation (15)):

$$\chi_{Al} = \frac{C_{Al,\,0} - C_{Al}}{C_{Al,\,0}} \tag{15}$$

The logarithmic form of Equation (15) is expressed as Equation (16):

$$k = \frac{1}{t}\ln\frac{1}{1 - \chi_{Al}} \tag{16}$$

where χ_{Al} is the dissolution rate of $NaAlO_2$ in the roasting clinkers at the time t, and $C_{Al,\,0}$ is the initial $NaAlO_2$ content in the roasting clinkers.

When the dissolution rate of $NaAlO_2$ was 50% (χ_{Al} = 0.5), the half-life of the first order reaction $t_{1/2}$ is given by Equation (17), whereas the half-life of the second order reaction $t'_{1/2}$ is written as Equation (18):

$$t_{1/2} = \frac{\ln 2}{k} \tag{17}$$

$$t'_{1/2} = \frac{1}{k\, C_{Al,\,0}} \tag{18}$$

The half-life results for first and second order reactions at different leaching temperatures are shown in Figure 8f. When the reaction order was one (n = 1), the half-life of the dissolution of $NaAlO_2$ within the leaching temperature of 303.15–363.15 K was 7.67–4.05 min. However, when the reaction order was two (n = 2), the half-life within the temperature of 303.15–363.15 K was 6.15–1.82 min. Combined with the experimental results shown in Figure 8a, the half-life showed high reliability in first order reaction, indicating that the dissolution of the $NaAlO_2$ was proportional to the reactant concentration.

The temperature dependence of the leaching reaction rate constant is expressed using the Arrhenius correlation (Equation (19)):

$$k = \mathrm{A} \cdot \exp(-E_a/RT) \tag{19}$$

where k is the overall reaction constant, A is the pre-exponential factor (min^{-1}), R is the ideal gas constant (8.3145 $J{\cdot}mol^{-1}{\cdot}K^{-1}$) and E_a ($J{\cdot}mol^{-1}$) is the apparent activation energy of the dissolution of $NaAlO_2$. The Arrhenius correlation can be rewritten in logarithmic form as Equation (20):

$$\ln k = \ln\mathrm{A} - \frac{E_a}{RT} \tag{20}$$

In general, when the activation energy is 40–300 $kJ{\cdot}mol^{-1}$, the reaction is controlled by the interface. However, when the activation energy is 8–20 $kJ{\cdot}mol^{-1}$, the reaction is controlled by the diffusion process [30]. Finally, the activation energy (E_a) calculated from the slope of Figure 9 was found to be 9.69 kJ·mol^{-1}, which is lower than the value of 20 kJ·mol^{-1}, thus confirming that the dissolution of $NaAlO_2$ was controlled by the diffusion process. This result is also similar to the result of the dissolution of $NaAlO_2$ mentioned in He's study, where the activation energy was reported to be 11.4010 kJ·mol^{-1} [30].

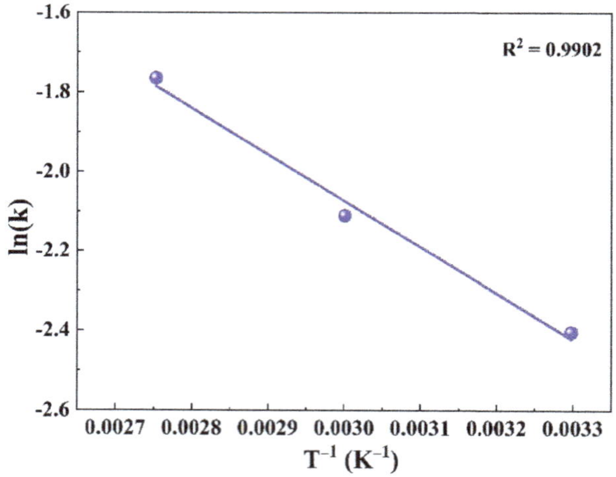

Figure 9. Relationship between the leaching rate constant k and temperature T.

3.5. Thermal Analysis and Product Characterization

In order to further evaluate the phase transition of the roasting process, the thermogravimetric and differential scanning calorimetry (TG-DSC) curves of the mixed materials of SAD and Na_2CO_3 (n(N/A) of 1.3) at a heating rate of 10 °C·min^{-1} are shown in Figure 10.

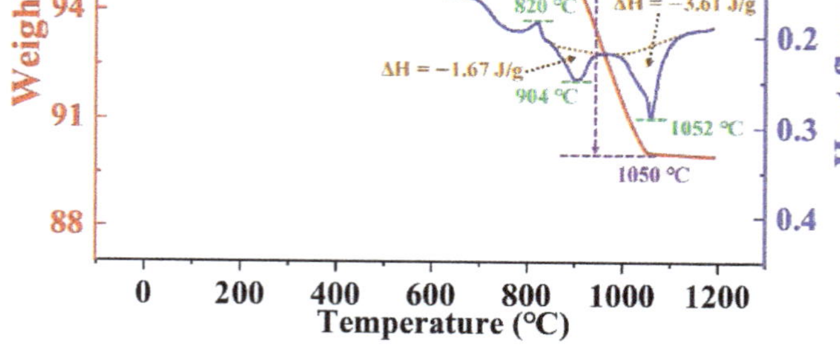

Figure 10. TG-DSC curves of SAD–Na_2CO_3 mixtures for the n(N/A) of 1.3.

The weight loss of the mixed materials can be divided into three stages: (1) from room temperature to 550 °C, the weight loss was caused by the evaporation of the attached water, and the weight was reduced by 1.24%; (2) from 550 °C to 1050 °C, CO_2 was released due to the reaction of Na_2CO_3 with the components in the SAD, resulting in a weight loss of 8.78 wt.%, which is also supported by the fact that Na_2CO_3 cannot undergo thermal decomposition within the temperature range of 550–1050 °C; (3) from 1050 °C to 1200 °C, the weight of the sample remained basically unchanged, indicating that the main reaction was completed.

The DSC curve showed three obvious endothermic peaks at 564 °C, 686 °C and 820 °C in the heating process of the mixed materials. Additionally, two obvious exothermic peaks at 904 °C and 1052 °C were observed. The enthalpy changes in the main reactions during the roasting process (see Section 3.1) are presented in Table 3, which indicate that the reactions (7)–(9) were endothermic, while the reactions (10) and (11) were exothermic. Previous studies have shown that Al_2O_3 and Na_2CO_3 react mainly at 500–700 °C, whereas Fe_2O_3 reacts with Na_2CO_3 at about 850 °C. Combining the thermodynamics calculations, XRD results and the TG-DSC analysis, the endothermic peaks at 564 °C, 686 °C and 820 °C should be attributed to reactions (7), (8) and (9), respectively. In addition, the exothermic peaks at 904 °C and 1052 °C corresponded to reactions (11) and (10), respectively, which were determined using the changes in enthalpy. The main chemical reactions of the roasting process were completed at around 1150 °C, which was consistent with the optimal roasting temperature derived in Section 3.2.

Table 3. Enthalpy changes in the reactions during the roasting process.

Reactions	ΔH at 500 °C ($kJ \cdot mol^{-1}$)	ΔH at 1000 °C ($kJ \cdot mol^{-1}$)
$Al_2O_3 + Na_2CO_3 \rightarrow 2NaAlO_2 + CO_2$	110.00	85.90
$MgAl_2O_4 + Na_2CO_3 \rightarrow 2NaAlO_2 + MgO + CO_2$	154.58	132.19
$Fe_2O_3 + Na_2CO_3 \rightarrow 2NaFeO_2 + CO_2$	135.38	90.93
$AlN + 0.5Na_2CO_3 + 0.75O_2 \rightarrow NaAlO_2 + 0.5CO_2 + 0.5N_2$	−452.605	−461.11
$SiO_2 + NaAlO_2 \rightarrow NaAlSiO_4$	−69.95	−38.98

The micromorphology of the roasting clinkers and leaching residues with the n(N/A) of 1.3 were observed using a scanning electron microscope (SEM). The energy dispersive spectroscopy (EDS) images are shown in Figure 11e–h. Figure 11a,b shows that the particles in the roasting clinkers were clustered and had a particle size of about 10–30 μm. The shape of the $NaAlO_2$ particles was smooth and irregular. The morphology of the MgO particles was agglomerated and spherical, with a particle size of 2–4 μm. Figure 11c shows that the $NaAlSiO_4$ in the leaching residues had a smooth block-shaped appearance, while $MgAl_2O_4$ was octahedral, as shown in Figure 11d.

The chemical compositions of the roasting clinkers and leaching residues under optimal conditions (roasting temperature of 1150 °C, Na_2CO_3/Al_2O_3 molar ratio of 1.3, roasting time of 1 h, leaching temperature of 90 °C, L/S ratio of 10 $mL \cdot g^{-1}$ and leaching time of 30 min) are presented in Table 4. The contents of N and Cl in the roasting clinkers reduced to 0.072 wt.% and 0.12 wt.%, respectively. Moreover, the AlN was oxidized to produce harmless N_2, and the chloride was evaporated to the gaseous phase, which was recovered after the condensation. The removals of N and Cl reached 98.93% and 97.14%, respectively. Therefore, the green toxification of SAD can be achieved in the roasting process without any pretreatment.

Table 4. Compositions of the roasting clinkers and the leaching residues under optimal conditions (%).

Roasting Clinkers	Al	O	F	Na	Ca	Si	Mg	Fe	N	K	Cl
Content (wt.%)	27.06	34.88	0.19	29.67	0.24	1.55	4.98	0.79	0.072	0.19	0.12
Leaching residues	Al	O	F	Na	Ca	Si	Mg	Fe	N	K	Cl
Content (wt.%)	10.64	43.63	0.87	10.10	1.36	7.52	21.27	4.26	0.04	0.06	0.03

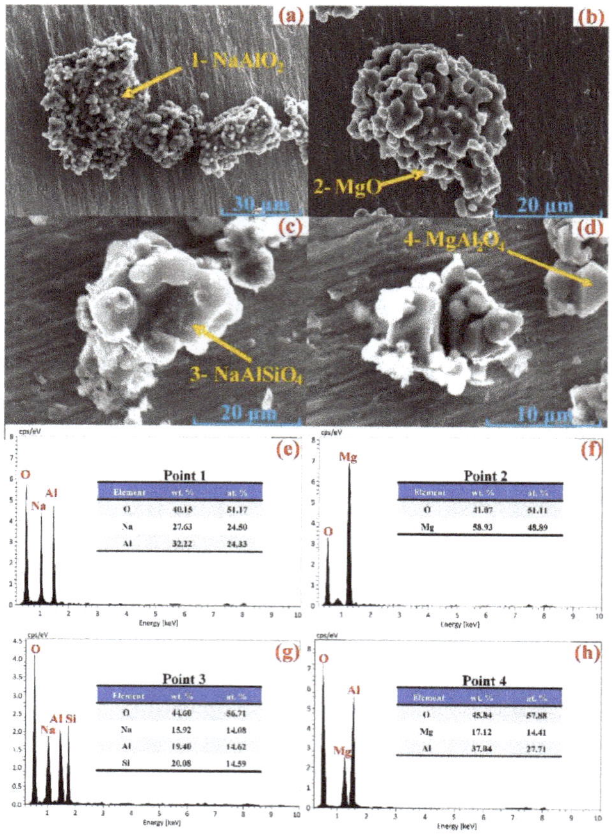

Figure 11. SEM micrographs of the roasting clinkers (**a**,**b**), leaching residues (**c**,**d**) and EDS (**e**–**h**).

4. Conclusions

SAD is a hazardous solid waste as well as an alumina-rich resource. The safe disposal and cost-effective recycling of SAD are serious challenges. A combined dry pressing and alkali roasting process was developed for the sustainable extraction of the valuable Al element from SAD. Based upon the results, the following conclusions can be drawn:

1. Both alkali additives (NaOH and Na_2CO_3) could react efficiently with SAD to obtain high recoveries of Al and Na. However, due to the deliquescence of the NaOH, the AlN reacted with water to rapidly release large amounts of NH_3. Therefore, the NaOH additive was not suitable for the dry pressing step.
2. The molar ratio of Na_2CO_3/Al_2O_3 and the roasting temperatures significantly affected the phase composition and leaching performance of the roasted products. Under optimal conditions, the recoveries of the Al and Na in the roasting clinkers reached the values of 90.79% and 92.03%, respectively.
3. The removal efficiencies of the N and Cl reached the values of 98.93% and 97.14%, respectively. The green detoxification and efficient extraction of the valuable Al from the SAD were simultaneously achieved in the roasting process. This process has the sustainable developing and practical application on SAD recovery.
4. The leaching kinetics showed that the $NaAlO_2$ dissolution in the roasting clinkers was a first order reaction and controlled by a layer diffusion process. The apparent activation energy was 9.69 $kJ{\cdot}mol^{-1}$. The experiments of desilication and crystal

seed decomposition can be conducted on synthesized metallurgical grade alumina in future.

Author Contributions: Conceptualization, H.L. and F.L.; methodology, M.X.; software, Z.W. and L.L.; investigation, R.Y. and J.H.; writing—original draft preparation, H.L.; writing—review and editing, H.Z. and F.L.; funding acquisition, F.L. All authors have read and agreed to the published version of the manuscript.

Funding: This research was funded by the National Key Research and Development Program of China, grant number 2019YFC1908403.

Institutional Review Board Statement: Not applicable.

Informed Consent Statement: Not applicable.

Data Availability Statement: Not applicable.

Conflicts of Interest: The authors declare no conflict of interest.

References

1. Abdulkadir, A.; Ajayi, A.; Hassan, M.I. Evaluating the Chemical Composition and the Molar Heat Capacities of a white Aluminum Dross. *Energy Procedia* **2015**, *75*, 2099–2105. [CrossRef]
2. Shen, H.; Liu, B.; Ekberg, C.; Zhang, S.G. Harmless disposal and resource utilization for secondary aluminum dross: A review. *Sci. Total Environ.* **2020**, *760*, 143968. [CrossRef] [PubMed]
3. Tsakiridis, P.E. Aluminium salt slag characterization and utilization—A review. *J. Hazard. Mater.* **2012**, *217–218*, 1–10. [CrossRef] [PubMed]
4. David, E.; Kopac, J. Hydrolysis of aluminum dross material to achieve zero hazardous waste. *J. Hazard. Mater.* **2012**, *209–210*, 501–509. [CrossRef] [PubMed]
5. Mostafa, M.; Ali, A. Hazardous aluminum dross characterization and recycling strategies: A critical review. *J. Environ. Manage.* **2018**, *223*, 452–468. [CrossRef]
6. Mahinroosta, M.; Allahverdi, A. A promising green process for synthesis of high purity activated-alumina nanopowder from secondary aluminum dross. *J. Clean. Prod.* **2018**, *179*, 93–102. [CrossRef]
7. Liu, F.Q.; Lv, H.; Zuo, Z.P.; Xie, M.Z.; Li, R.B.; Zhao, H.L. A denitrification-phase transition and protection rings (DPP) process for recycling electrolytic aluminum dross. *ACS Sustain. Chem. Eng.* **2021**, *9*, 13751–13760. [CrossRef]
8. Zuo, Z.P.; Lv, H.; Li, R.B.; Liu, F.Q.; Zhao, H.L. A new approach to recover the valuable elements in black aluminum dross. *Resour. Conserv. Recy.* **2021**, *174*, 105768. [CrossRef]
9. Liu, Y.; Leong, B.S.; Hu, Z.T.; Yang, E.H. Autoclaved aerated concrete incorporating waste aluminum dust as foaming agent. *Constr. Build. Mater.* **2017**, *148*, 140–147. [CrossRef]
10. Yoldi, M.; Fuentes-Ordoñez, E.G.; Korili, S.A.; Gil, A. Zeolite synthesis from aluminum saline slag waste. *Powder Technol.* **2020**, *366*, 175–184. [CrossRef]
11. Lv, H.; Xie, M.Z.; Shi, L.T.; Zhao, H.L.; Wu, Z.G.; Li, L.L.; Li, R.B.; Liu, F.Q. A novel green process for the synthesis of high-whiteness and ultrafine aluminum hydroxide powder from secondary aluminum dross. *Ceram. Int.* **2021**, *48*, 953–962. [CrossRef]
12. Meshram, A.; Singh, K.K. Recovery of valuable products from hazardous aluminum dross: A review. *Resour. Conserv. Recy.* **2018**, *130*, 95–108. [CrossRef]
13. Saravanakumar, R.; Ramachandran, K.; Laly, L.G.; Ananthapadmanabhan, P.V.; Yugeswaran, S. Plasma assisted synthesis of γ-alumina from waste aluminium dross. *Waste Manage.* **2018**, *77*, 565–575. [CrossRef] [PubMed]
14. Li, A.; Zhang, H.; Yang, H. Evaluation of aluminum dross as raw material for high-alumina refractory. *Ceram. Int.* **2014**, *40*, 12585–12590. [CrossRef]
15. Ewais, E.; Besisa, N. Tailoring of magnesium aluminum titanate based ceramics from aluminum dross. *Mater. Design* **2018**, *141*, 110–119. [CrossRef]
16. Ewais, E.M.M.; Khalil, N.M.; Amin, M.S.; Ahmed, Y.M.Z.; Barakat, M.A. Utilization of aluminum sludge and aluminum slag (dross) for the manufacture of calcium aluminate cement. *Ceram. Int.* **2009**, *35*, 3381–3388. [CrossRef]
17. Gil, A.; Korili, S.A. Management and valorization of aluminum saline slags: Current status and future trends. *Chem. Eng. J.* **2016**, *289*, 74–84. [CrossRef]
18. Murayama, N.; Maekawa, I.; Ushiro, H.; Miyoshi, T.; Shibata, J.; Valix, M. Synthesis of various layered double hydroxides using aluminum dross generated in aluminum recycling process. *Int. J. Miner. Process.* **2012**, *110*, 46–52. [CrossRef]
19. Mahinroosta, M.; Allahverdi, A. Enhanced alumina recovery from secondary aluminum dross for high purity nanostructured γ-alumina powder production: Kinetic study. *J. Environ. Manage.* **2018**, *212*, 278–291. [CrossRef]
20. Das, B.R.; Dash, B.R.; Tripathy, B.C.; Bhattacharya, I.N.; Das, S.C. Production of η-alumina from waste aluminium dross. *Miner. Eng.* **2007**, *20*, 252–258. [CrossRef]
21. Yang, Q.; Li, Q.; Zhang, G.F.; Shi, Q.; Feng, H.G. Investigation of leaching kinetics of aluminum extraction from secondary aluminum dross with use of hydrochloric acid. *Hydrometallurgy* **2019**, *187*, 158–167. [CrossRef]

22. Tsakiridis, P.E.; Oustadakis, P.; Agatzini-Leonardou, S. Aluminium recovery during black dross hydrothermal treatment. *J. Environ. Chem. Eng.* **2013**, *1*, 23–32. [CrossRef]
23. Nguyen, T.; Si, J.S.; Man, S.L. Development of a hydrometallurgical process for the recovery of pure alumina from black dross and synthesis of magnesium spinel. *J. Mater. Res. Technol.* **2020**, *9*, 2568–2577. [CrossRef]
24. Tripathy, A.K.; Mahalik, S.; Sarangi, C.K.; Tripathy, B.C.; Sanjay, K.; Bhattacharya, I.N. A pyro-hydrometallurgical process for the recovery of alumina from waste aluminium dross. *Miner. Eng.* **2019**, *137*, 181–186. [CrossRef]
25. Dash, B.R.; Das, B.R.; Tripathy, B.C.; Bhattacharya, I.N.; Das, S.C. Acid dissolution of alumina from waste aluminium dross. *Hydrometallurgy* **2008**, *92*, 48–53. [CrossRef]
26. David, E.; Kopac, J. Aluminum recovery as a product with high added value using aluminum hazardous waste. *J. Hazard. Mater.* **2013**, *261*, 316–324. [CrossRef]
27. Lv, H.; Zhao, H.L.; Zuo, Z.P.; Li, R.B.; Liu, F.Q. A thermodynamic and kinetic study of catalyzed hydrolysis of aluminum nitride in secondary aluminum dross. *J. Mater. Res. Technol.* **2020**, *9*, 9735–9745. [CrossRef]
28. Li, X.B.; Liu, J.H.; Wang, Y.L.; Zeng, L. Phase transformation of sodium hydrate alumino-silicate in alumina sintering process. *Chin. J. Nonferr. Met.* **2018**, *28*, 1225–1232. [CrossRef]
29. Xie, M.Z.; Zhao, H.L.; Wu, Z.G.; Liu, W.; Li, R.B.; Liu, F.Q. Study on kinetics of vacuum heat treatment process of the spent cathode carbon blocks from aluminum smelters. *J. Sustain. Metall.* **2020**, *6*, 715–723. [CrossRef]
30. He, L.Q.; Shi, L.; Huang, Q.Z.; Hayat, W.; Shang, Z.B.; Ma, T.F.; Wang, M.; Yao, W.D.; Huang, H.Y.; Chen, R. Extraction of alumina from aluminum dross by a non-hazardous alkaline sintering process: Dissolution kinetics of alumina and silica from calcined materials. *Sci. Total Environ.* **2021**, *777*, 146123. [CrossRef]

materials

MDPI

Review

Applications and Properties of Hemp Stalk-Based Insulating Biomaterials for Buildings: Review

Borja Martínez, Ernest Bernat-Maso and Lluis Gil *

Department of Strength of Materials and Structures in Engineering, Polytechnic University of Catalonia, 08222 Terrassa, Spain; borja.martinez@upc.edu (B.M.); ernest.bernat@upc.edu (E.B.-M.)
* Correspondence: lluis.gil@upc.edu

Abstract: There has been increasing interest in green and recyclable materials to promote the circular economy. Moreover, the climate change of the last decades has led to an increase in the range of temperatures and energy consumption, which entails more energy expenditure for heating and cooling buildings. In this review, the properties of hemp stalk as an insulating material are analyzed to obtain recyclable materials with green solutions to reduce energy consumption and reduce noise to increase the comfort of buildings. Hemp stalks are a low-value by-product of hemp crops; however, they are a lightweight material with a high insulating property. This study aims to summarize the research progress in materials based on hemp stalks and to study the properties and characteristics of the different vegetable binders that could be used to produce a bio-insulating material. The material itself and its microstructural and physical aspects that affect the insulating properties are discussed, as is their influence on durability, moisture resistance, and fungi growth. Research suggests using lignin-based or recyclable cardboard fiber to develop a bio-composite material from hemp stalk, but long-term stability requires further investigation.

Keywords: hemp stalk; bio-insulating material; sustainability; sound absorbing properties; thermal absorption properties; green composite material

Citation: Martínez, B.; Bernat-Maso, E.; Gil, L. Applications and Properties of Hemp Stalk-Based Insulating Biomaterials for Buildings: Review. *Materials* **2023**, *16*, 3245. https://doi.org/10.3390/ma16083245

Academic Editors: Lorena Zichella and Rossana Bellopede

Received: 21 March 2023
Revised: 14 April 2023
Accepted: 17 April 2023
Published: 20 April 2023

1. Introduction

Energy consumption has been increasing in recent decades, as has its consequent increase in greenhouse gas emissions into the atmosphere. In the building sector, climate conditioning consumes 45% of the total energy consumption in the building [1]. This effect is increased every year as long as more and more energy is needed to keep ideal temperatures. Due to the new requirements, studies are carried out to improve the energy performance of buildings, either through passive or active techniques. The demand for green buildings, "a building that, in its design, construction or operation, reduces or eliminates negative impacts, and can create positive impacts, on our climate and natural environment," [2] will increase substantially in the coming years. Through this type of technology, it is possible to reduce the carbon footprint of human activities and to achieve the objectives set in the Paris Agreements, to aim at "net-zero" emissions by 2050 [3].

A technique that can be explored in this field of research is the improvement of the performance of insulating materials based on green materials [4,5] with a positive carbon footprint [6]. Biomaterial or eco-friendly materials are materials that are produced and used in a way that minimizes harm to the environment. These materials are biodegradable, renewable, and sustainable, and have a reduced carbon footprint compared to traditional materials, reduce the amount of waste generated, and decrease the need for virgin materials. The use of eco-friendly materials is an important step towards achieving a more sustainable future. The European Industrial Hemp Association (EIHA) states that the hemp industry is an economically viable and socially responsible enterprise, as it contributes to restoring ecological balance and achieving decarbonization goals for a promising sustainable economy [7].

Hemp, *cannabis sativa* L., is a crop that can grow rapidly and yield up to four harvests in a single year. It is mainly produced for industrial or food purposes and is available in different varieties, each suited for specific applications. Hemp is a multi-functional crop with diverse applications across several fields. Despite its versatility, hemp cultivation declined globally in the 20th century due to the emergence of synthetic fibers that were cheaper to produce and possessed superior properties [8].

In recent years of the 21st century, hemp has aroused interest in the scientific community, considerably increasing the number of related publications in recent years (Figure 1). The figure shows the number of published articles only by ScienceDirect and MDPI to confirm the incremental tendency of scientific interest. However, the review takes into account articles published with other editors, considering the year of publication, methodology, and citations of each article.

Although it is primarily used in the textile industry, with the advent of the industrial era, other parts of the plant have been utilized in the market, leading to an increased range of products and applications [9]. These applications include fibers for textiles, seeds and oils for food, biomass fuel, construction materials, insulation for automotive textile parts, paper production from cellulose, and medicinal uses.

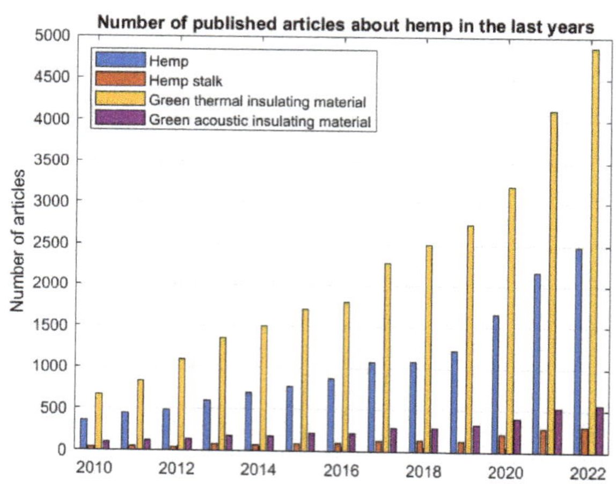

Figure 1. Number of articles published in ScienceDirect and MDPI about hemp and green insulation materials.

The remarkable versatility of hemp, increased farming productivity, rising demand resulting from emerging applications, and the growing need for eco-friendly materials have led to a surge in hemp farming in recent years. Despite the great farming tradition, new studies are still important to optimize the farming of hemp due to the numerous varieties of the crop. It is important to note that the usage and production of natural fibers are influenced by the customs and traditions of specific countries and regions. For example, China and Bangladesh, both located in Asia, have a rich history of the use of natural fibers, particularly jute, sisal, and coconut. Conversely, France and Belgium are renowned for being major producers of flax, whereas North America, despite its significant economic impact, does not possess the same level of prominence in relation to natural fibers. West Africa, Latin American countries, and India are the major oil palm cultivating countries [10–13]. In the case of hemp production, in Spain, according to data from the Ministry of Agriculture, in 2016, there were 61 ha devoted to hemp cultivation, and in 2020, that number increased to 510 ha [14]. This trend is also being reflected in the world, with Canada, the USA, China, and France being the countries with the highest number of hectares of hemp farms [15–17]. Industrial hemp production grew over 70% in Canada [18]. The US has increased its production from 9000 ha in 2016 to 93,000 in 2019 [19,20]. Furthermore, the estimation of the industrial hemp market presents gratifying results: the market size was estimated at $4.13b in 2021 and was expected to grow at a compound annual growth rate (CAGR) of 16.8% from 2022 to 2030 [21].

In addition to the innovations in the production of hemp, there are others related to less widespread applications, such as the use of fibers not only in the manufacture of textiles but also in construction, due to their mechanical properties when composite materials are manufactured from hemp fibers, obtaining a substitute material for synthetic fibers [22,23]. It also shows great potential in its use for biomass as a substitute for coal [18,24] or to produce activated carbon derived from hemp crops that can be utilized as an electrode material in a hybrid supercapacitor in order to produce a more environmentally friendly form of energy storage [25]. The reduction of CO_2 emissions can also involve the use of hemp, which can be utilized for its negative carbon footprint. Moreover, hemp biomass can also be processed into biochar, which has the ability to absorb various types of organic and inorganic pollutants in an environmentally friendly manner [26].

However, currently, there is a part of the plant that does not have a specific application, and most of such material is treated as a low-value by-product of hemp crop or crop waste. Despite the fact that hemp stalk is used in building applications or to breed animals, those applications do not cover all the offers of the material [24]. The hemp stalk is localized in the inner part of the stem and represents more than 50% of the entire plant by weight [27]. Despite not having any specific application that satisfies all the offers, it has interesting properties as an insulating material. In addition, it is a biomaterial that contains cellulose and has woody fibers. Hemp fiber is one of the vegetable fibers that contain the highest percentage of cellulose (70–74%) [22,28]. Although the hemp stalk has a lower amount of cellulose than the fibers, the percentage is still considerable (Table 1).

Table 1. Percentage of cellulose contained in hemp stalk.

Cellulose (%)	Hemicellulose (%)	Lignin (%)	Reference
44	18	28	[29]
50–60	15–20	20–30	[30]
34–44	31–37	19–28	[31]
49	25	25	[32]

The objective is to revalue a hemp by-product, hemp stalk (Figure 2) as the main material to develop a new biomaterial to use in green building will be reviewed, due to its insulating properties [33–35], positive carbon footprint, and its capability to produce a circular economy as a green material [36–38]. All these characteristics could be used to provide the stalk with new applications that add value to the material and, moreover, will increase the value of hemp farming [39]. So, in this way, more ecological materials will be produced.

Figure 2. Hemp stalk.

The main focus of this review is to cover the properties of hemp stalks and different green materials that can be used as a binder to develop a green insulating material based on hemp stalks. Recently published papers indicate that lignin-based resin, bio-epoxy resin [5], and recyclable cardboard fiber [40] are potential binding materials for developing a new bio-composite material based on hemp stalk. While this new bio-composite material based on hemp stalk and eco-friendly binding materials shows promise as a replacement for traditional inorganic insulating materials in building construction, further research is needed to improve its long-term stability. It is important to ensure that the material maintains its structural integrity and insulating properties over time to ensure its effectiveness and durability in real-world applications. Continued investigation and development of this material will be crucial to its success as a sustainable building material.

2. Insulating Material

Thermal/sound-insulating materials are materials that prevent the transfer of heat or sound. In the case of a building, the insulating materials improve the performance of energy consumption and provide a more comfortable area. The most used materials are rock wool and EPS (Expanded Polystyrene).

One of the most attractive properties of hemp is its insulating properties [41]. Moreover, in comparison with conventional materials, natural fibers have similar hygrothermal properties, and the process of retrofitting historical building envelopes is made more harmonious by their involvement [42].

2.1. Insulating Properties of Hemp Stalk

Hemp possesses exceptional insulating qualities, which are retained in the stalk. One of the primary advantages of using hemp stalk is its affordability, coupled with its insulation performance. However, ensuring its long-term stability as well as its chemical compatibility with various binders is crucial [43]. Since hemp is a particulate material, its mechanical properties, as well as thermal and acoustic insulation capabilities, are dependent on the sample's morphology. Therefore, the internal porosity of the material is the critical factor that determines its performance [44].

In the case of acoustic properties, the parameters of the hemp particles that influence its acoustic properties are the density of the particles, the bulk density, the thickness of the particles, and the shape factor [45]; this relationship is also shown in other biomaterials. On the other hand, the results also show that for low-frequency waves, acoustic insulation is not very effective due to the porosity of the material, although acoustic absorption can be increased in this range of frequencies by making a sandwich-type material.

Decreasing the stalk particle size improves the acoustic properties, the best case being those with an average length of 4 mm. Furthermore, the internal porosity can be predicted from the density of the particles [45,46]. At a microscopic level, hemp has a structure that brings on the absorption of acoustic waves due to the voids provided by the porosity of the material [33]. Hemp has a porosity of 78% with pores between 0.9–3 μm [47,48]. This study determines that the optimal properties for increasing the acoustic absorption of the stalk are for a particle to be 6 mm long with a density of 0.3 g/cm^3.

In the case of thermal insulation, it increases energy efficiency by reducing heat transfer with the outside and saving energy in the air conditioning of buildings. The porosity also contributes to its high thermal properties, although in this case, the properties of the matrix composite material are more influential properties for the final property of the biomaterial composite.

According to the hemp stalk's insulating properties, Table 2, hemp stalk is a green material that can be used in insulating applications, even though its properties are less competitive than commercial materials. The thermal conductivity of rockwool is 0.036–0.037 (W/m k) (28%, better performance than hemp stalk) [49,50] and EPS have 0.038 [50] (28%, better performance than hemp stalk). The acoustic absorption (α) of rockwool is 0.98–0.99 [51,52] (0.1 dB/dB, better performance than hemp stalk) and EPS 0.8–0.9 [53] (same performance of hemp stalk).

Table 2. Properties of hemp stalk.

Thermal Conductivity (W/m k)	Acoustic Absorption (α)	Density ρ_{shives} (kg/m^3)	Porosity ϕ_{inter}(%)	Reference
-	0.7–0.95	80–160	65–85	[45,46]
0.049–0.082	0.88–0.95	110–125	40	[47]
0.064–0.115	0.88–0.99	97–120	-	[54]
0.051	-	72	70–80	[55]

The elastic modulus of the hemp stalk (10–16 GPa) depends on its position in the stem of the plant according to the height at which it is located [56,57]. However, the area where the elastic modulus is greater is different among the species of hemp. The differences between the elastic modulus correspond to an increase or decrease in the size of the cell wall along the stem [56]. No studies have been found that prove whether these differences in cell size affect the insulating properties, detecting a research gap that could be the basis of interesting future research to increase hemp performance as raw material for insulating applications.

The properties shown by the stalk are suitable for use as both thermal and acoustic insulating material. However, it is a particle material, so it is necessary to use a vegetal binder to manufacture green composite material. In this insulating application, the mechanical resistance of the material is not a critical value; it is only necessary that it satisfy the minimum requirements and be a stable material.

2.2. Binder Materials for Hemp Stalk

This review focuses on green materials; however, there are not many studies with a 100% green material, so some results of non-green binder will be presented to study the behavior of the material and find out which green materials can obtain the best performance.

Different applications are being studied to use hemp stalk, taking advantage of its low price, such as using it as insulation in buildings and adding lime to form non-structural blocks [58,59]. Seeking a sustainable substitute for traditional walls, researchers carried out a study of the acoustic absorption properties of lime and hemp stalk walls, obtaining an average of between 40–50% acoustic absorption, obtaining better results with the less hydric binders and which can be manufactured on-site or placed as prefabricated units [60]. The main advantages of hempcrete are the insulating properties provided by the hemp stalk, and the lime binder provides protection against moisture, fungi, and fire [61]. In the same approach, adding hemp particles to the mortar as aggregates reduces its density, increases the insulating properties, and the material will increase the capability of CO_2 storage. Nevertheless, the mechanical properties decrease (maximum stress is reduced up to 30% when adding 8% hemp) [62–64]. Although cementitious matrices offer the benefits of affordability and adaptability, their use can result in chemical damage to hemp stalk [65]. In hot-dry regions where naturally ventilated buildings are preferred, the thermal design of walls based on compressed earth blocks (CEB) could be a viable and more ecological option [66]. Moreover, the thermal behavior of CEB was improved by 10% by incorporating 0.5% of date palm waste [67]. While compacted earth is a compelling environmentally-friendly material, it is crucial to thoroughly examine its dependability and longevity in the absence of long fibers [68].

Hemp stalk particles are also used as raw materials for compounds that are made up of wood particles. In this way, the manufacturing process mixes it with a binder to fabricate a material similar to a chipboard. The manufacture of this type of material consists of mixing the stalk with the binder material and applying pressure and temperature in a mold, then the adhesive cures and the material obtains the shape of the mold. The most commonly used binders are currently based on formaldehyde because of its mechanical properties, dynamic properties, abrasion resistance, and affordability [69]. Nevertheless, due to the fact that it is a toxic material in large quantities, its use has been decreasing in order to reduce formaldehyde. An intermediate solution is the use of 2 formaldehyde-based adhesives by partially substituting them for lignocellulose-based materials (wheat straw and pine and poplar particles), obtaining better results with PDMI (Polymeric Diphenylmethane Diisocyanate), obtaining better results by increasing the percentage of binding material [70,71]. PDMI shows better binding properties than UF (Urea formaldehyde), curing at a temperature of 180 °C and having pressure applied for 3 min [72]; these are the usual values in the industry. The process is also the same with vegetable agglomerate taking into account the curing temperature for each vegetable binder [73,74]. In this type of manufacturing, the structure, and size of the particles is also an important factor in the final properties of the material. If the particle size is very large, air gaps will be produced in the material as all the chips cannot be compacted together because the manufacturing process does not use a vacuum to prevent the air gaps. However, this problem can be solved by including saws and wood dust that occupy these holes together with the resin, thus increasing the mechanical properties of the material [75].

Another alternative is to completely eliminate formaldehyde-based resins by using synthetic ones. Although studies are being carried out to obtain vegetable resins that can achieve the regulatory requirements of the different applications, such as lignin-based wood adhesives [76,77], obtaining materials with great thermal properties, or vegetable proteins such as camellia protein, which is also a residue in the biodiesel production [78]. However, the main problem is the resistance to fire; that problem can be solved by adding a fire resistance coating. There are also studies to develop a sustainable, high-performance, and flame-retardant wood coating based on a curing agent of ammonium hydrogen phytate (AHP) [79,80]. Starch is also a good biobased binder for wood particles; for example, cassava starch binder can be used to elaborate low-density particleboard with excellent performance [81]. The fungi resistance is low; however, citric acid can be added to improve the fungal degradation by 10% [82].

Different innovative fabrication methods and renewable materials for thermal insulating applications are studied. However, the technology still needs more research to have a competitive price and solve different technical difficulties in using materials like vacuum insulating panels, aerogels, or nanocellulose. The research also proposes the use of recycled paper fiber as an insulating material based on cellulose [83]. To bind the different hemp particles in a material that can be lightweight, a manufacture method is needed that does not require applying high pressure to provide the maximum porosity in the material. Paper pulp fiber is proposed, which also has great thermal properties, and cellulose paper waste has a thermal conductivity value of 0.046–0.054 W/m K [84]. Among the different paper fibers, cardboard is a great option due to the facility for recycling and the mechanical/binding properties. Cardboard fibers have a mechanical resistance four times greater than eucalyptus fibers, measured from the binding [85,86]. Moreover, the length of the fiber is longer (cardboard 2.7 mm in average length and eucalyptus 0.76 mm [87]). Several layers of corrugated cardboard were tested, obtaining thermal conductivity values of 0.053 W/m K and a reduction of up to 80 dB in acoustic waves using several layers of corrugated cardboard [5]. In that case, the main problems were durability and moisture/fungi/fire resistance.

Table 3 summarizes the comparison of the performance advantages and disadvantages of bio-based wood adhesives for the stalk. Regardless of the type of bio-based adhesive used, it is crucial to assess their potential for wood composite application and comprehend their interaction with wood. This evaluation can provide valuable scientific insights to guide the development of adhesives in terms of mechanical strength, water and moisture resistance, and thermal and acoustic properties.

Table 3. Advantages and disadvantages of bio-binder for stalk [5,76–79,81–91].

Type of Bio-Binder	Advantages	Disadvantages
Lignin based	Recycle the secondary products produced in paper pulping industries Improve the modulus of elasticity Improve the thermal properties Improve the water resistance Good bonding strength	Need to add a catalyst material Increase the viscosity of adhesive Low fire resistance Reduce the curing rate Low porous structure Low level of substitution
Starch based	High level of substitution Good bonding strength Good film formation property	Low stability upon time Need a surface treatment to increase the water resistance Low fire resistance Slow drying process Poor water resistance Low fungal resistance
Plant protein based	Improve thermal stability Good adhesion strength High level of substitution	Poor water resistance Need a surface treatment to increase the water resistance Low porous structure Low fire resistance
Paper pulp based	Improve thermal/acoustic properties Good bonding strength Recyclable High porous structure High level of substitution	Need a surface treatment to increase the water resistance Slow drying process Poor fire resistance Low stability upon time Low fungal resistance

2.3. Acoustic Insulating Properties of Materials Based on Hemp Stalk

When a surface is contacted by sound waves, the energy is distributed into three categories: incident, reflected, and absorbed energy. In architectural acoustic design, it is helpful to utilize an average absorption coefficient that is assumed to rely solely on the physical attributes of the material. The sound absorption coefficient of any material is determined by the angle at which the sound wave hits the material and the frequency of the sound [92].

In the case of a composite material based on hemp stalks. The mechanism by which the materials absorb sound energy mainly involves three physical processes. Sound-absorbing composite materials have small holes that allow sound waves to access their interior, causing gas flow and friction. It triggers the conversion of a portion of the sound energy into heat energy, which leads to sound absorption. In the case of hemp stalks, when sound waves hit them, the viscous effects between air cavities attenuate some of the sound energy, converting it into heat. Moreover, the sound-absorbing composite materials have the ability to absorb certain sound waves through their own vibrations. Due to the force between chain segments, unique hollow structure, and large specific surface of hemp stalks, sound energy is attenuated and converted into heat and mechanical energy during propagation, resulting in an effective sound-absorbing effect [33,93,94]. The internal porosity of the composite is a crucial factor to consider in the sound absorption mechanism of a low-density insulation panel. The presence of voids, inner and outer spaces, as well as the density and thickness of the composites, directly influence sound absorption [95,96]. Low-density materials with more open structures exhibit lower absorption at low frequencies, while denser structures show better performance at higher frequencies (above 2000 Hz) [92]. Several studies investigating sound absorption in porous materials have found a direct relationship between thickness and low-frequency sound absorption. Increasing the thickness of the material leads to an increase in sound absorption at low frequencies. However, at higher frequencies, thickness has an insignificant effect on sound absorption [97]. Nevertheless, a sandwich-type material can increase acoustic absorption in low frequencies. The low porosity and high density of the material's surface cause sound waves to be reflected [44].

Table 4 shows the results obtained in different studies:

Table 4. Results of acoustic absorption of vegetal particles with different binders.

Fiber	Matrix	Eco-Friendly Material	Acoustic Absorption (α)	Pore Structure	Reference
Hemp stalk	Polycaprolactone	NO	0.6–0.9	Hollow microstructure	[33]
Hemp stalk	Lime	NO	0.6–0.9	Porous material (70–75%)	[46,60,98]
Hemp stalk	Portland cement & MgO-cement	NO	0.1–0.25	Low porosity	[99]
Hemp stalk	C2-H	NO	0.6–0.8	Porous material	[100]
Hemp stalk	Wheat starch	YES	0.7	Porous material (88–90%)	[89,101]
Sunflower stalk	Chitosan	YES	0.2	Low porosity	[102]
Sheep wool	Polypropylene	NO	0.3–0.6	Low porosity	[96]

The acoustic absorption values in Table 4 are lower than those presented by the hemp particles, which means that the composition of the binder and the new internal microstructure is a more significant parameter than the internal porosity or the particle size of the hemp stalk.

Based on the results presented in Table 4, it can be observed that a highly porous material provides good acoustic insulation as it contributes to the dissipation of sound waves, which is a more significant characteristic than sound wave reflection. Hemp stalk has been studied as an insulating material, but there are few studies on hemp due to the materials primarily used with inorganic binders that require the application of pressure and temperature to improve mechanical resistance, resulting in a denser material with low porosity that impairs its properties [103]. However, although a porous material performs better, a non-porous surface coating can be added to the sample to improve sound wave reflection. The use of certain binders can result in an elastic behavior of the composite material, allowing acoustical vibrations to be transmitted to the solid matrix. These elastic effects can have a significant impact on acoustic performance, causing both global effects on transmission and local effects on absorption [100].

2.4. Thermal Insulating Properties of Materials Based on Hemp Stalk

The main objective of thermal insulation is to enhance energy efficiency by limiting the transfer of heat through the building envelope. Insulation materials are designed to conduct heat poorly in order to minimize heat loss [104]. Heat conduction occurs due to the interaction between particles in a substance (solid, liquid, or gas) that results from particle movement. Hence, heat moves from more energetic particles to less energetic ones. Convection, on the other hand, refers to heat transfer between a solid surface and a fluid in motion, whereby heat is transferred through a combination of conduction from the solid to the fluid and bulk movement of fluid particles. Thermal insulation offers high thermal resistance, thereby retarding heat flow, primarily due to gases entrapped within the porous material structure [105]. Thermal insulation materials typically have low densities, which translates to high porosity. The insulation effect is largely attributed to the low thermal conductivity of still gases trapped within the voids of porous material [105]. The principal factors that affect thermal conductivity include raw materials, temperature, porosity, moisture content, and density. Other factors are airflow velocity and thickness [104]. For cellulose-based insulating materials, factors such as temperature, moisture content, and mass density are critical in determining the thermal conductivity value [42,83]. So, in this case, the environmental conditions (temperature, humidity) affect its insulation capacity.

In addition to acoustic insulation, the hemp stalk also has high thermal insulation properties, obtaining a coefficient of thermal conductivity of 0.05 W/m K (Table 2). In Europe, according to the DIN 4108, materials with a λ value lower than 0.1 W/m K may be classed as thermal insulating materials. Additionally, materials with thermal conductivity values lower than 0.03 W/m K are deemed highly effective as thermal insulators [104]. Moreover, its resistance to fire must be classified as at least one type E material according to the regulations. This means that the material should be able to withstand the attack of a small flame for a brief period without significant propagation of the flame. [106].

There are studies on that topic to develop new biomaterials using starch as a binder [107]. Straw and a geopolymer can be used to form an insulating material, obtaining thermal conductivity values of 0.101 W/m K [108,109], the biggest drawback being the resistance to water and fire when using green materials. Other studies show materials made with corn particles and epoxy resins obtain sufficient thermal conductivity values to be considered insulating materials [44].

Table 5 shows the results obtained in different studies:

Table 5. Results of thermal absorption of vegetal particles with different binders.

fiber	Matrix	Eco-Friendly Material	Thermal Conductivity (W/m K)	Reference
Hemp stalk	Lime	NO	0.08–0.13	[60,110,111]
Hemp stalk	Portland cement & MgO-cement	NO	0.08–0.115	[98,99,112]
Hemp stalk	Wheat starch	YES	0.06–0.07	[89,101]
Hemp stalk	Cassava starch	YES	0.026	[113]
Hemp stalk	Reactive vegetable protein	YES	0.078	[114]
Sunflower stalk	Chitosan	YES	0.056–0.058	[102]
Corn stalk	Rice huck ashes	YES	0.06–0.08	[107]
Bamboo powder	Bio-glues	YES	0.10–0.20	[115]
Corn stalk	Epoxy	NO	0.10	[44]
Sheep wool	Polypropylene	NO	0.06–0.10	[96]
Flax stalk	Lignin & and biobased epoxy	NO	0.074	[77]

In addition to using vegetable resins, it is also proposed to replace the particles of pine and other common trees with crop waste, such as hemp stalk. The influence of the starch-stalk ratio on the properties of the material is studied, and it shows that by increasing the hemp particle ratio, the mechanical properties are reduced due to the increase in porosity, which decreases the load transfer capacity. Nevertheless, it improves thermal performance [101]. The thermal insulating performance of materials is improved by higher porosity and density. Unlike acoustic properties, the thermal properties of the matrix material play a crucial role in determining the final properties of the composite material.

At a mean temperature of 24 °C, the apparent thermal conductivity of hemp stalk was tested for various densities and found to increase with rising temperature. While wood-based fiberboards are utilized as thermal insulation materials due to their low density and high thermal resistance, their porous internal structures make them sensitive to environmental changes [116]. Consequently, their thermal conductivity increases by roughly 50% as the temperature increases from 10 to 60 °C [117].

An increase in relative moisture in materials can lead to a decrease in their thermal conductivity and make them more prone to mold formation. For composites made of jute, flax, hemp shives, and fibers, a relative air humidity of 70% leads to a relative material humidity of 5–10%. Additionally, it has been demonstrated that an increase in material moisture from 0–10% results in an increase in thermal conductivity [118].

2.5. Carbon Storage Properties

Although the different studies show high insulating properties, they are not the only important characteristic of these materials. Developing a biomaterial based on hemp stalk can increase the value of the hemp crop, produce more environmentally friendly materials, and also result in materials that act as CO_2 accumulators [2].

The hemp absorbs CO_2; meanwhile, it is growing through the photosynthesis process. Moreover, due to very rapid growth, the biomass accumulated by hemp crops has a significant effect in absorbing atmospheric CO_2; however some of the carbon stored in this biomass is turned back into the atmosphere as CO_2 due to the biodegradation of leaves and roots. Assuming a concentration of 0.5 kg of carbon per kg of dry matter, it can be calculated that 1.84 kg of CO_2 is sequestered per kg of dry hemp through photosynthesis during the plant's growth. This means that a ton of dry hemp can store 325 kg of CO_2, taking into account the amount of emissions during the farming as the diesel consumption, transportation of the seeds, etc [119].

Hempcrete is a sustainable building material made from hemp stalks, lime, and water, and the lime in the material also takes part in the CO_2 sequestration due to the carbonation process of the lime. During the carbonation process, the lime reacts with the CO_2 present in the air, resulting in the conversion of calcium hydroxide ($Ca(OH)_2$) into calcium carbonate ($CaCO_3$) [120]. The carbonation increases the mechanical resistance of the hempcrete, and, additionally, the absorption of CO_2 during this process can have significant implications for the environmental impact of this product [121]. Hempcrete can store 300 kg of CO_2 per m^3 [122–124]. With these two properties, we have a new material that reduces the energy consumption of the building due to the thermal insulation capacities and also stores CO_2, obtaining a material that would passively help to obtain buildings with zero emissions [125].

Based on these examples, it can be concluded that the binder used in the composite material has a significant impact on its carbon footprint. The use of recyclable materials with low impact on the carbon footprint is proposed for the binder material.

2.6. Durability

In order to develop new biocomposite insulating materials, it is necessary to ensure the durability conditions to ensure the commercial product can be reached. However, there are only a few papers that study the durability of long-term tests in those materials. The longevity of bio-composites can be impacted by various forms of biological degradation, such as mold growth, as well as environmental factors like fluctuations in temperature and humidity [105].

Investigating the water absorption characteristics of biocomposites is crucial, given the weak water resistance of biomaterial. For biocomposites used outdoors in construction applications, their water absorbency is a critical factor affecting their mechanical properties and dimensional stability [126].

Hemp fibers also present low stability against contact with water, which sees their Young's modulus reduced by 50% after immersion in water [127]. Moreover, using a 100% vegetable material increases the risk of mold growing in high humidity conditions [115]. Mold growth can be relieved by increasing the pH of the material, as is the case with hempcrete, which has anti-fungal properties [61].

In the case of VFRCM (vegetal fabric reinforced cementitious matrix), the issue arises from the alkaline hydrolysis resulting from the production of $Ca(OH)_2$ during cement hydration. The calcium adsorption is pH dependent, and the pectin contained in fibers can react with calcium ions in an alkaline environment [128]. Moreover, a reduction in compressive strength has been observed in hempcrete when using $Ca(OH)_2$ treated hemp stalks compared to untreated ones [129,130]. To incorporate a material that is rich in Al_2O_3 and SiO_2 can mitigate the degradation of the hemp fibers. This material should have the ability to consume $Ca(OH)_2$ during mortar hydration [131]. Alternatively, the fiber can undergo physical or chemical treatment. To extend these findings to vegetable binders, it is crucial to ensure compatibility between the hemp stalk and the vegetal binder to achieve optimal results.

To prevent the degradation of the biomaterials, a solution is to introduce a coating in order to protect the material against environmental effects. To increase the water resistance of these materials, surface coatings such as NaOH, silane, and epoxy have been studied, which reduce the moisture absorption of the fibers [132–135].

In one research work, a vegetal coating is proposed by coating a bamboo particle with pine resin, obtaining a hydrophobic material after treatment; in addition, after the coating, bamboo particles become a stable material against the environment [136]. Nevertheless, the pine resin worsens the fire resistance properties. Arabic gum also presents good results as a coating in cases of humidity and fire, although it cannot be used in direct contact with water since Arabic gum is soluble in water, so when the layer is in direct contact, the protective coating is removed until the material is uncoated [137]. Although these coatings increase the durability of the material, no studies have been found that investigate how the insulating properties behave when these coatings are added.

The coating would not only have the function of protecting the material, but also of stabilizing the interior moisture. Since moisture has a great impact on the insulating properties, as seen in the previous sections.

3. Conclusions and Further Research Interests

According to the results presented in the different studies of the last decades, hemp is positioned as a green material with great insulating properties that could replace some inorganic commercial materials to improve the insulating performance of buildings and build green buildings. These data can be corroborated by the increase in interest in the scientific community, with the increase in research each year, and also with the increase in the hemp industry in the business sector.

- Varieties and positions on the stem can cause differences in the elastic modulus of the hemp stalk of up to 60%. However, no studies have been found to confirm whether these differences in cell size impact the insulating properties of hemp.
  This research gap presents an interesting opportunity for future studies to explore ways to enhance the performance of hemp as a raw material for insulating applications. With the farming of the different varieties of hemp, it is necessary to study which variety produces the hemp stalk with the best insulating properties and in which part of the stem it is localized.
- It is suggested to use a binder that aids in creating a low-density and porous material as the microstructure of the composite material plays a significant role in determining its insulating properties.
  It is still necessary to delve into some issues in order to develop new green insulating materials, such as the study of more green binders that do not need to apply pressure or temperature during the manufacturing process. In this way, the manufacturing cost is reduced, and lightweight materials with high porosity would be produced that would stimulate the insulating properties of the composite material.
  In this way, paper pulp-based binder shows great opportunities for the development of future research.

- Despite all the studies, further research is still needed on the use of a 100% plant-based composite material, including the binder. However, among those presented, starch stands out.
  By using starch as a binder, an acoustic absorption (α) of 0.7 and a thermal conductivity of 0.03 W/m K have been achieved, which are respectively 0.2 db/db lower and 15% higher than conventional materials.
- It has been observed that only a few studies have investigated the acoustic properties, fire resistance, fungi growth, and long-term durability of 100% green composite materials.
- A ton of dry hemp can store 325 kg of CO_2.
  Hemp is exhibited as a material capable of storing CO_2 and producing a renewable material with a circular economy. The binder also affects the carbon footprint of the composite material. To prioritize the carbon footprint, a recycled biomaterial such as cardboard fiber can be chosen as a binder.
- It is necessary to protect the biomaterial against ambient conditions.
  The main problem for a biocomposite material is degradation over time. So in order to produce a commercial product it will be necessary to ensure the stability of the material in ambient conditions. Nevertheless, no research has been found that investigates the influence of using an eco-friendly coating on the insulation properties of a composite material.
  To protect the biocomposite material against ambient degradation, some vegetable coating, such as colophony, shows a great performance. Arabic gum is proposed as an effective solution for fire protection, but it only provides protection against moisture and not direct contact with water.

In this review, a large amount of research related to the development of new hemp stalk-based composite materials has been compiled. Much research has been done, and it is arousing increasing interest in the scientific community for its performance and sustainability reasons. Therefore, it is recommended that these issues continue to be investigated in order to develop a new competitive material capable of replacing current inorganic materials.

Author Contributions: Conceptualization, B.M., E.B.-M. and L.G.; methodology, B.M., E.B.-M. and L.G.; software, B.M.; validation, B.M., E.B.-M. and L.G.; formal analysis, B.M.; investigation, B.M.; resources, B.M.; data curation, B.M.; writing—original draft preparation, B.M.; writing—review and editing, E.B.-M. and L.G.; visualization, L.G.; supervision, E.B.-M. and L.G.; project administration, L.G.; funding acquisition, L.G. All authors have read and agreed to the published version of the manuscript.

Funding: This research was funded by ACCIÓ grant number VALSUCANEM ACE014/20/000024.

Institutional Review Board Statement: Not applicable.

Informed Consent Statement: Not applicable.

Data Availability Statement: All the data is available within the manuscript.

Conflicts of Interest: The authors declare no conflict of interest.

References

1. UNEP. Buildings and Climate Change: Current Status, Challenges and Opportunities. *DG Environment News Alert Dervice*; United Nations Environment Programme: Nairobi, Kenya, 2007; p. 1.
2. Jami, T.; Karade, S.R.; Singh, L.P. A review of the properties of hemp concrete for green building applications. *J. Clean. Prod.* **2019**. *239*, 117852. [CrossRef]
3. Delbeke, J.; Runge-Metzger, A.; Slingenberg, Y.; Werksman, J. The paris agreement. *Towards a Climate-Neutral Europe: Curbing the Trend*; Routledge: London, UK, 2019; pp. 24–45. [CrossRef]
4. Ahmed, A.T.; Islam, M.Z.; Mahmud, M.S.; Sarker, M.E.; Islam, M.R. Hemp as a potential raw material toward a sustainable world: A review. *Heliyon* **2022**, *8*, e08753. [CrossRef] [PubMed]
5. Asdrubali, F.; D'Alessandro, F.; Schiavoni, S. A review of unconventional sustainable building insulation materials. *Sustain. Mater. Technol.* **2015**, *4*, 1–17. [CrossRef]

6. Zampori, L.; Dotelli, G.; Vernelli, V. Life cycle assessment of hemp cultivation and use of hemp-based thermal insulator materials in buildings. *Environ. Sci. Technol.* **2013**, *47*, 7413–7420. [CrossRef] [PubMed]
7. Barbash, V.A.; Yashchenko, O.V.; Yakymenko, O.S.; Zakharko, R.M.; Myshak, V.D. Preparation of hemp nanocellulose and its use to improve the properties of paper for food packaging. *Cellulose* **2022**, *29*, 8305–8317. [CrossRef]
8. Amaducci, S.; Scordia, D.; Liu, F.H.; Zhang, Q.; Guo, H.; Testa, G.; Cosentino, S.L. Key cultivation techniques for hemp in Europe and China. *Ind. Crops Prod.* **2015**. *68*, 2–16. [CrossRef]
9. Ranalli, P.; Venturi, G. Hemp as a raw material for industrial applications. *Euphytica* **2004**, *140*, 1–6. [CrossRef]
10. Biagiotti, J.; Puglia, D.; Kenny, J.M. A Review on Natural Fibre-Based Composites-Part I. *J. Nat. Fibers* **2004**, *1*, 37–68. [CrossRef]
11. Sanjay, M.R.; Arpitha, G.R.; Naik, L.L.; Gopalakrishna, K.; Yogesha, B. Applications of Natural Fibers and Its Composites: An Overview. *Nat. Resour.* **2016**, *7*, 108–114. [CrossRef]
12. Mwaikambo, L.Y. Review of the history, properties and application of plant fibres *Afr. J. Sci. Technol. (Ajst) Sci. Eng. Ser.* **2006**, *7*, 120–133.
13. Peças, P.; Carvalho, H.; Salman, H.; Leite, M. Natural Fibre Composites and Their Applications: A Review *J. Compos. Sci.* **2018**, *2*, 66. [CrossRef]
14. Nota de Prensa del Ministerio de Agricultura de España. Available online: https://www.mapa.gob.es/es/prensa/ultimas-noticias/la-superficie-dedicada-al-c%C3%A1%C3%B1amo-se-multiplica-por-ocho-en-los-%C3%BAltimos-cinco-a%C3%B1os/tcm:30-560562#:~:text=La%20superficie%20dedicada%20al%20cultivo,solicitud%20%C3%BAnica%20de%20la%20PAC (accessed on 9 September 2022).
15. Horner, J.; Milhollin, R.; Roach, A.; Morrison, C.; Schneider, R. *Comparative Analysis of the Industrial Hemp Industry*; University of Missouri: Columbia, MO, USA, 2019.
16. Rupasinghe, H.V.; Zheljazkov, V.D.; Davis, A.; Kumar, S.K.; Murray, B. Industrial Hemp (*Cannabis sativa* subsp. *sativa*) as an Emerging Source for Value-Added Functional Food Ingredients and Nutraceuticals. *Molecules* **2020**, *25*, 4078. [CrossRef]
17. Schluttenhofer, C.; Yuan, L. Challenges towards Revitalizing Hemp: A Multifaceted Crop. *Trends Plant Sci.* **2017**, *22*, 917–929. [CrossRef]
18. Parvez, A.M.; Lewis, J.D.; Afzal, M.T. Potential of industrial hemp (*Cannabis sativa* L.) for bioenergy production in Canada: Status, challenges and outlook. *Renew. Sustain. Energy Rev.* **2021**, *141*, 110784. [CrossRef]
19. Dhoubhadel, S.P. Challenges, Opportunities, and the Way Forward for the U.S. Hemp Industry. *J. Gender Agric. Food Secur.* **2021**, *19*, 79–96.
20. Mark, T.; Shepherd, J.; Olson, D.; Snell, W.; Proper, S.; Thornsbury, S. Economic Viability of Industrial Hemp in the United States: A Review of State Pilot Programs. *J. Gender Agric. Food Secur.* **2020**, *1*, 1–22.
21. Advanced Materials. *Industrial Hemp Market Size, Share & Trends Analysis Report by Product (Seeds, Fiber, Shivs), by Application (Animal Care, Textiles, Personal Care), by Region (North America, APAC), and Segment Forecasts, 2022–2030*; Technical Report; Grand View Research: San Francisco, CA, USA, 2022.
22. Manaia, J.P.; Manaia, A.T.; Rodriges, L. Industrial Hemp Fibers: An Overview. *Fibers* **2019**, *7*, 106. [CrossRef]
23. Santoni, A.; Bonfiglio, P.; Fausti, P.; Marescotti, C.; Mazzanti, V.; Mollica, F.; Pompoli, F. Improving the sound absorption performance of sustainable thermal insulation materials: Natural hemp fibres. *Appl. Acoust.* **2019**, *150*, 279–289. [CrossRef]
24. Marrot, L.; Candelier, K.; Valette, J.; Lanvin, C.; Horvat, B.; Legan, L.; DeVallance, D.B. Valorization of Hemp Stalk Waste Through Thermochemical Conversion for Energy and Electrical Applications. *Waste Biomass Valorization* **2022**, *13*, 2267–2285. [CrossRef]
25. Mandal, S.; Hu, J.; Shi, S.Q. A comprehensive review of hybrid supercapacitor from transition metal and industrial crop based activated carbon for energy storage applications. *Mater. Today Commun.* **2023**, *34*, 105207. [CrossRef]
26. Hosny, M.; Fawzy, M.; Eltaweil, A.S. Phytofabrication of bimetallic silver-copper/biochar nanocomposite for environmental and medical applications. *J. Environ. Manag.* **2022**, *316*, 115238. [CrossRef] [PubMed]
27. Colinart, T.; Glouannec, P.; Chauvelon, P. Influence of the setting process and the formulation on the drying of hemp concrete. *Constr. Build. Mater.* **2012**, *30*, 372–380. [CrossRef]
28. Gümüşkaya, E.; Usta, M.; Balaban, M. Carbohydrate components and crystalline structure of organosolv hemp (*Cannabis sativa* L.) bast fibers pulp. *Bioresour. Technol.* **2007**, *98*, 491–497. [CrossRef] [PubMed]
29. Vignon, M.R.; Garcia-Jaldon, C.; Dupeyre, D. Steam explosion of woody hemp chènevotte. *Int. J. Biol. Macromol.* **1995**, *17*, 395–404. [CrossRef] [PubMed]
30. Evrard, A.; Erde, L.T. Transient Hygrothermal Behavior of Lime-Hemp Materials. Ph.D. Thesis, Ecole Polytechnique de Louvain Unité d' Architecture Transient, Louvain, Belgium, 2015.
31. Thomsen, A.B.; Rasmussen, S.; Bohn, V.; Nielsen, K.V.; Thygesen, A. *Hemp Raw Materials: The Effect of Cultivar, Growth Conditions and Pretreatment on the Chemical Composition of the Fibres*; Risø National Laboratory: Roskilde, Denmark, 2005; Volume 1507, pp. 6–30.
32. Zhang, X.; Guo, J.; Liu, Y.; Hao, X.; Ji, X.; Yang, Q. Biochemical preparation of hydrophobic and lipophilic nanocellulose from hemp stalk. *Mater. Today Chem.* **2023**, *27*, 101346. [CrossRef]
33. Su, S.; Gao, Y.; Zhou, X.; Xiong, X.; Wang, Y.; Lyu, L. Structure of Waste Hemp Stalks and Their Sound Absorbing Properties. *Polymers* **2022**, *14*, 4844. [CrossRef]
34. Ingrao, C.; Giudice, A.L.; Bacenetti, J.; Tricase, C.; Dotelli, G.; Fiala, M.; Siracusa, V.; Mbohwa, C. Energy and environmental assessment of industrial hemp for building applications: A review. *Renew. Sustain. Energy Rev.* **2015**. *51*, 29–42. [CrossRef]

35. Charai, M.; Sghiouri, H.; Mezrhab, A.; Karkri, M. Thermal insulation potential of non-industrial hemp (*Moroccan cannabis sativa* L.) fibers for green plaster-based building materials. *J. Clean. Prod.* **2021**, *292*, 126064. [CrossRef]
36. Adesina, I.; Bhowmik, A.; Sharma, H.; Shahbazi, A. A review on the current state of knowledge of growing conditions, agronomic soil health practices and utilities of hemp in the United States. *Agriculture* **2020**, *10*, 129. [CrossRef]
37. Karche, T.; Singh, M.R. The application of hemp Cannabis sativa l. for a green economy: A review. *Turk. J. Bot.* **2019**, *43*, 710–723. [CrossRef]
38. Guessasma, S.; Bassir, D.; Hedjazi, L. Influence of interphase properties on the effective behaviour of a starch-hemp composite. *Mater. Des.* **2015**, *65*, 1053–1063. [CrossRef]
39. Kirilovs, E.; Gusovius, H.J.; Kukle, S.; Emsins, J. Performance of Fibreboards Made from Wetpreserved Hemp. *Mater. Sci. Text. Cloth. Technol.* **2014**, *8*, 65. [CrossRef]
40. Martinez, B.; Gil, L.; Bernat, E. Study of an insulating hemp-based bio-material: Mechanical, thermal and acoustic properties. *Mater. Compuestos* **2022**, *7*. [CrossRef]
41. Liao, J.; Zhang, S.; Tang, X. Sound Absorption of Hemp Fibers (*Cannabis Sativa* L.) Based Nonwoven Fabrics and Composites: A Review. *J. Nat. Fibers* **2022**, *19*, 1297–1309. [CrossRef]
42. Jerman, M.; Palomar, I.; Kočí, V.; Černý, R. Thermal and hygric properties of biomaterials suitable for interior thermal insulation systems in historical and traditional buildings. *Build. Environ.* **2019**, *154*, 81–88. [CrossRef]
43. Bokhari, S.M.; Chi, K.; Catchmark, J.M. Structural and physico-chemical characterization of industrial hemp hurd: Impacts of chemical pretreatments and mechanical refining. *Ind. Crops Prod.* **2021**, *171*, 113818. [CrossRef]
44. Binici, H.; Aksogan, O.; Demirhan, C. Mechanical, thermal and acoustical characterizations of an insulation composite made of bio-based materials. *Sustain. Cities Soc.* **2016**, *20*, 17–26. [CrossRef]
45. Glé, P.; Gourdon, E.; Arnaud, L. Modelling of the acoustical properties of hemp particles. *Constr. Build. Mater.* **2012**, *37*, 801–811. [CrossRef]
46. Glé, P.; Gourdon, E.; Arnaud, L. Acoustical properties of materials made of vegetable particles with several scales of porosity. *Appl. Acoust.* **2011**, *72*, 249–259. [CrossRef]
47. Kosiński, P.; Brzyski, P.; Tunkiewicz, M.; Suchorab, Z.; Wiśniewski, D.; Palczyński, P. Thermal Properties of Hemp Shives Used as Insulation Material in Construction Industry. *Energies* **2022**, *15*, 2461. [CrossRef]
48. Balčiūnas, G.; Vėjelis, S.; Lekūnaitė, L.; Kremensas, A. Assessment of structure influence on thermal conductivity of hemp shives composite. *Environ. Eng. Manag. J.* **2016**, *15*, 699–705. [CrossRef]
49. Ducoulombier, L.; Lafhaj, Z. Comparative study of hygrothermal properties of five thermal insulation materials. *Case Stud. Therm. Eng.* **2017**, *10*, 628–640. [CrossRef]
50. Al-Homoud, M.S. Performance characteristics and practical applications of common building thermal insulation materials. *Build. Environ.* **2005**, *40*, 353–366. [CrossRef]
51. Wang, C.N.; Torng, J.H. Experimental study of the absorption characteristics of some porous fibrous materials. *Appl. Acoust.* **2001**, *62*, 447–459. [CrossRef]
52. Tang, X.; Yan, X. Acoustic energy absorption properties of fibrous materials: A review. *Compos. Part Appl. Sci. Manuf.* **2017**, *101*, 360–380. [CrossRef]
53. de Moraes, E.G.; Sangiacomo, L.; P. Stochero, N.; Arcaro, S.; R. Barbosa, L.; Lenzi, A.; Siligardi, C.; Novaes de Oliveira, A.P. Innovative thermal and acoustic insulation foam by using recycled ceramic shell and expandable styrofoam (EPS) wastes. *Waste Manag.* **2019**, *89*, 336–344. [CrossRef]
54. Delhomme, F.; Hajimohammadi, A.; Almeida, A.; Jiang, C.; Moreau, D.; Gan, Y.; Wang, X.; Castel, A. Physical properties of Australian hurd used as aggregate for hemp concrete. *Mater. Today Commun.* **2020**, *24*, 100986. [CrossRef]
55. Sinka, M.; Sahmenko, G.; Korjakins, A.; Radina, L.; Bajare, D. Hemp thermal insulation concrete with alternative binders, analysis of their thermal and mechanical properties. *Iop Conf. Ser. Mater. Sci. Eng.* **2015**, *96*, 012029. [CrossRef]
56. Li, X.; Wang, S.; Du, G.; Wu, Z.; Meng, Y. Variation in physical and mechanical properties of hemp stalk fibers along height of stem. *Ind. Crops Prod.* **2013**, *42*, 344–348. [CrossRef]
57. Khan, M.M.R.; Chen, Y.; Laguë, C.; Landry, H.; Peng, Q.; Zhong, W. Compressive properties of Hemp (*Cannabis sativa* L.) stalks. *Biosyst. Eng.* **2010**, *106*, 315–323. [CrossRef]
58. Brzyski, P.; Gładecki, M.; Rumińska, M.; Pietrak, K.; Kubiś, M.; Łapka, P. Influence of hemp shives size on hygro-thermal and mechanical properties of a hemp-lime composite. *Materials* **2020**, *13*, 5383. [CrossRef]
59. Lagouin, M.; Magniont, C.; Sénéchal, P.; Moonen, P.; Aubert, J.E.; Laborel-préneron, A. Influence of types of binder and plant aggregates on hygrothermal and mechanical properties of vegetal concretes. *Constr. Build. Mater.* **2019**, *222*, 852–871. [CrossRef]
60. Kinnane, O.; Reilly, A.; Grimes, J.; Pavia, S.; Walker, R. Acoustic absorption of hemp-lime construction. *Constr. Build. Mater.* **2016**, *122*, 674–682. [CrossRef]
61. Magwood, C. *Essential Hempcrete*; Endeavour Centre: Peterborough, ON, Canada, 2016.
62. Valle-Zermeño, R.D.; Aubert, J.E.; Laborel-Préneron, A.; Formosa, J.; Chimenos, J.M. Preliminary study of the mechanical and hygrothermal properties of hemp-magnesium phosphate cements. *Constr. Build. Mater.* **2016**, *105*, 62–68. [CrossRef]
63. Seng, B.; Magniont, C.; Lorente, S. Characterization of a precast hemp concrete block. Part II: Hygric properties. *J. Build. Eng.* **2019**, *24*, 100579. [CrossRef]

64. Selvaranjan, K.; Navaratnam, S.; Gamage, J.C.P.H.; Thamboo, J.; Siddique, R.; Zhang, J.; Zhang, G. Thermal and environmental impact analysis of rice husk ash-based mortar as insulating wall plaster. *Constr. Build. Mater.* **2021**, *283*, 122744. [CrossRef]
65. Mendizabal, V.; Martínez, B.; Mercedes, L.; Bernat-Maso, E.; Gil, L. Vegetal-FRCM Failure under Partial Interaction Mechanism. *Appl. Sci.* **2022**, *12*, 12964. [CrossRef]
66. Hema, C.; Messan, A.; Lawane, A.; Soro, D.; Nshimiyimana, P.; van Moeseke, G. Improving the thermal comfort in hot region through the design of walls made of compressed earth blocks: An experimental investigation. *J. Build. Eng.* **2021**, *38*, 102148. [CrossRef]
67. Nusrat, J.; Jeff, C.; Badr, A.; Rafal, L.A.M.; Karyono, K. Thermophysical Properties of Sawdust and Coconut Coir Dust Incorporated Unfired Clay Blocks. *Constr. Mater.* **2022**, *66*, 234–257. [CrossRef]
68. Laibi, A.B.; Poullain, P.; Leklou, N.; Gomina, M.; Sohounhloué, D.K.C. Influence of the kenaf fiber length on the mechanical and thermal properties of Compressed Earth Blocks (CEB). *Ksce J. Civ. Eng.* **2018**, *22*, 785–793. [CrossRef]
69. Sonnenschein, M.F.; Wendt, B.L.; Sonnenschein, G.F. Interfacial factors affecting polymeric diphenylmethane diisocyanate/wood bond strength. *J. Appl. Polym. Sci.* **2005**, *98*, 449–455. [CrossRef]
70. Cülfük, M.; Karakuş, K.; Atar, İ.; Bacşboğa, İ.H.; Bozkurt, F.; Mengeloğlu, F. Determination of some properties of particleboards produced with two different adhesives. In Proceedings of the International Symposium on New Horizons in Forestry, Isparta, Turkey, 18–20 October 2017.
71. Hasan, K.M.; Horváth, P.G.; Alpár, T. Thermomechanical Behavior of Methylene Diphenyl Diisocyanate-Bonded Flax/Glass Woven Fabric Reinforced Laminated Composites. *ACS Omega* **2021**, *6*, 6124–6133. [CrossRef] [PubMed]
72. Papadopoulos, A.N. Property Comparisons and Bonding Efficiency of UF and PMDI Bonded Particleboards as Affected by Key Process Variables. *BioResources* **2006**, *1*, 201–208. [CrossRef]
73. Kremensas, A.; Kairyte, A.; Vaitkus, S.; Vejelis, S.; Balčiunas, G. Mechanical performance of biodegradable thermoplastic polymer-based biocomposite boards from hemp shivs and corn starch for the building industry. *Materials* **2019**, *16*, 845. [CrossRef] [PubMed]
74. Le, A.T.; Gacoin, A.; Li, A.; Mai, T.H.; Rebay, M.; Delmas, Y. Experimental investigation on the mechanical performance of starch-hemp composite materials. *Constr. Build. Mater.* **2014**, *61*, 106–113. [CrossRef]
75. Mirski, R.; Derkowski, A.; Dziurka, D.; Dukarska, D.; Czarnecki, R. Effects of a chipboard structure on its physical and mechanical properties. *Materials* **2019**, *12*, 3777. [CrossRef]
76. Mansouri, N.E.E.; Pizzi, A.; Salvadó, J. Lignin-based wood panel adhesives without formaldehyde. *Holz als Roh-und Werkstoff* **2007**, *65*, 65–70. [CrossRef]
77. Essid, S.; Hegde, V.J.; Mahieu, A.; Bizet, L.; Leblanc, N.; Saouab, A. Comparison of the properties of flax shives based particleboards prepared using binders of bio-based lignin and partially bio-based epoxy resin. *Int. J. Adhes. Adhes.* **2021**, *109*. [CrossRef]
78. Zhu, X.; Wang, D.; Li, N.; Sun, X.S. Bio-Based Wood Adhesive from Camelina Protein (a Biodiesel Residue) and Depolymerized Lignin with Improved Water Resistance. *ACS Omega* **2017**, *2*, 7996–8004. [CrossRef]
79. Song, F.; Liu, T.; Fan, Q.; Li, D.; Ou, R.; Liu, Z.; Wang, Q. Sustainable, high-performance, flame-retardant waterborne wood coatings via phytic acid based green curing agent for melamine-urea-formaldehyde resin. *Prog. Org. Coatings* **2022**, *162*, 106597. [CrossRef]
80. Zhao, X.; Liu, T.; Ou, R.; Hao, X.; Fan, Q.; Guo, C.; Sun, L.; Liu, Z.; Wang, Q. Fully Biobased Soy Protein Adhesives with Integrated High-Strength, Waterproof, Mildew-Resistant, and Flame-Retardant Properties. *Acs Sustain. Chem. Eng.* **2022**, *10*, 6675–6686. [CrossRef]
81. Monteiro, S.; Martins, J.; Magalhães, F.D.; Carvalho, L. Low density wood-based particleboards bonded with foamable sour cassava starch: Preliminary studies. *Polymers* **2016**, *8*, 354. [CrossRef]
82. Hazim, M.; Amini, M.; Hashim, R.; Sulaiman, N.S.; Mohamed, M.; Sulaiman, O. Environmentally friendly binder. *BioResources* **2020**, *15*, 4234–4248.
83. Abu-Jdayil, B.; Mourad, A.H.; Hittini, W.; Hassan, M.; Hameedi, S. Traditional, state-of-the-art and renewable thermal building insulation materials: An overview. *Constr. Build. Mater.* **2019**, *214*, 709–735. [CrossRef]
84. Aditya, L.; Mahlia, T.M.; Rismanchi, B.; Ng, H.M.; Hasan, M.H.; Metselaar, H.S.; Muraza, O.; Aditiya, H.B. A review on insulation materials for energy conservation in buildings. *Renew. Sustain. Energy Rev.* **2017**, *73*, 1352–1365. [CrossRef]
85. Chen, Y.; Wan, J.; Dong, X.; Ma, Y. Fiber properties of eucalyptus kraft pulp with different carboxyl group contents. *Cellulose* **2013**, *20*, 2839–2846. [CrossRef]
86. Tarrés, Q.; Pèlach, M.À.; Alcalà, M.; Delgado-Aguilar, M. Cardboard boxes as raw material for high-performance papers through the implementation of alternative technologies: More than closing the loop. *J. Ind. Eng. Chem.* **2017**, *54*, 52–58. [CrossRef]
87. Sarikaya, E.; Çallioğlu, H.; Demirel, H. Production of epoxy composites reinforced by different natural fibers and their mechanical properties. *Compos. Part Eng.* **2019**, *167*, 461–466. [CrossRef]
88. Ferdosian, F.; Pan, Z.; Gao, G.; Zhao, B. Bio-based adhesives and evaluation for wood composites application. *Polymers* **2017**, *9*, 70. [CrossRef]
89. Le, A.T.; Gacoin, A.; Li, A.; Mai, T.H.; Wakil, N.E. Influence of various starch/hemp mixtures on mechanical and acoustical behavior of starch-hemp composite materials. *Compos. Part Eng.* **2015**, *75*, 201–211. [CrossRef]

90. Jafari, V.; Labafzadeh, S.R.; Jeihanipour, A.; Karimi, K.; Taherzadeh, M.J. Construction and demolition lignocellulosic wastes to bioethanol. *Renew. Energy* **2011**, *36*, 2771–2775. [CrossRef]
91. Orlova, A.M. Furfural-based Binder for Composite Construction Materials. *Procedia Eng.* **2016**, *153*, 527–530. [CrossRef]
92. Nandanwar, A.; Kiran, M.C.; Varadarajulu, K.C. Influence of Density on Sound Absorption Coefficient of Fibre Board. *Open J. Acoust.* **2017**, *7*, 73876. [CrossRef]
93. Zhu, X.; Kim, B.J.; Wang, Q.; Wu, Q. Recent Advances in the Sound Insulation Properties of Bio-based Materials. *BioResources* **2013**, *9*. [CrossRef]
94. Sagartzazu, X.; Hervella-Nieto, L.; Pagalday, J.M. Review in sound absorbing materials. *Arch. Comput. Methods Eng.* **2008**, *15*, 311–342. [CrossRef]
95. Cao, L.; Fu, Q.; Si, Y.; Ding, B.; Yu, J. Porous materials for sound absorption. *Compos. Commun.* **2018**, *10*, 25–35. [CrossRef]
96. Ilangovan, M.; Navada, A.P.; Guna, V.; Touchaleaume, F.; Saulnier, B.; Grohens, Y.; Reddy, N. Hybrid biocomposites with high thermal and noise insulation from discarded wool, poultry feathers, and their blends. *Constr. Build. Mater.* **2022**, *345*, 128324. [CrossRef]
97. Ibrahim, M.; Melik, R. Physical Parameters Affecting Acoustic Absorption Characteristics of Fibrous Materials. *Proc. Math. Phys. Soc. Egypt* **1978**, *46*, 125–130.
98. Zakrevskaya, L.V.; Gavrilenko, A.; Andreeva, K.; Lubin, P.; Kapush, I.; Udin, I. Wall materials based on complex binders and organic aggregate. *IOP Conf. Ser. Mater. Sci. Eng.* **2020**, *896*, 012083. [CrossRef]
99. Stevulova, N.; Kidalova, L.; Cigasova, J.; Junak, J.; Sicakova, A.; Terpakova, E. Lightweight composites containing hemp hurds. *Procedia Eng.* **2013**, *65*, 69–74. [CrossRef]
100. Abbas, M.S.; Gourdon, E.; Glé, P.; McGregor, F.; Ferroukhi, M.Y.; Fabbri, A. Relationship between hygrothermal and acoustical behavior of hemp and sunflower composites. *Build. Environ.* **2021**, *188*, 107462. [CrossRef]
101. Bourdot, A.; Moussa, T.; Gacoin, A.; Maalouf, C.; Vazquez, P.; Thomachot-Schneider, C.; Bliard, C.; Merabtine, A.; Lachi, M.; Douzane, O.; et al. Characterization of a hemp-based agro-material: Influence of starch ratio and hemp shive size on physical, mechanical, and hygrothermal properties. *Energy Build.* **2017**, *153*, 501–512. [CrossRef]
102. Mati-Baouche, N.; De Baynast, H.; Lebert, A.; Sun, S.; Lopez-Mingo, C.J.S.; Leclaire, P.; Michaud, P. Mechanical, thermal and acoustical characterizations of an insulating bio-based composite made from sunflower stalks particles and chitosan. *Ind. Crops Prod.* **2014**, *58*, 244–250. [CrossRef]
103. Crini, G.; Lichtfouse, E.; Chanet, G.; Morin-Crini, N. Applications of hemp in textiles, paper industry, insulation and building materials, horticulture, animal nutrition, food and beverages, nutraceuticals, cosmetics and hygiene, medicine, agrochemistry, energy production and environment: A review. *Environ. Chem. Lett.* **2020**, *18*, 1451–1476. [CrossRef]
104. Hung Anh, L.D.; Pásztory, Z. An overview of factors influencing thermal conductivity of building insulation materials. *J. Build. Eng.* **2021**, *44*, 102604. [CrossRef]
105. Schritt, H.; Pleissner, D. Recycling of organic residues to produce insulation composites: A review. *Clean. Waste Syst.* **2022**, *3*, 100023. [CrossRef]
106. *UNE-EN 13501-2:2009+A1:2010*; Clasificación en Función del Comportamiento Frente al Fuego de los Productos de Construcción y Elementos Para la Edificación. Parte 2: Clasificación a Partir de Datos Obtenidos de los Ensayos de Resistencia al Fuego Excluidas; Asociacion Espanola de Normalizacion y Certificacion: Madrid, Spain, 2010.
107. Srikanth, G.; Fernando, A.; Selvaranjan, K.; Gamage, J.C.P.H.; Ekanayake, L. Development of a plastering mortar using waste bagasse and rice husk ashes with sound mechanical and thermal properties. *Case Stud. Constr. Mater.* **2022**, *16*, e00956. [CrossRef]
108. Wei, K.; Lv, C.; Chen, M.; Zhou, X.; Dai, Z.; Shen, D. Development and performance evaluation of a new thermal insulation material from rice straw using high frequency hot-pressing. *Energy Build.* **2015**, *87*, 116–122. [CrossRef]
109. Liu, L.; Zou, S.; Li, H.; Deng, L.; Bai, C.; Zhang, X.; Wang, S.; Li, N. Experimental physical properties of an eco-friendly bio-insulation material based on wheat straw for buildings. *Energy Build.* **2019**, *201*, 19–36. [CrossRef]
110. Ahlberg, J.; Ahlberg, J.; Georges, E.; Norlén, M. The potential of hemp buildings in different climates and the hempcrete building system. *Tve* **2014**, *14*.
111. Sutton, A.; Black, D.; Pete, W. *Hemp Lime: And introduction to Low-Impact Building Materials*; BRE Information Papers; University of Bath: Bath, UK, 2011; pp. 1–6.
112. Balčiunas, G.; Vejelis, S.; Vaitkus, S.; Kairyte, A. Physical properties and structure of composite made by using hemp hurds and different binding materials. *Procedia Eng.* **2013**, *57*, 159–166. [CrossRef]
113. Canto, A.; Batista, M.; Sanchez, J.; Moreno, M.; James, A. Aislante térmico a base de materiales orgánicos. *Revista de Iniciación Cientfica* **2008**, *4*, 48–51. [CrossRef]
114. Sassoni, E.; Manzi, S.; Motori, A.; Montecchi, M.; Canti, M. Novel sustainable hemp-based composites for application in the building industry: Physical, thermal and mechanical characterization. *Energy Build.* **2014**, *77*, 219–226. [CrossRef]
115. Nguyen, D.M.; Grillet, A.C.; Bui, Q.B.; Diep, T.M.H.; Woloszyn, M. Building bio-insulation materials based on bamboo powder and bio-binders. *Constr. Build. Mater.* **2018**, *186*, 686–698. [CrossRef]
116. Manohar, K.; Ramlakhan, D.; Kochhar, G.; Haldar, S. Biodegradable fibrous thermal insulation. *J. Braz. Soc. Mech. Sci. Eng.* **2006**, *28*, 45–47. [CrossRef]
117. Troppová, E.; Švehlík, M.; Tippner, J.; Wimmer, R. Influence of temperature and moisture content on the thermal conductivity of wood-based fibreboards. *Mater. Struct. Constr.* **2015**, *48*, 4077–4083. [CrossRef]

118. Korjenic, A.; Petránek, V.; Zach, J.; Hroudová, J. Development and performance evaluation of natural thermal-insulation materials composed of renewable resources. *Energy Build.* **2011**, *43*, 2518–2523. [CrossRef]
119. Pervaiz, M.; Sain, M.M. Carbon storage potential in natural fiber composites. *Resour. Conserv. Recycl.* **2003**, *39*, 325–340. [CrossRef]
120. Ashraf, W. Carbonation of cement-based materials: Challenges and opportunities. *Constr. Build. Mater.* **2016**, *120*, 558–570. [CrossRef]
121. Grist, E.R.; Paine, K.A.; Heath, A.; Norman, J.; Pinder, H. The environmental credentials of hydraulic lime-pozzolan concretes. *J. Clean. Prod.* **2015**, *93*, 26–37. [CrossRef]
122. Jami, T.; Kumar, S. Assessment of Carbon Sequestration of Hemp Concrete. In Proceedings of the International Conference on Advances in Construction Materials and Systems, Chennai, India, 3 September 2017. [CrossRef]
123. Ip, K.; Miller, A. Life cycle greenhouse gas emissions of hemp–lime wall constructions in the UK. *Resour. Conserv. Recycl.* **2012**, *69*, 1–9. [CrossRef]
124. Arrigoni, A.; Pelosato, R.; Melià, P.; Ruggieri, G.; Sabbadini, S.; Dotelli, G. Life cycle assessment of natural building materials: The role of carbonation, mixture components and transport in the environmental impacts of hempcrete blocks. *J. Clean. Prod.* **2017**, *149*, 1051–1061. [CrossRef]
125. Sartori, I.; Hestnes, A.G. Energy use in the life cycle of conventional and low-energy buildings: A review article. *Energy Build.* **2007**, *39*, 249–257. [CrossRef]
126. Stevulova, N.; Cigasova, J.; Purcz, P.; Schwarzova, I.; Kacik, F.; Geffert, A. Water absorption behavior of hemp hurds composites. *Materials* **2015**, *8*, 2243–2257. [CrossRef]
127. Bollino, F.; Giannella, V.; Armentani, E.; Sepe, R. Mechanical behavior of chemically-treated hemp fibers reinforced composites subjected to moisture absorption. *J. Mater. Res. Technol.* **2023**, *22*, 762–775. [CrossRef]
128. Sedan, D.; Pagnoux, C.; Chotard, T.; Smith, A.; Lejolly, D.; Gloaguen, V.; Krausz, P. Effect of calcium rich and alkaline solutions on the chemical behaviour of hemp fibres. *J. Mater. Sci.* **2007**, *42*, 9336–9342. [CrossRef]
129. Števulová, N.; Terpáková, E.; Čigášová, J.; Junák, J.; Kidalová, L. Chemically Treated Hemp Shives as a Suitable Organic Filler for Lightweight Composites Preparing. *Procedia Eng.* **2012**, *42*, 948–954. [CrossRef]
130. Chabannes, M.; Garcia-Diaz, E.; Clerc, L.; Bénézet, J.C. Effect of curing conditions and Ca(OH)2-treated aggregates on mechanical properties of rice husk and hemp concretes using a lime-based binder. *Constr. Build. Mater.* **2016**, *102*, 821–833. [CrossRef]
131. Wei, J.; Ma, S.; Thomas, D.G. Correlation between hydration of cement and durability of natural fiber-reinforced cement composites. *Corros. Sci.* **2016**, *106*, 1–15. [CrossRef]
132. Sepe, R.; Bollino, F.; Boccarusso, L.; Caputo, F. Influence of chemical treatments on mechanical properties of hemp fiber reinforced composites. *Compos. Part Eng.* **2018**, *133*, 210–217. [CrossRef]
133. Liu, M.; Silva, D.A.S.; Fernando, D.; Meyer, A.S.; Madsen, B.; Daniel, G.; Thygesen, A. Controlled retting of hemp fibres: Effect of hydrothermal pre-treatment and enzymatic retting on the mechanical properties of unidirectional hemp/epoxy composites. *Compos. Part Appl. Sci. Manuf.* **2016**, *88*, 253–262. [CrossRef]
134. Liu, M.; Thygesen, A.; Summerscales, J.; Meyer, A.S. Targeted pre-treatment of hemp bast fibres for optimal performance in biocomposite materials: A review. *Ind. Crops Prod. Prod.* **2017**, *108*, 660–683. [CrossRef]
135. Sandrine, U.B.; Isabelle, V.; Hoang, M.T.; Maalouf, C. Influence of chemical modification on hemp-starch concrete. *Constr. Build. Mater.* **2015**, *81*, 208–215. [CrossRef]
136. Su, N.; Fang, C.; Zhou, H.; Tang, T.; Zhang, S.; Fei, B. Hydrophobic treatment of bamboo with rosin. *Constr. Build. Mater.* **2021**, *271*, 121507. [CrossRef]
137. Atgié, M.; Garrigues, J.C.; Chennevière, A.; Masbernat, O.; Roger, K. Gum Arabic in solution: Composition and multi-scale structures. *Food Hydrocoll.* **2019**, *91*, 319–330. [CrossRef]

MDPI
St. Alban-Anlage 66
4052 Basel
Switzerland
Tel. +41 61 683 77 34
Fax +41 61 302 89 18
www.mdpi.com

Materials Editorial Office
E-mail: materials@mdpi.com
www.mdpi.com/journal/materials

www.ingramcontent.com/pod-product-compliance
Lightning Source LLC
Chambersburg PA
CBHW061805210926
51850CB00006B/617
9783036577104